Fault Diagnosis of Dynamic Systems

Teresa Escobet · Anibal Bregon ·
Belarmino Pulido · Vicenç Puig
Editors

Fault Diagnosis of Dynamic Systems

Quantitative and Qualitative Approaches

Springer

Editors
Teresa Escobet
Research Center for Supervision, Safety
and Automatic Control (CS2AC)
Universitat Politècnica de Catalunya (UPC)
Terrassa, Spain

Belarmino Pulido
Department of Computer Science
University of Valladolid
Valladolid, Spain

Anibal Bregon
Department of Computer Science
University of Valladolid
Segovia, Spain

Vicenç Puig
Research Center for Supervision, Safety
and Automatic Control (CS2AC)
Universitat Politècnica de Catalunya (UPC)
Terrassa, Spain

ISBN 978-3-030-17727-0 ISBN 978-3-030-17728-7 (eBook)
https://doi.org/10.1007/978-3-030-17728-7

This Springer imprint is published by the registered company Springer Nature Switzerland AG
The registered company address is: Gewerbestrasse 11, 6330 Cham, Switzerland

Preface

Nowadays, physical and software systems, designed and built through engineering processes, are everywhere: our home and our office are full of electronic devices, our factories are almost fully automatized, we have cars and trucks full of complex electronic systems, etc. Hence, it is required that these systems work as much as possible and as safer as possible. For these tasks, automated diagnosis is mandatory because for most devices, it is almost impossible to obtain the necessary experience to build knowledge-based systems before they become obsolete, or there are enough number of variants for the same mechanism that it is not possible to adjust existing solutions. In those cases where a lot of data are available, it would be possible to learn models using data-driven techniques. Although pure data-driven approaches can learn models from complex data collections, it is also clear that they usually lack the explanation capabilities inherent to other techniques. For all those reasons, model-based reasoning, built upon correct behavior models, seems as a fairly suitable alternative to expert systems and data-driven models.

For almost 30 years, two large research communities (the artificial intelligence and the control theory) have made considerable advances in model-based diagnosis, mostly using different techniques and reasoning assumptions. It has not been until 15 years ago that both communities started the necessary cooperation to understand each other and to produce the required symbiosis to produce new techniques taking the best from both sides.

The Aim of the Book

The aim of this book is to provide a glimpse of the fundamental issues and techniques used in both fields, known as DX and FDI communities. The book has been organized as a collection of chapters that could be taught in a 30 h seminar, but containing several examples or case studies, so that students can compare different solutions to the same problem. These chapters made up the first part of the book.

Being aware of the importance of interdisciplinary work, this book also includes a collection of methods that are not purely model-based but that can be used to improve or to collaborate with other techniques trying to solve more complex scenarios. Presenting these techniques and their application to well-known case studies should be considered as a second part of the book.

The Object of Study

The needs for increased performance, safety, and reliability of engineering systems provide the motivation for developing Integrated Systems Health Management (ISHM) methodologies that include efficient fault detection, diagnosis, and recovery mechanisms to reduce downtime and increase system availability through the life of the system. This can be achieved by monitoring system behavior during operation, automatically detecting and isolating faults and degradations, and recovering to a normal operating mode when faults occur.

Expected Audience

The book represents a guide for students, researchers, and engineers starting to work on fault diagnosis and willing to have a reference guide on the main concepts and standard approaches on model-based fault diagnosis. Readers with experience on one of the two main communities will also find it useful to learn the fundamental concepts of the other community and the synergies between both communities.

The book is also open to researchers or academics who are already familiar with the standard approaches, since they will find a collection of advanced approaches with more specific and advanced topics or with application to different domains.

Finally, engineers and researchers looking for transferable fault diagnosis methods will also find valuable insights in the book.

The Content

Chapter 1 reviews the main concepts of fault diagnosis of dynamic systems establishing a common framework with unified terminology and introduces the methodologies of the different diagnosis approaches considered.

Chapter 2 presents different modeling approaches to be used to perform process monitoring, fault detection, and fault isolation. In this chapter, several application examples are introduced, which will be applied to illustrate the diagnosis schemes presented and studied in this book.

Chapter 3 presents a formal definition of fault detection and isolation in terms of consistency relation. After that, the definition of structural analysis based on bipartite graphs is introduced for analyzing the structural properties of dynamical systems and computing the structured residuals. These structured residuals provide the interaction between signals and allow finding analytical redundancy, which can be exploited for fault diagnosis.

The FDI approach, that relied on the use of system and control theory to develop methods for model-based diagnosis, is presented in Chap. 4. The key concept is to exploit the analytic redundancy concept that is based on generating fault indicators, named residuals, from the comparison of the available measurements from sensors and their estimation using models. The chapter will present the methods that are available from the literature to generate residuals (parity equations/space, state observers and parameter estimation), and illustrate how to use them to carry out the fault detection and isolation tasks.

Chapter 5 introduces the DX approach to model-based diagnosis. First, the main ideas of Consistency-based Diagnosis (CBD) methodology, together with the logical formalization provided by Reiter's work, are presented. The formalization of the set of minimal diagnoses is then elaborated from the set of minimal conflicts. Second, it provides a description of the General Diagnostic Engine (GDE), which is the de facto computational paradigm. The chapter finishes by extending the basic CBD approach with predictive fault models, while retaining the essential no exoneration assumption. Both theoretical and practical views will be illustrated using the classical polybox example provided in Chap. 2.

Chapter 6 analyzes the main problems found by the DX community approach: the online computation of minimal conflicts by means of an ATMS-like dependency-recording engine and the need for an extension to deal with dynamic systems diagnosis. To cope with the first problem, different approaches have been proposed. Some of them were extensions of the original GDE, while others can be categorized as topological methods, which are described in the first part of the chapter. The second part copes with the second problem, dynamics: we review the whole set of proposals made to extend Reiter's formalization and the GDE to dynamic systems. The classical polybox example and the three tanks example provided in Chap. 2 will be used to illustrate all the concepts.

For more than 20 years, both the DX and FDI communities developed techniques and theories to approach model-based diagnosis from apparently two widely separated points of view. In Chap. 7, both FDI and DX are revised with the aim of establishing a logical framework allowing comparison. Moreover, the hypotheses underlying the two approaches are clearly stated. The two approaches are then shown equivalent under some assumptions. A practical comparison is provided using the classical polybox example and potential synergies are explained.

Multivariate statistical process control is a data-driven monitoring method, with fault detection and isolation capabilities, based on the principles of Principal Component Analysis (PCA). It extends the PCA concept to establish a reference model build with data recorded during normal operation conditions of the process. Then, the consistency of process observation with respect to this statistical model is evaluated in order to detect possible faults. Chapter 8 presents the fundamentals of PCA and describes its extension for process monitoring including fault detection and isolation.

Chapter 9 is concerned with the diagnosis of systems modeled as Discrete-Event Systems (DES). DES model dynamic systems with discrete changes through "events". The chapter discusses the relevance of this formalism and shows how the diagnosis problem is defined and how it can be solved. Because the problem is computationally hard, several techniques that can be used to alleviate the computational cost are also presented. Finally, the chapter focuses on the problem of diagnosability, which is the question whether in a system modeled with a given DES the occurrence of a fault can always be detected and identified in a bounded amount of time.

In these previous chapters, the classical approaches to model-based diagnosis were introduced. However, real-life problems require to include additional techniques, while retaining the basic ideas presented in the first part of this book.

In the second part, the book starts addressing the problem of model uncertainty. In the previous approaches, the models used to reason about a system behavior assumed that the model is reasonably close to reality. However, the difference between the model estimation and the system observations does not always point out to fail. In real-life applications, model uncertainties in the model parameters should be considered making the diagnosis problem harder.

Chapter 10 will review the passive approach when considering the nominal model plus the uncertainty on every parameter bounded by intervals. This type of uncertainty modeling provides a type of models known as *interval models*. Noise will be also be considered to be unknown-but-bounded and modeled in a deterministic framework. This chapter also reviews the different approaches that can be used to identify interval models for fault detection. Finally, the chapter presents the application of some of the approaches presented in the chapter using the two-tank case study introduced in Chap. 2.

The objective of Chap. 11 is twofold. First, standard estimation strategies for fault detection and isolation of a class of nonlinear systems are briefly discussed along with some state-of-the-art references. This first part refers to the classical residual-based framework and involves such techniques like parameter estimation, parity relation, and observers. A particular attention is focused on the last group of approaches and a brief nonlinear observer review is provided, ending with a discussion about the robustness issue. The remaining part of the chapter is devoted to the direct fault estimation strategies.

Another option to deal with model uncertainty is to use stochastic models. Diagnosis using stochastic methods is a widely researched topic for which many different approaches have been proposed. Rather than view stochastic diagnosis from the perspective of a heterogeneous set of methodologies, in Chap. 12 we provide a unifying framework based on the theory of Bayesian filtering and the computational framework of factor graphs. We illustrate the framework by comparing and contrasting several models on the tank benchmark.

Chapter 13 provides a quick review of research in fault detection and isolation of hybrid systems, and then develops a structural diagnosis approach for them. Hybrid systems are characterized by continuous behaviors that are interspersed with discrete mode changes in the system, making the analysis of their behaviors quite complex. The work presented in this chapter adopts a comprehensive approach to mode detection and to diagnoser design using analytic redundancy methods to detect the operating mode of the system even in the presence of system faults.

Chapter 14 presents the main ideas of constraint-driven fault diagnosis (CDD). CDD allows to obtain the causes of why a correctly designed hardware system does not work as expected. CDD problems would be automatically solved using consistency and search techniques that are traditional methods to solve Constraint Satisfaction Problems (CSPs) and Constraint Optimization Problems (COPs). For certain CSPs/COPs, the computational complexity is very high; therefore, the complexity reduction is the cornerstone issue in solving CSPs/COPs, especially when there is a large number of constraints and/or wide domains of the variables.

During software development and maintenance, the costs associated with bug isolation and repair represent a substantial percentage of (human and financial) resources. Improving the detection and isolation of defects would allow a significant saving of the final cost of the software production. Bugs do not only appear in the development phase, but also in the maintenance, considering that many of the developed software products will be changed during its useful life to be able to adapt to new requirements. Chapter 15 discusses in detail some of the promising techniques to isolate faults in software-intensive systems: model-based debugging.

The fault detection and later diagnosis of abnormal behaviors of business processes are crucial from the strategic point of view of the organizations, since their proper working is an essential requirement. Unexpected faults can provide undesirable halts in the processes, thereby causing cost increase and production decrease. Therefore, and due to the complexity of business process models, Chap. 16 shows how to apply diagnosis techniques at different stages of the business process life cycle with the goal of identifying and isolating faults both at design-time and run-time.

Together with diagnosis, prognostic is a necessary task for system reconfiguration and fault-adaptive control in complex systems. Chapter 17 presents the main ideas of prognostics and its related concepts, including end of life and remaining useful life. Afterward, the ideas underlying the prognostics process are presented and the different sources of uncertainty that appear, and make the prognostics process challenging, are discussed. Finally, the chapter focuses on model-based prognostics and gives the reader a thorough perspective of the prognostics process using dynamic models of systems, with an emphasis on physics-based models.

A large majority of today's critical and complex systems constitute embedded electronic modules for monitoring, control, and enhanced functionality. However, it has been found that these modules are often also the first element in the system to fail. Studying and analyzing the performance degradation of such critical electronic systems enhance operational safety and reliability, avoid catastrophic failures, and reduce maintenance costs over the life of the system. The development of prognostics methodologies for electronics has become more important as more electrical systems are being used in a wide range of application fields. These methodologies are presented in Chap. 18.

Terrassa, Spain
Segovia, Spain
Valladolid, Spain
Terrassa, Spain
December 2018

Teresa Escobet
Anibal Bregon
Belarmino Pulido
Vicenç Puig

Contents

Part II Advanced Approaches

Acronyms

AB	Abnormal predicate
AI	Artificial intelligence
ALS	Advances life support system
AOC	Abnormal operating conditions
ARMA	Autoregressive moving average
ARR	Analytical redundancy relations
ATMS	Assumption-based truth maintenance system
AWRS	Advanced water recovery system
BDC	Business data constraint
BDD	Binary decision diagrams
BFS	Breadth-first search
BG	Bond graph
BILP	Binary integer linear programming
BM	Behavioral model
BN	Bayesian network
BPM	Business process model
BPMN	Business process modeling notation
BPMS	Business process management systems
CARC	Context analytical redundancy constraint
CBD	Consistency-based diagnosis
CDD	Constraint-driven fault diagnosis
CFG	Control flow graph
CN	Context network
CoMSS	Complement of an MSS
COP	Constraint optimization problems
CPT	Conditional probability table
CS	Context set
CSP	Constraint satisfaction problems
DAG	Directed acyclic graph
DbC	Design by contract

DBN	Dynamic Bayesian network
DES	Discrete-event system
DM	Dulmage–Mendelsohn
DNNF	Disjunction negation normal form
DOC	Diagnosis of overdetermined CSPs
DRE	Dependency register engine
DTP	Disjunctive temporal problem
DX	Diagnosis
EOL	End of life
ETTF	Estimated time to failure
FD	Fault diagnosis
FDI	Fault detection and identification
FG	Factor graph
FM	Feature model
FS	Fault signature
FSM	Fault signature matrix
FTC	Fault tolerant control
GARR	Global ARR
GB	Gröbner basis algorithm
GDE	General diagnostic engine
GLR	Generalized likelihood ratio
HBG	Hybrid bond graph
HMM	Hidden Markov models
HMSO	Hybrid minimal structurally overdetermined
HSO	Hybrid structurally overdetermined
ISO	International organization for standardization
KF	Kalman filter
LGM	Linear Gaussian model
LPV	Linear parameter varying
LTI	Linear time-invariant
MA	Moving average
MAP	Maximum aposteriori
MBD	Model-based diagnosis
MCC	Minimal conflict contexts
MEC	Minimal evaluation chain
MEM	Minimal evaluation model
MHS	Minimal hitting set
MMSE	Maximum mean-squared error
MSD	Minimal structurally determined
MSO	Minimal structurally overdetermined
MSS	Maximal satisfiable subsets
MTES	Minimal test equation support
MUS	Minimal unsatisfiable subsets
NCSP	Numeric overconstrained CSPs
NOC	Normal operating conditions

OM	Observational model
OS	Observed signatures
PAIS	Process-aware information system
PBDM	Physics-based degradation modeling
PC	Possible conflict
PCA	Principal component analysis
PGM	Probabilistic graphical model
PHM	Prognostics and health management
PLS	Partial least squares
PSO	Minimal structurally overdetermined
QR	Qualitative reasoning
RC	Relevant contexts
RO	Reverse osmosis
RUL	Remaining useful life
SAT	Boolean satisfiability problem
SCAP	Sequential causality assignment procedure
SD	System description
SFL	Spectrum-based fault localization
SHM	Systems health management
SIS	Sequential importance sampling
SMBD	Stochastic model-based diagnosis
SPE	Square prediction error
SPRT	Sequential probability ratio test
SSA	Static single assignment
TCG	Temporal causal graph
TES	Test equation support
TS	Test support
UIO	Unknown input observe
WFM	Workflow management

Chapter 1
Introduction

Joaquim Armengol, María Jesús de la Fuente and Vicenç Puig

1.1 Introduction

Nowadays, physical and software systems, designed and built through engineering processes and software, are everywhere: our home and our office are full of electronic devices, our factories are almost fully automatized, we have cars and trucks full of complex electronic systems, almost every electronic system contains hundreds or thousands of lines of code, our computers run operating systems made up of hundreds of small programs, etc. Hence, it is required that these systems work as expected and as safer as possible. For these tasks, automated diagnosis is mandatory because for most devices, it is almost impossible to obtain necessary experience to build knowledge-based systems before they become obsolete, or there are enough number of variants for the same mechanism that it is not possible to adjust existing solutions. In those cases where a lot of data are available, it would be possible to learn models using data-driven techniques. Although pure data-driven approaches can learn models from complex data collections, it is also clear that they usually lack the explanation capabilities inherent to other techniques. For all those reasons

J. Armengol
Departament d'Enginyeria Elèctrica, Electrònica i Automàtica, Universitat de Girona,
Girona, Spain
e-mail: joaquim.armengol@udg.cat

M. J. de la Fuente
Departament de Ingeniería de Sistemas y Automática, Universidad de Valladolid,
Valladolid, Spain
e-mail: mjfuente@eii.uva.es

V. Puig (✉)
Research Center for Supervision, Safety and Automatic Control (CS2AC), Unversitat Politècnica de Catalunya (UPC),
Terrassa, Spain
e-mail: vicenc.puig@upc.edu

T. Escobet et al. (eds.), *Fault Diagnosis of Dynamic Systems*,
https://doi.org/10.1007/978-3-030-17728-7_1

model-based reasoning, built upon correct behavior models, seem as a fairly suitable alternative to expert systems and data-driven models.

For almost 30 years two huge research communities (the Artificial Intelligence (AI) and the control theory) have made considerable advances in model-based diagnosis, mostly using different techniques and reasoning assumptions. It has not been until 15 years ago that both communities started the necessary co-operation to understand each other and to produce the required symbiosis to produce new techniques taking the best from both sides.

In this book, we provide a glimpse of the fundamental issues and techniques used in both fields, known as DX and FDI communities, organized as a collection of chapters that could be taught in a 3–4 h seminar each, but containing several examples or case studies, so that students can compare different solutions to the same problem. These chapters made up the first part of the book.

Being aware of the importance of interdisciplinary work, this book also includes a collection of methods that are not purely model-based but that can be used to improve or to collaborate with other techniques trying to solve more complex scenarios. Presenting these techniques and their application to well-known case studies should be considered as a second part of the book, where also more advanced topics and methods are introduced to motivate the students go to further in the research on fault diagnosis and related techniques. Additionally, we include several software tools that can be used to implement these fundamental concepts in small projects or case studies.

1.2 Preliminary Concepts

1.2.1 System Concept

François Cellier starts his classical book about modeling [2] with some basic definitions that will be used in the rest of the book like "system", "model", and "simulation". Many of those definitions are useful for this book, including the definition of "system", which is the first one. Cellier copies this definition from Brian Gaines [3]:

> A system is what is distinguished as a system. At first sight, this looks to be a nonstatement. Systems are whatever we like to distinguish as systems. Has anything been said? Is there any possible foundation here for a systems science? I want to answer both these questions affirmatively and show that this definition is full of content and rich in its interpretation.
>
> Let me first answer one obvious objection to the definition above and turn it to my advantage. You may ask, "What is peculiarly systemic about this definition?" "Could I not equally well apply it to all other objects I might wish to define?", e.g., a rabbit is what is distinguished as a rabbit. "Ah, but," I shall reply, "my definition is adequate to define a system but yours is not adequate to define a rabbit." In this lies the essence of systems theory: that to distinguish some entity as being a system is a necessary and sufficient criterion for its being a system, and this is uniquely true for systems, whereas to distinguish some entity as being anything else is a necessary criterion to its being that something but not a sufficient one.

So, applying this definition, the only thing that has to be done in order to define a system is to mark a border between the system and the rest of the universe hence dividing the universe in two parts: the system and the rest. Usually, there are interactions between the system and the rest of the universe: events that happen in the rest of the universe affect the system (these are called *inputs* to the system) and events that happen in the system affect the rest of the universe (these are called *outputs* of the system). An example could be a tank: the walls of the tank mark clearly which part of the universe is the tank and which is not, and there may be input and output flows or heat exchanges between the tank and the rest of the universe.

A technological system is a system designed so as to accomplish one, or several, missions. For instance, in the case of the tank, the mission can be to store water or to store hot water. Usually, there can be distinguished subsystems in the system. These subsystems interact with each other and are called components of the system. For instance, in the case of the tank, one of the components is the tank itself, but there can be others like pipes or valves. This concept of system is not only applicable to physical systems, but to software systems too. For instance, a program is made up of routines or it is organized as objects that interact through methods. These are the components of the system, while the data introduced to perform a computation are the inputs, and the results provided are the outputs (that can be stored or transferred to other systems (program)).

Components are technological devices that introduce constraints between variables. For instance, the variations of the level in the component tank are related, due to the law of conservation of mass, to the input and output flows: if the input flow is greater than the output flow, then the level increases, and vice versa. This can be expressed with a mathematical model:

$$A\frac{dh(t)}{dt} = q_{in}(t) - q_{out}(t), \tag{1.1}$$

where A is the section of the tank, $h(t)$ is the level at time t, and $q_{in}(t)$ and $q_{out}(t)$ are, respectively, the input and the output flows.

In this case, the model is called *dynamic*, because the time plays an important role: to know the level at a specific time point is not enough knowing the flows at that time point, it is necessary to have more information like past levels and flows. On the other side, a model is called *static* when it allows to compute the value of a variable in a time instant only knowing the values of other variables at the same time instant. For instance, the flow in a pipe and the pressures at both ends of the pipe are so closely related that if, at a time point, the value of two of these variables is known, the third can be calculated:

$$q(t) = k\,(p_1(t) - p_2(t))\,, \tag{1.2}$$

where $q(t)$ is the flow in the pipe, $p_1(t)$ and $p_2(t)$ are the pressures at both ends of the pipe and k is a factor of proportionality that depends on the characteristics of the pipe.

Anyway, the values of the variables may change across time, so the evolution of the variables can be graphically represented and these graphical representations are called trajectories. Sometimes, graphics represent the evolution of one or several variables with respect to time and other times they represent the evolution of one variable with respect to other variables and the time is implicit.

In any case, we usually want to control systems: make that the variables follow a trajectory such that they reach a desired value. For instance, making that the level of the tank is a desired one. To do so, sensors and actuators are needed. Sensors are devices necessary to measure the value of the variables. For instance, a level sensor measures level and a temperature sensor (a thermometer) measures temperature. Once the actual value of a variable is known, some action is to be taken in order to bring it to the desired value. This is done by means of the actuators, for instance, by means of input or output valves in the example of the tank. The strategy to control the system can be automatically applied by an automatic controller. For instance, in the case of the tank, the controller may open or close input and output valves depending on the actual level of the tank and in order to bring and maintain the level at the desired value.

So, essentially, a system (or a subsystem) transforms energy or information. For instance, a tank converts the energy due to the pressure of the input flow to potential energy, which is related to level. At the same time, level is the cause of the pressure at the output pipe, which is translated into an output flow (kinetic energy). When these variables are measured through a sensor, information is obtained (a signal), so a sensor converts energy to information. The controller uses this information to decide what to do, which is information as well. Finally, the actuator closes the loop and provides energy to the system according to the information given by the controller.

In the case of software or information systems, they perform some kind of algorithmic transformation of thc information provided as input. For instance, a web service is able to accept demands and to produce results. Its behavior is limited by the amount of resources provided, the amount of memory, and the time constraints to finish a transaction.

1.2.2 Fault Concept

Sometimes, the controller cannot achieve its goals because there is a *malfunction* in the system, in the sensors or in the actuators. These abnormal behaviors are called *faults*. So, a fault is a deviation from the normal, acceptable, usual, and standard behavior. Faults may have multiple causes: errors in the phase of the system design, errors in the phase of the system implementation, errors during operation, etc. In many cases, these errors are due to humans, but sometimes, faults are due to the deterioration of the components across time due to aging (use, wear) or due to damages produced by previous faults. In the case of software systems, the source of faults is a wrong design (i.e., wrong specifications) or a bug in the code. There are no faults due to aging, but maintenance can create inconsistencies.

The consequences of having faults are varied as well. Sometimes, the system can still work but with a worse performance, for instance, wasting energy or wasting raw materials, or producing lower quality products or at a lower rate. In all these cases, consequences are economic, but additionally, there can be damages to the environment or to humans. The effects of software faults can be catastrophic as well, since these programs can be part of a control system as in aerospace vehicles or satellites, but can also cause malfunctions in any system whose software controller or model is faulty.

There are many different faults and they can be classified in different ways. One classification derives from the magnitude (size) of the fault and recalls its definition. The definition states that a fault is a deviation from the acceptable behavior, so if a deviation from the usual behavior is acceptable, then it is not a fault. Sometimes, a deviation of 1% is unacceptable so it is considered a fault, and some other times a deviation of 10% is acceptable so it is not considered a fault. On the other side, if the fault is as big as to permanently interrupt the system's ability to perform a required function under specific operating conditions, it is rather called a *failure*. For instance, if the tank has a hole in its wall, it can still store water, with a more or less important leakage. If an earthquake splits the tank into several pieces, it cannot store water so the catastrophic fault is better called a failure. Faults in information systems are different, since any fault will cause sooner or later a malfunction.

Physical faults can also be classified depending on the localization of the faults. For instance, if the inflow to a wastewater treatment plant is beyond the maximum design flow, the quality of the output water will be lower than desired. This is a fault, but it is a fault neither of the system, nor of the sensors, nor of the actuators, changing them for new ones would not solve the problem. This is called an *external fault*. System, sensors, and actuators perform well, but the interactions between the system and the environment are not compatible with the goals. Hence, a fault is called *internal* when there is a fault in the system, in a sensor or in an actuator. Examples of internal faults are: a leakage in the tank (system fault), a 5% permanent error in the measurement of the sensor (sensor fault), or a valve that cannot be totally closed (actuator fault).

Another way to classify physical faults is depending on temporal aspects. There are faults that appeared suddenly at a specific time point, for instance, when an electric wire got disconnected. This is an *abrupt fault* and can be modeled as a step. Other faults, for instance, due to aging, appear slowly and the time point in which they are big enough to be considered as faults is not clear. These are called *incipient* or evolutionary faults and usually are modeled using ramps, exponential functions or parabolas. Finally, there are *intermittent* faults, they are not permanent and can be modeled with pulses, perhaps with different widths and heights. In software systems, faults are usually permanent.

Finally, faults can be classified depending on the way that they affect the behavior of the system. In an *additive* fault, the changes at the output of the system depend on the magnitude of the fault and do not depend on the inputs. For instance, a faulty thermometer can give measurements 2 °C above the real temperature. This is called an offset and can be represented (and corrected) by means of an addition. If the measurement of the faulty thermometer is 10% greater than the real temperature, it is called a *multiplicative* fault. In this case, the changes at the output of the faulty component depend on the magnitude of the fault and of the input of the component. In this case, the fault can be modeled (and corrected) by means of a product (a gain). Many faults cannot be classified as neither additive nor multiplicative because they are neither purely additive nor purely multiplicative.

1.2.3 Fault Diagnosis and Tolerance Concepts

Hence, when a system is controlled it would be desirable that, even in the presence of faults, the system continues in operation as long as possible, provided both efficiency and security remain acceptable. This is called *Fault Tolerant Control* (FTC) and is one of the goals of the automation. Its aim is to make the system stable and retain an acceptable performance. There are different techniques that can be used depending on the size of the fault. Robust control intends to design a single controller that performs well even when there are small differences between the real behavior of the components and the one that has been used for the design. This is called a *passive* technique. If faults are larger, *active* techniques are needed. One active technique is adaptive control: the parameters of the controller are retuned for the new situation or, even, the structure of the controller is changed (a new control algorithm is used). Sometimes, this is not enough and is necessary to reconfigure the system in order to keep it in normal operation. For instance, if there is a leakage in a pipe, an alternative path is selected closing and opening the adequate valves. In this case, it is said that the system has been accommodated to the fault, which is a way to handle the fault. If this option cannot be used, perhaps the system can still work but with different goals, so instead of working in normal operation it then works in degraded operation. If this is not possible, the system must be brought to a safe stop.

As it has been seen, many tasks are related to dealing with faults. First one, a necessity to start other tasks, is *fault detection*, i.e., to determine the existence of a fault. If a fault has been detected, if there is a fault, then there is *fault diagnosis*. In fact, fault detection is considered as the first stage of fault diagnosis. A diagnostic is the set of components that explain the fault. On one side fault diagnosis involves *fault isolation*, that intends to determine characteristics of the fault like kind, location, root cause and which is the faulty component. And, on the other side, *fault identification* intends to determine characteristics like the size and the type of the fault or the time when it has occurred.

Once the fault has been detected and diagnosed, *fault tolerance* is the task that proposes solutions like, for instance, the ones related to fault handling. So, there are *monitoring systems* that make a surveillance of other systems and *supervision systems* that, in addition, propose solutions. In order to propose solutions, an analysis of the situation is necessary, the possibilities of achieving a given objective when a fault occurs must be assessed. But, these possibilities will exist only if the system has been designed to allow the achievement of a given objective not only in normal operation but also in given faulty situations, i.e., these possibilities will exist only if the system has been provided with the necessary means like hardware architecture or software mechanisms. In other words, if there is a leakage in a pipe, an alternative path can be used if it exists, and it exists if it was foreseen when the system was designed. The best analysis is useless without a proper design.

1.2.4 Fault Diagnosis Performance Assessment

This book presents many different methods for fault diagnosis, with different needs and different results. There are different ways to compare them. In the case of the fault detection methods, the output of the method is the activation of an alarm when a fault is detected, so a good fault detection method activates the alarm when there is a fault, this is a true positive, and does not activate the alarm when there is no fault, this is called a true negative. But sometimes, there are errors. It can happen that the method raises the alarm when the system is healthy, i.e., there is no fault. This is called a *false alarm* or a false positive. On the contrary, if there is a fault and the method does not activate the alarm, it is called a *missed alarm*, a false negative or an undetected fault.

A method is more reliable than another method, if it has less detection errors. This can be measured in several ways:

- *Sensitivity*. Indicates the ratio of times the faults are detected

$$Sensitivity = \frac{Detection}{Fault} = \frac{TruePositive}{TruePositive + FalseNegative}.$$

  The goal is that sensitivity approaches to one: all faults are detected.

- *Specificity*. Indicates the goodness of the method when it does not activate the alarm

$$Specificity = \frac{NoDetection}{NoFault} = \frac{TrueNegative}{TrueNegative + FalsePositive}.$$

  The goal is that specificity approaches to one: when the system is healthy there are no alarms.

- *False Positive Rate*. Indicates the false alarm ratio, i.e., the percentage of false fault indications.

$$FalsePositiveRate = \frac{Detection}{NoFault} = \frac{FalsePositive}{FalsePositive + TrueNegative} = 1 - Specificity.$$

  The goal is False Positive Rate approaches to zero: no false positives.

- *False Negative Rate*. Indicates the missed alarm ratio, i.e., the percentage of missed fault detections.

$$FalseNegativeRate = \frac{NoDetection}{Fault} = \frac{FalseNegative}{FalseNegative + TruePositive} = 1 - Sensitivity.$$

The goal is False Negative Rate approaches to zero: no false negatives.

1.2.5 Robustness, Detectability, Isolability, and Redundancy Concepts

Many times detection errors are caused by *uncertainty*. Noise is a source of uncertainty: the real value of a variable is unknown and a measurement, which can differ from the real value, is used instead. Modeling errors are another source of uncertainty: the model of a system is a mathematical expression that represents the behavior of the real system, but it is only an approximation to keep it simple. These models contain parameters (a length, a mass, an electrical resistance, etc.) which are considered as constants but maybe they are not due to the effects of as, e.g., environmental conditions (temperature or atmospheric pressure). This is another source of uncertainty. These unknown and uncontrolled inputs, like ambient temperature or atmospheric pressure, are called *disturbances* and act on the system and affect its behavior, so they are a source of uncertainty. Uncertainties and disturbances must be taken into account when designing a fault detection method in order to minimize its effects. In this case, the method is said to be *robust*. If a method raises an alarm due to a disturbance, it means that it is not robust in front of that disturbance. Similar things can happen in the presence of noise or modeling errors. Robustness is a crucial factor when comparing different fault detection methods.

Methods can also be compared when trying to detect different faults: there can exist faults that are *detectable* for some methods and *undetectable* for others. Sometimes, a method is more sensitive than another one to a specific fault while for other faults the comparison can differ substantially. A related measurement is the *detection time*. Faults must be detected as soon as possible in order to minimize their effects, so perhaps two methods both are able to detect a specific fault, but one of them is better because it needs much less time than the other. Or, finally, one method is better than another one because obtains similar results with simpler algorithms so with less computation effort and computation power.

Similarly, fault diagnosis methods can obtain different results as well, so they can be compared. Concerning diagnosis errors, a false positive consists in labeling as faulty a component which is healthy, i.e., working normally. On the other side, a False Negative consists in labeling a faulty component as normal. Methods can also be compared regarding *diagnosability* or *isolability*: if a fault can be diagnosed using a method and cannot be diagnosed using another method, the former method is better for that specific fault. Or maybe both methods can diagnose the fault but one needs less time or less computation effort.

Finally, the concept of *redundancy* is introduced. Suppose that there is a sensor in a system in order to measure a variable. How can we know if the sensor is working properly or not? One way is to use two sensors instead. If both sensors give similar measurements, probably they are working well. The exception is that both have the same fault so both provide similar erroneous measurements. If measurements are clearly different, then one or both sensors are faulty, so it is easy to detect that there is a fault, but how can we know which sensor is faulty, i.e., how can we diagnose the fault? One way is using three sensors and through voting: if the three sensors provide similar measurements, probably all of them are working correctly; if two sensors give similar measurements, probably the third one is faulty. This simple technique is called *physical redundancy* and can be used for sensors, for actuators, and for systems. Main drawbacks are at least economical due to the cost of the redundant devices, but in many cases, it is a question of space or weight. An alternative is *analytical redundancy*, which will be described in the next sections.

1.3 Overview of Methods

In literature, several classifications for FDI techniques can be found depending on the authors. Fault detection and isolation are highly complex task and many solutions have been proposed. These solutions are studied from different fields such as process control, artificial intelligence, statistics or electronics. Also, these techniques have been applied to industry, aeronautics, electronic components, and image processing, among others.

According to [4], the fault detection and isolation methods are divided into two groups: the model-free methods and the model-based methods, depending on the need of using a mathematical model (usually first-principles, state space or input–output models) or not. Another classical classification divided the FDI methods in three categories: model-based methods, knowledge-based methods, and data-driven (or model-free) methods. Finally, a most recent classification published is proposed in the reviews of [7–9]. The authors divide these techniques into three classes: quantitative-based models, qualitative-based models, and process history-based methods.

All fault diagnosis methods (data-driven, analytical, or knowledge-based) have their advantages and disadvantages, so that no single approach is best for all applications. Usually, the best process monitoring scheme employs even different methods to detect and isolate the possible faults in the system.

1.3.1 Model-Based Approaches

Model-based approaches are based on the comparison between a measured signal, the actual plant output, and its estimation calculated in terms of an explicit mathematical model of the system in nominal operation conditions (Fig. 1.1). The difference is called residual, these residuals should be zero-valued when the system is in normal operation and should diverge from zero when a fault occurs in the system. So, the faults are detected by setting a (fixed o variable) threshold on the residual. In a more

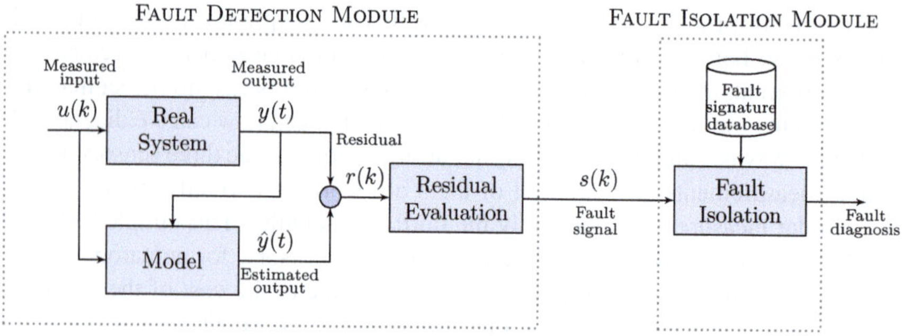

Fig. 1.1 Stages of model-based fault diagnosis

general and conceptual structure the model-based methods consist of two stages: residual generation and residual evaluation, also called, decision-making stage.

The AI approach to model-based diagnosis usually relies upon qualitative models to cope with parameter uncertainty, making the solution less precise. But, at the same time, the fault detection stage is more simple because it needs to be check for a discrepancy between the estimated and the observed variable. Only recently AI approaches have started to include the concept of residual, instead of discrepancy.

Most of the model-based fault diagnosis methods rely on the concept of *analytical redundancy*. In contrast to *physical redundancy*, where measurements from parallel sensors are compared to each other, now sensory measurements are compared to analytically computed values of the respective variable. Such computations use present and/or previous measurements for other variables and the mathematical plant model describing their nominal relationship to the measurable variable.

The difference between the different methods of analytical redundancy is in terms of the mathematical model used to generate the residuals. There are two main research areas that working in this field of model-based fault diagnosis approaches:

- *FDI community* uses methods from the control engineering field.
- *DX community* uses reasoning methods from the artificial intelligence field.

These two communities although face the problem of fault diagnosis using models, use different nomenclature, concepts, assumptions, and techniques to implement the model-based approaches for fault diagnosis. Some of these techniques are described in this book.

In the FDI community, the different techniques used are:

- *Observers* generate residuals from the output prediction error using some form of observer (as, e.g., Luenberger observer, full rank observer, or unknown input observer).
- *Kalman filters* assume a stochastic description of the uncertainty. In this case, the innovation (prediction error) of the Kalman Filter can be used as a fault detection residual; its mean is zero if there is not a fault and becomes nonzero in the presence of faults. Since the innovation sequence is white, it is easy to use statistical

techniques as the Sequential Probability Ratio Test (SPRT) [1] or the Generalized Likelihood Ratio (GLR) [5] methods to detect the faults.
- *Parity equations* or consistency relations obtained by direct manipulation of the state space or the input–output model of the system.
- Parameter estimation, using the identification algorithms to identify a linear model of the system.

However, as it has been realized, there is fundamental equivalence between parity relations and observer-based designs, in that both techniques produce identical residuals if the generators have been designed for the same specification.

All previous techniques are well developed for linear systems but there are extensions to the nonlinear systems. Several extensions exist depending on how the nonlinear model of the system is build as, e.g., nonlinear first-principles models, neural networks, Takagi–Sugeno (fuzzy), or Linear Parameter Varying (LPV) models.

From the point of view of the DX community, the models are oriented to diagnosis, and also there are several approaches:

- *Methods based on consistency*: The system is modeled by a finite set of interconnected components. Each component is modeled using a local model of correct behavior. Fault diagnosis is based on checking the consistency of the local models with the observations providing a collection of conflicts, and from those to choose the best candidates.
- *Method based on qualitative and semi-qualitative models*: The system dynamics are represented by means of qualitative or semi-qualitative models obtained from an abstraction of the first-principles dynamics. Fault diagnosis is based on logic reasoning using this type of models to check the consistency of the observed behavior and the one predicted using the models.

In general, the problem of model-based fault diagnosis can be stated as the determination of faults of a system from the comparison of available measurements with a priori information represented by the system mathematical model, through the generation of fault indicative signals (residuals) and its analysis. So, we have a mathematical model of the system and construct the residuals as the difference between the available measure and the model output. If this residual is zero there is not a fault, if it is different from zero there is a conflict, i.e., a fault. The better the model used to represent the dynamic behavior of the system, the better will be the chance of improving the reliability and performance in diagnosing faults. However, is it the model accuracy enough that the residual will be zero in nominal conditions? In most real cases, the answer is no because a perfect and complete mathematical model of a physical system is never available. Usually, the parameters vary with time in an uncertain manner (uncertainties), there are modeling errors, and the characteristics of the disturbances and noise are unknown. Thus, there is always a mismatch between the actual process and its mathematical model even if there are no faults. Uncertainty and disturbances constitute a source of false and missed alarms which can worsen the fault detection and isolation system performance to such an extent that it may even become totally useless.

Hence, there is a need to develop robust fault diagnosis algorithms. The robustness of a fault diagnosis system means that it must be insensitive or even invariant to uncertainties and unmodeled disturbances whilst at the same time sensitive to faults. Both, the faults and the uncertainty affect the residual and discrimination between their effects is difficult. There are two ways to overcome this problem and to increase the robustness of the FDI scheme: to use robust residual generation or to use robust residual evaluation.

- *Robust residual generation*: The heart of the model-based fault diagnosis is the generation of residuals. So, the first approach to increase the robustness of the fault diagnosis system is to generate robust residuals following the *active* approach. This is achieved by using a mathematical model of the monitored system that includes all kinds of uncertainties that can occur in practice and affect the behavior of the system in order to build a set of transformed residuals as insensitive as possible to their effect while still sensitive to the fault effect. There exist robust extensions of the classical model-based methods to deal with uncertainty leading to robust observers or robust parity equations.
- *Robust residual evaluation*: The other alternative is achieving robustness in the decision-making stage by considering the effect of uncertainty in the generation of the detection threshold (*passive robustness*). The goal in this case is to minimize the false and missing alarm rates due to the effects of the modeling uncertainty and unknown disturbances will have on the residuals. This can be achieved in several ways, e.g., using adaptive thresholds, i.e., thresholds that change with the dynamic of the system, using thresholds generated by fuzzy logic, using statistical decisions or using set-based approaches, i.e., generate envelopes, etc.

1.3.2 Data-Driven Methods

These kinds of methods based only on data have a lot of names in the literature; they are referred as model-free approaches, process history-based methods, data-based methods, data mining methods, or instance-based methods. The techniques grouped in this category only exploit experimental data to design the fault diagnosis system. These methods are specially indicated for processes where the mathematical models do not exist or are incomplete, imprecise, or very difficult to calculate. When the dimensionality (number of variables) or complexity (distributed, nonlinear, variant systems) makes infeasible the use of other techniques, and when exists or is feasible to get a case base (examples) of previously documented experiences to infer a model.

There are some tasks to carried out in the design of fault diagnosis systems when the data-driven approach is used:

- *Preprocessing:* The aim of this step is to transform the data in such a way, that it can be more effectively processed by the actual model. The usual steps are the handling of missing data, outliers detection and replacement, filtering, normalization, etc. The importance of this step is critical in the behavior of the final system, and it requires a large amount of manual work and expert knowledge about the underlying system.

- *Exploratory data analysis, transformation, and feature extraction:* Extract information from raw data or transform the data to get a better representation. Typically, the data measured in the process industry are strongly colinear. This results from the partial redundancy in the measurements collected from the plant, so these environments are often called data rich but information poor. In this case, only informative variables are required. Redundant information unnecessarily increases the model complexity, which has often negative effects on the model training and performance. There are two ways to deal with the colinearity problem. One way is by transforming the input variables into a new reduced space with less colinearity: feature extraction. Another way to handle colinearity is to select a subset of the variables which is less colinear and at the same time are the most significant ones: feature selection.
- *Model selection, construction, and validation:* It is necessary to select the model to be trained taking into account the available data. There is no unified theoretical approach for this task and thus the model type and its parameters are often selected in an ad hoc manner for each method. After finding the optimal model structure and training the model, the trained system has to be evaluated on independent data once again, i.e., the model must be validated. In this task, different questions need to be taken into account: Are the assumptions made on available data true?. How representative is the available data? Is the model consistent with the actual and future data?. Here is the generalization problem, i.e., the designed fault diagnosis system must act well with data not used in the model construction and validation.
- *Model exploitation:* To use the method designed for the detection and diagnosis of faults in the plant.

All these tasks are usually done in an iterative way, e.g., after the standardization and missing value treatment which are usually performed only once, an outlier removal and feature selection are repeatedly applied until the model developer considers the data as being ready to be used for the training and evaluation of the actual model. Once the model is constructed, if the validation is not adequate, the designer can change the data, the type of model, and so on until a good model is obtained for the final purpose: to detect and isolate faults in the real plant.

This procedure can be performed in a qualitative way, like expert systems, or in a quantitative way. In this case, there are also two possibilities:

1. *Computational methods:* those obtained from methods developed in the area of computer science or Artificial Intelligence, and also can be considered as knowledge-based approach:

    - *Pattern recognition methods:* use the association between data patterns and fault classes without an explicit modeling of the internal process state or structure. Examples include clustering techniques, classification methods, case-based reasoning, etc. They are black box methods that learn the pattern based entirely on data in the training sessions.

- *Soft computing methods* that uses a combination of data and heuristic knowledge, as neural networks, fuzzy logic, Support Vector Machines (SVM), etc. These techniques can be used in different forms in order to detect faults, just as a data-driven method, in other words, to use the networks, the SVM techniques to classify the data in normal or faulty data. Or to use the approaches, as a model-based techniques, that uses the neural networks, the fuzzy logic, the SVM to build a nonlinear model of the system.

2. *Statistical models:* where a probabilistic behavior is assumed in data as in the case of:

   - *Parametric models:* where a predefined function specified by a set of parameters is assumed as a model: a distribution function, regression models, multivariate statistical techniques as Partial Least Squares (PLS) or Principal Component Analysis (PCA), etc. These last techniques compare the past trends of the process under control with the current state of the plant in order to detect and diagnose faults.
   - *Nonparametric models:* data follows a distribution function but this is neither predefined or parametrized: histograms, spectrum analysis (fault-specific frequencies in sound, vibration, etc).

The problems with these methods driven by data are that it is necessary data in normal as in faulty situations, in order to determine the different fault classes, data that not always is available in real systems. As consequence, these methods are only able to detect known faults, in other words, faults that data are available, for new faults or faults that have not enough data it is possible to detect them, but not to isolate them. Also, these methods are based on the capacity of generalization of the system, in other words, that they can learn all the behavior of the system just with a few data, and must respond adequately with new data is available.

1.3.3 Qualitative Models and Search Strategies

This name is given by [8], but other authors classify this type of methods as *method based on knowledge*. In this group, two main classes are grouped together: on the one hand, the methods based on qualitative methods such as knowledge-based expert systems, and on the other hand, the search strategies such as topographic and symptomatic searches. Topographic searches find malfunctions using a template of normal operation and symptomatic searches try to find symptoms to direct the search to the fault location.

- *Expert systems*. Expert systems have been used in the area of fault diagnosis [6]. Expert systems are knowledge-based techniques which are closer in style to human problem-solving, and are used to imitate the reasoning of human experts when diagnosing the faults. The experience from a domain expert can be formulated in terms of rules, which can be combined with the knowledge from first-principles or a structural description of the system for diagnosing faults. Also, expert systems are able to capture human diagnostic associations that are not readily translated into mathematical or causal models. This methodology has a great advantage, that is a consolidated approach and there are methodologies and working systems to implement it. However, it has also some disadvantages: First, related to experience: one of the main difficulties of applying expert systems is the knowledge acquisition step, which is the step of collecting adequate knowledge from domain experts and translating it into computer programs. Domain experts may not be available for unique operation scenarios and for new plants. Or, when the domain experts are available they may not understand or be able to explain clearly how they solve a problem. Expert systems are application specific. Developing an effective expert system can be time-consuming and costly for large-scale systems. Second, related to classification methods: there is not available knowledge from new faults, and multiple symptoms can occur with multiple faults, and sometimes, it is difficult to distinguish multiple faults from a fault with similar symptoms. And finally, related to software: the correctness and completeness of the information stored in the knowledge base specifies the performance achievable by the expert system, to benefit from new experience and knowledge the knowledge base also needs to be updated periodically, with induces the problem of maintenance of the base, in other words, to maintain always the consistency of the base.

- *Causal models.* Another technique based on knowledge is the causal analysis that uses the concept of causal modeling of fault-symptom relationships. As the signed direct graph (digraphs) that is a qualitative model-based approach, where the process variables are represented by nodes and its relationships (constraints, i.e., the equations that relations the variables) are represented by arcs, a heuristic search for the graph provides the cause of the fault. Another example is the Fault Tree Analysis (FTA) that is a top-down, deductive failure analysis in which an undesired state of a system is analyzed using Boolean logic to combine a series of lower level events.

Acknowledgements This work has been partially funded by the Spanish State Research Agency (AEI) and the European Regional Development Fund (FEDER) through the projects MASCONTROL (ref. MINECO DPI2015-67341-C2-2-R), (ref. MINECO DPI2016-78831-C2-2-r) DEOCS (ref. MINECO DPI2016-76493) and SCAV (ref. MINECO DPI2017-88403-R). This work has also been partially funded by AGAUR of Generalitat de Catalunya through the grants 2017 SGR 01551/2017 SGR 482 and by Agència de Gesti6 d'Ajuts Universitaris i de Recerca.

References

1. Baseville, M.: Detecting changes in signals and systems—a survey. Automatica **24**(3), 309–326 (1988)
2. Cellier, F.: Continuous System Modeling. Springer, Berlin (1991)
3. Gaines, B.: General systems research: quo vadis. Gen. Syst.: Yearb. Soc. Gen. Syst. Res. **24**, 1–9 (1979)

4. Gertler, J.: Fault Detection and Diagnosis in Engineering Systems. Marcel Dekker, New York (1998)
5. Pouliezos, A.D., Stavrakakis, G.: Real Time Fault Monitoring of Industrial Processes. Kluwer Academic Publishers, Dordrecht (1994)
6. Tzafestas, S., Singh, M., Schmidt, G. (eds.): System Fault Dianostics, Realiability and Related Knowledge-Based Approaches. D. Reidel Publishing Company, Dordrecht (1987)
7. Venkatasubramanian, V., Rengaswamy, R., Kavuri, S.N.: A review of process fault detection and diagnosis. Part II: qualitative models and search strategies. Comput. Chem. Eng. **27**(3), 313–326 (2003)
8. Venkatasubramanian, V., Rengaswamy, R., Kavuri, S.N., Yin, K.: A review of process fault detection and diagnosis. Part I: quantitative model-based methods. Comput. Chem. Eng. **27**(3), 291–311 (2003)
9. Venkatasubramanian, V., Rengaswamy, R., Kavuri, S.N., Yin, K.: A review of process fault detection and diagnosis. Part III: process history based methods. Comput. Chem. Eng. **27**(3), 327–346 (2003)

Chapter 2
Case Studies and Modeling Formalism

Teresa Escobet, Belarmino Pulido, Anibal Bregon and Vicenç Puig

2.1 Introduction

In this chapter, several application examples are introduced, which will be used to illustrate the diagnosis schemes presented and studied in this book. The objective of introducing these application examples in the first part of this book, immediately after the introduction, is to provide the reader with the useful application background and understandings of some basic technical concepts in the process monitoring and fault diagnosis field.

To illustrate the component-oriented, first-order logic approach, typical in the artificial intelligence community proposal for model-based diagnosis, the classical polybox example will be used. This is a static system made up of three multipliers and two adders, with several interconnected inputs/outputs that helps to clarify the logic-based approach and also provides interesting interactions to explain conflicts and their relation to structurally independent *Analytical Redundancy Relations* (ARRs).

For dynamic systems, we will use a collection of tanks sequentially connected through pipes at the bottom part of the tanks. With these connections and different

T. Escobet (✉)
Research Center for Supervision, Safety and Automatic Control (CS2AC), Universitat Politècnica de Catalunya (UPC), Terrassa, Spain
e-mail: teresa.escobet@upc.edu

B. Pulido
Departamento de Informática, Universidad de Valladolid, Valladolid, Spain
e-mail: belar@infor.uva.es

A. Bregon
Departamento de Informática, Universidad de Valladolid, Segovia, Spain
e-mail: anibal@infor.uva.es

V. Puig
Research Center for Supervision, Safety and Automatic Control (CS2AC), Universitat Politècnica de Catalunya (UPC), Terrassa, Spain
e-mail: vicenc.puig@upc.edu

T. Escobet et al. (eds.), *Fault Diagnosis of Dynamic Systems*,
https://doi.org/10.1007/978-3-030-17728-7_2

sets of available sensors, we can illustrate all the concepts related to dynamic systems, structural analysis, diagnosability, sensor location, etc. Even more, we can introduce basic concepts for hybrid and discrete event systems diagnosis.

After introducing these simple examples, other more elaborated benchmarks that can be used by students to evaluate the techniques described in this book are introduced.

Different modeling approaches can be used to perform process monitoring, fault detection, and fault isolation. We can have models for static systems or for dynamic systems, equation or object-oriented, we can have quantitative or qualitative models, component-oriented or system-oriented, etc. Special attention must be also paid about how the different approaches consider fault modeling, either implicitly or explicitly in the equations. In this chapter, we will provide a glimpse of the most used techniques for model-based diagnosis.

2.2 Case Studies

We use some static and dynamic systems to present the fundamentals concepts related to structural analysis and model-based diagnosis.

2.2.1 Static Case Studies

The first static system is the polybox example taken from [7] and given in Fig. 2.1. This system has three multipliers, M1, M2, and M3 (operator $\times$), and two adders, A1 and A2 (operator $+$). Each one of the five components might be faulty.

The behavior of this system is given by the following equations in normal operation:

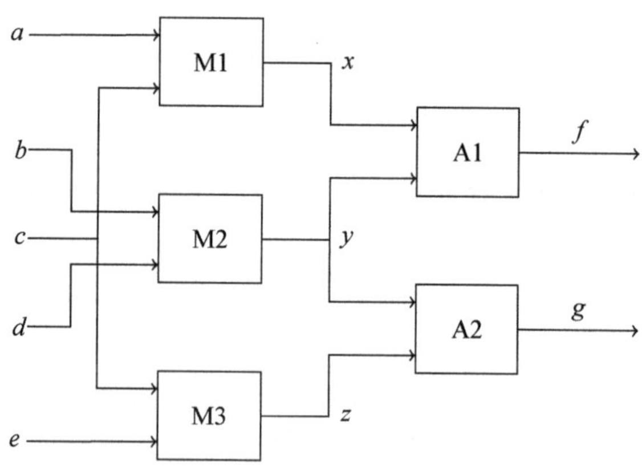

Fig. 2.1 The polybox system

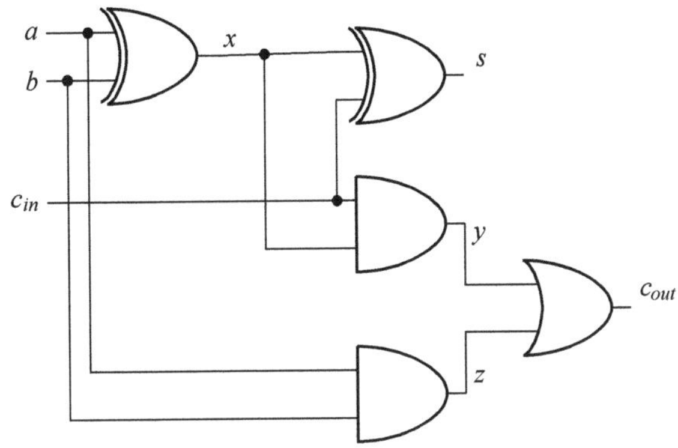

Fig. 2.2 A full adder

$$
\begin{aligned}
\text{M1}\colon x &= a \times c,\\
\text{M2}\colon y &= b \times d,\\
\text{M3}\colon z &= c \times e,\\
\text{A1}\colon f &= x + y,\\
\text{A2}\colon g &= y + z.
\end{aligned}
\tag{2.1}
$$

This system has five inputs, a, b, c, d, and e, and two outputs f and g; all these variables are observable. There are also three unobserved variables, x, y, and z.

The second static system taken from [22] is the binary full adder given in Fig. 2.2. This system has two XOR gates, two AND gates, and one OR gate, XOR1, XOR2, AND1, AND2, and OR1, respectively. The five components may be faulty.

In normal operation, the system is described by the following equations:

$$
\begin{aligned}
\text{XOR1}\colon x &= \text{XOR}(a, b),\\
\text{XOR2}\colon s &= \text{XOR}(x, c_{in}),\\
\text{AND1}\colon y &= \text{AND}(c_{in}, x),\\
\text{AND2}\colon z &= \text{AND}(a, b),\\
\text{OR1}\colon c_{out} &= \text{OR}(y, z).
\end{aligned}
\tag{2.2}
$$

This system has three inputs, a, b, and c_{in}, and two outputs s and c_{out}; all these variables are observable. There are also three unobserved variables, x, y, and z.

2.2.2 *Continuous System*

Regarding continuous systems, the three-tank system taken from [9] and shown in Fig. 2.3 will be used as the running example to illustrate some of the techniques covered in the book.

The system is made up of three identical tanks, $\{T_1, T_2, T_3\}$. All three tanks have the same physical features such as the ame height and cross-sectional area, A. There

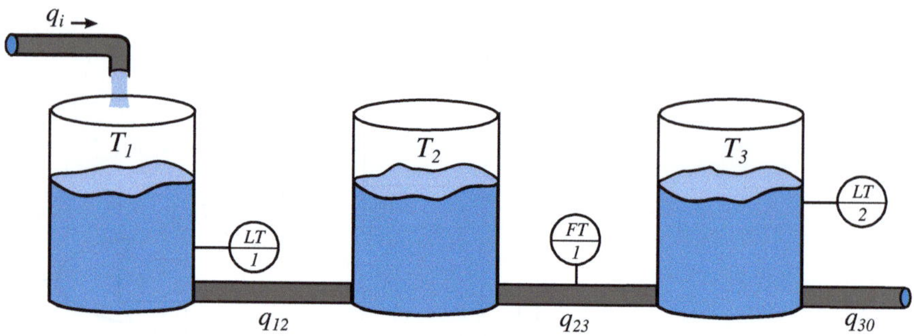

Fig. 2.3 Schematic representation of three-tank system

is a measured input flow q_i for tank T_1, which is drained into T_2 via a pipe q_{12}. A similar process gets the flow from T_2 to T_3 via pipe q_{23}. Finally, there is an output flow q_{30} from T_3. The system has three sensors measuring the level in thanks T_1 and T_3 (level transducers $LT1$ and $LT2$, respectively), and another sensor measuring the flow through pipe q_{23} (flow transducer $FT1$).

The following three dynamic system equations model the behavior of the system in normal conditions. The change in the level in each tank, $\dot{h}_{T_j}(t)$ for $j = 1, \ldots, 3$, is computed according to mass balances:

$$\dot{h}_{T_1}(t) = \frac{q_i(t) - q_{12}(t)}{A}, \tag{2.3}$$

$$\dot{h}_{T_2}(t) = \frac{q_{12}(t) - q_{23}(t)}{A}, \tag{2.4}$$

$$\dot{h}_{T_3}(t) = \frac{q_{23}(t) - q_{30}(t)}{A}. \tag{2.5}$$

Flows between tanks, $\{q_{12}(t), q_{23}(t), q_{30}(t)\}$, are modeled as

$$q_{12}(t) = S_{p_1}\text{sign}\big(h_{T_1}(t) - h_{T_2}(t)\big)\sqrt{2g \mid h_{T_1}(t) - h_{T_2}(t) \mid}, \tag{2.6}$$

$$q_{23}(t) = S_{p_2}\text{sign}\big(h_{T_2}(t) - h_{T_3}(t)\big)\sqrt{2g \mid h_{T_2}(t) - h_{T_3}(t) \mid}, \tag{2.7}$$

$$q_{30}(t) = S_{p3}\sqrt{2gh_{T_3}(t)}, \tag{2.8}$$

being S_{p_i} the cross-sectional area of the pipes. The three-tank system parameters are shown in Table 2.1.

In the case study, six possible faults are considered: three-tank leakages and three pipe blockages. Parameters $Stuck_{q_{jk}}$ and $Leakage_{T_j}$ are used to model pipe blockages and tank leakages, respectively, being the new mass balance:

$$\dot{h}_{T_1}(t) = \frac{q_i(t) - q_{12}(t) - q_{f_1}(t)}{A}, \tag{2.9}$$

Table 2.1 Parameter values for the three-tank system

Parameters	Symbol	Values	
Cross-sectional area of the tanks	A	0.154	$[m^2]$
Gravity coefficient	g	9.81	$[m/s^2]$
Cross-sectional area of the pipe q_{12}	S_{p1}	2.3×10^{-5}	$[m^2]$
Cross-sectional area of the pipe q_{23}	S_{p2}	3.0×10^{-5}	$[m^2]$
Cross-sectional area of the pipe q_{30}	S_{p3}	2.25×10^{-5}	$[m^2]$
Max. height of tanks	$h_{T_i}^{max}$	1.20	[m]

$$\dot{h}_{T_2}(t) = \frac{q_{12}(t) - q_{23}(t) - q_{f_2}(t)}{A}, \tag{2.10}$$

$$\dot{h}_{T_3}(t) = \frac{q_{23}(t) - q_{30}(t) - q_{f_3}(t)}{A}. \tag{2.11}$$

Flow between tanks, $\{q_{12}(t), q_{23}(t), q_{30}(t)\}$, and leakage flows, $\{q_{f_1}(t), q_{f_2}(t), q_{f_3}(t)\}$, are modeled as

$$q_{12}(t) = \left(1 - Stuck_{q_{12}}\right) S_{p_1} \operatorname{sign}\left(h_{T_1}(t) - h_{T_2}(t)\right) \sqrt{2g \mid h_{T_1}(t) - h_{T_2}(t) \mid}, \tag{2.12}$$

$$q_{f_1}(t) = Leakage_{T_1} \sqrt{2g h_{T_1}(t)}, \tag{2.13}$$

$$q_{23}(t) = \left(1 - Stuck_{q_{23}}\right) S_{p_2} \operatorname{sign}\left(h_{T_2}(t) - h_{T_3}(t)\right) \sqrt{2g \mid h_{T_2}(t) - h_{T_3}(t) \mid}, \tag{2.14}$$

$$q_{f_2}(t) = Leakage_{T_2} \sqrt{2g h_{T_2}(t)}, \tag{2.15}$$

$$q_{30}(t) = \left(1 - Stuck_{q_{30}}\right) S_{p_3} \sqrt{2g h_{T_3}(t)}, \tag{2.16}$$

$$q_{f_3}(t) = Leakage_{T_3} \sqrt{2g h_{T_3}(t)}. \tag{2.17}$$

In this way, we can deal with both additive and multiplicative faults in the model by just setting the corresponding parameters to 0 or 1. Notice that models in normal conditions, Eqs. (2.3)–(2.5) and (2.6)–(2.8), are equal to abnormal model, Eqs. (2.9)–(2.11) and (2.12), (2.14), (2.16), being $q_{f_j} = 0$ and $Stuck_{q_{jk}} = 0$.

It is assumed that the measured variables are h_{T_1}, h_{T_3} and q_{23} and the input flow q_i is known. It is also assumed that the initial water level in the three tanks is zero, but this can be configured for each diagnosis scenario. Table 2.1 lists the parameters value used for simulation.

In the following, the variable t in parentheses will be omitted when no confusion can arise.

2.3 System Modeling

A system is made of a set of components and a set of sensors, which provide a set of observations. The (normal) behavior model of the system expresses the constraints that link its descriptive variables. It is given by a set of relations, the formal expression of which depends on the type of knowledge (analytical, qualitative, production rules, numerical tables, etc.). In the classical DX approaches to diagnosis usually, there is a component-oriented description of the system behavior. On the other hand, in the classical FDI view, the model is stated at a system level, where the equations establish relations between input, output, and control variables in the system.

In most of the diagnostic methodologies presented in this book, a component-oriented model of the system will be required. Following the classification given by [26], the model-based methodologies can be classified as qualitative or quantitative. In component-oriented model, the model is developed based on the physical fundamentals of the process components. Quantitative models use fundamental mathematical relations between the inputs and the outputs of each component, in contrast to qualitative models where these relationships are expressed in terms of qualitative functions between different components of the system. The most popular qualitative models in diagnosis are: structural models, bond graph, and causal graphs. These approaches depend upon the knowledge from experts in both the normal and fault situations. In this chapter, we introduce three modeling methodologies: Differential-algebraic equations, consistency-based diagnosis, and causal graphs.

2.3.1 Differential-Algebraic Equations (DAEs)

Quantitative models can be given as a set of Differential-Algebraic Equations (DAEs). In this case, the system model can be defined as $\mathcal{M}(SM, \mathcal{U}, \mathcal{Y}, \mathcal{X}, \Theta)$, where

- SM is the system model or set of DAEs, defined over a collection of known and unknown variables;
- $\mathcal{U}$ is a *set of inputs*;
- $\mathcal{Y}$ is a *set of outputs*;
- $\mathcal{X}$ is a *set of state* and intermediate, i.e., unknown variables;
- Θ is the *set of model parameters* including possible faults.

Example 2.1 The polybox system in Fig. 2.1 is a static system and can be formulated as a set of Algebraic Equations (AEs):

$$
\begin{aligned}
\mathcal{M}(SM, \mathcal{U}, \mathcal{Y}, \mathcal{X}, \Theta) &= \\
SM &= \{\mathrm{M1} : x = a \times c, \\
&\quad\ \mathrm{M2} : y = b \times d, \\
&\quad\ \mathrm{M3} : z = c \times e, \\
&\quad\ \mathrm{A1} : f = x + y, \\
&\quad\ \mathrm{A2} : g = y + z\} \\
\mathcal{U} &= \{a, c, b, d, e\} \\
\mathcal{Y} &= \{f, g\} \\
\mathcal{X} &= \{x, y, z\} \\
\Theta &= \{\mathrm{M1}, \mathrm{M2}, \mathrm{M3}, \mathrm{A1}, \mathrm{A2}\}.
\end{aligned}
$$

Example 2.2 The full adder given in Fig. 2.2 can be formulated as a set of nonlinear AE:

$$\begin{aligned}
\mathcal{M}(SM, \mathcal{U}, \mathcal{Y}, \mathcal{X}, \Theta) = \\
SM &= \{\text{X1} : x = \text{XOR}(a, b), \\
&\quad \text{X2} : s = \text{XOR}(x, c_{in}), \\
&\quad \text{A1} : y = \text{AND}(c_{in}, x), \\
&\quad \text{A2} : z = \text{AND}(a, b), \\
&\quad \text{O1} : c_{out} = \text{OR}(y, z)\} \\
\mathcal{U} &= \{a, b, c_{in}\} \\
\mathcal{Y} &= \{s, c_{out}\} \\
\mathcal{X} &= \{x, y, z\} \\
\Theta &= \{\text{X1}, \text{X2}, \text{A1}, \text{A2}, \text{O1}\}.
\end{aligned}$$

Example 2.3 In the case of three-tank system (see Fig. 2.3), the model can be summarized as

$$\begin{aligned}
e1 : \quad & \dot{h}_{T_1} = \frac{q_i - q_{12} - q_{f_1}}{A}, \\
e2 : \quad & \dot{h}_{T_2} = \frac{q_{12} - q_{23} - q_{f_2}}{A}, \\
e3 : \quad & \dot{h}_{T_3} = \frac{q_{23} - q_{30} - q_{f_3}}{A}, \\
e4 : \quad & q_{12} = \left(1 - Stuck_{q_{12}}\right) S_{p_1} \operatorname{sign}\left(h_{T_1} - h_{T_2}\right) \sqrt{2g \mid h_{T_1} - h_{T_2} \mid}, \\
e5 : \quad & q_{f_1} = Leakage_{T_1} \sqrt{2g h_{T_1}}, \\
e6 : \quad & q_{23} = \left(1 - Stuck_{q_{23}}\right) S_{p_2} \operatorname{sign}\left(h_{T_2} - h_{T_3}\right) \sqrt{2g \mid h_{T_2} - h_{T_3} \mid}, \\
e7 : \quad & q_{f_2} = Leakage_{T_2} \sqrt{2g h_{T_2}}, \\
e8 : \quad & q_{30} = \left(1 - Stuck_{q_{30}}\right) S_{p_3} \sqrt{2g h_{T_3}}, \\
e9 : \quad & q_{f_3} = Leakage_{T_3} \sqrt{2g h_{T_3}}.
\end{aligned}$$

The observational model, which relates internal and measured variables, is given by the following equations[1]:

$$\begin{aligned}
e10 &: h_{T_{1,obs}} = h_{T_1}, \\
e11 &: h_{T_{3,obs}} = h_{T_3}, \\
e12 &: q_{23,obs} = q_{23}.
\end{aligned}$$

In our system model, SM, the set of state variables is $\mathcal{X}$, and contains those variables whose evolution changes over time. We make explicit the relation between the state variables, h_{T_i} in our example, and their derivatives: dh_{T_i}, through equations:

[1] For the sake of simplicity, we assume that sensor models are the identity operator.

$$e13 : h_{T_1} = \int \dot{h}_{T_1} dt, \tag{2.18}$$

$$e14 : h_{T_2} = \int \dot{h}_{T_2} dt, \tag{2.19}$$

$$e15 : h_{T_3} = \int \dot{h}_{T_3} dt. \tag{2.20}$$

Using the model notation based on DAEs introduced in the previous section, we can summarize the model as follows:

$$\begin{aligned}
&\mathcal{M}(SM, \mathcal{U}, \mathcal{Y}, \mathcal{X}, \Theta) = \\
&\quad SM \quad = \{\text{e1}, \ldots, \text{e15}\}, \\
&\quad \mathcal{U} \quad = \{q_i\}, \\
&\quad \mathcal{Y} \quad = \{h_{T_{1,obs}}, h_{T_{3,obs}}, q_{23,obs}\}, \\
&\quad \mathcal{X} \quad = \{h_{T_1}, h_{T_2}, h_{T_3}, q_{12}, q_{23}, q_{30}, q_{f_1}, q_{f_2}, q_{f_3}, \dot{h}_{T_1}, \dot{h}_{T_2}, \dot{h}_{T_3}\}, \\
&\quad \Theta \quad = \{A, S_p, Stuck_{q_{12}}, Stuck_{q_{23}}, Stuck_{q_{30}}, Leakage_{T_1}, \\
&\qquad\qquad Leakage_{T_2}, Leakage_{T_3}\}.
\end{aligned}$$

A typical set of SM may be formulated in the temporal domain as a state-space model:

$$BM : \dot{\mathbf{x}}(t) = f(\mathbf{x}(t), \mathbf{u}(t), \theta), \tag{2.21}$$

$$OM : \mathbf{y}(t) = g(\mathbf{x}(t), \mathbf{u}(t), \theta), \tag{2.22}$$

where BM is the behavioral model, OM is the observational model, f and g are real functions over $\Re$, $\mathbf{x}(t) \in \Re^{n_x}$, $\mathbf{u}(t) \in \Re^{n_u}$, $\mathbf{y}(t) \in \Re^{n_y}$, and $\theta \in \Re^{n_\theta}$. BM describes the way the system evolves in time, as a consequence of the system inputs, while OM describes the measurements which are available [25].

We can also obtain an equivalent state-space representation to the set of DAEs for the three-tank system as in Eqs. (2.21) and (2.22). In this case, we have three-state variables $\{h_1, h_2, h_3\}$. If we use the nominal behavior model, without parameters used to model the faults:

$$\begin{aligned}
\dot{h}_{T_1} &= f_1(h_{T_1}, h_{T_2}, q_i), \\
\dot{h}_{T_2} &= f_2(h_{T_1}, h_{T_2}, h_{T_3}), \\
\dot{h}_{T_3} &= f_3(h_{T_2}, h_{T_3}), \\
h_{T_{1,obs}} &= g_1(h_{T_1}), \\
h_{T_{3,obs}} &= g_2(h_{T_3}), \\
q_{23,obs} &= g_3(h_{T_2}, h_{T_3}),
\end{aligned}$$

where

$$\begin{aligned}
f_1(\cdot) : \dot{h}_{T_1} &= (q_i - q_{12})/A \\
&= \left(q_i - (1 - Stuck_{q_{12}}) S_{p_1} \operatorname{sign}(h_{T_1} - h_{T_2}) \sqrt{2g \mid h_{T_1} - h_{T_2} \mid}\right)/A,
\end{aligned}$$

$$
\begin{aligned}
f_2(\cdot): \dot{h}_{T_2} &= (q_{12} - q_{23})/A \\
&= \Big(\big(1 - Stuck_{q_{12}}\big) S_{p_1} \operatorname{sign}\big(h_{T_1} - h_{T_2}\big)\sqrt{2g \mid (h_{T_1} - h_{T_2} \mid} - \\
&\quad \big(1 - Stuck_{q_{23}}\big) S_{p_2} \operatorname{sign}\big(h_{T_2} - h_{T_3}\big)\sqrt{2g \mid h_{T_2} - h_{T_3} \mid}\Big)/A, \\
f_3(\cdot): \dot{h}_{T_3} &= (q_{23} - q_{30})/A \\
&= \Big(\big(1 - Stuck_{q_{23}}\big) S_{p_2} \operatorname{sign}\big(h_{T_2} - h_{T_3}\big)\sqrt{2g \mid (h_{T_2} - h_{T_3} \mid)} - \\
&\quad \big(1 - Stuck_{q_{30}}\big) S_{p_3} \sqrt{2g \mid h_{T_3} \mid}\Big)/A,
\end{aligned}
\tag{2.23}
$$

$$
\begin{aligned}
g_1(\cdot): h_{T_{1,obs}} &= h_{T_1}, \\
g_2(\cdot): h_{T_{3,obs}} &= h_{T_3}, \\
g_3(\cdot): q_{23,obs} &= q_{23} \\
&= \big(1 - Stuck_{q_{23}}\big) S_{p_2} \operatorname{sign}\big(h_{T_2} - h_{T_3}\big)\sqrt{2g \mid h_{T_2} - h_{T_3} \mid}.
\end{aligned}
$$

2.3.2 Logical Models for AI Approaches

Consistency-based diagnosis, CBD, is the main formulation of the artificial intelligence community. It was developed for static, time-invariant parameter systems diagnosis. Its main features are as follows:

- It is component-oriented: the system is made up of a collection of connected components that only interact through their terminals.
- System behavior is obtained from the component models and the system structure, which describes how components are connected.
- It uses only correct behavior models, which are local.

CBD is usually applied to systems made up of a collection of components, which are interconnected, and the global behavior is obtained by local estimations made by each component model which is later propagated through the component terminals to other components. The propagation paths are described in the system structure. Typical examples are physical devices made up of transistors, resistances, valves, pumps, etc. Magnitudes in the models (such as mass, pressure, or voltage) are linked to the terminals. Terminals are modeled as ideal components, but it is possible to model faults in the connections by modeling them as physical elements. For instance, we can provide a model for an electric wire if we want to associate a potential fault to an open wire.

Main assumption in CBD is the *no function in structure*: local models must make no assumption about its environment, potential connections, etc. Local models must contain only relations among inputs, internal variables, and terminal outputs.

In real-world applications, it can be difficult to comply with this assumption. In those cases, it is possible to lump together a set of components into a super-component, with a higher level of abstraction, but still satisfying the no function in structure principle.

This is an essential point in the CBD reasoning process, because it is the only way to make assumptions of correct behavior only local to the component. This assumption of correct behavior will be made explicit in the local model by means of an special predicate $AB(\cdot)$, which states that the component is behaving abnormal or not. The nominal behavior will be stated as $\neg AB(c)$, being c a component type, and this assumption can later be retracted while reasoning about inconsistencies. In that way, a typical model in CBD makes explicit the assumption in the left-hand side of the formula, while the constraints used to express the model are in the right-hand side of the formula, as in

$$ADD(c) \wedge \neg AB(c) \Rightarrow out(c) = in_1(c) + in_2(c).$$

The reader should notice that the relation between the correct assumption and the constraints is done by means of $\Rightarrow$ and not through the biconditional, $\Leftrightarrow$, thus excluding exoneration from the reasoning process, and maintaining the soundness of the inference. This issue will be deeply explained in Chap. 5.

Reiter, in his seminal paper "A Theory of diagnosis from First Principles" [22], proposed a general formal theory for CBD. Such formalization uses the language of first-order logic and it is still the conceptual framework for CBD. It defines precisely what is a diagnosis problem and its set of solutions: a system and its diagnoses.

Unfortunately, there is no general logic theory universally accepted to model dynamic systems.

Definition 2.1 (*System*) A system is a triplet $\{SD, COMP, OBS\}$, where

- SD, the system description, is a finite set of first-order sentences.
- $COMP$, the system components, is a finite set of symbols of constants.
- OBS, the system observations, is a finite set of first-order sentences.

In Reiter's theory, a system specifies a diagnosis problem:

SD defines the system structure and provides the models of correct behavior of its components. Connections are modeled with equality relations over the values of the signals of their terminals. The correct behavior assumption is made explicit with the $\neg AB(x)$ literal.

$COMP$ provides the names of the system components.

OBS states the observations of the system for the posed diagnosis problem. They are modeled with equality relations between constant values—the observed values—and the values of the signals of the terminals were observations take place. In other model-based approaches, it is called the observational model.

Example 2.4 The Polybox system can be written using first-order logic formula as

$$
\begin{aligned}
COMP &= \{\text{M1, M2, M3, A1, A2}\}\\
SD &= \{ADD(\text{c}) \wedge \neg AB(\text{c}) \Rightarrow out(\text{c}) = in_1(\text{c}) + in_2(\text{c}),\\
&\quad MULT(\text{c}) \wedge \neg AB(\text{c}) \Rightarrow out(\text{c}) = in_1(\text{c}) \times in_2(\text{c}),\\
&\quad MULT(\text{M1}), MULT(\text{M2}), MULT(\text{M3}), ADD(\text{A1}), ADD(\text{A2}),\\
&\quad out(\text{M1}) = in_1(\text{A1}), out(\text{M2}) = in_2(\text{A1}),\\
&\quad out(\text{M2}) = in_1(\text{A2}), out(\text{M3}) = in_2(\text{A2})\}, \text{ for c} \in \mathcal{COMP}\\
OBS &= \{in_1(\text{M1}), in_2(\text{M1}), in_1(\text{M2}), in_2(\text{M2}),\\
&\quad in_1(\text{M3}), in_2(\text{M3}), out(\text{A1}), out(\text{A2})\}.
\end{aligned}
$$

Example 2.5 The full adder system can be written using first-order logic formula as

$$
\begin{aligned}
COMP &= \{\text{X1, X2, A1, A2, O1}\}\\
SD &= \{XORgate(\text{c}) \wedge \neg AB(\text{c}) \Rightarrow out(\text{c}) = in_1(\text{c}) \oplus in_2(\text{c}),\\
&\quad ANDgate(\text{c}) \wedge \neg AB(\text{c}) \Rightarrow out(\text{c}) = in_1(\text{c}) \wedge in_2(\text{c}),\\
&\quad ORgate(\text{c}) \wedge \neg AB(\text{c}) \Rightarrow out(\text{c}) = in_1(\text{c}) \vee in_2(\text{c}),\\
&\quad XORgate(\text{X1}), XORgate(\text{X2}), ANDgate(\text{A1}),\\
&\quad ANDgate(\text{A2}), ORgate(\text{O1}),\\
&\quad out(\text{X1}) = in_1(\text{X2}), out(\text{X1}) = in_2(\text{A1}),\\
&\quad out(\text{A1}) = in_1(\text{O1}), out(\text{A2}) = in_2(\text{O1})\}, \text{ for c} \in \mathcal{COMPS}\\
OBS &= \{in_1(\text{X1}), in_2(\text{X1}), in_2(\text{X2}), in_1(\text{A1}),\\
&\quad in_1(\text{A2}), in_2(\text{A2}), out(\text{X2}), out(\text{O1})\}.
\end{aligned}
$$

Example 2.6 The three-tank system cannot be directly written using First-Order Logic (FOL), because it requires a explicit representation for time. There is no general extension that includes time, as it will explained later in Chap. 6. However, we could represent the instantaneous relations between variables in the model using the following FOL sentences:

$$
\begin{aligned}
COMPS &= \{\text{T}_1, \text{T}_2, \text{T}_3, q_{12}, q_{23}, q_{30}\}\\
SD &= \{\text{Tank(T)} \wedge \neg AB(\text{T}) \Rightarrow \dot{h}(T) = \tfrac{q_{in}(T) - q_{out}(T)}{A},\\
&\quad \text{Pipe(P)} \wedge \neg AB(\text{P}) \Rightarrow\\
&\quad q(P) = S_p \text{sign}\big(h_{left}(P) - h_{right}(P)\big)\sqrt{2g \mid h_{left}(P) - h_{right}(P) \mid},\\
&\quad \text{Tank(T1), Tank(T2), Tank(T3)},\\
&\quad \text{Pipe}(q_{12}), \text{Pipe}(q_{23}), \text{Pipe}(q_{30}),\\
&\quad q_{out}(T_1) = q_{in}(T_2), q_{out}(T_2) = q_{in}(T_3),\\
&\quad h_{left}(q_{12}) = h_{T_1}, h_{right}(q_{12}) = h_{T_2}, h_{left}(q_{23}) = T_2,\\
&\quad h_{right}(q_{23}) = T_3, h_{left}(q_{30}) = T_3, h_{right}(q_{30}) = 0,\\
OBS &= \{q_i, h_{T_1}, h_{T_3}, q_{23}\}.
\end{aligned}
$$

To comply with the DX only correct behavior, component-oriented approach, we have decided to consider pipes as isolable components: $Pipe(q_{12})$, $Pipe(q_{23})$, and $Pipe(q_{30})$. The reader should notice the straightforward relation of the parameters in Θ used to model faults, and the $\neg AB(\cdot)$ predicates in these sentences. For instance, $\neg AB(T_i)$ is related to potential leakage faults in tanks T_i, while $\neg AB(P_{xy})$ makes reference to potential fault blockages in pipes q_{xy}.

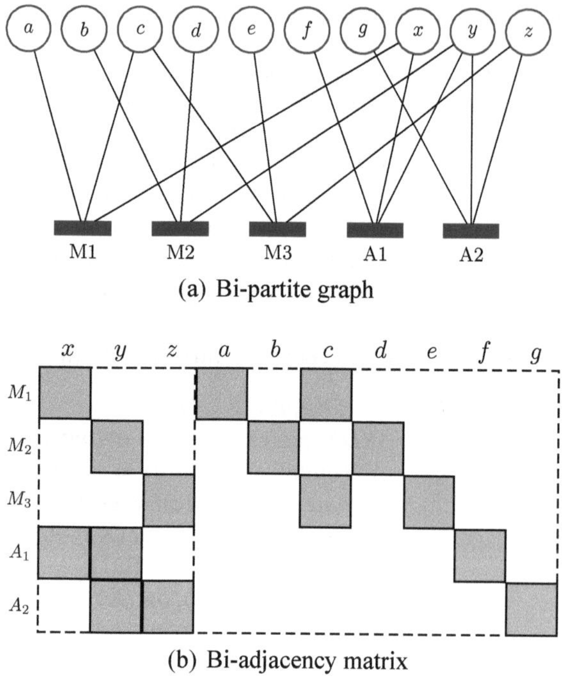

Fig. 2.4 Bipartite graph and bi-adjacency matrix for the polybox model

2.3.3 Structural Graph

The model can also be given by an structural model. The structural model contains only the information of which variables are contained in each constraint. The system's structural model can be represented by a bipartite graph $\mathcal{G} = (\mathcal{E} \bigcup \mathcal{V}, \mathcal{A})$ where $\mathcal{E}$ is the set of system equations or constraints, $\mathcal{V}$ is the set of variables, and $\mathcal{A}$ is a set of arcs such that $a(i,j) \in \mathcal{A}$ iff variable $v_j \in \mathcal{V}$ appears in relation $e_i \in \mathcal{E}$. A bipartite graph can also be represented by the bi-adjacency matrix which crosses model relations in rows and model variables in columns.

Example 2.7 The polybox system modeled with an structural model. The model has five constraints $\mathcal{E} = \{M1, M2, M3, A1, A2\}$ and ten variables $\mathcal{V} = \{x, y, z, a, b, c, d, e, f, g\}$. The bipartite graph and corresponding bi-adjacency matrix are shown in Fig. 2.4.

Example 2.8 The three-tank system can be structurally modeled considering normal or abnormal behavior conditions.

- Structural models in normal conditions are

$$\mathcal{E} = \{e1, e2, e3, e4, e6, e8, e10, e11, e12, e13, e14, e15\},$$
$$\mathcal{V} = \{h_{T_1}, h_{T_2}, h_{T_3}, q_{12}, q_{23}, q_{30}, \dot{h}_{T_1}, \dot{h}_{T_2}, \dot{h}_{T_3}, h_{T_{1,obs}}, h_{T_{3,obs}}, q_{23,obs}, q_i\},$$

and the bipartite graph and corresponding bi-adjacency matrix are shown in Fig. 2.5.

- Structural model in abnormal conditions:

$$\begin{aligned}\mathcal{E} &= \{e1, e2, e3, e4, e5, e6, e7, e8, e9, e10, e11, e12, e13, e14, e15\},\\ \mathcal{V} &= \{h_{T_1}, h_{T_2}, h_{T_3}, q_{12}, q_{23}, q_{30}, \dot{h}_{T_1}, \dot{h}_{T_2}, \dot{h}_{T_3}, q_{f_1}, q_{f_2}, q_{f_3}, Stuck_{q12}, Stuck_{q23}, Stuck_{q30},\\ &\quad Leakage_{T_1}, Leakage_{T_2}, Leakage_{T_3}, h_{T_{1,obs}}, h_{T_{3,obs}}, q_{23,obs}, q_i\},\end{aligned}$$

with the bi-adjacency matrix shown in Fig. 2.6.

2.3.4 *Modeling with Bond Graphs*

Bond graphs are labeled, directed graphs that present a topological, domain-independent, lumped-parameter, energy-based methodology for modeling the dynamic behavior of physical systems [13, 24]. The main advantages of bond graphs as a modeling language can be summarized as follows:

- It is an intuitive modeling approach based on the notion of systems topology; therefore, system models can be directly linked to system structure. The explicit modeling of topology (unlike the state-space formalism) provides additional information to establish causal relations among system variables.
- It is a generic energy-based framework modeling language that applies across multiple physical domains, such as mechanical, electrical, fluid, and thermal.
- The standard mathematical models of dynamic system behavior, e.g., the state-space and I/O formulations, can be systematically derived from bond graph models.

System models are represented as graphs, with nodes of the graph representing components or submodels, and edges, called *bonds*, representing ideal[2] energy transfer paths between the components and submodels. Bonds are drawn as half arrows ($\rightharpoonup$), and specified by a bond number i, i.e., $\overset{i}{\rightharpoonup}$. Dynamic behavior of the system is defined as a function of energy exchange between components of a system [13]. The state of the system is defined by the distribution of energy among its components at any particular time. Energy exchange between components is expressed in terms of *power*, i.e., the rate of transfer of energy between system components. Power is the product of two conjugated variables [5]: *effort* (e) and *flow* (f). Examples of effort and flow variables are force and velocity in mechanical systems, voltage and current in electrical networks, and pressure and volume or mass flow rate in hydraulic systems.

Bond graphs include a small set of domain-independent primitive elements [5] that interface with bonds through *ports*. Hence, bond graphs contain one-port, two-port, and multi-port elements. One-port elements model energy *storage elements* as capacitances (C:C, such that $\dot{e}_i = \frac{1}{C} f_i$) and inertias (I:$I$, such that $\dot{f}_i = \frac{1}{I} e_i$), energy

[2]These connections neither generate nor dissipate energy.

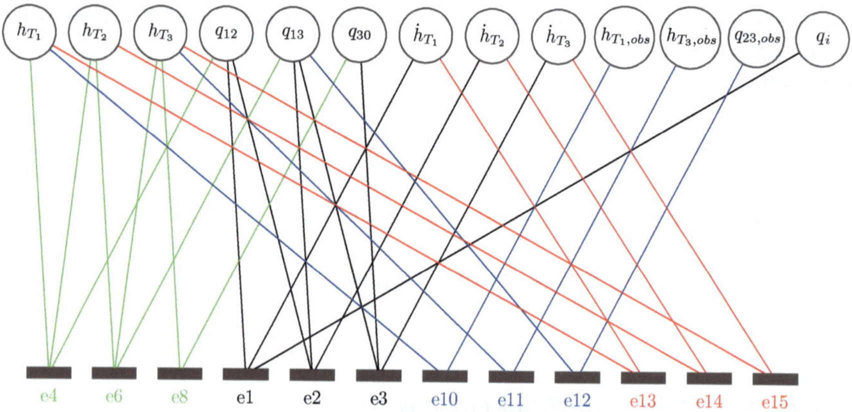

(a) Bi-partite graph

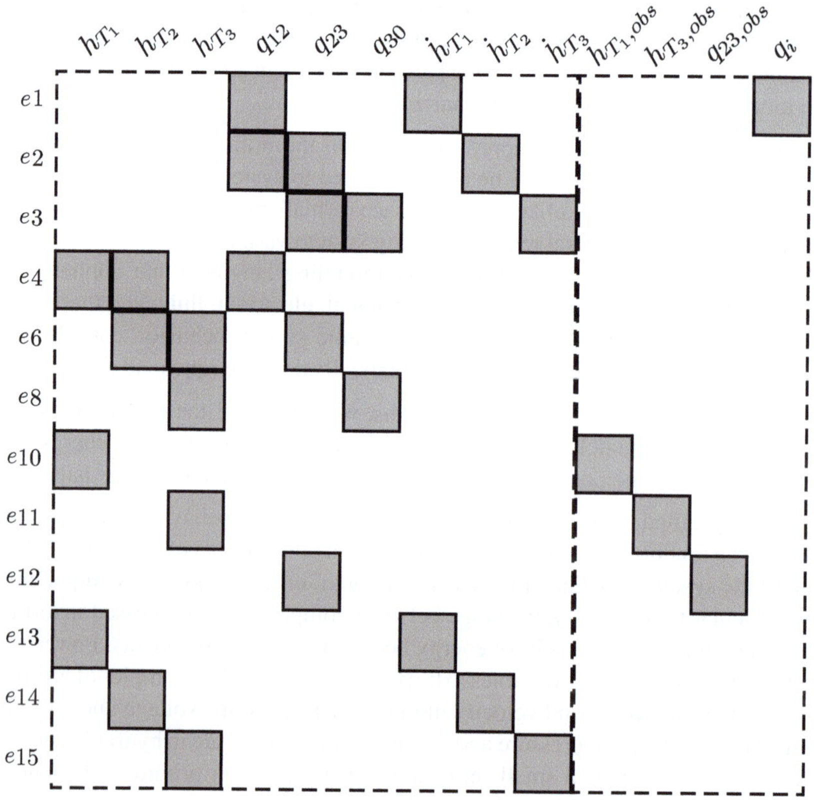

(b) Bi-adjacency matrix

Fig. 2.5 Bipartite graph and bi-adjacency matrix for the three-tank model in normal operation

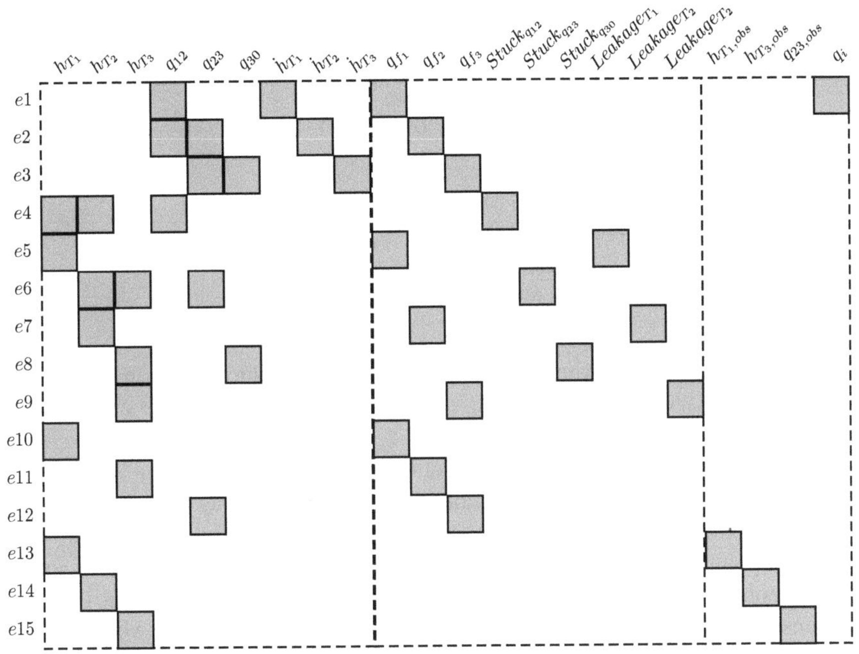

Fig. 2.6 Bi-adjacency matrix for the three-tank model in abnormal operation

dissipation elements as resistors (R:R, such that $e_i = Rf_i$), and energy *source elements* as sources of effort (Se:u, such that $e_i = u$) and flow (Sf:u, such that $f_i = u$). Two-port elements model energy *conversion elements* as transformers (TF:n, such that $e_i = ne_j$ and $f_j = nf_i$) and gyrators (GY:r, such that $e_i = rf_j$ and $e_j = rf_i$). Finally, multi-port elements, the 0-junction and the 1-junction, represent ideal energy connections for one- and two-port elements. 0-junctions model points of common effort, where all efforts are equal and $\sum f = 0$, while 1-junctions model points of common flow, where all flows are equal and $\sum e = 0$. The direction of the bonds determines the signs of the efforts and the flows in 0- and 1-junctions, respectively.

Other than bonds and BG elements, *signal links* can be present in a bond graph model. Signal links are drawn as full arrows ($\rightarrow$) and represent information transfer pathways. Nonlinearities in the models are introduced using *modulated elements*. Modulated elements are denoted with an M prefixed to the BG element name, and their parameters are algebraic functions of other system variables or external signal. For example, MR:R denotes a modulated resistor, or MSe:u denotes a modulated source of effort.

Signal links are also used to explicitly model sensors in the system as effort detectors (De:u, such that $u = e_i$) and flow detectors (Df:u, such that $u = f_i$).

Building system models in the bond graph framework follow an intuitive approach, where energy-related functions are associated with different parts of the system and translated into bond graph elements. The system structure is analyzed to establish

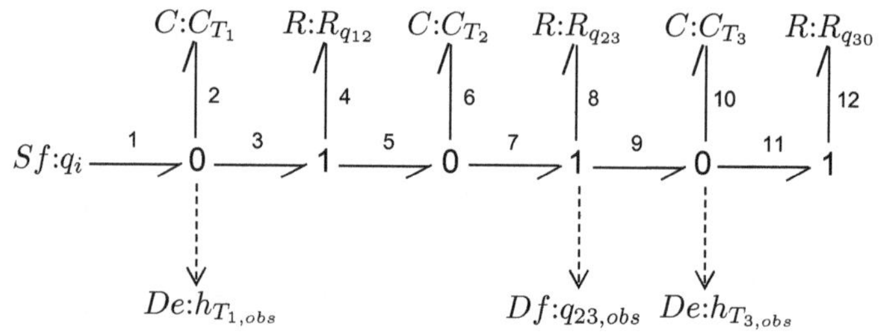

Fig. 2.7 Bond graph model for the running example

the energy flow paths between the system components, which are then represented as bonds. Conservation of energy at connections between multiple elements are modeled by the appropriate junctions.

Figure 2.7 shows the bond graph model for the three-tank system in Fig. 2.3. The tanks are modeled as capacitors that hold fluid, and the pipes as resistances to flow. 0-and 1-junctions represent the common effort (i.e., pressure) and common flow (i.e., flow rate) points in the system, respectively. Measurement points shown as D_e and D_f components are connected to junctions, but in this case we represented the output to the sensors with a dashed line (instead of using a signal link) because we considered that the sensors are implicitly included in the BG model.

2.3.4.1 Causality Assignment

Causality of a bond is depicted by the causal stroke (|) on one end of a bond, with the BG element near the causal stroke imposing flow on the opposite element of the bond. The causal assignment procedure takes into account the four different types of causal constraints [5]: fixed causality (S_e, S_f), constrained causality (0-junction, 1-junction), preferred causality (C, I), and indifferent causality (R). Preferred causality, i.e., causality for storage elements, can be placed in either integral or derivative causality. Figure 2.8 shows the possible causal assignments for each one of the bond graph elements.

There exist well-defined methods for causal assignments to bonds [1], and these assignments can be employed for equation generation, and for testing the correctness of the bond graph model. The Sequential Causal Assignment Procedure (SCAP) [13] systematically assigns causality in a bond graph model. Figure 2.9 shows the bond graph model of the three-tank system after the causality assignment process. Bond 1 imposes flow to the 0-junction (because of the Sf element). Since we are using integral causality in the capacitor elements, bond 2 imposes effort to the 0-junction; consequently, bond 3 must impose flow to the 0-junction. The process is similar for the rest of the bonds in the BG model.

Causal analysis of the bond graph model provides the computational relations between the effort and flow variables associated with individual bonds as a set of constituent equations. Constituent equations defined for components and junctions

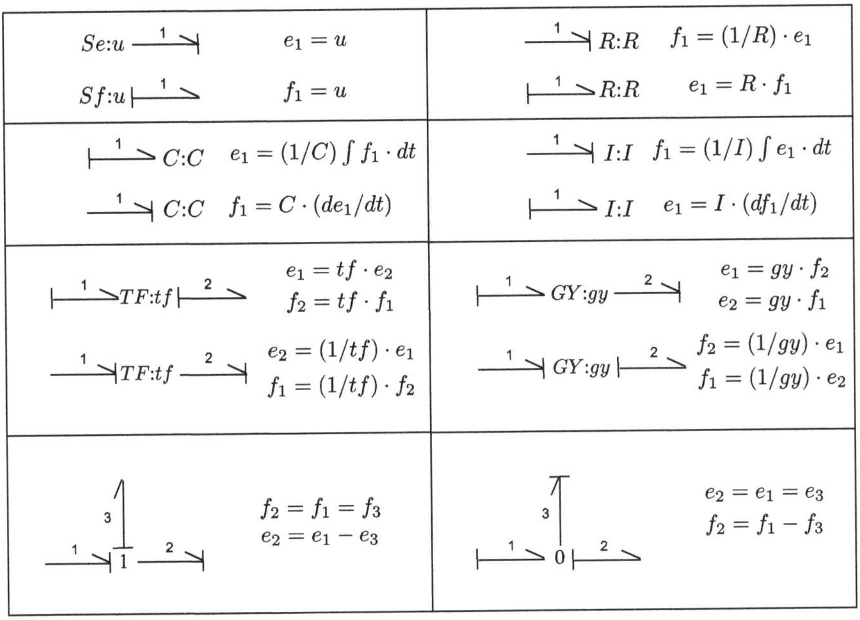

Fig. 2.8 Possible causal assignments and their corresponding constituent equations for the set of bond graph elements

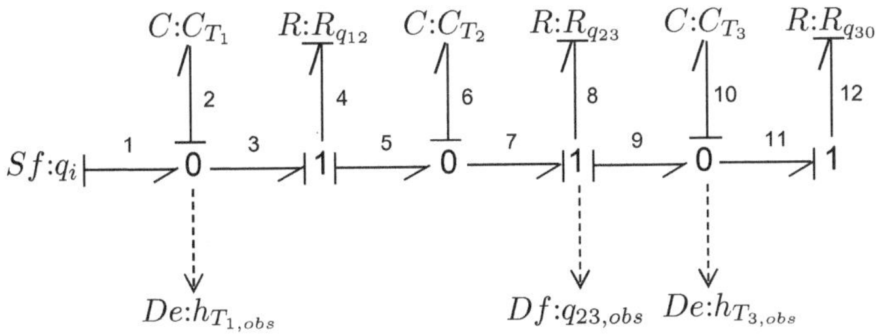

Fig. 2.9 Bond graph model for the running example with causality assigned

can be combined to derive state equation or input/output models of system behavior. Details of the derivation process can be found in [13, 23]. As an example, the constituent equation of capacitor C_{T_1} is given by the equation: $e_2 = \frac{1}{C_{T_1}} \int f_2 \cdot dt$, where e_2 and f_2 correspond to the pressure in the tank, and net flow rate into the tank (i.e., the difference between input and output flow rate), respectively. Similarly, the constituent equations for the first of the 1-junctions may be expressed as $f_3 = f_4 = f_5$ and $e_4 = e_3 + e_5$. Following causal chains, the constituent equations can be algebraically composed to derive the state equations of the system. In the bond graph model, the effort variables for capacities and the flow variables for inertias in integral causality define the set of state variables.

2.4 Other Case Studies

In the literature, other fault diagnosis case studies have been proposed and will be used in some of the chapters of this book. In the following, these case studies are briefly described providing some references where the benchmark is more detailed. Associated with these benchmarks and in general in fault diagnosis, some performance indexes are typically used for assessing and comparing the performance of different methods.

Most of these benchmarks can be found in the IFAC group "Industrial Application of Advanced FDI/FTC Technology" web site.[3]

2.4.1 Servoactuator

This benchmark is based on real industrial servoactuator of a Lublin sugar factory in Poland. This benchmark has been the developed in the context of the DAMADICS European research training network specially designed to compare and contrast a wide range of fault diagnosis methods.

The actuator is considered as an assembly of devices consisting of a control valve, a spring-and-diaphragm pneumatic servomotor, and a positioner. The control valve acts on the flow of the fluid passing through the pipeline installation. A servomotor carries out a change in the position of the control valve plug, thus acting on fluid flow rate. A spring-and-diaphragm pneumatic servomotor is a compressible fluid powered device in which the fluid acts upon the flexible diaphragm, to provide linear motion of the servomotor stem. The positioner is a device applied to eliminate the control-valve-stem miss-positions produced by the external or internal sources such as, e.g., friction among other.

In [2], the case study and benchmark has been described in detail. This reference was part of a special issue in Control Engineering journal devoted to this case study. In this special issue, several methods were proposed and compared using several performance indexes.

The benchmark uses a complete set of real and simulated fault scenarios as a common platform for testing developed in MATLAB/Simulink and for comparing fault diagnosis methods considering a wide range of faults.

2.4.2 Two-Tank System

This benchmark is based on a two-tank system that was proposed in the context of the CHEM European project.

---

[3]https://tc.ifac-control.org/6/4/working-groups.

The benchmark problem concerns the two coupled tanks connected by a pipe. The main aim of the two tanks is to provide a continuous water flow to a consumer. The fist tank is filled by a pump up to a reference water level. The water level in this tank is controlled by a PI level controller acting to the inlet flow provided by the pump. The water flow between the tanks can be controlled by a valve using an on/off controller in order to keep the water level of the second tank in a pre-established band. The quantity of water outflow to a consumer supplied by the second tank is simulated by the position of output valve. Leakage in both tanks can be induced using valves available in each tank.

In [24], this case study is presented in detail not only presenting the mathematical model but also the structural analysis using bond graphs. A platform developed in MATLAB/Simulink has also been developed for comparing fault diagnosis methods considering a wide range of faults.

2.4.3 *Wind Turbine*

This benchmark is based on a wind turbine that was proposed in the context of an international competition that had several editions in several international conferences presented for the first time in [17] in its simpler version. A more complex and realistic one has been presented in [16] using the FAST high-fidelity simulator.

The benchmark model deals with the wind turbine on a system level, and it includes sensor, actuator, and system faults, namely, faults in the pitch system, the drive train, the generator, and the converter system. Since it is a system-level model, converter and pitch system models are simplified because these are controlled by internal controllers working at higher frequencies than the system model. The model represents a three-bladed pitch-controlled variable-speed wind turbine with a nominal power of 4.8 MW.

In [18], the solutions and results obtained in the international competition by several teams were presented and compared. This comparison relies on additional test data in which the faults occur in different operating conditions than in the test data used for the FDI design.

2.4.4 *Tennessee Eastman Challenge*

The Tennessee Eastman process challenge is a realistic simulation of a chemical plant which is widely accepted as a benchmark for control and monitoring studies. The process, which is described in detail in [10], has two products from four reactants. Additionally, an inert and a by-product are also present making a total of eight components. The process allows a total 52 measurements out of which 41 are of process variables and 11 are manipulated variables. Reference [10] defined 20 process faults and an additional valve fault.

This benchmark has been widely used for testing data-driven methods. See as, e.g., [27] among other. The Simulink models can be found in web site.[4]

2.4.5 ADAPT Electrical Power System

This benchmark is based on the Electrical Power System (EPS) testbed in the ADAPT lab at NASA Ames Research Center. It introduces a number of specifications including a standardized fault catalog, a common set of metrics, a library of modular and standardized test scenarios, a test protocol, and an evaluation algorithm for processing diagnostic data and calculating the performance metrics. The testing procedure is usually scenario-based, where each scenario may have faults injected into the system. To diagnose faults, each diagnostic algorithm has access to real-time data from the ADAPT EPS. Moreover, a standardized output scheme is enforced on the diagnostic algorithms to ensure the generation of common datasets for the calculation of metrics. The data from the testbed and this output of the diagnostic system are saved to a database, and the diagnostic algorithm performance is evaluated according to a predefined set of metrics.

Two international competitions based on this benchmark has been proposed [15, 19].

2.5 Simulation of System Models

As previously mentioned, a major difference between the FDI and the DX approaches to model-based diagnosis is their different focuses on system dynamics. While the FDI approach is entirely concerned with dynamic systems, the DX approach was mainly concerned in its early stages with static systems diagnosis. While in the FDI field there is no doubt about how to model dynamic behavior, either using continuous or discrete time models, or using a state-space representation or the input/output model approach, on the other hand, there is a vast catalog of definitions regarding time in the DX field, and there is neither a general consensus nor a general theory.

For instance, from a FDI point of view, if we focus on the state-space representation for a nonlinear dynamic system, we can state the model as

$$\dot{\mathbf{x}}(t) = \mathbf{f}(\mathbf{x}(t), \mathbf{u}(t), \mathbf{v}(t)), \qquad (2.24)$$

$$\mathbf{y}(t) = \mathbf{g}(\mathbf{x}(t), \mathbf{u}(t), \mathbf{w}(t)), \qquad (2.25)$$

where $\mathbf{x} \in \Re^{n_x}$, $\mathbf{u} \in \Re^{n_u}$, and $\mathbf{y} \in \Re^{n_y}$ are the state, input, and output vectors, respectively, $\mathbf{v} \in \Re^{n_v}$ $\mathbf{w} \in \Re^{n_w}$ represents the process and measurement noise vectors,

[4]https://depts.washington.edu/control/LARRY/TE/download.html.

respectively, and $\mathbf{f}(\cdot)$ and $\mathbf{g}(\cdot)$ are nonlinear functions. The dimension of a vector a is denoted by n_a.

The general model described by Eqs. (2.24) and (2.25) can be implemented as a simulation or a state observer model as follows:

$$\dot{\hat{\mathbf{x}}}(t) = \mathbf{f}(\hat{\mathbf{x}}(t), \mathbf{u}(t), \mathbf{v}(t)) + \mathbf{k}(\mathbf{y}(t) - \hat{\mathbf{y}}(t)), \quad (2.26)$$

$$\hat{\mathbf{y}}(t) = \mathbf{g}(\hat{\mathbf{x}}(t), \mathbf{u}(t), \mathbf{w}(t)), \quad (2.27)$$

where $\hat{\mathbf{x}}$ and $\hat{\mathbf{y}}$ are the estimated state and output variables, respectively, and $\mathbf{k}$ is the observer gain, which filters out the difference between the measured and estimated variables, minimizing the error, $\mathbf{e}_y(t) = \mathbf{y}(t) - \hat{\mathbf{y}}(t)$, considering a given design criterion. Depending on the computed value for $\mathbf{k}$, if $\mathbf{k} = 0$, we have the simulation model for the system, but if $\mathbf{k} \neq 0$ we have a general state observer model for the same system [20].

On the other hand, in the DX field, main issue in modeling dynamics comes from the representation of the system state variables, and the influence in the current state from previous states (assuming that the diagnosis of combinatorial circuits, i.e., with no state, can be seen as a sequence of static diagnosis processes). In other words, the main difference between static and dynamic system models is the presence of equations or constraints linking state variables and their derivatives [11, 12, 21]. Following Dressler's proposal [11], we can distinguish two kinds of constraints in the model: *differential* constraints, those used to model dynamic behavior—*inter-state constraints* in Dressler's terminology—and *instantaneous* constraints, those used to model static or instantaneous relations between system variables—*intra-state constraints* in Dressler's terminology. As shown by Frisk et al. [12] the fact that differential constraints are explicitly represented or not does not introduce genuinely new structural information.

Differential constraints can be interpreted in two ways, depending on the selected causality assignment. In integral causality, constraint is solved by computing $x(k)$ using $x(k-1)$ and $\dot{x}(k-1)$. In derivative causality, it is assumed that the derivative, $\dot{x}(k)$, can be computed based on present and past samples, being x known. Integral causality usually implies using simulation, and it is the preferred approach in the DX field. Derivative causality is the preferred approach in the FDI approach. Both have been demonstrated to be equivalent for numerical models, assuming adequate sampling rates and precise approximations for derivative computation are available, and assuming initial conditions for simulation are known, according to Chantler et al. [6].

In quantitative terms, we should be able to somehow represent the relation between a state variable and its derivative as

$$x(k) = x(k-1) + \Delta t\, \dot{x}(k-1).$$

If we can compute $\dot{x}(k-1)$, then we can perform integration and to obtain $x(k)$ by simulation. If we can observe the variables $x(k)$, we can compute $\dot{x}(k)$ and later estimate the next state $\dot{x}(k+1) = x(k) + \Delta t\, \dot{x}(k)$.

These concepts can be easily implemented for our three-tank system. If we take the set of DAEs representing system dynamics in Sect. 2.2.2, we can obtain easily the simulation code written directly in MATLAB: http://www.infor.uva.es/~belar/SoftwareCPCs/threeTanksSimulation.m.

2.6 Modeling Discussion

In model-based diagnosis, the modeling stage is of paramount importance. Depending on the available knowledge about the system, we can have first-principles mathematical models from physics, state-space analytical models, expertise-based rules, qualitative models, input/output relations, black box models as in neural networks, etc. What it is clear is the need for a model to be fed with available observations and the ability of the model to do estimations that must be checked against observations to detect inconsistencies.

The different approaches illustrated in this book will use some kind of analytical or qualitative model. The model can be designed from a system-level perspective, which is the usual approach in the FDI community [3], or designed from a component-level perspective [8, 22], which is the usual approach in the DX community. There are also a lot of methods that use an intermediate, structural model [14], which only states the relations between model variables, without specifying the relation. Finally, there are other methods that rely upon graphical models such as bond graphs [5] or some kind of causal graphs [4].

Some types of models are better suited for some techniques, such as causal graphs for qualitative reasoning, first-order logic formulas for pure consistency-based diagnosis or state-space models for state observers or analytical redundancy relations computation. But other techniques are more broadly used, and they can be even transformed from one type to other or just apply a different level of abstraction. That is the case of the structural models obtained from a set of differential-algebraic equations.

In this chapter, we have presented different case studies: some of them are well suited for the logical formulation in consistency-based diagnosis and can be seen as combinatorial, state-less systems. This is the case of the widely used polybox example. Other systems are more suited to the most typical dynamic systems analyzed in FDI approaches. We will use through this book a three-tank system made up of tanks and pipes.

Different techniques will make explicit or not the correct behavior assumptions in the models. While consistency-based diagnosis makes explicit the assumption by means of local models, the FDI approach does not and usually applies an exoneration assumption which is not used in consistency-based diagnosis as a default.

What is clear is that whatever the system, it is necessary to take into account the needs for models from the design stage, because the presence or absence of observations, or the control-oriented design, for instance, defines what can be expected from the diagnosis stage. At the same time, the precision in the models also determines what can be expected from the diagnosis process. As pointed out in different seminal model-based diagnosis articles, diagnosis will be as good or as precise as the available models.

Acknowledgements This work has been partially funded by the Spanish State Research Agency (AEI) and the European Regional Development Fund (ERFD) through the projects DEOCS (ref. MINECO DPI2016-76493) and SCAV (ref. MINECO DPI2017-88403-R). This work has also been partially funded by AGAUR of Generalitat de Catalunya through the Advanced Control Systems (SAC) group grant (2017 SGR 482) and by Agència de Gestió d'Ajuts Universitaris i de Recerca.

References

1. Antic, D., Vidojkovic, B., Mladenovic, M.: An introduction to bond graph modelling of dynamic systems. In: 4th International Conference on Telecommunications in Modern Satellite, Cable and Broadcasting Services, vol. 2, pp. 661–664 (1999)
2. Bartys, M., Patton, R., Syfert, M., de las Heras, S., Quevedo, J.: Introduction to the DAMADICS actuator FDI benchmark study. Control. Eng. Pract. **14**(6), 577–596 (2006)
3. Blanke, M., Kinnaert, M., Lunze, J., Staroswiecki, M.: Diagnosis and Fault-Tolerant Control. Springer, Berlin (2006)
4. Bousson, K., Trave-Massuyes, L., Zimmer, L.: Causal model-based diagnosis of dynamic systems. LAAS Report No 94231 (1994)
5. Broenink, J.: Introduction to physical systems modelling with bond graphs. SiE Whitebook on Simulation methodologies (1999)
6. Chantler, M., Daus S. Vikatos, T., Coghill, G.: The use of quantitative dynamic models and dependency recording engines. In: Proceedings of the Seventh International Workshop on Principles of Diagnosis (DX-96), pp. 59–68. Val Morin, Quebec, Canada (1996)
7. Davis, R.: Diagnostic Reasoning Based on Structure and Behavior. In: D.G. Bobrow (ed.) Qualitative Reasoning About Physical Systems, pp. 347–410. Elsevier, Amsterdam. https://doi.org/10.1016/B978-0-444-87670-6.50010-8 (1984)
8. De Kleer, J., Williams, B.C.: Diagnosing multiple faults. Artif. Intell. **32**(1), 97–130 (1987)
9. Ding, S.X.: Case study and application examples. Data-driven Design of Fault Diagnosis and Fault-tolerant Control Systems, pp. 11–21. Springer, Berlin (2014)
10. Downs, J., Vogel, E.: A plant-wide industrial process control problem. Comput. Chem. Eng. **17**(3), 245–255 (1993)
11. Dressler, O.: On-line diagnosis and monitoring of dynamic systems based on qualitative models and dependency-recording diagnosis engines. In: Proceedings of the Twelfth European Conference on Artificial Intelligence (ECAI-96), pp. 461–465. Budapest, Hungary (1996)
12. Frisk, E., Dustegor, D., Krysander, M., Cocquempot, V.: Improving fault isolability properties by structural analysis of faulty behavior models: application to the damadics benchmark problem. In: Proceedings of SAFEPROCESS-2003. Washington, DC, USA (2003)
13. Karnopp, D., Rosenberg, R., Margolis, D.: System Dynamics, A Unified Approach, 3rd edn. John Wiley, Hoboken (2000)
14. Krysander, M., Åslund, J., Nyberg, M.: An efficient algorithm for finding minimal overconstrained subsystems for model-based diagnosis. IEEE Trans. Syst., Man, Cybern.-Part A: Syst. Hum. **38**(1), 197–206 (2008)
15. Kurtoglu, T., Narasimhan, S., Poll, S., Garcia, D., Kuhn, L., de Kleer, J., van Gemund, A., Provan, G., Feldman, A.: First International Diagnostic Competition–DXC'09. In: Proceedings of the 20th International Workshop on Principles of Diagnosis (DX'09). Stockholm, Sweden (2009)

16. Odgaard, P.F., Johnson, K.E.: Wind turbine fault detection and fault tolerant control-an enhanced benchmark challenge. In: American Control Conference (ACC), 2013, pp. 4447–4452. IEEE, Washington, DC, USA (2013)
17. Odgaard, P.F., Stoustrup, J., Kinnaert, M.: Fault tolerant control of wind turbines-a benchmark model. In: Proceedings of the 7th IFAC Symposium on Fault Detection, Supervision and Safety of Technical Processes, vol. 1, pp. 155–160. Barcelona, Spain (2009)
18. Odgaard, P.F., Stoustrup, J., Kinnaert, M.: Fault-tolerant control of wind turbines: a benchmark model. IEEE Trans. Control. Syst. Technol. **21**(4), 1168–1182 (2013)
19. Poll, S., de Kleer, J., Feldman, A., Garcia, D., Kurtoglu, T., Narasimhan, S.: Second international diagnostics competition–DXC'10. In: Proceedings of the 21st International Workshop on Principles of Diagnosis (DX'10). Philadelphia, Pennsylvania, USA (2010)
20. Puig, V., Quevedo, J., Escobet, T., Meseguer, J.: Toward a better integration of passive robust interval-based FDI algorithms. In: Proceedings of the 6th IFAC Symposium on Fault Detection, Supervision and Safety of Technical Processes, SAFEPROCESS06. Beijing, China (2006)
21. Pulido, B., Bregon, A., Alonso-González, C.: Analyzing the influence of differential constraints in Possible Conflict and ARR computation. In: Meseguer, P., Mandow, L., Gasca, R.M. (eds.) Current Topics in Artficial Intelligence. CAEPIA 2009 Selected Papers. Springer, Berlin (2010)
22. Reiter, R.: A theory of diagnosis from first principles. Artif. Intell. **32**(1), 57–95 (1987)
23. Roychoudhury, I., Daigle, M., Biswas, G., Koutsoukos, X., Mosterman, P.: A method for efficient simulation of hybrid bond graphs. In: Proceedings of the International Conference on Bond Graph Modeling and Simulation, ICBGM07, pp. 177–184. San Diego, CA, USA (2007)
24. Samantaray, A.K., Bouamama, B.O.: Model-based Process Supervision: A Bond Graph Approach. Advances in Industrial Control. Springer Science & Business Media, Berlin (2008)
25. Staroswiecki, M.: Quantitative and qualitative models for fault detection and isolation. Mech. Syst. Signal Process. **14**(3), 301–325 (2000)
26. Venkatasubramanian, V., Rengaswamy, R., Yin, K., Kavuri, S.N.: A review of process fault detection and diagnosis: Part i: quantitative model-based methods. Comput. Chem. Eng. **27**(3), 293–311 (2003)
27. Yin, S., Ding, S.X., Haghani, A., Hao, H., Zhang, P.: A comparison study of basic data-driven fault diagnosis and process monitoring methods on the benchmark tennessee eastman process. J. Process Control **22**(9), 1567–1581 (2012)

Part I
Standard Approaches

Chapter 3
Structural Analysis

Erik Frisk, Mattias Krysander and Teresa Escobet

3.1 Introduction

Fault diagnosis (FD) aims at carefully identifying which fault can be hypothesized to be the cause of monitoring events. In general, when addressing the FD problem, two strategies can be found in the literature: hardware redundancy based on the use of physical redundancies, and software or logic redundancies based on the use of software/intelligent sensors or models combining the information provided by sensor measurements, actuator commands, and system knowledge.

A powerful instrument for determining the detectability/isolability properties of a system is performing a structural analysis. This technique allows to assess whether the number and place of sensors are adequate to comply with the specifications of diagnosis. Structural analysis is well suited for computer support, and all examples in this chapter have been performed using the freely available MATLAB toolbox [7] that can be downloaded from https://faultdiagnosistoolbox.github.io.

In this chapter, a formal definition of fault detection and isolation is given in terms of consistency relation. After that, the definition of structural analysis based on bi-partite structure graphs is introduced for analyzing the structural properties of dynamical systems and computing the structured residuals. These structured residuals provide the interaction between signals and allow finding the analytical redundancies, which can be exploited for fault diagnosis.

E. Frisk (✉) · M. Krysander
Department of Electrical Engineering, Linköping University, SE-581 83 Linköping, Sweden
e-mail: erik.frisk@liu.se

M. Krysander
e-mail: mattias.krysander@liu.se

T. Escobet
Research Center for Supervision, Safety and Automatic Control (CS2AC), Universitat Politècnica de Catalunya, Terrassa, Spain
e-mail: teresa.escobet@upc.edu

T. Escobet et al. (eds.), *Fault Diagnosis of Dynamic Systems*,
https://doi.org/10.1007/978-3-030-17728-7_3

3.2 Background

This section will introduce basic concepts for structural analysis, all diagnosis-related notions will be introduced in the corresponding section.

3.2.1 *Structural Models*

A system can be said to consist of a set of components and sensors. Its dynamic behavior is described by a model, i.e., a set of constraints that describes the behavior and relations among all model variables. The formal expression of the model depends on the type of knowledge (analytical, qualitative, production rules or numerical tables, etc.). It generally relies on a component-based description, which associates a set of constraints to each component. The structural analysis described later in this chapter will use a structural representation of the model. As mentioned in Sect. 3.1, the structural representation is a coarse model description that only considers which variables that are included in each model relation/equation.

Let $\mathcal{E}$ be the set of constraints in the model and $\mathcal{V}$ the set of variables, then the structural model can be represented by the bi-partite graph $\mathcal{G} = (\mathcal{E} \bigcup \mathcal{V}, \mathcal{A})$ where $\mathcal{A}$ is a set of arcs between nodes in the two node sets $\mathcal{E}$ and $\mathcal{V}$. An arc $(e_i, v_j) \in \mathcal{A}$ iff variable $v_j \in \mathcal{V}$ appears in the model relation $e_i \in \mathcal{E}$. A common way to visualize the structural model is with the so-called *bi-adjacency matrix* of the graph. The bi-adjacency matrix is a matrix with equally many rows and columns as constraints and variables, respectively. One element (i, j) of the bi-adjacency matrix is empty when the variable v_j is not included in constraint e_i. The bi-adjacency matrix of a graph representing the structure of a model will also simply be called structural matrix.

The set of variables $\mathcal{V}$ can be partitioned into unknown, $\mathcal{X} \subset \mathcal{V}$, and known, $\mathcal{Z} \subset \mathcal{V}$, variables, i.e., $\mathcal{V} = \mathcal{X} \bigcup \mathcal{Z}$. For analysis purposes, the part of the structure related to unknown variables will be most important and the subgraph only including constraints and unknown variables will be called the *reduced structure graph.*

3.2.2 *Dulmage–Mendelsohn's Decomposition and Matchings*

A key tool when analyzing structural models is the Dulmage–Mendelsohn's decomposition [4], used for diagnosis in [2, 12]. The general Dulmage–Mendelsohn's decomposition is illustrated in Fig. 3.1 where, by a suitable reordering of constraints and variables, the bi-adjacency matrix of a reduced structural graph is converted to a block triangular form.

The subgraph G^- with node sets E^- and X^- represents the underdetermined part of the model, G^0 with node sets E^0 and X^0 the exactly determined part, and G^+ with node sets E^+ and X^+ the overdetermined part. The overdetermined part con-

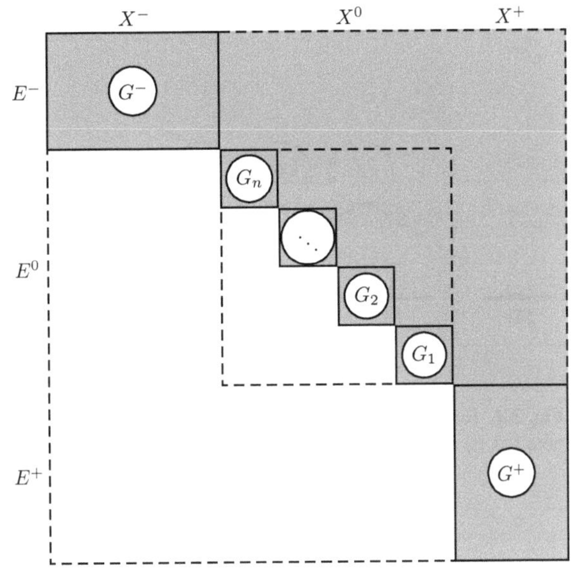

Fig. 3.1 Dulmage–Mendelsohn's decomposition with over- and underdetermined part, the just-determined part with the finer structure in the Hall components

tains redundancy, and can, therefore, be used for diagnosis. In the exactly determined part, there is a finer structure of Hall components, here denoted G_i. Note that the Dulmage–Mendelsohn's components $G^-, G_1, \dots G_n, G^+$ are unique, but the triangular structure may vary, for example, if the Hall components are independent. The notation $()^+$ will be used as an operator with a slight abuse of notation.

A key concept used frequently in the following sections is *matching* [8].

Definition 3.1 (*Matching*) For a bi-partite graph $\mathcal{G} = (\mathcal{E} \bigcup \mathcal{V}, \mathcal{A})$, a matching $\Gamma \subset \mathcal{A}$ is a set of edges with no common vertices. A matching is said to be *maximal* if it contains the largest number of edges. A matching is said to be *perfect* if all vertices in the graph are matched, i.e., is included in an edge in Γ. A matching is said to be complete with respect to a node set of all nodes are matched.

Matchings have many uses in fault diagnosis which will be explored in the upcoming sections, for example, the structural rank of a matrix equals the size of a maximal matching. Also, a matching can, in this context, be interpreted as if equation e_i is matched with the variable x_j, then e_i can be interpreted as a mechanism for solving for x_j if all other variables are known. This means that a matching can be useful for finding a causal interpretation of the model, i.e., a computational order for the variables which will be further explored in Sect. 3.4.3.

As an example, consider the polybox model presented in Sect. 2.2. The model has five constraints $\mathcal{E} = \{M1, M2, M3, A1, A2\}$ and ten variables partitioned in unknown variables $\mathcal{X} = \{x, y, z\}$, known inputs $\mathcal{U} = \{a, b, c, d, e\}$, and known outputs $\mathcal{Y} = \{f, g\}$, collected as $\mathcal{V} = \{\mathcal{X}, \mathcal{U}, \mathcal{Y}\}$. The reduced bi-partite graph and corresponding bi-adjacency matrix (or structural matrix) are shown in Fig. 3.2.

Using this example, different matchings can be defined on a given bi-partite graph. These matchings are given by the following set of disjoint edges:

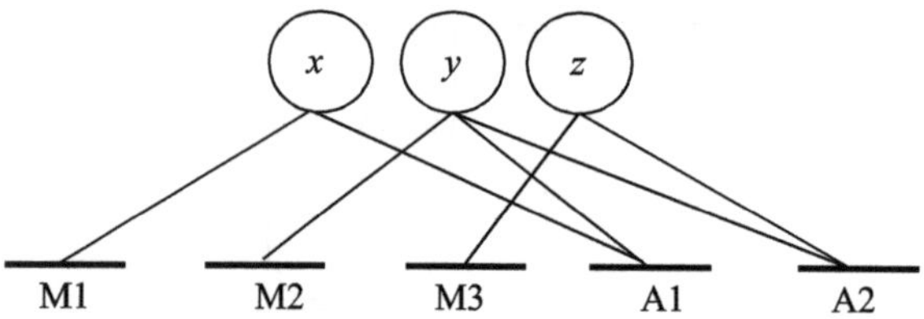

(a) Bi-partite graph

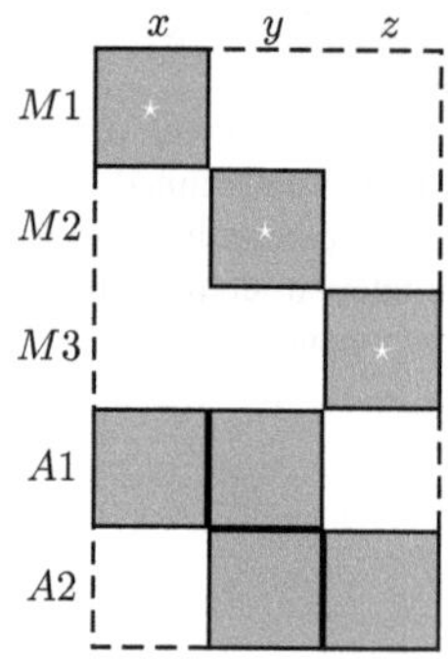

(b) Bi-adjacency matrix

Fig. 3.2 Reduced bi-partite graph and bi-adjacency matrix for the polybox model. A matching, noted as Γ_1 in (3.1), is indicated by the white stars

$$
\begin{aligned}
\Gamma_1 &= \{(M1, x), (M2, y), (M3, z)\} \\
\Gamma_2 &= \{(M1, x), (M2, y), (A2, z)\} \\
\Gamma_3 &= \{(M1, x), (M3, z), (A1, y)\} \\
\Gamma_4 &= \{(M1, x), (M3, z), (A2, y)\} \\
\Gamma_5 &= \{(M2, y), (M3, z), (A1, x)\} \\
\Gamma_6 &= \{(M2, y), (A1, x), (A2, z)\}.
\end{aligned}
\tag{3.1}
$$

Note that, in the bi-adjacency matrix, a matching is represented by selecting at most one edge in each row and in each column.

In this example, the number of constraints is greater to the number of unknown variables. In this case, the constraints not involved in the perfect matching are denoted as redundant relations.

Definition 3.2 (*Redundant Relation*) Let a bi-partite graph $\mathcal{G} = (\mathcal{E} \bigcup \mathcal{X}, \mathcal{A})$ represent the structure of a model and $\Gamma \subseteq \mathcal{A}$ a complete matching with respect to unknown variables $\mathcal{X}$. Then a *Redundant Relation* (RR) is a relation in $\mathcal{E}$ which is not involved in the complete matching Γ.

RRs are not needed to determine any of the unknown variables. Every RR produces an *Analytical Redundant Relation* (ARR) when the unknown variables involved in the RR are replaced by their formal expression, as will be seen in Sect. 3.4.

3.3 Fault Isolability Analysis

The set of faults that are detectable and isolable from one another in a system depend on the physics of the system and how it is interconnected as well as what measurements that are available. Fault detectability and isolability are system properties that put limits on the diagnosis performance that can be achieved with any diagnosis system. Thus, it may be beneficial to carry out a diagnosability analysis before starting the design of a diagnosis system. Structural methods have proven useful for analyzing diagnosability properties of a model and this section is a brief introduction to this topic.

One way to model a fault is to add a new fault variable f, to the model such that $f = 0$ in the fault-free case and $f \neq 0$ in the faulty case. In the faulty case, the value of f is unknown and might vary over time. With a slight abuse of notation f will both be referred to as fault signal and its corresponding fault. Now, after introducing faults in the model M the variables $\mathcal{V}$ can be partitioned into unknown variables $\mathcal{X}$, known variables $\mathcal{Z}$, and faults $\mathcal{F}$, i.e., $\mathcal{V} = \mathcal{X} \cup \mathcal{Z} \cup \mathcal{F}$. Without loss of generality, it is possible to assume that a single fault can only violate one equation. If a fault signal f appears in more than one equation, we simply replace f in the equations with a new variable x_f and add equation $f = x_f$ which then will be the only equation violated by this fault. Let $e_f \in M$ be the equation that might be violated by a fault $f \in \mathcal{F}$.

A small linear system will be used as a running example to introduce concepts, although the results will be equally applicable to large-scale, nonlinear, differential–algebraic models. Consider a fifth-order linear system of ordinary differential equations

$$\begin{aligned}
e_1 &: \quad \dot{x}_1 = -x_1 + x_2 + x_5 \\
e_2 &: \quad \dot{x}_2 = -2x_2 + x_3 + x_4 \\
e_3 &: \quad \dot{x}_3 = -3x_3 + x_5 + f_1 + f_2 \\
e_4 &: \quad \dot{x}_4 = -4x_4 + x_5 + f_3 \\
e_5 &: \quad \dot{x}_5 = -5x_5 + u + f_4 \\
e_6 &: \quad y_1 = x_4 + f_5 \\
e_7 &: \quad y_2 = x_5,
\end{aligned} \tag{3.2}$$

where x_i are the state variables, u a known control signal, y_i sensor signals, and f_i the faults we want to detect and isolate. This means that $\mathcal{X} = \{x_1, \ldots, x_5\}$, $\mathcal{Z} = \{u, y_1, y_2\}$, and $\mathcal{F} = \{f_1, \ldots, f_5\}$. In the example $e_{f_1} = e_{f_2} = e_3$, $e_{f_3} = e_4$, $e_{f_4} = e_5$, and $e_{f_5} = e_6$. The structure of the model is shown in Fig. 3.3 where the three different parts correspond to unknown variables, faults, and known variables from left to right.

The main problem that will be described in this section is how a model structure as the one shown in Fig. 3.3 can be utilized to find which faults in the model that are detectable and isolable.

3.3.1 *Fault Detectability Analysis*

A fault f is said to be detectable if there exists an observation that is consistent with fault mode f and inconsistent with the no-fault mode. Let $\mathcal{O}_f$ denote the set of

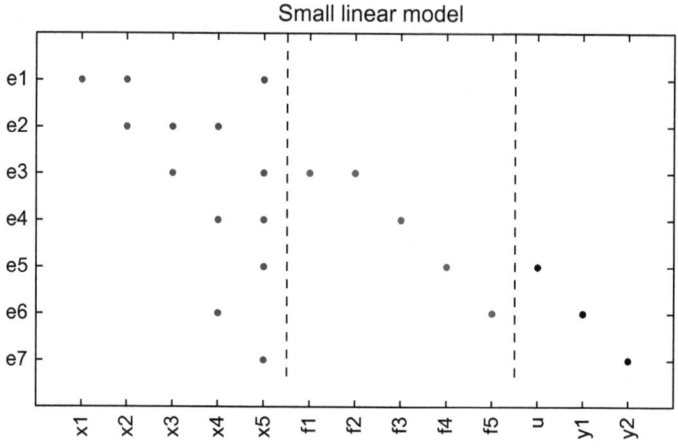

Fig. 3.3 Structural model of the small linear example

observations consistent with a mode f for a given model M and let NF denote the fault-free mode. Detectability can then formally be defined as [14]:

Definition 3.3 A fault mode f is detectable in a model M if $\mathcal{O}_f \setminus \mathcal{O}_{NF} \neq \emptyset$.

From a structural viewpoint this means that a detectable fault affects an equation in the structurally overdetermined part of M, thus structural detectability can be defined as [2]:

Definition 3.4 A fault f is structurally detectable in a model M if $e_f \in M^+$.

Returning to running example the correspondence between detectable faults and structurally detectable faults can be illustrated as follows. The faults f_3, f_4 and f_5 are the detectable faults and residuals capable of detecting them are, for example,

$$\begin{aligned} \{e_4, e_6, e_7\}: \quad & r_1 = \dot{y}_1 + 4y_1 - y_2 = f_3 + \dot{f}_5 + 4f_5 \\ \{e_5, e_7\}: \quad & r_2 = \dot{y}_2 + 5y_2 - u = f_4, \end{aligned} \tag{3.3}$$

which in fact spans the space of all linear residual generators for this model. The set of equations indicate which equations that have been used to derive the corresponding residual generator. The expression in only known variables is the computational form of the residual and the expression in fault variables is the internal form of the residual generator indicating the fault sensitivity. Thus, faults f_1 and f_2 are not detectable.

The Dulmage–Mendelsohn's decomposition of the model $M = \{e_1, \ldots, e_7\}$ is shown in Fig. 3.4a where it can be seen that the structurally overdetermined part of the model is $M^+ = \{e_4, e_5, e_6, e_7\}$. The equations $e_{f_3} = e_4$, $e_{f_4} = e_5$, and $e_{f_5} = e_6$ corresponding to the detectable faults f_3, f_4 and f_5 belong to M^+, but not the equations corresponding to the other faults. This implies according to Definition 3.4 that the

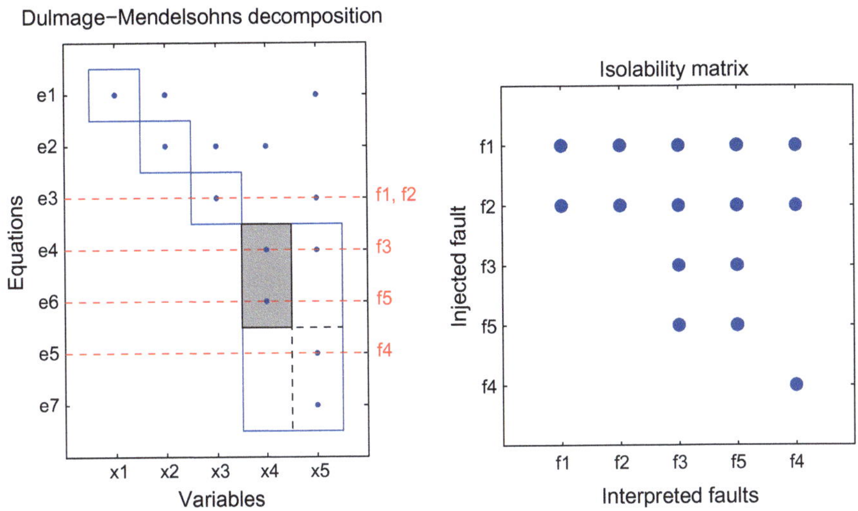

(a) A canonical decomposition. (b) Fault isolability matrix.

Fig. 3.4 Diagnosis analysis results for the model in Fig. 3.3

detectable faults, f_3, f_4, and f_5, are the structurally detectable faults in M which is in accordance with the analytical result above.

3.3.2 *Fault Isolability Analysis*

A fault f_i is said to be isolable from a fault f_j if there exists an observation that is consistent with fault mode f_i and inconsistent with fault mode f_j and a formal definition is as follows [14].

Definition 3.5 A fault mode f_i is isolable from a fault mode f_j in a model M if $\mathcal{O}_{f_i} \setminus \mathcal{O}_{f_j} \neq \emptyset$.

Detection is a special case of isolation, i.e., a fault is detectable if the fault is isolable from the no-fault mode. By noting this similarity, it holds that a fault f_i, isolable from f_j, can violate a monitorable equation in the model describing the behavior of the process having a fault f_j. The set of equations valid with a fault f_j is $M \setminus \{e_{f_j}\}$ and the monitorable part of these equations is, in the generic case, equal to $(M \setminus \{e_{f_j}\})^+$. This motivates the following structural characterization of isolability [10, 11].

Definition 3.6 A fault f_i is structurally isolable from f_j in a model M if

$$e_{f_i} \in (M \setminus \{e_{f_j}\})^+. \tag{3.4}$$

To illustrate the definition, consider the case of determining which faults are isolable from sensor fault f_5 in the running example. The overdetermined part of $M \setminus \{e_{f_5}\}$ where $e_{f_5} = e_6$ is $\{e_5, e_7\}$, i.e., the last two equations in Fig. 3.4a. Only fault f_4 belongs to these equations, thus only f_4 is structurally isolable from f_5 according to the definition.

Single fault isolability $\mathcal{I}$ for a model with faults $\mathcal{F}$ is given by the set of ordered fault pairs (f_i, f_j) such that f_i is structurally isolable from f_j, i.e.,

$$\mathcal{I} = \{(f_i, f_j) \in \mathcal{F} \times \mathcal{F} | f_i \text{ is structurally isolable from } f_j\}.$$

The faults belonging to $(M \setminus \{e_{f_j}\})^+$ are the faults isolable from f_j. Thus, single fault isolability can be computed as

$$\mathcal{I} = \cup_{f_j \in \mathcal{F}} \{(f_i, f_j) | e_{f_i} \in (M \setminus \{e_{f_j}\})^+\}$$

making one Dulmage–Mendelsohn's decomposition for each fault in the model.

Single fault isolability $\mathcal{I}$ can be visualized with a so called fault isolability matrix [10]. The isolability matrix for the running example is shown in Fig. 3.4b. Both rows and columns correspond to the faults and a dot in position (f_i, f_j) means that f_i is not isolable from f_j, i.e., $(f_i, f_j) \notin \mathcal{I}$. Unique isolation of all faults corresponds to the identity matrix. The dots in row f_i correspond to the interpreted faults under the condition that fault f_i is the injected fault. The part of the fault isolability matrix corresponding to detectable faults are symmetric, in the example faults f_3, f_4, f_5.

The interpretation of the isolability matrix in Fig. 3.4b is that the faults f_1 and f_2 are not isolable from any other fault since these faults are not even detectable. The block corresponding to f_3 and f_5 indicates that these faults are not isolable from each other but isolable from all other faults. Finally, f_4 is uniquely isolable.

3.3.3 Canonical Isolability Decomposition of the Overdetermined Part

The Dulmage–Mendelsohn's decomposition reveals the overdetermined part of a model. The overdetermined part can be decomposed into a finer decomposition based on isolability properties of the model and this finer decomposition is described next.

Consider a model where the overdetermined part of the model is equal to the model itself, i.e., $M = M^+$. Such model can be partitioned into a number of equation sets M_k as shown in Fig. 3.5 such that fault f_i and fault f_j are structurally isolable if and only if the faults belong to different equations sets M_i. The white parts of the matrix have only zero entries. The partitioning consists of m equation sets M_i in the figure, the n first equation sets M_i correspond to the n blocks (M_i, X_i), for $i \in \{1, 2, \ldots, n\}$ in the matrix and the $m - n$ last equation sets M_i contain only one equation each and are not related to any variables.

Fig. 3.5 Canonical decomposition of the overdetermined part of a model

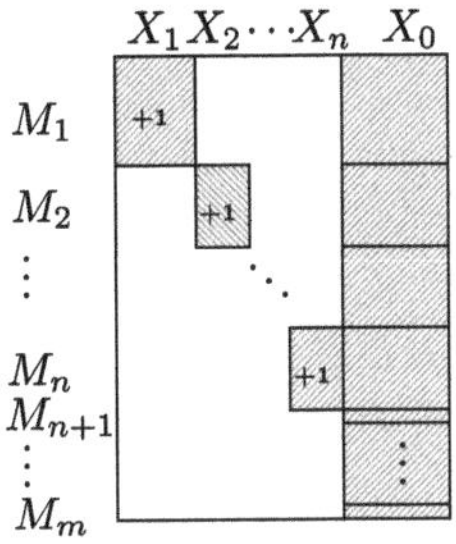

This partitioning can be computed as follows. Given an equation $e \in M$, the equation set M_i including equation e can be computed as

$$M_i = M \setminus (M \setminus \{e\})^+. \tag{3.5}$$

To understand the relation between this decomposition and structural isolability note the similarity between (3.5) and Definition 3.6.

Let e in (3.5) be e_f. The faults in $(M \setminus \{e_f\})^+$ are the faults isolable from f according to Definition 3.6, which means that the faults in the complement set M_i contain all faults that are not isolable from f. Thus faults in the same set are not isolable from one another and faults in different sets are isolable from each other. More details of the decomposition are given [11].

Consider the running example again and the decomposition in Fig. 3.4a. First, note that this decomposition only involves the overdetermined part, i.e., $\{e_4, e_5, e_6, e_7\}$. There is one block with more than one equation and that is the gray block $\{e_4, e_6\}$. This block contains the faults f_3 and f_5. These faults are therefore not isolable from each other which also is in agreement with Fig. 3.4b. Then there are two equation sets $\{e_5\}$ and $\{e_7\}$ with cardinality one. Since the fault f_4 has its own block it means that it is uniquely isolable. This is also confirmed in Fig. 3.4b.

3.3.4 Case Study—The Three-Tank System

Finally, the diagnosability analysis tools are applied to one of the benchmark examples, i.e., the three-tank system. The structural model of the three-tank system is shown in Fig. 3.6a, in which the observed variables $h_{T_1,obs}$, $h_{T_3,obs}$ and $q_{23,obs}$ are $LT1$, $LT2$, and $FT1$, respectively. Note that all dynamics of the model are separated from the algebraic part by writing the dynamics with so-called differential constraints e_{13}, e_{14}, e_{15} in the particular form

$$\frac{d}{dt}h_{T_i} = \dot{h}_{T_i}.$$

This way of representing dynamics is useful in structural analysis. In the structural model special entries in the matrix indicate a differential constraint, a D for $\dot{h}_{T_i}$, i.e., if solving for this variable differentiation of h_{Ti} is required, and I for h_{Ti} because solving for this variable requires integration of $\dot{h}_{T_i}$.

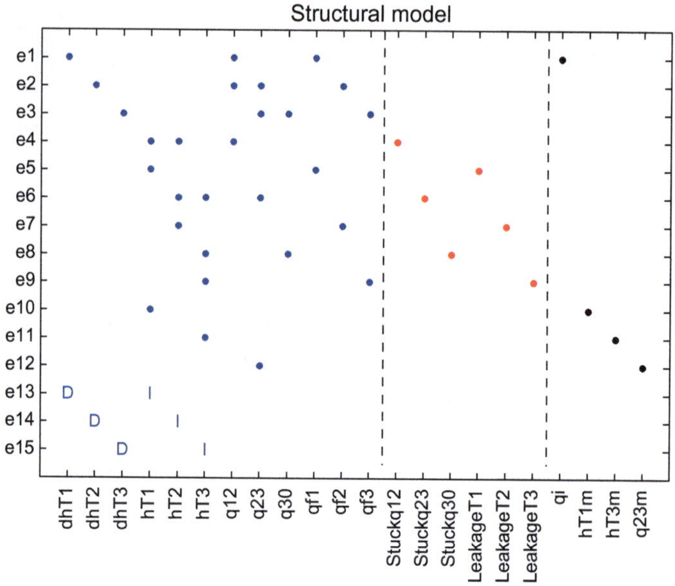

(a) Structural model of the three tank example.

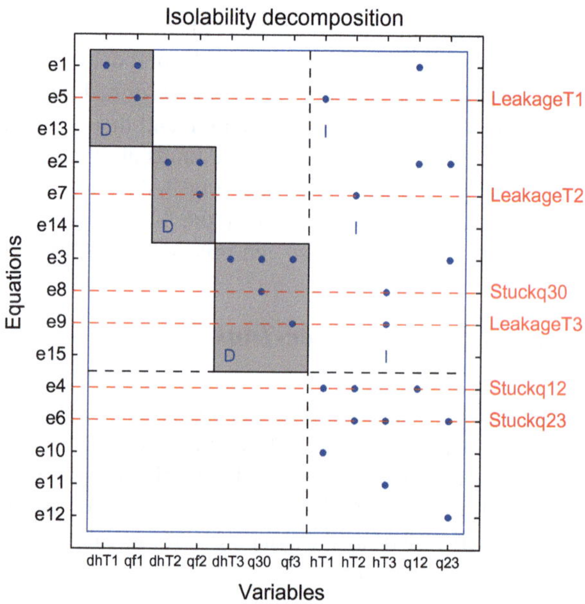

(b) Structural decomposition.

Fig. 3.6 Structural model and decomposition of the three-tank system

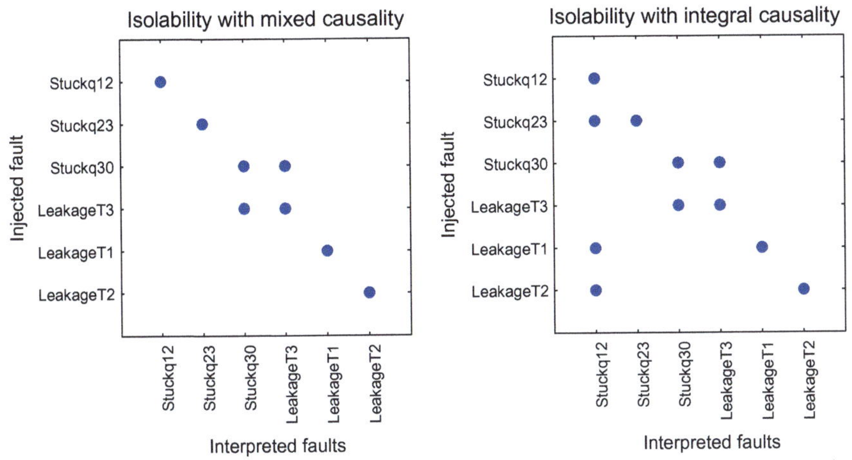

(a) Fault isolability matrix

(b) Possible fault isolability using integration based residual methods.

Fig. 3.7 Fault isolability matrix for the three-tank example

The decomposition of the model is shown in Fig. 3.6b. The complete model is structurally overdetermined; thus, all faults are structurally detectable. The equation partitioning includes three gray blocks and five single equation blocks corresponding to the five last equations. This concludes that all faults are uniquely isolable except for the two faults Stuck_{q30} and Leakage_{T_3} both contained in the same block, thus not isolable from each other.

Single fault isolability is summarized in the fault isolability matrix in Fig. 3.7a. In this analysis, dynamics is not considered, i.e., the additional information that there are D:s and I:s in the structural model, but there are extensions that take this into account [5, 6]. For example standard observer based techniques solve differential equations by integration only, i.e., with integral causality. Under this restriction of causality, a reduced structural isolability can be computed and the result is shown in Fig. 3.7b. This figure shows that Stuck_{q23}, Leakage_{T_1}, Leakage_{T_2} is not isolable from Stuck_{q12} using residual generators based on integration.

3.4 Testable Submodels

The issue of residual generation using structural analysis has been studied by several authors [1]. Structural analysis allows to determine the set of constraints or equations from which residuals can be generated and it can provide a computation sequence to be used. The main objective of this section is to present some algorithms to find all the testable subsystems.

3.4.1 Basic Definitions

From the linear system represented by the model (3.2), the residuals given in (3.3) has been derived by algebraic elimination of the known variables. Both equations are called analytical redundancy relations or parity relations and can be used to check if u, y_1 and y_2 are consistent with the model (3.2). This means that the equation sets given in (3.3) are testable submodels. Also, this example shows the fault influences on the residuals determining detectability and isolability capability of the residuals. These sets of faults are called *Test Supports* (TSs).

Performing the Dulmage–Mendelsohn's canonical decomposition of a model M, the Structural Overdetermined (SO) part if not being the empty set, i.e., $M^+ \neq \varnothing$ is a testable submodel. This motivates us to define the following structural characterization of testable submodels [12].

Definition 3.7 A non-empty set of equations $M \neq \varnothing$ is a *Proper Structurally Overdetermined* (PSO) set if $M = M^+$.

Definition 3.8 A PSO set is a *minimal structurally overdetermined* (MSO) set if no proper subset is a PSO set.

For linear systems, the dimension of the residual space quantifies the degree of redundancy of a model, and next the corresponding structural property is defined.

Definition 3.9 The degree of structural redundancy of a model M is defined as

$$\varphi(M) = |M^+| - |X^+|.$$

Example 3.1 To illustrate these definitions we consider the model in (3.2) and its corresponding Dulmage–Mendelsohn's canonical decomposition in Fig. 3.4a. In this model:

- $M^- = \varnothing$, $M^0 = \{e_1, e_2, e_3\}$, and $M^+ = \{e_4, e_5, e_6, e_7\}$.
- M^+ is a PSO with a redundancy

$$\varphi(M) = |M^+| - |X^+| = 4 - 2 = 2 > 0.$$

- M^+ is not an MSO set, because there are PSO subsets. In this example, there are three MSO sets

$$\begin{aligned}
MSO_1 &: \{e_5, e_7\} \text{ with } \varphi\ (MSO_1) = 2 - 1 = 1\\
MSO_2 &: \{e_4, e_6, e_7\} \text{ with } \varphi\ (MSO_2) = 3 - 2 = 1\\
MSO_3 &: \{e_4, e_5, e_6\} \text{ with } \varphi\ (MSO_3) = 3 - 2 = 1.
\end{aligned}$$

Notice that an MSO set is a PSO set with structural degree of redundancy 1. We can also conclude that each MSO is a testable submodel. MSO and PSO models characterize model redundancy but faults are not taken into account. The residuals presented in (3.3) have two parts the computational form and the internal form, describing the fault influence. As shown below, the three MSOs are characterized by a set of equations and, also, are sensitive to a set of faults:

$$\begin{aligned} \text{Faults}(MSO_1) &= \{f_4\} \\ \text{Faults}(MSO_2) &= \{f_3, f_5\} \\ \text{Faults}(MSO_3) &= \{f_3, f_4, f_5\}. \end{aligned} \tag{3.6}$$

Now, let $F(M)$ denote the set of faults that influence any of the equations in M. Then, since a PSO set exactly characterizes a set of equations that can be used to form a test, a formal definition is then given by [13].

Definition 3.10 Given a model $\mathcal{M}$ and a set of faults $\mathcal{F}$, a non-empty subset of faults $\zeta \subseteq \mathcal{F}$ is a test support if there exists a PSO set $M \subseteq \mathcal{M}$ such that $F(M) = \zeta$.

It is also important to characterize sets of equations that can be used to form a test.

Definition 3.11 An equation set M is a *Test Equation Support* (TES) if

1. $F(M) \neq \emptyset$,
2. M is a PSO set, and
3. for any $M' \not\supseteq M$ where M' is a *PSO* set it holds that $F(M') \not\supseteq F(M)$.

Here it is interesting to consider Minimal Test Support (MTS) and Minimal Test Equation Support (MTES).

Definition 3.12 Given a model, a test support is a *minimal test support* (MTS) if no proper subset is a test support.

Definition 3.13 A TES M is a minimal TES (MTES) if there exists no subset of M that is a TES.

Example 3.2 Let's continue with the example above to illustrate these definitions. In this model:

- F(MSO_1), F(MSO_2), and F(MSO_3) are test supports.
- MSO_3 is not a TES set, because F(MSO_3) $\supset$ F(MSO_1) $\cup$ F(MSO_2).
- MSO_1 and MSO_2 are MTES sets.

3.4.2 *MSO Algorithm*

In [13], it is presented a procedure to find MSO sets. It is based on examining the set M of constraints of a proper structurally over-constrained structure graph. The proposed function is based on a top-down approach in the sense that we start with the entire model and then reduce the size of the model step by step until an MSO set remains. The function `FindMSO`, computes the set of all MSO sets contained in the model. Assume without lack of generality that the input model fulfills $M = M^+$.

```
function M_MSO = FindMSO(M)
    if φ(M) = 1
        M_MSO := {M}
    else
        M_MSO := ∅
        for each e ∈ M
            M' := (M \ {e})^+
            M_MSO := M_MSO ∪ FindMSO(M')
        end
    end
end
```

To illustrate the steps in this function, consider the polybox system given in Sect. 1.3.3. The reduced structural representation of the polybox system is given in Table 3.1.

The structural redundancy of $M = \{e_1, \ldots, e_5\}$ is $\varphi(M) = 2$ and the for-loop is entered in the function `FindMSO`. First e_1 is removed and the overdetermined part of the new set $M \setminus \{e_1\} = \{e_2, e_3, e_4, e_5\}$ is $(M \setminus \{e_1\})^+ = \{e_2, e_3, e_5\}$. The set M' in the algorithm becomes $(M \setminus \{e_1\})^+ = \{e_2, e_3, e_5\}$. In this case $\varphi\left(M'\right) = 1$, and the equation set is saved as an MSO in $\mathcal{M}_{MSO}$. Then e_2 is removed, and the MSO set $(M \setminus \{e_2\})^+ = \{e_1, e_3, e_4, e_5\}$ is found. Next, e_3 is removed, and $M' = (M \setminus \{e_3\})^+ = \{e_1, e_2, e_4\}$. Next, e_4 is removed, and the MSO set is $\{e_2, e_3, e_5\}$, the same MSO is found once again. Finally, e_5 is removed, and the MSO set is $\{e_1, e_2, e_4\}$, already found before.

In the approach above, the same MSO set can be found more than once which decreases efficiency. To avoid this situation, and guarantee that each MSO set is only found once, equivalence classes defined on the set of equations is useful. Let R be a relation on the set of equations M such that $(e', e) \in R$ if

$$e' \notin (M \setminus \{e\})^+. \tag{3.7}$$

Table 3.1 Bi-adjacency matrix for the reduced structural model of polybox system

Equations	Unknown		
	x	y	z
e_1	×		
e_2		×	
e_3			×
e_4	×	×	
e_5		×	×

Table 3.2 Equivalent classes in the polybox system

Lumped structure	y
$M_1 = \{e_1, e_4\}$	×
$M_2 = \{e_2\}$	×
$M_3 = \{e_3, e_5\}$	×

This implies that $(M \setminus \{e\})^+ = (M \setminus \{e'\})^+$, i.e., removing e and e' and then extracting the overdetermined part results in the same set of equations. Since removing an equation and then computing the overdetermined part is a key step in the function `FindMSO` (line 7), equivalence classes can be used to avoid finding the same MSO set more than once. The set M can be partitioned into m equivalence classes M_i. The equivalence classes can then be lumped together forming a reduced structure. Full details of this algorithm are described in [12]. Note also that this partitioning of equations is the same as explained in Sect. 3.3.3.

We illustrate this using the polybox example. The lumped structure is given in Table 3.2. Notice that the three MSOs are obtained removing one of the equations that belong to the same class.

3.4.3 *Residual Generation Based on Matching*

Each MSO set comprises at least one constraint that can be used as an *analytical redundancy relation*(ARR).

Definition 3.14 Let M be a model and $\mathcal{O}$ the set of observation of this model. If $r(z) = 0$ for all $z \in \mathcal{O}$ then $r(z) = 0$ is an ARR for M.

The ARRs are obtained choosing a complete matching on unknown variables of the MSO sets.

According to Example 3.1, without considering faults, a complete matching of each MSO is defined on a given bi-partite graph as illustrated in Fig. 3.8 by the bold edges. The matching is given by the following set of disjoint edges:

$$\begin{aligned} \Gamma_1 &= \{(e_6, x_4), (e_7, x_5)\} \\ \Gamma_2 &= \{(e_7, x_5)\}, \end{aligned} \tag{3.8}$$

and the redundant equations e_4 and e_5 allow to compute the residuals used in (3.3).

Choosing a complete matching on unknown variables of the MSOs allows to rearrange the equations of M and to rewrite for designing the residual generator.

Definition 3.15 A model taking only known variables z as input, and generating a scalar signal r as output, is a residual generator for the model $M(z, \mathbf{o}, \mathbf{x})$, if for all z consistent with $M(z, \mathbf{x})$, it holds that $\lim_{t\to\infty} r(z) = 0$.

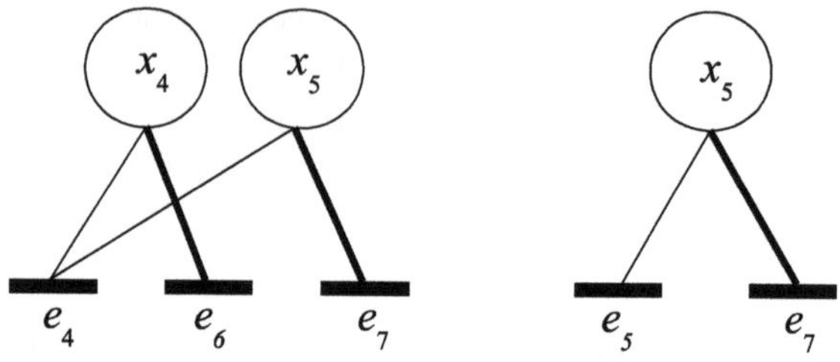

Fig. 3.8 MSO_1 and MSO_2 complete matchings

But not all the potential MSOs allow to generate residuals and not all matchings can lead to a residual. This matter is linked with the conditions under which a variable substitution can be actually performed. A given relation can be interpreted in a causal way, and the substitution of variables in one ARR can be performed when having consistent causalities. Classical non-invertible constraints are nonlinear functions and conditional functions, and we can include, also, in this group the differential constraints.

Taking into account differential constraints, in the model of the linear system presented in (3.2) five more equation must be considered

$$e_{7+i} : \dot{x}_i = \frac{dx_i(t)}{dt},$$

where $i = 1, \ldots, 5$ which, in fact, expresses the relations between a variable and its derivative. Its Dulmage–Mendelsohn's decomposition is shown in Fig. 3.9, where D and I are used to mark the derivative and integral variables in the differential constraint.

In this new case, there are two MSOs:

$$MSO_1 : \{e_4, e_6, e_7, e_{11}\}$$
$$MSO_2 : \{e_5, e_7, e_{12}\},$$

and, considering integral causality, the complete matching are given by the following set of disjoint edges:

$$\Gamma_1 = \{(e_6, x_4), (e_7, x_5), (e_4, \dot{x}_4)\}$$
$$\Gamma_2 = \{(e_7, x_5), (e_5, \dot{x}_5)\}$$

derived from Fig. 3.10. The redundant equations are in this case e_{11} and e_{12}. Taking into account causal relations, the complete matching gives the computational sequence of the variables to generate the residuals. Figure 3.11 shows a computational graph for the residuals obtained for integral causality.

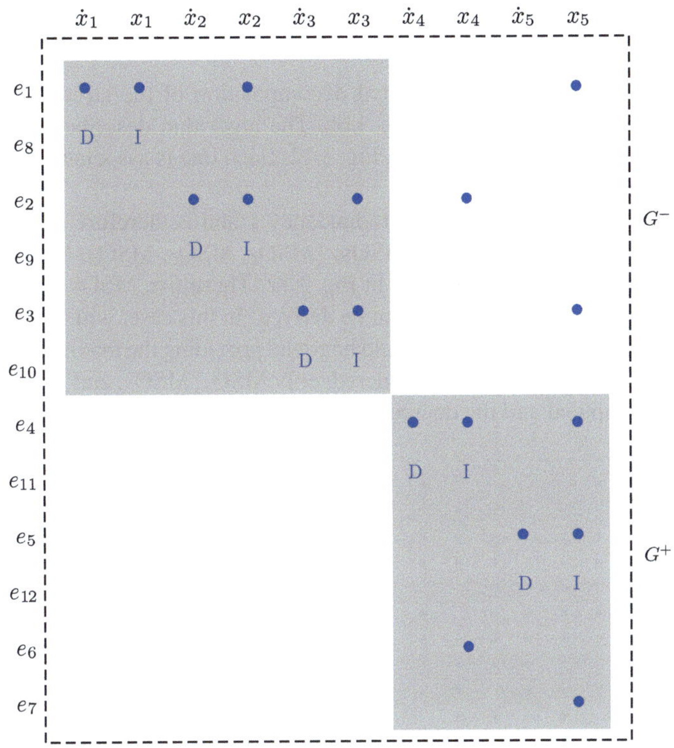

Fig. 3.9 Dulmage–Mendelsohn's decompositions

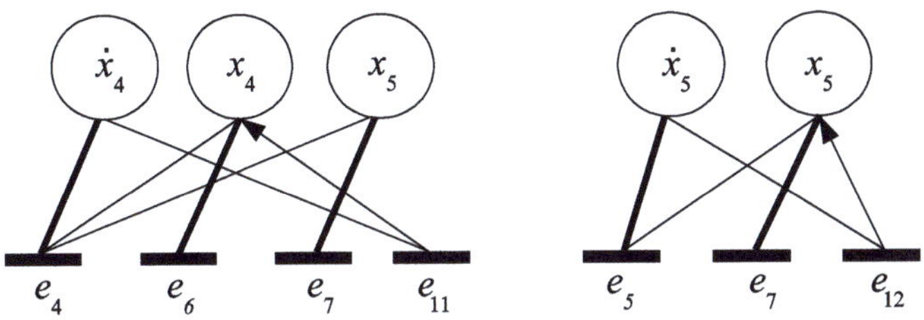

Fig. 3.10 MSO_1 and MSO_2 complete matchings with derivative variables

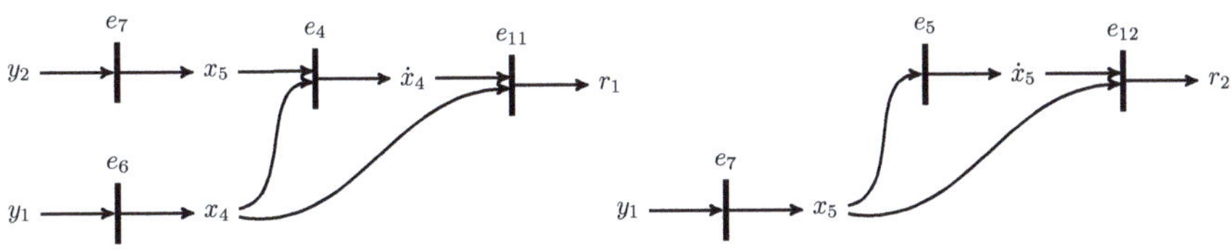

Fig. 3.11 Computational residuals sequence

3.4.4 Case Study—The Three-Tank System

The Dulmage–Mendelsohn's canonical decomposition of the structural model of the three-tank system is given in Fig. 3.6b. The algorithm described in Sect. 3.4.2 determines the 17 MSO sets listed in Fig. 3.12. Each one is associated with a set of faults.

The MTES have in this example redundancy 1 and is therefore also MSO sets. The MTES are given by the set of MSOs: $\{\text{MSO}_1, \text{MSO}_2, \text{MSO}_5, \text{MSO}_8, \text{MSO}_9\}$, that have been marked in gray color in Fig. 3.12. These five MSOs provide the set of constraints from which an ARR can be derived. In this case, without considering constraints restrictions, five ARRs could be found providing the maximum isolability [12]. When integral causality is considered, only MSO_1, MSO_2, and MSO_5 could be structurally computed and the degree of isolability decreases. Figure 3.13 shows the

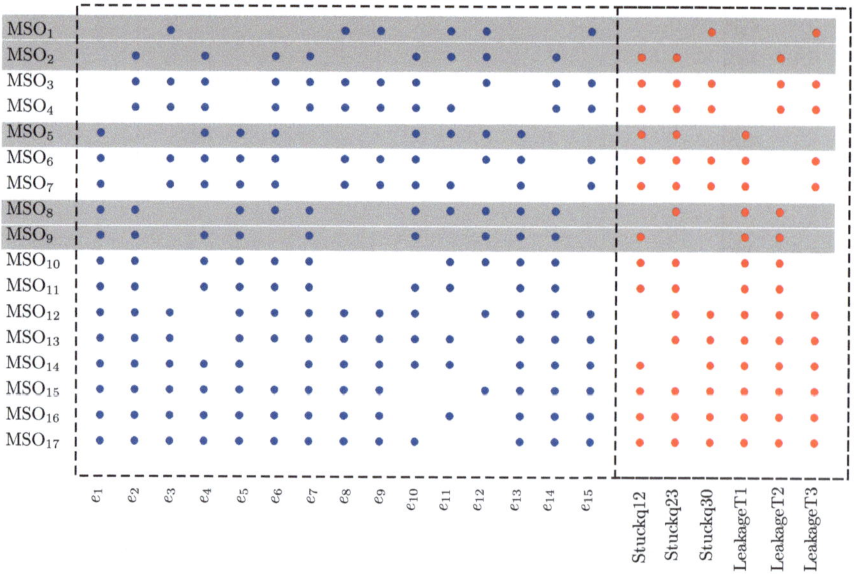

Fig. 3.12 MSOs computed in the tree tank system

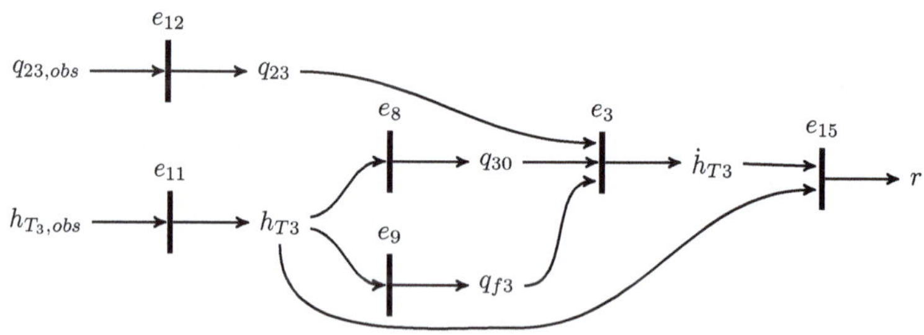

Fig. 3.13 Residual obtained from MSO_1 when integral causality is considered

residual obtained from causal matching of the constraints in MSO_1 when integral causality is considered.

3.5 Sensor Placement

The number and location of sensors significantly influences the expected fault detection and isolation performance. The objective of this section is to describe an approach how to, from a given model and a specified detectability and isolability performance specification, compute a characterization of all possible sets of sensors that meet the requirements. This presentation describes the main ideas and main results, for full details, proofs, optimizations, and further properties, see [11].

3.5.1 The Basic Sensor Placement Problem

To illustrate the fundamental problem, consider the small linear example model introduced in (3.2) but without the two sensors. With no sensors, there is no redundancy and the faults are not detectable. Then, a question is which sensors should be added to detect and isolate the faults? Before answering this, let's introduce some notation and let $\mathcal{F}$ denote the set of faults. A detectability performance specification is then a set $F_{det} \subseteq \mathcal{F}$ specifying the detectability requirement and an isolability requirement is a set $\mathcal{I}$ of ordered pairs $(f_i, f_j) \in F_{det} \times F_{det}$, meaning that f_i is isolable from f_j. Note that we assume that all faults that are included in the isolability specification $\mathcal{I}$ are also required to be detectable.

Since the fault isolability capability always increases when adding new sensors (for the moment, consider new sensors to be fault free), there are minimal elements in the family of sensor sets that achieves a certain level of fault isolability. Therefore, *minimal sensor set* is defined as follows.

Definition 3.16 (*Minimal sensor set*) Let $\mathcal{S}$ be the set of possible sensor locations, i.e., the set of measurable variables, and let S be a multiset defined on $\mathcal{S}$. Then S is a minimal sensor set, with respect to a given detectability and isolability specification, if adding the sensors in S fulfills the specification and all proper subsets of S do not.

Note that S is a multiset, which is similar to a set but allows for multiple instances of a member. Multisets are used since it may be necessary to add more than one sensor measuring the same variable if also the new sensors can fail.

Going back to the small linear model, it can be verified, using methods from Sect. 3.3, that there are five minimal sensor sets that achieve maximal fault isolation: $\{x_1, x_3\}$, $\{x_1, x_4\}$, $\{x_2, x_3\}$, $\{x_2, x_4\}$, and $\{x_3, x_4\}$. Thus, adding sensors measuring the variables in any of these sets, or a superset of the variables, achieves maximum fault isolability. Now, it could be the case that the new sensors may also become faulty. If maximum fault isolability is desired also for faults in the new sensors, there are nine minimal sensor sets where one sensor set is two sensors measuring x_1 and one for x_3, i.e., the multiset $S = \{x_1, x_1, x_3\}$ is a minimal sensor set.

3.5.2 A Structural Approach

This section describes the approach. A general assumption here is that the model does not contain any underdetermined part. This means, for example, that the model can be simulated with a unique solution given an initial state. Without loss of generality, it is also assumed that (1) no fault affects more than one equation and (2) that possible sensors measure a function of 1 unknown variable. Both these assumptions can be fulfilled by introducing additional variables. For example, consider the case where a sensor measures some function $h(x)$ rather than just a single variable. Then, introduce a new variable x_{new}, an equation $x_{new} = h(x)$, and then the new measurement equation will then be $y = x_{new}$.

The approach consists of three steps: (1) how to select sensors to achieve detectability, (2) how to rewrite isolability requirements as detectability requirements, and then (3) put it all together.

3.5.2.1 Sensor Placement for Detectability

Consider again the small example model introduced in (3.2) and fault f_3. To make this fault detectable, according to Definition 3.4, an additional sensor is needed such that equation $e_{f_3} = e_4$ is in the overdetermined part of the model. It is straightforward to verify that f_3 becomes detectable if and only if any one of the variables $\{x_1, x_2, x_4\}$ are measured. To understand why exactly these variables give detectability of f_3, consider first Fig. 3.14a where the model structure is shown. The model structure is here already in triangular form, but in general, the Dulmage–Mendelsohn's decomposition has to be done first to get the upper-triangular form.

The strong components are denoted with b_i and, for example, block b_1 is said to be connected with b_2 via a nonzero element in position $(1, 2)$ and in a similar fashion is b_2 connected to b_4. Such connections in the structure graph then imply an *ordering* among the strong components, e.g., $b_2 \leq b_1$ and $b_5 \leq b_2$. This ordering is a partial order and the Hasse diagram is shown in Fig. 3.14b. The interpretation of the Hasse diagram is that if there is an edge *from* b_j to *to* b_i this means that $b_j \leq b_i$. Note that it is a partial order, i.e., not all strong components are ordered. For example, there is no order between b_3 and b_4. It can be shown that this ordering makes it possible to state exactly which equations, in an exactly determined model, that become overdetermined when adding a sensor. Each block b_i is directly related to an equation set M_i, given by the Dulmage–Mendelsohn's decomposition.

Theorem 3.1 ([11]) *Let M be an exactly determined set of equations, b_i a strongly connected component in M with equations M_i, and $e \notin M$ an equation corresponding*

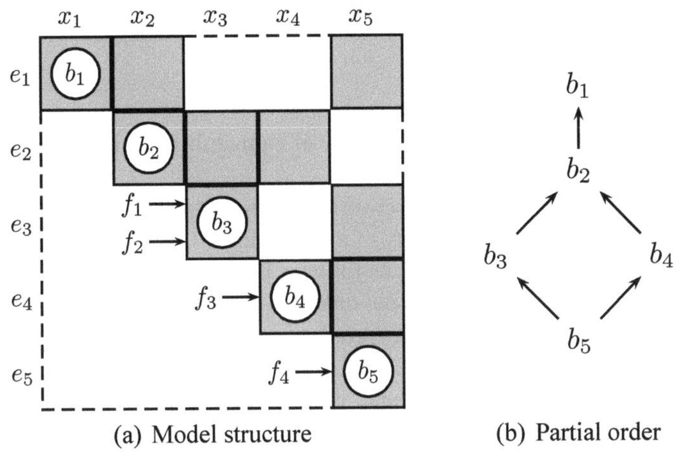

Fig. 3.14 Model structure and Hasse diagram of the block partial order over the strongly connected components for the small example model

to measuring any variable in b_i. *Then*

$$(M \cup \{e\})^+ = \{e\} \cup (\cup_{b_j \leq b_i} M_j). \tag{3.9}$$

It follows from Theorem 3.1 that measuring a variable in a block ordered higher than the block the fault enters achieves detectability. To use this result for sensor selection, let $\mathcal{P} \subseteq \mathcal{X}$ be a set of possible sensor locations and introduce the set

$$D(f_i) = \{x | b_i \leq b_j, x \in X_j \cap \mathcal{P}\}, \tag{3.10}$$

where X_j is the set of variables corresponding to block b_j, and b_i the block that is influenced by the fault f_i. The set $D(f_i)$ is then the set of variables such that measuring *any* variable in the set achieves detectability of f_i. Consider again the fault f_3 in the small example that affects an equation in b_4 in Fig. 3.14. To make f_3 detectable, *any* variable in b_1, b_2, or b_4 should be measured, i.e., $D(f_3) = \{x_1, x_2, x_4\}$.

Further, this means that to find sensors that achieve detectability of a *set* of faults, compute detectability sets $D(f_i)$ for each fault f_i and select a sensor set that has a non-empty intersection with each $D(f_i)$. This means that a solution measures at least one variable that makes each fault detectable. Finding the sensor set thus corresponds to finding a *hitting set* for the sets $D(f_i)$. To obtain minimal solutions, a minimal hitting set algorithm [9, 15] can be directly applied to find all minimal sensor sets that achieves detectability. For the example model, the corresponding detectability sets for all faults are

$$D(f_1) = D(f_2) = \{x_1, x_2, x_3\}, \quad D(f_3) = \{x_1, x_2, x_4\}, \quad D(f_4) = \mathcal{X}$$

and minimal sensor sets that achieve detectability of all faults are then $\{x_1\}$, $\{x_2\}$, and $\{x_3, x_4\}$.

3.5.2.2 Sensor Placement for Isolability of Detectable Faults

This section describes how to find sensors such that detectable faults are also isolated. Achieving maximum isolability of a set of single faults $\mathcal{F}$ can be divided into $|\mathcal{F}|$ subproblems, one for each fault, as follows. For each fault $f_j \in \mathcal{F}$, find all measurements that make the maximum possible number of faults $f_i \in \mathcal{F} \setminus \{f_j\}$ isolable from f_j. The solution to the isolability problem will then be obtained by combining the results from all subproblems.

Now, to show how this can be done, each subproblem will be formulated as a detectability problem. Assume that M is a model, including sensors such that all faults are detectable, and M_S represents a set of equations describing an additional sensor set S, e.g., if one sensor is added measuring a variable x, then M_S consists of the equation $y = x$. Given the sensor set S, a fault f_i is isolable from f_j in the model $M \cup M_S$ if

$$e_{f_i} \in ((M \setminus \{e_{f_j}\}) \cup M_S)^+$$

according to Definition 3.6. By introducing $M' = M \setminus \{e_{f_j}\}$, this can be written as

$$e_{f_i} \in (M' \cup M_S)^+ \tag{3.11}$$

which according to Definition 3.4 means that f_i is structurally detectable in $M' \cup M_S$. Hence, the maximum possible number of faults $f_i \in \mathcal{F} \setminus \{f_j\}$ are *isolable* from f_j in $M \cup M_S$ if the maximum possible number of faults $f_i \in \mathcal{F} \setminus \{f_j\}$ are structurally *detectable* in the model $(M \setminus \{e_{f_j}\}) \cup M_S$. Thus, each isolability subproblem can be formulated as a detectability problem.

To outline the solution of one such subproblem, again consider the small linear example. Assume that we have added sensors measuring $\{x_3, x_4\}$ such that all faults are detectable. Furthermore, assume that these sensors can be faulty and denote these new sensor faults f_5 and f_6, respectively. A row permuted structure of the obtained model $M = \{e_1, e_2, \ldots, e_7\}$ is shown in Fig. 3.15. Consider now the subproblem associated with fault f_1, i.e., isolate as many faults as possible from f_1. The set M' in (3.11) is equal to $M' = M \setminus \{e_{f_1}\} = M \setminus \{e_3\}$. The subproblem is then to, given the model M', find the minimal additional sensors S such that as many of the faults $f_2, f_3, \ldots, f_6$ as possible become detectable in $M' \cup M_S$.

The faults can, depending on which equations they violate, be divided into the following three types: faults that do not violate any equation in M', faults that violate equations in the structurally overdetermined part $(M')^+$, and faults that violate other equations in M', i.e., $M' \setminus (M')^+$. In the example, we have that $(M')^+ = \{e_4, e_5, e_7\}$ and $M' \setminus (M')^+ = \{e_1, e_2, e_6\}$ which is equal to the structurally just-determined part of M'. This implies that f_2 is not included in M'; f_3, f_4, and f_6 belong to the structurally overdetermined part; and f_5 to the structurally just-determined part. Fault f_2 is not

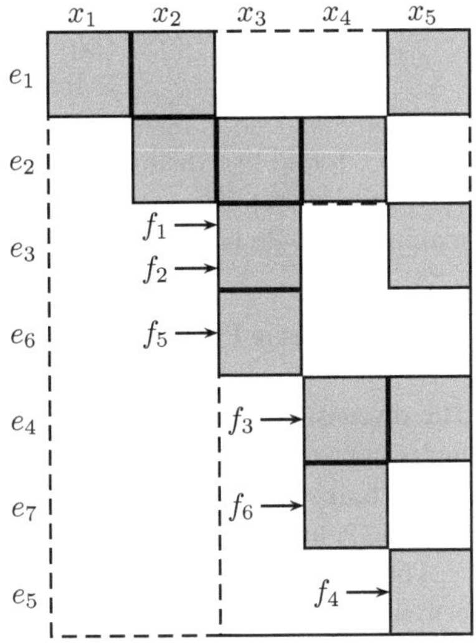

Fig. 3.15 Block structure of the example in Sect. 3.5.1 extended with measurements of x_3 and x_4

included in M' and cannot be structurally detectable in $M' \cup M_S$ for any sensor set S. This implies that f_2 is not isolable from f_1 with any sensor addition and this also follows from the fact that these two faults violate the same equation. The faults f_3, f_4, and f_6 in the structurally overdetermined part $(M')^+$ are according to Definition 3.4 structurally detectable in M' and require no additional measurements. Fault f_5 in the just-determined part is not detectable, but f_5 can become detectable in $M' \cup M_S$ if S is appropriately selected.

By applying the approach from Sect. 3.5.2.1 to the structurally just-determined part of M', i.e., the subgraph of M' defined by the node sets $\{e_1, e_2, e_6\}$ and $\{x_1, x_2, x_3\}$, we get that $D(f_5) = \{x_1, x_2, x_3\}$. Hence, one of the variables in the detectability set $\{x_1, x_2, x_3\}$ must be measured to make the faults $\mathcal{F} \setminus \{f_1, f_2\}$ detectable in $M' \cup M_S$ and this implies that all faults in $\mathcal{F} \setminus \{f_1, f_2\}$ are isolable from f_1 in $M \cup M_S$. The solution to the subproblem related to fault f_1 will be the computed detectability set. The next result formalizes the solution of a subproblem.

Theorem 3.2 *Let M be a set of equations with no structurally underdetermined part, $\mathcal{F}$ a set of structurally detectable faults in M, $\mathcal{P} \subseteq \mathcal{X}$ the set of possible sensor locations, and M_S the equations added by adding the sensor set S.*

For an arbitrary fault f_j, let M^0 be the just-determined part of $M \setminus \{e_{f_j}\}$, F^0 the set of faults contained in M^0, and $\mathcal{D}$ detectability sets for faults F^0 in M^0 according to procedure in Sect. 3.5.2.1.

Then, the maximum possible number of faults $f_i \in \mathcal{F} \setminus \{f_j\}$ is structurally isolable from f_j in $M \cup M_S$ if and only if S has a non-empty intersection with all non-empty sets in $\mathcal{D}$.

For the example shown in Fig. 3.15, the families of detectability sets of the different subproblems are

$$\begin{aligned}&\{\{x_1, x_2, x_3\}\} \text{ for } f_1, f_2, \text{ and } f_5\\&\{\{x_1, x_2, x_4\}\} \text{ for } f_3 \text{ and } f_6\\&\varnothing \text{ for } f_4.\end{aligned} \tag{3.12}$$

We have found two distinct detectability sets and the minimal hitting sets are $\{x_1\}$, $\{x_2\}$, and $\{x_3, x_4\}$. These sets are the minimal additional measurements that achieve maximum single fault isolability.

3.5.2.3 Sensor Placement for Both Detectability and Isolability

The discussion above shows how isolability can be achieved in a model where all faults are structurally detectable. Next, the approach will be extended to models where faults may not be structurally detectable in the original model. Again, the solution is first outlined for the small linear example.

The faults in the model, without any sensors, are not detectable and the objective is to find all minimal sensor sets that maximize fault detectability and isolability. It is previously shown that the minimal sets of measurements to achieve full detectability are $\{x_1\}$, $\{x_2\}$, and $\{x_3, x_4\}$.

If we start by adding the first set, i.e., a sensor measuring x_1 described by an equation e_s, a new model $M \cup \{e_s\}$ is obtained where all faults are detectable. Since all faults are detectable, the previously described method to achieve maximum isolability can be applied to the model $M \cup \{e_s\}$. The minimal sensor sets that solve this problem are $\{x_3\}$ and $\{x_4\}$. By combining this result with the fact that a sensor measuring x_1 has been added to obtain detectability, it follows that $\{x_1, x_3\}$ and $\{x_1, x_4\}$ are two possible sensor sets that achieve maximum detectability and isolability. To compute all minimal sensor sets that achieve maximum isolability, we also have to investigate the solutions when we decide to measure $\{x_2\}$ or $\{x_3, x_4\}$ to obtain full detectability. By solving one isolability problem for each of the minimal sensor sets that achieves full detectability, we get that the minimal sensor sets are $\{x_1, x_3\}$, $\{x_1, x_4\}$, $\{x_2, x_3\}$, $\{x_2, x_4\}$, and $\{x_3, x_4\}$ which are the same sets as in Sect. 3.5.1.

The above procedure for the small linear example can be directly extended to any model. All operations needed have polynomial complexity except for the minimal hitting set algorithm which is NP-hard. This means that the worst case might be intractable; however, this is a direct consequence of the objective to find *all* minimal sensor sets. If the objective is to find not all, but small sensor sets, approximate hitting set algorithms [3] can be applied which makes large problems tractable. Note however that well-formed models of physical systems typically have a structure which makes the computations less demanding. Also, and this is maybe the most important aspect, the number of measurable signals in the specification P is a prime indicator of the complexity. Thus, it is not primarily the number of equations or number of faults, but rather the user-specified sensor specification that controls the complexity.

To conclude the sensor placement description, consider again the three-tank example from Chap. 2. In Sect. 3.3, it was shown that with the original set of three sensors, full fault isolability was not possible since faults Leakage_{T_3} and Stuck_{q30} could not be isolated. Now, remove the three sensors and assume that possible sensors are height

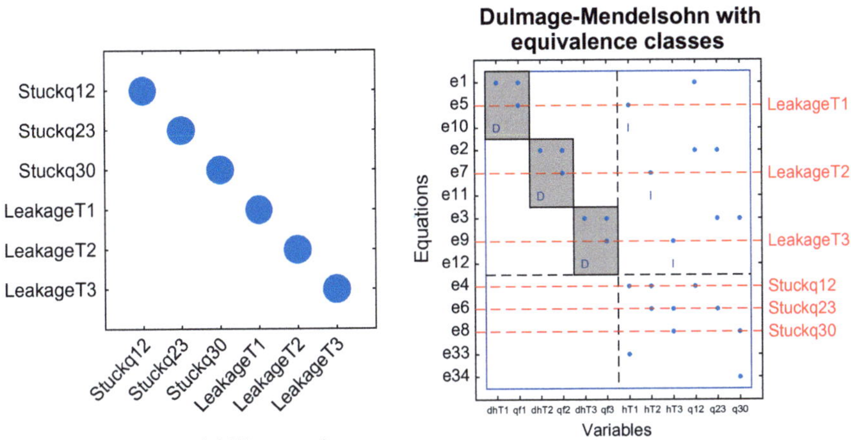

(a) Isolability matrix (b) Dulmage-Mendelsohn

Fig. 3.16 Isolability matrix and detailed Dulmage–Mendelsohn's decomposition, with a canonical decomposition of the overdetermined part, resulting from the sensor placement analysis of the three-tank system

and flow sensors. With the presented approach, there are five minimal sensor sets where 1 is of size 2 and 4 of size 3. The smallest sensor set is $\mathcal{S}_1 = \{h_{T_1}, q_{30}\}$, i.e., measuring the height of the first tank and the flow out of the last tank gives information on the complete process. Figure 3.16a shows the isolability matrix, which should be compared to the corresponding isolability matrix in Fig. 3.7a with the original sensors. Figure 3.16b shows the Dulmage–Mendelsohn's decomposition, with canonical decomposition of the overdetermined part, which confirms that all faults are isolable from each other since they all are in different equivalence classes.

3.6 Summary and Discussion

In this chapter, we have introduced structural analysis for diagnosis system design and model analysis. We have introduced the properties of fault detectability and isolability of a model. Both allow checking the limit on the diagnosis performance. The model structure has also been used for determining the redundant set of constraints or equations from which analytical redundancy relations or sequential residual generators can be designed. Computation sequences for the sequential residual generators have been provided. And finally, structural models have been used for characterizing all possible sets of sensors that meet a specified detectability and isolability performance specification.

References

1. Armengol, J., Bregón, A., Escobet, T., Gelso, E., Krysander, M., Nyberg, M., Olive, X., Pulido, B., Travé-Massuyès, L.: Minimal structurally overdetermined sets for residual generation: a comparison of alternative approaches. IFAC Proc. Vol. **42**(8), 1480–1485 (2009)
2. Blanke, M., Kinnaert, M., Lunze, J., Staroswiecki, M.: Diagnosis and Fault-Tolerant Control. Springer, Berlin (2006)
3. De Kleer, J.: Hitting set algorithms for model-based diagnosis. In: 22nd International Workshop on Principles of Diagnosis (DX-11), Murnau, Germany (2011)
4. Dulmage, A.L., Mendelsohn, N.S.: Coverings of bipartite graphs. Can. J. Math. **10**, 517–534 (1958)
5. Frisk, E., Bregon, A., Åslund, J., Krysander, M., Pulido, B., Biswas, G.: Diagnosability analysis considering causal interpretations for differential constraints. IEEE Trans. Syst. Man Cybern. Part A: Syst. Hum. **42**(5), 1216–1229 (2012)
6. Frisk, E., Krysander, M., Åslund, J.: Analysis and design of diagnosis systems based on the structural differential index. In: IFAC World Congress, Toulouse, France (2017)
7. Frisk, E., Krysander, M., Jung, D.: A toolbox for analysis and design of model based diagnosis systems for large scale models. IFAC-PapersOnLine **50**(1), 3287–3293 (2017). 20th IFAC World Congress
8. Harary, F.: Graph Theory. Addison-Wesley Publishing Company, Reading (1969). ISBN 0-201-41033-8
9. Kleer, J.D., Williams, B.: Diagnosing multiple faults. Artif. Intell. **32**, 97–130 (1987)
10. Krysander, M.: Design and analysis of diagnosis systems using structural methods. Ph.D. thesis, Linköpings Universitet (2006)
11. Krysander, M., Frisk, E.: Sensor placement for fault diagnosis. IEEE Trans. Syst. Man Cybern. Part A: Syst. Hum. **38**(6), 1398–1410 (2008)
12. Krysander, M., Åslund, J., Nyberg, M.: An efficient algorithm for finding minimal overconstrained subsystems for model-based diagnosis. IEEE Trans. Syst. Man Cybern. Part A: Syst. Hum. **38**(1), 197–206 (2008)
13. Krysander, M., Åslund, J., Frisk, E.: A structural algorithm for finding testable sub-models and multiple fault isolability analysis. In: 21st International Workshop on Principles of Diagnosis (DX-10), Portland, Oregon, USA, pp. 17–18 (2010)
14. Nyberg, M., Frisk, E.: Residual generation for fault diagnosis of systems described by linear differential-algebraic equations. IEEE Trans. Autom. Control **51**(12), 1995–2000 (2006)
15. Reiter, R.: A theory of diagnosis from first principles. Artif. Intell. **32**, 57–95 (1987)

Chapter 4
FDI Approach

Vicenç Puig, María Jesús de la Fuente and Joaquim Armengol

4.1 Introduction

Model-based Fault Detection and Isolation (FDI) of dynamic systems is based on the use of models (*analytical redundancy*) to check the consistency of observed behaviors. This consistency check is based on computing the difference between the predicted value from the model and the real value measured by the sensors. Then, this difference, known as *residual*, is compared with a threshold value (zero in the ideal case). When the residual is greater than the threshold, it is considered that there is a fault in the system. Otherwise, it is considered that either the system is working properly or, if it is faulty, the fault cannot be detected. This is denoted as residual evaluation. Due to the presence of noise, disturbances, and model errors, the residuals are never zero, even if there is no fault. Therefore, the detection decision requires testing the residual against thresholds, obtained empirically or by theoretical considerations. Also the desensitizing of the residual from the noise, the disturbances, and the model errors while maximizing fault sensitivity is the goal of the robust design of the detection and diagnosis algorithms. Fault detection is followed by the fault isolation procedure which intends to distinguish a particular fault from others.

V. Puig (✉)
Research Center for Supervision, Safety and Automatic Control (CS2AC), Unversitat Politècnica de Catalunya (UPC),
Terrassa, Spain
e-mail: vicenc.puig@upc.edu

M. J. de la Fuente
Departamento de Ingeniería de Sistemas y Automática, Universidad de Valladolid,
Valladolid, Spain
e-mail: mjfuente@eii.uva.es

J. Armengol
Departament d'Enginyeria Elèctrica, Electrònica i Automàtica, Universitat de Girona,
Girona, Spain
e-mail: joaquim.armengol@udg.cat

T. Escobet et al. (eds.), *Fault Diagnosis of Dynamic Systems*,
https://doi.org/10.1007/978-3-030-17728-7_4

While a single residual is sufficient to detect faults, a set (or a vector) of residuals is required for fault isolation [13]. If a fault can be distinguished from other faults using a residual set, then it is said that this fault is isolable.

Model-based FDI started with the seminal works of Clark [7] and others. Since then, a huge amount of research has been carried out. As a result, today there exists a set of methods that conform the basis of this field and can be considered as the foundations of more advanced methods. Such methods are parity methods [13], observer methods [5], and parameter estimation methods [18, 19]. Relations between such methods have been established by several authors and are now well understood [13]. On the other hand, in the literature, two different approaches to construct residual sets with the desired isolability properties are described.

One approach is based on designing a vector of *structured residuals* [13]. Each residual is designed to be sensitive to a subset of faults, while remaining insensitive to the remaining faults. The design procedure consists of two steps—the first step is to specify the sensitivity and insensitivity relationships between residuals and faults according to the assigned isolation task, while the second one is to design a set of residual generators according to the previous relationships. The fault isolation problem then consists of a separate threshold test for each residual using a decision table.

An alternative way of achieving the isolability of faults is by designing a vector of *directional residuals* [13], which lies in a fixed and fault-specified direction in the residual space, in response to a particular fault.

4.2 Fault Detection

The principle of model-based fault detection is to test whether the measured inputs and outputs from the system are consistent with the model of the faultless system. The system to be monitored can be described by a linear uncertain dynamic model in discrete-time and state-space form as follows:

$$\begin{aligned} \mathbf{x}(k+1) &= \mathbf{A}(\boldsymbol{\theta})\mathbf{x}(k) + \mathbf{B}(\boldsymbol{\theta})\mathbf{u}(k) + \mathbf{w}(k) \\ \mathbf{y}(k) &= \mathbf{C}(\boldsymbol{\theta})\mathbf{x}(k) + \mathbf{D}(\boldsymbol{\theta})\mathbf{u}(k) + \mathbf{v}(k), \end{aligned} \tag{4.1}$$

where $\mathbf{y}(k) \in \Re^{n_y}$, $\mathbf{u}(k) \in \Re^{n_u}$, $\mathbf{x}(k) \in \Re^{n_x}$ are the system output, input, and the state-space vectors, respectively; $\mathbf{w}(k) \in \Re^{n_x}$ and $\mathbf{v}(k) \in \Re^{n_y}$ are the disturbances and noises; the state, input, output, and direct transmission matrices are $\mathbf{A}(\boldsymbol{\theta}) \in \Re^{n_x \times n_x}$, $\mathbf{B}(\boldsymbol{\theta}) \in \Re^{n_x \times n_u}$, $\mathbf{C}(\boldsymbol{\theta}) \in \Re^{n_y \times n_x}$, and $\mathbf{D}(\boldsymbol{\theta}) \in \Re^{n_y \times n_u}$, respectively; $\boldsymbol{\theta} \in \Re^{n_\theta}$ is the vector of parameters corresponding to the normal operation conditions.

The system in Eq. (4.1) can, alternatively, be expressed in input–output form using the shift operator q^{-1} and assuming zero initial conditions:

$$\mathbf{y}(k) = \mathbf{M}(q^{-1}, \boldsymbol{\theta})\mathbf{u}(k), \tag{4.2}$$

where $\mathbf{M}(q^{-1}, \boldsymbol{\theta})$ is given by

$$\mathbf{M}(q^{-1}, \boldsymbol{\theta}) = \mathbf{C}(\boldsymbol{\theta})(q\mathbf{I} - \mathbf{A}(\boldsymbol{\theta}))^{-1}\mathbf{B}(\boldsymbol{\theta}) + \mathbf{D}(\boldsymbol{\theta}).$$

If the measurements are inconsistent with the model of the faultless system, the existence of a fault can be proved. The residual usually describes the consistency check between the real behavior measured by the sensors, $y(k)$, and the predicted value, $\hat{y}(k)$, using the model:

$$r_i(k) = y_i(k) - \hat{y}_i(k) \quad i = 1, \ldots, n_y, \tag{4.3}$$

where n_y is the number of output sensors.

The fault detection task consists of deciding if a residual given by Eq. (4.3) is violated at a given instant or not by generating a *fault signal* s_i according to:

$$s_i(k) = \begin{cases} 0, & \text{if } |r_i(k)| < \tau_i \text{ (no fault)} \\ 1, & \text{if } |r_i(k)| \geq \tau_i \text{ (fault)} \end{cases}, \tag{4.4}$$

where τ_i is the threshold associated to the ith residual.

The different methods of FDI depend on the different models used to calculate the residual. As was explained before, the most classical models used to form the FDI point of view are: state observers (or Kalman filters in case of stochastic systems), parity equations, and parameter estimation methods that are going to be described in the following sections. In these classical methods, the model obtained is a linear one, but if the system is nonlinear, the FDI methods can be extended using nonlinear models obtained by: neural networks, neuro-fuzzy systems, nonlinear observers, fuzzy-observers, and so on [5].

4.2.1 Observers

As discussed in the introduction, fault detection can be based on the use of observers. The basic idea here is to estimate the outputs of the system from the measurements (or a subset of measurements) by using either Luenberger observer in a deterministic setting or a Kalman filter in a stochastic setting. Then, the (weighted) output estimation error (or innovations in the stochastic case) is used as a residual.

A state observer with Luenberger structure can be used to obtain the output estimation $\hat{\mathbf{y}}(k)$, in the residual (4.3):

$$\begin{aligned} \hat{\mathbf{x}}(k+1) &= \mathbf{A}\hat{\mathbf{x}}(k) + \mathbf{B}\mathbf{u}(k) + \mathbf{L}\left(\mathbf{y}(k) - \hat{\mathbf{y}}(k)\right) \\ \hat{\mathbf{y}}(k) &= \mathbf{C}\hat{\mathbf{x}}(k), \end{aligned} \tag{4.5}$$

where $\mathbf{L}$ is the observer gain, designed to stabilize the observer matrix $(\mathbf{A} - \mathbf{LC})$ and to guarantee desired fault detection performance. So, the estimation error dynamics and also the residual are

$$\begin{aligned} \mathbf{e}(k+1) &= \mathbf{x}(k+1) - \hat{\mathbf{x}}(k+1) = (\mathbf{A} - \mathbf{LC})\mathbf{e}(k) \\ \mathbf{r}(k) &= \mathbf{C}(k)\mathbf{e}(k). \end{aligned} \tag{4.6}$$

Here, the state error vanishes asymptotically and the residual in ideal conditions when there is not a fault is zero.

A system with all possible faults can be described by state-space model as

$$\begin{aligned} \mathbf{x}(k+1) &= \mathbf{A}(\boldsymbol{\theta})\mathbf{x}(k) + \mathbf{B}(\boldsymbol{\theta})\mathbf{u}(k) + \mathbf{F}(\boldsymbol{\theta})\mathbf{f}(k) \\ \mathbf{y}(k) &= \mathbf{C}(\boldsymbol{\theta})\mathbf{x}(k) + \mathbf{D}(\boldsymbol{\theta})\mathbf{u}(k) + \mathbf{H}(\boldsymbol{\theta})\mathbf{f}(k), \end{aligned} \tag{4.7}$$

where $\mathbf{f}(k) \in \Re^{n_f}$ are fault vectors, and the fault entry matrices **F** and **H** are known. Introducing this process equations into the observer equation according to (4.5) leads the state error and the residuals as

$$\begin{aligned} \mathbf{e}(k+1) &= (\mathbf{A} - \mathbf{LC})\mathbf{e}(k) + \mathbf{Ff}(k) - \mathbf{LHf}(k) \\ \mathbf{r}(k) &= \mathbf{C}(k)\mathbf{e}(k) + \mathbf{Hf}(k). \end{aligned} \tag{4.8}$$

It can be seen here that the residuals depend solely on faults because the state deviations $\mathbf{e}(k)$ are asymptomatically vanished.

The predicted output of the observer (4.5), expressing the system matrices in observer canonical form, can also be expressed in input/output form

$$\hat{\mathbf{y}}(k) = \mathbf{G}_u(q)\mathbf{u}(k) + \mathbf{G}_y(q)\mathbf{y}(k), \tag{4.9}$$

where

$$\begin{aligned} \mathbf{G}_u(q) &= \mathbf{C}(q\mathbf{I} - \mathbf{A} + \mathbf{LC})^{-1}\mathbf{B} \\ \mathbf{G}_y(q) &= \mathbf{C}(q\mathbf{I} - \mathbf{A} + \mathbf{LC})^{-1}\mathbf{L}. \end{aligned} \tag{4.10}$$

So the residuals can also be calculated through this input/output form.

4.2.2 Parity Equations

Parity equations can be seen as particular cases of the observer (4.5) [13]. In case the observer gain is taken equal to zero ($\mathbf{L} = 0$), the observer becomes an *interval simulator*, since the output prediction is based only in the inputs and previous output predictions, and output prediction becomes

$$\hat{\mathbf{y}}(k) = \mathbf{G}_u(q)\mathbf{u}(k), \tag{4.11}$$

while the residual is given by

$$\mathbf{r}(k) = \mathbf{y}(k) - \hat{\mathbf{y}}(k) = \mathbf{y}(k) - \mathbf{G}_u(q)u(k). \tag{4.12}$$

According to [13], Eq. (4.12) corresponds with an *ARMA primary parity equation* or *residual*. This is an *open-loop approach*. Interval simulation requires solving the optimization problem following the same strategy as in the case of the interval observer but using the system matrices (4.1). In order to reduce computing complexity, as in the observer case, a time window could also be used. In this case, this approach is known as ℓ-order ARMA parity equation [32].

On the other hand, in case that the observer gain is designed such that the poles are at the origin (*deadbeat observer*), the observer becomes an *interval predictor*, since the output prediction is based only on measured inputs and outputs and follows the real system output after the minimum number of samples. The prediction equation is moving average (MA) and follows a *closed-loop approach*. Thus, the corresponding residuals are called *MA primary parity equations* or *residuals* [13].

Assuming additive faults, the input–output relation given by (4.11) should be modified in the following way

$$\mathbf{y}(k) = \mathbf{G}_u(q)\mathbf{u}(k) + \mathbf{S}_f(q)\mathbf{f}(k), \tag{4.13}$$

where $\mathbf{S}_f(q)$ is fault transfer function and $\mathbf{f}(k)$ is the fault. Then, the residual (4.12) can alternatively be expressed as

$$\mathbf{r}(k) = \mathbf{y}(k) - \mathbf{G}_u(q)\mathbf{u}(k) = \mathbf{S}_f(q)\mathbf{f}(k), \tag{4.14}$$

where the first part of the residual is the computational form that depends only on the inputs and outputs of the system and the second part is the internal form that shows how the residuals only depend on the faults.

The parity equations can also be calculated with the state-space model following the procedure defined by Chow and Willsky [6]. The output estimation (4.11) can also be expressed in state-space form as

$$\hat{\mathbf{y}}(k) = \mathbf{C}\left(\mathbf{A}^k\mathbf{x}(0) + \sum_{i=0}^{k-1}\left(\mathbf{A}^{k-1-i}\mathbf{B}\mathbf{u}(i)\right)\right) + \mathbf{D}\mathbf{u}(k). \tag{4.15}$$

The output estimation for a time window of length $p+1$, with $(p \leq n_x)$, and n_x the number of system states, are summarized as

$$\mathbf{Y}(k) = \mathbf{O}\mathbf{x}(k-p) + \mathbf{T}\mathbf{U}(k), \tag{4.16}$$

where

$$\mathbf{Y}(k) = \left(\mathbf{y}(k-p) \ldots \mathbf{y}(k-1)\ \mathbf{y}(k)\right)^t \tag{4.17}$$

$$\mathbf{O} = \begin{pmatrix} \mathbf{C} \\ \mathbf{CA} \\ \vdots \\ \mathbf{CA}^p \end{pmatrix} \tag{4.18}$$

$$\mathbf{T} = \begin{pmatrix} \mathbf{D} & 0 \cdots & 0 & 0 \\ \mathbf{CB} & D \cdots & 0 & 0 \\ \vdots & \ddots & & \vdots \\ \mathbf{CA}^{p-1}\mathbf{B} & \cdots & \mathbf{CB} & \mathbf{D} \end{pmatrix} \tag{4.19}$$

$$\mathbf{U}(k) = \left(\mathbf{u}(k-p) \cdots \mathbf{u}(k-1)\ \mathbf{u}(k)\right)^t \tag{4.20}$$

Hence, (4.16) describes the $(p+1)$ input and output signals and the initial state vector $\mathbf{x}(k-p)$ over a time interval of length $(p+1)$, thus forming a *temporal redundancy*. The matrix $\mathbf{O}$ is the observability matrix and the $\mathbf{T}$ matrix is a Toeplitz matrix.

Now, the primary residual can be calculated as

$$\mathbf{e}(k) = \mathbf{Y}(k) - \mathbf{TU}(k) = \mathbf{Ox}(k-p), \tag{4.21}$$

where the first part of the residual is the computational form and the second one is the internal form. But as $\mathbf{e}(k)$ depends on the state vector $\mathbf{x}(k-p)$, this dependency must be eliminated, obtaining a transformed residual as

$$\mathbf{r}(k) = \mathbf{W}(\mathbf{Y}(k) - \mathbf{TU}(k)), \tag{4.22}$$

where the transformation W must satisfy the condition:

$$\mathbf{WO} = 0, \tag{4.23}$$

by satisfying this condition the residual is decoupled from the state. The dimension of W is $1x(p+1)n_y$ with n_y the number of outputs. If the order of A is n_x, the matrix $\mathbf{O}$ has order $n_x(p+1)n_y$. Through the Eq. (4.23), n_x elements of $\mathbf{W}$ are determined, however, the remaining $(p+1)n_y - n_x$ can be chosen freely.

Assuming faults in the system, i.e., considering the state-space model representation of Eq. (4.7) and following a procedure similar to the one presented in this section, the internal form of the parity equation is obtained as

$$\mathbf{e}(k) = \mathbf{Ox}(k-p) + \mathbf{T_f f}(k), \tag{4.24}$$

where $\mathbf{T_f}$ is a Toeplitz matrix similar to $\mathbf{T}$. Thus, the transformed residual is

$$\mathbf{r}(k) = \mathbf{W}(\mathbf{Y}(k) - \mathbf{TU}(k)) = \mathbf{WT_f f}(k), \tag{4.25}$$

where as before the transformation matrix $\mathbf{W}$ must satisfy the condition $\mathbf{WO} = 0$ and the new residual only depends on the faults.

4.2.3 Parameter Estimation

Alternative to the observer/parity space approach presented in previous sections, fault detection via parameter estimation relies on the principle that possible faults in the process can be associated with specific physical parameters of the system such as length, mass, speed, drag coefficient, viscosity, resistances, capacities, etc. And the faults which make themselves noticeable in these physical process constants are, therefore, also expressed in terms of the model parameters of that system. So, if the physical process coefficients which indicate process faults are not directly measurable, an attempt can be made to determine their changes via the changes in the process model parameters. So, the following procedure is available [17]:

- Establishment of the process model for the measurable input and output variables, usually derived from first principles (mass, energy and momentum balances, the physical-chemical equations, etc.), it can be written in terms of differential equations, such that their parameters θ are expressed in dependence of the physical coefficients, p

$$y(k) = f(u(k), \theta), \tag{4.26}$$

  where $\mathbf{y}$ and $\mathbf{u}$ are the measurable output and input variables and θ are the unmeasured model parameters to estimate.
- Determination of the relationship between the model parameters θ_i and the physical coefficients p_j, i.e., $\theta = g(p)$.
- Estimation of the model parameters θ_i from the inputs and outputs of the process $\mathbf{u(k)}$, $\mathbf{y(k)}$ by an estimation procedure:

$$\hat{\theta}(k) = h\Big(y(1), \ldots, y(k), u(1), \ldots, u(k)\Big). \tag{4.27}$$

- Calculation of the process coefficients in terms of the model parameters:

$$\hat{p}(k) = g^{-1}(\hat{\theta}(k)). \tag{4.28}$$

- Now the decision on whether a fault has occurred can be taken based either on the changes in the physical coefficients Δp_j or on the changes of the model parameters $\Delta\theta_i$.

Hence, the basis of this class of methods is the combination of theoretical modeling and parameter estimation of continuous time models. A necessary requirement of this procedure is, however, the existence of the inverse relationship Eq. (4.28). However, for some classes of nonlinear systems, the physical coefficients can be estimated directly. In these cases, there exists a set of differential equations based on first principles that define the process model. Since some of the equations are linear with respect to the physical parameters to be estimated, there is no need for the existence of that inverse relationship [11, 12].

4.3 Robustness

4.3.1 Introduction

As stated above, analytical redundancy is based on the use of models. However, when building a model of a dynamic process to monitor its behavior, there is always some mismatch between the modeled and the real behavior due to the fact that some effects are neglected, there are nonlinearities that are linearised in order to simplify the model, there are parameters which have different values in different units of the same component (tolerance), there are errors in the parameters (or in the structure) of the model that are introduced in the model calibration process, etc. so, there is uncertainty in the model due to the modeling errors. This uncertainty is called *structured* if it affects only to the values of the parameters of the model and not its structure, and *nonstructured* if it affects the structure of the model. Anyway, it can affect the results of the fault detection procedure by generating false alarms or undetected faults.

In FDI literature, a fault detection algorithm able to handle the uncertainty is called *robust*. The *robustness* of an FDI algorithm is the degree of sensitivity to faults compared to the degree of sensitivity to uncertainty [5]. The goal is to maximize the detectability and isolability of faults while minimizing the effects of uncertainty and disturbances [22].

Research on robust fault diagnosis methods has been very active in the FDI community in recent years. One of the most developed families of approaches, called *active*, is based on generating residuals which are insensitive to uncertainty, while at the same time being sensitive to faults. This approach has been extensively developed by several researchers using different techniques: unknown input observers, robust parity equations, H_∞, etc. In [5], there is an excellent survey of this active approach.

On the other hand, there is a second family of approaches, called *passive*, which enhances the robustness of the fault detection system at the decision-making stage [5] by propagating the uncertainty to the residuals and generating an *adaptive threshold*. Seminal papers suggesting this approach are [16] in the time domain and [8] in the frequency domain. This passive approach has been developed in recent times by several researchers and is still under research, see for example some recent results in [1, 3, 10, 14, 23, 27–29, 31], among others. This approach has also been integrated with *Qualitative Reasoning* tools (coming from AI community) like *CA*∼*EN* [9, 33], *SQualTrack* [3] or *MOSES* [30]. For a more detailed review, the reader is referred to [2].

As it was defined in the introduction, the residual vector, known also as *analytical redundant relation* (ARR), is the difference between measured $\mathbf{y}(k)$ and predicted system outputs $\hat{\mathbf{y}}(k)$

$$\mathbf{r}(k) = \mathbf{y}(k) - \hat{\mathbf{y}}(k). \tag{4.29}$$

Ideally, the residuals should only be affected by the faults. However, the presence of disturbances, noise, and modeling errors causes the residuals to become nonzero, and thus interferes with the detection of faults.

Therefore, the fault detection procedure must be robust against these undesired effects [5]. In case of modeling a dynamic system using an interval model, the predicted output is described by a set that can be bounded at any step by an interval

$$\hat{y}_i(k) \in [\underline{\hat{y}_i}(k), \overline{\hat{y}_i}(k)], \tag{4.30}$$

in a non-faulty case. Such interval is computed independently for each output (neglecting couplings between outputs) as follows:

$$\underline{\hat{y}_i}(k) = \min_{\theta \in \Theta}(\hat{y}_i(k, \boldsymbol{\theta})) \text{ and } \overline{\hat{y}_i}(k) = \max_{\theta \in \Theta}(\hat{y}_i(k, \boldsymbol{\theta})). \tag{4.31}$$

It can be computed using the algorithm based on numerical optimization presented in [26]. Then, the fault detection test is based on propagating the parameter uncertainty to the residual, and checking if

$$\mathbf{y}(k) \in \left[\underline{\hat{\mathbf{y}}}(k) - \sigma, \overline{\hat{\mathbf{y}}}(k) + \sigma\right], \tag{4.32}$$

where σ is the noise bound. Equivalently, this test can be formulated in terms of the residual, checking if

$$0 \in \left[\underline{\mathbf{r}}(k), \overline{\mathbf{r}}(k)\right] = \mathbf{y}(k) - \left[\underline{\hat{\mathbf{y}}}(k) - \sigma, \overline{\hat{\mathbf{y}}}(k) + \sigma\right], \tag{4.33}$$

holds or not. In case it does not hold, a fault can be indicated. This is called the *direct test*, in which effective thresholds are found using intervals to bound the uncertainty in parameters and measurements [2, 10, 24, 25].

Alternatively, the *inverse test* consists of checking if there exists some parameter value in the parameter uncertainty set Θ such that model (4.2) is consistent with the system measurements. More formally, to check if

$$\exists \boldsymbol{\theta} \in \Theta \mid \hat{\mathbf{y}}(k, \boldsymbol{\theta}) \in \left[\mathbf{y}(k) - \sigma, \mathbf{y}(k) + \sigma\right] \tag{4.34}$$

holds or not. In the case that this condition is not satisfied, a discrepancy between measurements and the model is detected and a fault should be indicated. This test can be implemented with the parameter estimation algorithms used in the error-bounding approach [21].

The direct test is related to the use of parity equation or observer methods while the inverse test is related to parameter estimation methods. According to [19], parity equations and observer approaches are more suitable for additive faults while parameter estimation approach is better suited for multiplicative (parametric) faults.

4.3.2 Active Robustness: Disturbance Decoupling

If the system is affected by disturbances **d** and faults **f**, the model (4.1) is modified as

$$\begin{aligned} \mathbf{x}(k+1) &= \mathbf{A}\mathbf{x}(k) + \mathbf{B}\mathbf{u}(k) + \mathbf{E}\mathbf{d}(k) + \mathbf{F}\mathbf{f}(k) \\ \mathbf{y}(k) &= \mathbf{C}\hat{\mathbf{x}}(k) + \mathbf{G}\mathbf{d}(k) + \mathbf{H}\mathbf{f}(k)\,. \end{aligned} \tag{4.35}$$

In this case, using a Luenberger observer to estimate the state-space of the system gives the estimation error $\mathbf{e}(k+1)$ and the residual $\mathbf{r}(k)$ as

$$\begin{aligned} \mathbf{e}(k+1) &= (\mathbf{A} - \mathbf{L}\mathbf{C})\,\mathbf{e}(k) + \mathbf{E}\mathbf{d}(k) + \mathbf{F}\mathbf{f}(k) \\ \mathbf{r}(k) &= \mathbf{C}\mathbf{e}(k) + \mathbf{G}\mathbf{d}(k) + \mathbf{H}\mathbf{f}(k)\,. \end{aligned} \tag{4.36}$$

As it is possible to see here, if the observer is well-defined $e(k) \to 0$, and the residuals depend on the faults and on the disturbances. So in this case, the Luenberger observer gives a residual not decoupled with the disturbances, so the residual is not robust. In order to increase the robustness, it is necessary to design a residual decoupled with the disturbances, this can be done using an unknown input observer (UIO) [5].

The system including faults and disturbances (4.35) can also be expressed in input–output form as follows:

$$\mathbf{y}(k) = \mathbf{G}_u(q)u(k) + \mathbf{S}_f(q)\mathbf{f}(k) + \mathbf{S}_d(q)\mathbf{d}(k), \tag{4.37}$$

where $\mathbf{G}_u(q)$ is a transfer function matrix with ARMA functions of the shift operator q as elements. Similarly, $\mathbf{S}_f(q)$ and $\mathbf{S}_d(q)$ are transfer function matrices that represent the sensitivity to the faults and the disturbances, respectively. Now, the residuals are

$$\mathbf{e}(k) = \mathbf{y}(k) - \mathbf{G}_u(q)u(k) = \mathbf{S}_f(q)\mathbf{f}(k) + \mathbf{S}_d(q)\mathbf{d}(k). \tag{4.38}$$

This is a vector of residuals that as before is called *ARMA primary residuals* and the Eq. (4.38) is called *ARMA primary parity equation*, also the middle expression of this equation gives the computational form of these residuals while the right-hand side provides the internal form. These primary residuals depend on faults and disturbances, so they are not robust. However, additional residuals can be generated from them by linear transformation:

$$\mathbf{r}(k) = \mathbf{W}(q)\mathbf{e}(k), \tag{4.39}$$

where $\mathbf{W}(q)$ is a transfer function matrix, chosen so the residual will be independent to the disturbances, i.e., the $\mathbf{W}(q)$ matrix should satisfy: $\mathbf{W}(q)\mathbf{S_d}(q) = 0$, and the residuals only are affected by the faults.

However, in many practical situations, the perfect decoupled observer or parity equation does not exist, and the solution, in this case, is to find an optimal approximation due to a performance criterion that takes into account the sensitivity with respect to the disturbances $\mathbf{d}$ and the sensitivity to the faults $\mathbf{f}$ as

$$\min J = \min \frac{||\mathbf{S}_d(q)||}{||\mathbf{S}_f(q)||}. \tag{4.40}$$

4.3.3 Passive Robustness: Envelope Generators

An actual system or a part of it can be represented by a model described by the following nonlinear discrete-time equation:

$$\mathbf{y}(k) = \mathbf{f}(\mathbf{y}(k-1), \ldots, \mathbf{y}(k-n), \mathbf{u}(k-1), \ldots, \mathbf{u}(k-m), \boldsymbol{\theta}), \tag{4.41}$$

where $\mathbf{y}(k) \in \Re^{n_y} \ldots \mathbf{y}(k-n) \in \Re^{n_y}$ are the outputs of the system at instants $k \ldots k-n$, $\mathbf{f}$ is a vector of continuous functions, $\mathbf{u}(k-1) \in \Re^{n_u} \ldots \mathbf{u}(k-m) \in \Re^{n_u}$ are the inputs at instants $k-1 \ldots k-m$, and $\boldsymbol{\theta} \in \Re^{n_\theta}$ is a vector of parameters.

An Analytical Redundancy Relation (ARR) is an algebraic constraint deduced from the system model which contains only measured variables. An ARR for Eq. (4.41) is

$$\tilde{\mathbf{y}}(k) = \hat{\mathbf{y}}(k), \tag{4.42}$$

where $\tilde{\mathbf{y}}(k)$ is the measured output of the system at instant k and $\hat{\mathbf{y}}(k)$ is the analytically predicted output of the model at instant k:

$$\hat{\mathbf{y}}(k) := \mathbf{f}(\tilde{\mathbf{y}}(k-1), \ldots, \tilde{\mathbf{y}}(k-n), \tilde{\mathbf{u}}(k-1), \ldots, \tilde{\mathbf{u}}(k-m), \boldsymbol{\theta}), \tag{4.43}$$

being $\tilde{\mathbf{u}}(k)$ the measurement of the input at instant k.

An ARR is used to check the consistency of the observations with respect to the system model. Therefore, a fault is detected when

$$\tilde{\mathbf{y}}(k) \neq \hat{\mathbf{y}}(k), \tag{4.44}$$

or equivalently

$$\mathbf{r}(k) \neq \mathbf{0}, \tag{4.45}$$

where

$$\mathbf{r}(k) = \hat{\mathbf{y}}(k) - \tilde{\mathbf{y}}(k), \tag{4.46}$$

is the so called *residual* of the ARR.

The main problem is that the measured output $\tilde{\mathbf{y}}(k)$ and the computed output $\hat{\mathbf{y}}(k)$ are seldom the same because the model is, by definition, inaccurate, i.e., it is an approximated representation of the system. This is the consequence of the uncertainties of the system and the procedure of systems' modeling. Moreover, the measurements $\tilde{\mathbf{y}}(k)$ and $\tilde{\mathbf{u}}(k)$ are approximations to the real values of the variables $\mathbf{y}(k)$ and $\mathbf{u}(k)$ due to uncertainties in sensors: noise, errors in the analog to digital conversion, bias, drift, nonlinearities, inaccuracies due to calibration, etc.

Therefore, the uncertainty has to be considered. In some methods, the uncertainty is taken into account when the behavior of the actual system is compared with the behavior of the model, so a fault is indicated when the difference is larger than a threshold τ:

$$||\mathbf{r}(k)|| > \tau. \tag{4.47}$$

An important difficulty is to determine the size of τ. If it is too small, faults are indicated even when none exist, i.e., there are false alarms. On the other side, if the threshold is too large, the amount of missed alarms, or undetected faults, increases.

4.3.3.1 Consistency Test

Applying the principle of analytical redundancy, but taking into account the uncertainty of measurements and parameters, a fault is detected when

$$\left(\forall\tilde{\mathbf{y}}(k) \in \tilde{\mathbf{Y}}(k)\right) \left(\forall\hat{\mathbf{y}}(k) \in \hat{\mathbf{Y}}(k)\right) \mathbf{r}(k) \neq \mathbf{0}, \tag{4.48}$$

where $\tilde{\mathbf{Y}}(k)$ is the range of possible values that the output $\mathbf{y}(k)$ can take (it is obtained from the measurement $\tilde{\mathbf{y}}(k)$ and from the properties of the sensor) and $\hat{\mathbf{Y}}(k)$ is the range of possible values of the output of the model, which is an interval because the output of the model depends on interval inputs and parameters.

Therefore, at each time step, the range of a function in a parameter space has to be computed. Range computation is a task related to global optimization [15, 20], which usually requires considerable computational effort. The computational cost can be lowered by computing this range in an iterative way so at each iteration an outer approximation, closer to the exact range than the one obtained in the previous iteration, is obtained. This iterative procedure can stop when Eq. (4.48) is fulfilled because the fault has already been detected.

The problem is that this iterative procedure never stops either when there is no fault or when the fault cannot be detected using this method, for instance, because there is a fault but

$$\left(\exists\tilde{\mathbf{y}}(k) \in \tilde{\mathbf{Y}}(k)\right) \left(\exists\hat{\mathbf{y}}(k) \in \hat{\mathbf{Y}}(k)\right) \mathbf{r}(k) = \mathbf{0}, \tag{4.49}$$

due to dynamics, for instance.

A way to stop this procedure, in this case, is computing also an inner approximation to $\hat{\mathbf{Y}}(k)$. This is described in the following.

Both approximations, the outer one and the inner one, can be computed using Modal Interval Analysis. Considering the modal interval *-extensions $\mathbf{r}^*(k)$ of the residual functions $\mathbf{r}(k)$ to the proper intervals $\hat{\mathbf{Y}}(k)$ and $\tilde{\mathbf{Y}}(k)$, Eq. (4.48) is equivalent, in accordance with the Semantic Theorems, to the interval relation

$$[0, 0] \nsubseteq \mathbf{r}^*(k), \tag{4.50}$$

and Eq. (4.49) is equivalent to the interval relation

$$[0, 0] \subseteq \mathbf{r}^*(k). \tag{4.51}$$

Remark that, in this case, $\mathbf{r}^*(k)$ is the range of $\mathbf{r}(k)$ for the domains $\hat{\mathbf{Y}}(k)$ and $\tilde{\mathbf{Y}}(k)$. Using the necessary inner and outer roundings of $\mathbf{r}^*(k)$, Eq. (4.48) is true if

$$[0, 0] \nsubseteq Outer\left(\mathbf{r}^*(k)\right), \tag{4.52}$$

and Eq. (4.49) is true if

$$[0, 0] \subseteq Inner\left(\mathbf{r}^*(k)\right). \tag{4.53}$$

The f^* algorithm [3] is used to compute these approximations in an iterative way. The execution is stopped either when Eq. (4.52) is satisfied, hence a fault has been detected, or when Eq. (4.53) is true, in which case the ARR is consistent and a faulty or healthy behavior cannot be assured, so if there is a fault, it cannot be detected.

4.3.3.2 Window Consistency

Consider a discrete first-order model,

$$y(k) = a\,y(k-1) + b\,u(k-1), \tag{4.54}$$

where a and b are uncertain model parameters such that a belongs to the interval A and b belongs to the interval B.

At step k, the range of $\hat{\mathbf{Y}}(k)$ has to be computed. And, at step $k+1$, the range to be computed is $\hat{\mathbf{Y}}(k+1)$, being

$$y(k+1) = a\,y(k) + b\,u(k). \tag{4.55}$$

Two situations must be distinguished in this case. A first case is when the system is considered time variant, in which the values of a and b at steps k and $k+1$ can be considered different:

- $a_k \neq a_{k+1}$.
- $b_k \neq b_{k+1}$.

In this case, the ranges of $\hat{\mathbf{Y}}(k)$ and $\hat{\mathbf{Y}}(k+1)$ can be computed independently.

A second case is when the system is considered time-invariant, i.e., there is uncertainty in the values of a and b, but it is known that these values are the same at step k and at step $k+1$. In this case, there is a dependency between Eqs. (4.54) and (4.55) which is made explicit by substituting $y(k)$ into $y(k+1)$:

$$y(k+1) = a\,(a\;y(k-1) + b\;u(k-1)) + b\;u(k)\,. \tag{4.56}$$

In order to distinguish $y(k+1)$ of Eq. (4.56) from $y(k+1)$ of Eq. (4.55), the concept of *window length* (w) is introduced:

$$y(k+1|k-1) = a\,(a\;y(k-1) + b\;u(k-1)) + b\;u(k)\,, \tag{4.57}$$

or, in general

$$\mathbf{y}(k|k-w) = \mathbf{f}_w(\mathbf{y}(k-w), \mathbf{u}(k-w), \ldots, \mathbf{u}(k-1), \boldsymbol{\theta})\,, \tag{4.58}$$

which constitutes a *sliding time window*.

This second case, time-invariant systems, is more difficult to handle than the first one and is the case considered below.

The window consistency allows to determine the consistency of a set of system measurements (inputs and output) between two sampling times with respect to the predicted behavior in the same time interval. The distance between the two considered sampling times is called window length. Then, for a window length w, in accordance with Eq. (4.48), a fault is detected when

$$\begin{aligned}\left(\forall\tilde{\mathbf{y}}(k)\in\tilde{\mathbf{Y}}(k)\right)\left(\forall\hat{\mathbf{y}}(k|k-w)\in\hat{\mathbf{Y}}(k|k-w)\right)\\ \left(\forall\tilde{\mathbf{u}}(k-w)\in\tilde{\mathbf{U}}(k-w)\right)\ldots\\ \left(\forall\tilde{\mathbf{u}}(k-1)\in\tilde{\mathbf{U}}(k-1)\right)(\forall\boldsymbol{\theta}\in\Theta)\,\mathbf{r}_w(k)\neq\mathbf{0},\end{aligned} \tag{4.59}$$

where $\tilde{\mathbf{U}}(k)$ is defined similarly to $\tilde{\mathbf{Y}}(k)$ and is the range of possible values that the input $\mathbf{u}(k)$ can take (it is obtained from the measurement $\tilde{\mathbf{u}}(k)$ and from the properties of the sensor),

$$\mathbf{r}_w(k) = \hat{\mathbf{y}}(k|k-w) - \tilde{\mathbf{y}}(k) \tag{4.60}$$

and

$$\hat{\mathbf{y}}(k|k-w) = \mathbf{f}_w(\tilde{\mathbf{y}}(k-w), \tilde{\mathbf{u}}(k-w), \ldots, \tilde{\mathbf{u}}(k-1), \boldsymbol{\theta}) \tag{4.61}$$

is the corresponding predicted behavior for a window length of w.

Notice that, for two different window lengths w_1 and w_2, it can happen that with w_1 the fault is not detected but with w_2 the fault is detected. In this case, it can be assured that there is a fault because detecting it with one window length is a sufficient condition to assure it. Consequently, a fault is detected if

$$\begin{aligned}\left(\forall\tilde{\mathbf{y}}(k)\in\tilde{\mathbf{Y}}(k)\right)\left(\forall\hat{\mathbf{y}}(k|k-w_1)\in\hat{\mathbf{Y}}(k|k-w_1)\right)\\ \left(\forall\tilde{\mathbf{u}}(k-w_1)\in\tilde{\mathbf{U}}(k-w_1)\right)\dots\\ \left(\forall\tilde{\mathbf{u}}(k-1)\in\tilde{\mathbf{U}}(k-1)\right)(\forall\boldsymbol{\theta}\in\Theta)\,\mathbf{r}_{w_1}(k)\neq\mathbf{0}\\ \vee\dots\vee\\ \left(\forall\tilde{\mathbf{y}}(k)\in\tilde{\mathbf{Y}}(k)\right)\left(\forall\hat{\mathbf{y}}(k|k-w_n)\in\hat{\mathbf{Y}}(k|k-w_n)\right)\\ \left(\forall\tilde{\mathbf{u}}(k-w_n)\in\tilde{\mathbf{U}}(k-w_n)\right)\dots\\ \left(\forall\tilde{\mathbf{u}}(k-1)\in\tilde{\mathbf{U}}(k-1)\right)(\forall\boldsymbol{\theta}\in\Theta)\,\mathbf{r}_{w_n}(k)\neq\mathbf{0}.\end{aligned}\tag{4.62}$$

The fault detection results obtained using several window lengths are generally better, i.e., there are less missed alarms, than the ones obtained using a single window length, whatever is the length in the latter case. The best window length not only depends on the dynamics of the system but also on the type of fault to be detected.

4.4 Fault Isolation

The *actual fault signature* of the system $\mathbf{s}(k) = [s_1(k), s_2(k), \dots s_n(k)]$, obtained as a result of the fault detection phase (see (4.4)) is provided to the fault isolation module which will try to isolate the fault and give a diagnosis (see Fig. 4.1).

The actual fault signature is compared against the *theoretical Fault Signature Matrix* (FSM) that binary codifies the influence on every residual, in the set of considered residuals $r_1, r_2, \dots r_{n_r}$, of every fault in set of considered faults $f_1, f_2, \dots f_{n_f}$. Thus, this matrix has as many rows as residuals and as many columns as considered faults. An element FSM_{ij} of this matrix is equal to 1 means that the jth fault appears in the expression of the ith residual. Otherwise, it is equal to 0. Assuming classical FDI fault hypotheses, i.e., single faults and no-compensation (exoneration), fault isolation will consist in looking for a column of the FSM that matches the actual fault signature $\mathbf{s}(k)$ (see Fig. 4.1). Therefore, this classic approach in the FDI community is also known as Column Reasoning [13].

In the literature, there exists two different approaches to construct residual sets with the desired isolability properties. One approach is based on designing a vector of *structured residuals* [13]. Each residual is designed to be sensitive to a subset of faults, while remaining insensitive to the remaining faults. The design procedure consists of two steps: the first step is to specify the sensitivity and insensitivity

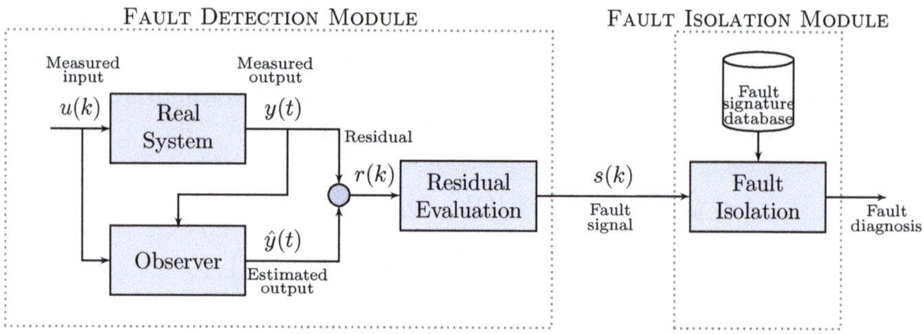

Fig. 4.1 Conceptual diagram of the model-based fault detection and isolation approach

relationships between residuals and faults according to the assigned isolation task, while the second is to design a set of residual generators according the previous relationships. The fault isolation problem then consists of a separate threshold test for each residual using a decision table. An alternative way of achieving the isolability of faults is by designing a vector of *directional residuals* [13], which lies in a fixed and fault-specified direction in the residual space, in response to a particular fault. The fault isolation problem consists of determining which of the known fault directions, called fault signatures, the generated residual vector lies the closest to.

Another alternative to obtain the set of residuals to be used for fault isolation is using the structural analysis [4] that allows to combine model equations with measurements to obtain residual expressions as well as the FSM matrix.

4.4.1 Structured Residuals

A set of structured residuals can be designed using the method proposed by [13], where each individual residual is designed to be sensitive to a single fault while remaining insensitive to the rest of faults.

For this purpose, let us consider a set of primary residuals each one given by (4.3). Assuming additive faults, the input–output relation given by (4.13) should be used. Then, the residual used is the one expressed in (4.14) that can be modified as

$$\mathbf{r}(k) = \mathbf{W}(q)\mathbf{S}_f(q)\mathbf{f}(k) \tag{4.63}$$

that shows how the residuals respond to the faults. The set of residuals can be structured by imposing the desired fault response as

$$\mathbf{r}(k) = \mathbf{Z}(q)\mathbf{f}(k), \tag{4.64}$$

where $\mathbf{Z}(q)$ is desired fault transfer function matrix. Comparing Eqs. (4.63) and (4.64), the design equation for the filter $\mathbf{W}(q)$ can be derived:

$$\mathbf{W}(q)\mathbf{S}(q) = \mathbf{Z}(q). \tag{4.65}$$

Assuming that the number of residuals n_r is equal to the number of considered faults n_f (i.e., $\mathbf{S}(q)$ is a square fault transfer matrix), the solution of design Eq. (4.65) can be found by direct inversion:

$$\mathbf{W}(q) = \mathbf{Z}(q)\mathbf{S}^{-1}(q), \tag{4.66}$$

what requires that $rank(\mathbf{S}_f(q)) = n_r = n_f$. The set of structured residuals (4.63) can alternatively be designed using a bank of dedicated observers in two forms: *Dedicated Observer Scheme* (DOS) where each observer is sensitive just to one fault and *Generalized Observer Scheme* (GOS) where each observer is sensitive to all the faults except one.

Fault isolation using structured residuals is based on the idea of *pattern matching*, which consists of comparing the theoretical *fault signatures* (FS) corresponding to all considered faults with the computed observed fault signature obtained from the residuals. In general, FS contain the expected evolution of the residual immediately after that a fault occurs (known also as symptom), which are derived off-line from the structural properties of the residuals. FS can include both analytic and heuristic information generated to produce distinctive pattern for each particular fault. FS are stored as matrices in the fault diagnosis database and used to match to the trends of online residuals using pattern matching methods. Finally, a decision logic algorithm proposes the most probable fault candidate.

Fault patterns are organized according to a theoretical FSM. This interpretation assumes that the occurrence of f_j always affect a system and is observable at least in one residual r_i, hypothesis known as fault exoneration, and that f_j is the only fault affecting the monitored system. Several FS matrices can be considered in the evaluation task: Boolean fault signal activation ($FSM01$), fault signal signs ($FSMsign$) and fault residual sensitivity ($FSMsensit$).

Each element of the $FSMsensit$ is computed with the following equation:

$$FSMsensit_{ij}(k) = \begin{cases} \frac{S_{r_i,f_j}(q^{-1})}{max(\mathcal{R}_i(k))-r_i^0} u_a(k-k_0), & \text{if } r_i^0 \geq 0 \text{ and } k \geq k_0, \\ \frac{S_{r_i,f_j}(q^{-1})}{min(\mathcal{R}_i(k))-r_i^0} u_a(k-k_0), & \text{if } r_i^0 < 0 \text{ and } k \geq k_0, \\ 0, & \text{if } S_{r_i,f_j} = 0 \text{ or } k < k_0, \end{cases} \tag{4.67}$$

where u_a is a unitary abrupt step input, S_{r_i,f_j} is the sensitivity associated to the nominal residual r_i^0 regarding the fault hypothesis f_j, and k_0 is the fault occurrence time instant. As a consequence of the fault residual sensitivity time dependence, $FSMsensit_{ij}$ dynamically evolves since the fault occurrence time instant.

Besides, $FSM01$ and $FSMsign$ matrix can be easily derived from $FSMsensit$ (4.67) by applying the following conversions:

$$FSM01_{ij}(k) = \begin{cases} 1 & \text{if } FSM01_{ij}(k) \neq 0, \\ 0 & \text{if } FSM01_{ij}(k) = 0. \end{cases} \tag{4.68}$$

$$FSMsign_{ij}(k) = \begin{cases} 1 & \text{if } FSMsensit_{ij}(k) > 0, \\ -1 & \text{if } FSMsensit_{ij}(k) < 0, \\ 0 & \text{if } FSMsensit_{ij}(k) = 0. \end{cases} \tag{4.69}$$

The decision logic algorithm starts when the first residual is activated and lasts T_w time instants or till all fault hypotheses except one are rejected. Fault hypotheses are rejected when an unexpected activation signal has been observed according to those fault hypotheses. Rejection is based on using the results of $factor01_j(k)$ and $factorsign_j(k)$. If any of these factors are null for a given fault hypothesis, it will be rejected. Every factor represents some kind of a filter, suggesting a set of possible fault hypotheses. At the end of the time window T_w, for each non-rejected fault hypothesis f_j, a fault isolation indicator ρ_j could be determined, for instance, although other logic formula could be considered:

$$\rho_j(k) = \max_{p \in [k - T_w/\Delta t, k]} (factor01_j(p) \times |factorsign_j(p)|, |factorsensit_j(p)|). \tag{4.70}$$

A set of fault candidates with their corresponding fault isolation indicator is provided as the final diagnosis result, so that the greatest fault isolation indicator will determine the diagnosed fault. In the case of nonisolable faults, they will all have a high fault isolation indicator.

4.4.2 Directional Residuals

Considering the vector of primary residuals (4.3) and the fault sensitivity introduced in (4.13), fault isolation is based on looking for the fault signature stored in the columns of (4.13) that presents the closest direction with (4.3). This is done by computing the normalized projections, ψ_j, onto each sensitivity vector (column of (4.13)) as

$$\psi_j = \frac{\mathbf{r}^T \mathbf{s}_j}{|\mathbf{r}||\mathbf{s}_j|}, \tag{4.71}$$

for $j = 1, \ldots, n_f$, where n_f is the number of candidate faults.

Then, the largest projection will allow to determine the fault candidate that better matches with the observed residual (4.3).

$$\psi_k = \max(\psi_1, \ldots, \psi_{n_f}). \tag{4.72}$$

4.5 Application Example

In this section, the case study based on the three tank system example proposed in Chap. 2 will be used to illustrate some of the methods presented in this chapter.

The goal is to detect using the nominal (non-faulty) model whether the system is faulty or not (Fault Detection). And, in case that the fault is detected, to discover which is the fault that is present in the system (Fault Isolation). First, to perform this task, it is assumed that only the measurements of the water levels are available using the direct residuals (primary residuals obtained comparing the measurements against the system model). The obtained results are compared when considering only the available measurements (h_{T_1}, h_{T_3}, and q_{23}) proposed in Chap. 2 and the structured residuals derived in Chap. 3 for the tank system case study.

4.5.1 Faults and Uncertainties

As explained in Sect. 4.4, the results of the fault detection task are used for the fault isolation task by comparing the actual fault signature with the theoretical Fault Signature Matrix (FSM). This matrix has as many rows as residuals and as many columns as considered faults. In this case, it is assumed that only the measurements of the three water levels are available, so there are three residuals and the matrix has three rows. On the other side, the faults that are taken into account are as follows:

- f_1: level sensor of tank 1.
- f_2: level sensor of tank 2.
- f_3: level sensor of tank 3.
- f_4: leakage in tank 1.
- f_5: leakage in tank 2.
- f_6: leakage in tank 3.
- f_7: clogging between tank 1 and tank 2.
- f_8: clogging between tank 2 and tank 3.
- f_9: clogging at the output of tank 3.

Since there are nine considered faults, the fault signature matrix has nine columns (see Table 4.1). The cells of the matrix are equal to one if the fault affects the residual and zero otherwise. As the level of tank i at time t $h_i(t)$ is computed from the initial levels $h_i(0)$ and the input flow $q_i(t)$, if a fault in the sensor of tank j appears, only the residual of tank j will be affected. This way the three first columns of the matrix are built. The six remaining columns contain no 0 because a leakage in one tank affects all the levels and so does a clogging.

From this matrix, it can be observed that sensor faults can be isolated, but there is no way to distinguish a leakage from a clogging or to indicate where it is. More information is needed in this case. As discussed in Sect. 4.4, this information can

Table 4.1 Fault signature matrix for the primary residuals with binary information

	f_1	f_2	f_3	f_4	f_5	f_6	f_7	f_8	f_9
r_1	1	0	0	1	1	1	1	1	1
r_2	0	1	0	1	1	1	1	1	1
r_3	0	0	1	1	1	1	1	1	1

Table 4.2 Fault signature matrix for the primary residuals with sign information

	f_1	f_2	f_3	f_4	f_5	f_6	f_7	f_8	f_9
r_1	± 1	0	0	-1	-1	-1	$+1$	$+1$	$+1$
r_2	0	± 1	0	-1	-1	-1	-1	$+1$	$+1$
r_3	0	0	± 1	-1	-1	-1	-1	-1	$+1$

be added to the FSM by means of signs: $+1$ indicates that the real level is above the predicted value, -1 indicates that the real level is below the predicted value and ± 1 indicates that the measured level can be above or below the predicted value depending on the fault. This is the case of the sensor faults: when a sensor is faulty the measurement provided by the sensor can be a number greater than or less than the real value. So, the FSM with signs is presented in Table 4.2.

Now, all faults can be detected and diagnosed, but there is still no way to know, when there is a leakage, where it is. To deal with this problem, more information would be needed in this case or to use a set of structured residuals as the ones presented in Chap. 3 designed specifically to isolate these faults.

Noise in measurements and parametric uncertainties in the model are considered by assuming an unknown-but-bounded description as follows:

$$\begin{aligned} h_i &\quad \text{within the interval } H_i, \\ q_i &\quad \text{within the interval } Q_I, \\ s_i &\quad \text{within the interval } S_i, \\ c_{ii+1} &\quad \text{within the interval } C_{ii+1}. \end{aligned}$$

The way how these bounds can be obtained from the measurements of the system is detailed in Chap. 10.

4.5.2 Results

Several scenarios including some of the faults defined above will be presented in this section to illustrate fault detection and isolation:

- The system is not faulty. There are neither cloggings nor leakages.
- The system is faulty because of a leakage. The system is non-faulty until $t = 5000$ s and from this time on, there is a leakage in tank 2 ($Leakage_{T_2}$) of approximately 10% of the water inflow.
- The system is faulty because of a clogging. The system is non-faulty until $t = 5000$ s and from this time on, there is a clogging between tanks 1 and 2 of approximately 10% of the nominal flow. In this case, parameter $Stuck_{q_{12}}$ is not just 0 or 1 as it was introduced in Chap. 2.

In all cases, 10000 s are simulated. As it is shown in Fig. 4.2, the input flow is $150\,\text{cm}^3/\text{s} \pm 1\%$ to use a linearized model of the system around this operating point to generate parity equations in ARMA form (4.37).

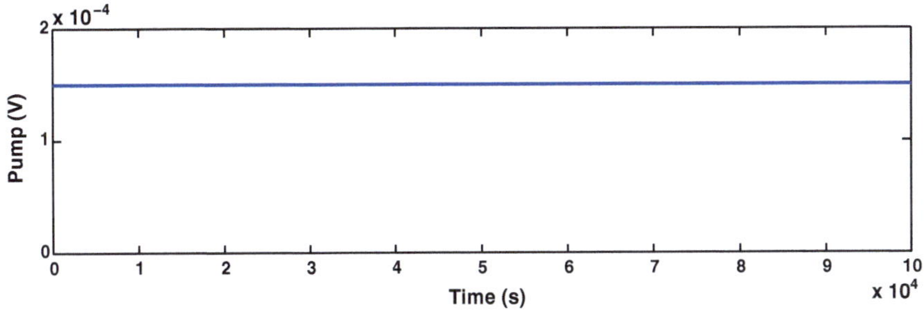

Fig. 4.2 Pump input

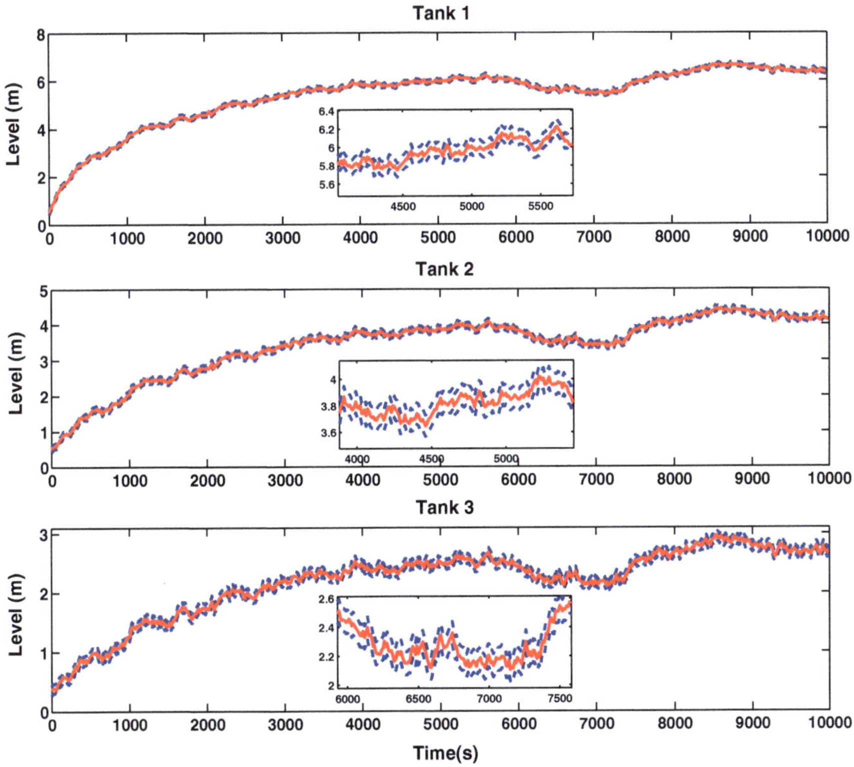

Fig. 4.3 Envelopes without faults

Non-faulty system. If the system is non-faulty, according to (4.49), the experimental values (solid line) will be inside the band (dashed lines), as shown in Fig. 4.3, in which a set of pseudo-experimental measurements has been created by means of a real-numbered simulation using the model of the system without faults and starting from the initial water levels $h_1(0) = 0.51$ m, $h_2(0) = 0.50$ m and $h_3(0) = 0.35$ m. As it can be observed from Fig. 4.3, the system is only affected by noise and

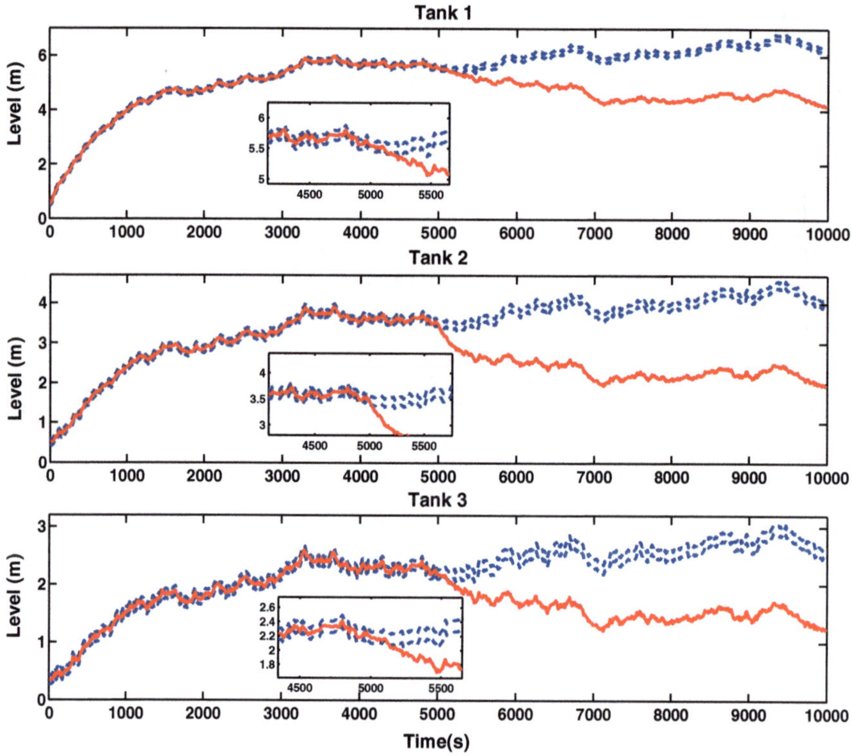

Fig. 4.4 Envelopes with leakage in tank 2

parametric uncertainty and the model estimation can properly follow the system during the healthy functioning of the system.

Faulty system. If the system is faulty, according to (4.48), the fault is detected when a measurement is outside its outer estimated envelope. Then, two different procedures can be used to generate the residual: (i) residual generation by means of a difference between the value(s) of output(s) from the model and the real measured value(s) of output(s) given by the sensor. (ii) structural residual generation, where each individual residual is designed to be sensitive to a single fault whilst remaining insensitive to the rest of faults.

In the first considered fault scenario, a leakage in tank 2 is introduced at time instant $k = 5000$ s and it remains until the end of the simulation. Figure 4.4 shows the interval of the level estimation in this case. Since the fault is simulated after time instant $k = 5000$ s, before this time instant, the system is only affected by the effect of parametric uncertainty and noises. Consequently, the bounds that are obtained at the first 5000 time instants show only the effect of both uncertainties. As it can be seen from Fig. 4.4, after $k = 5000$ s, the fault can be detected when a measurement is outside its outer estimated envelope. This point can also be seen in Fig. 4.5, where

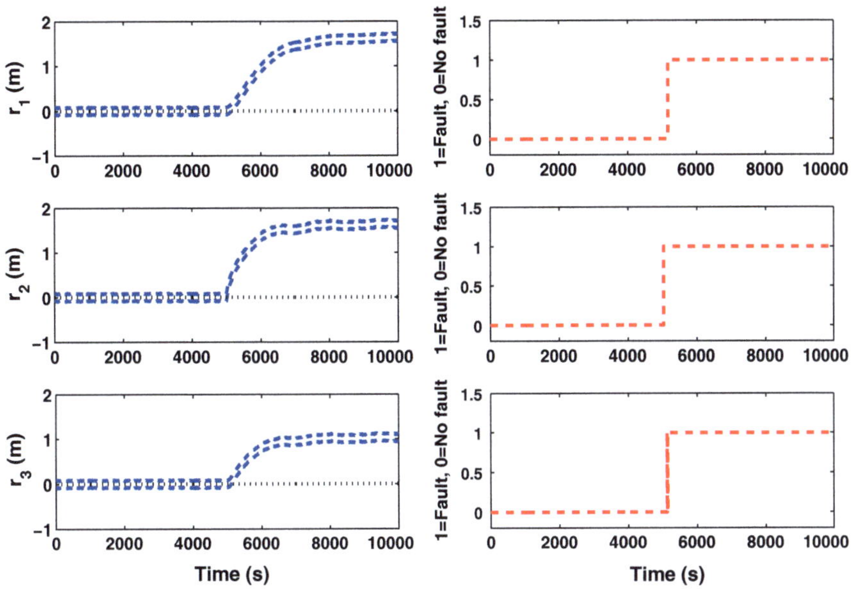

Fig. 4.5 Primary residual envelopes for the case of leakage in tank 2

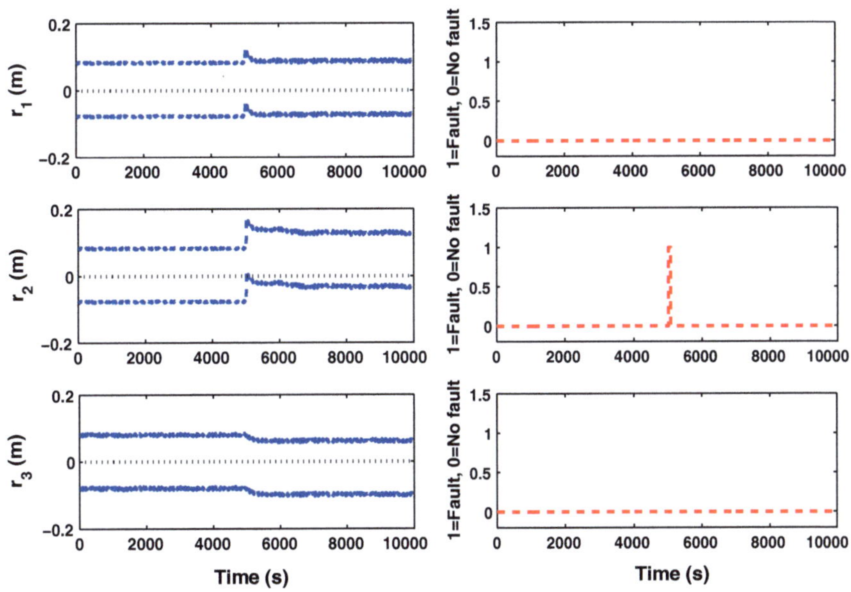

Fig. 4.6 Structured residuals envelopes for the case of leakage in tank 2

the primary residuals are generated by means of a difference between the predicted output and the real measured values of outputs.

It can be observed from Fig. 4.5 that zero is not included between the upper and lower bounds of the residual set. Then, the existence of the fault will be proved. Moreover, the fault indication is shown on the right side of Fig. 4.5. In Fig. 4.5, the value equal to one means the occurrence of the fault.

However, one major drawback of primary residuals is related to the inability to distinguish all the faults as discussed using binary and sign information as shown in Tables 4.1 and 4.2. This can also be seen in Fig. 4.5, after fault occurrence, all three residuals are violated in order to detected the fault. On the other hand, using structural residuals designed in Chap. 3 that are sensitive to a single fault whilst remaining insensitive to the rest of faults, the leakage in tank 2 can be isolated. This can be seen in Fig. 4.6 where the residual r_2 is designed to be sensitive to this fault while the other residuals are insensitive. Thus, not only the leakage fault can be detected, but also, based on results presented in Fig. 4.6 can be isolated.

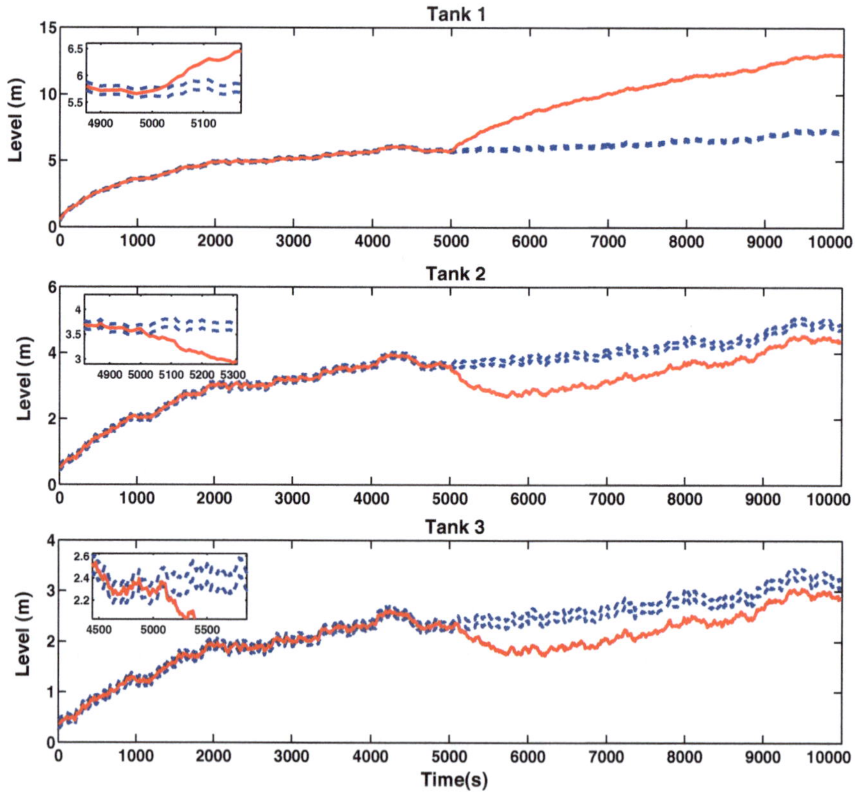

Fig. 4.7 Envelopes and exact measurements with clogging between tanks 1 and 2

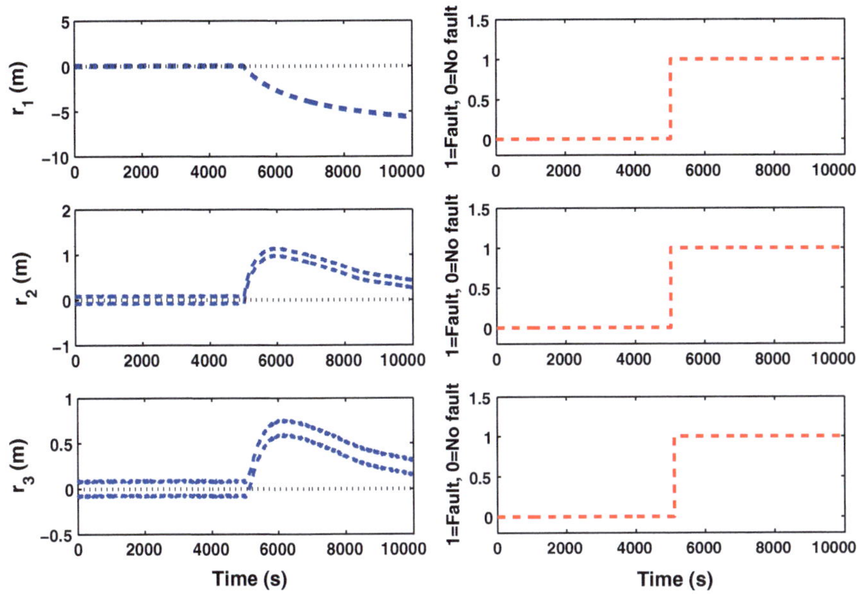

Fig. 4.8 Residual envelopes for the case of clogging between tanks 1 and 2

A second fault scenario is analyzed consisting of a clogging between tanks 1 and 2. Similar to the the leakage fault, the occurrence of the fault is simulated after $k = 5000$ s. The level estimation envelopes that are obtained are shown in Fig. 4.7.

As it can be seen in Fig. 4.7, the fault occurrence can be detected when the measurements are outside of the prediction interval. Or, as analogously shown in Fig. 4.8, when zero is not included in the residual interval.

As in the case of the leakage in tank 2, Fig. 4.8 show that for the case of clogging between tanks 1 and 2, all the primary residuals are violated because of the effect of the considered fault. Thus, although the fault can be detected, it cannot be isolated as previously discussed. However, the isolation of this fault is possible using the structural residuals generated in Chap. 3 designed to separate all the leakages and cloggings. Figure 4.9 shows the evaluation of the structured residuals. As it can be seen from Fig. 4.9, the influence of the considered fault can be observed in residuals r_2 and r_3, but not in residual r_1, that allow the isolation. Then, by considering the structural residual not only this fault can be detected, but it can also be isolated.

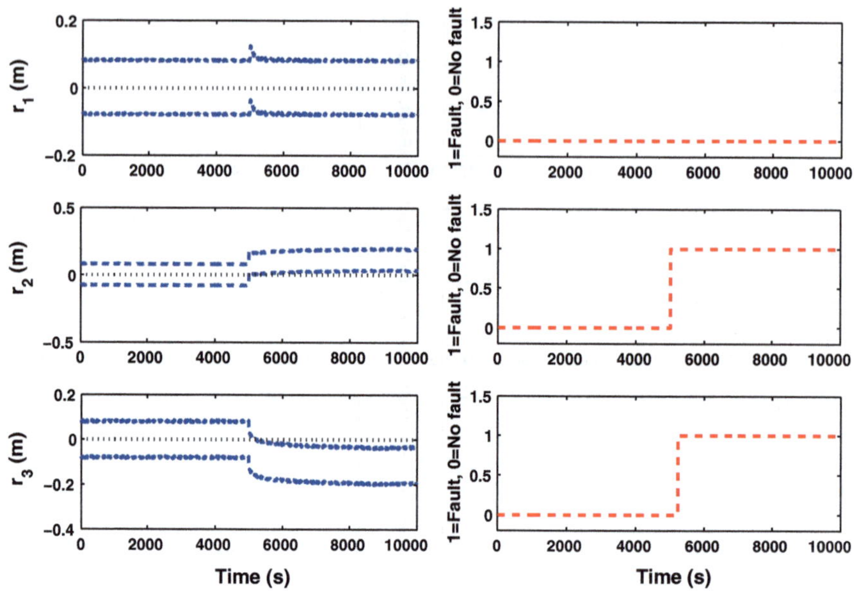

Fig. 4.9 Residual envelopes for the case of clogging between tanks 1 and 2

Acknowledgements This work has been partially funded by the Spanish State Research Agency (AEI) and the European Regional Development Fund (FEDER) through the projects MASCON-TROL (ref. MINECO DPI2015-67341-C2-2-R), (ref. MINECO DPI2016-78831-C2-2-r), DEOCS (ref. MINECO DPI2016-76493) and SCAV (ref. MINECO DPI2017-88403-R). This work has also been partially funded by AGAUR of Generalitat de Catalunya through the grants 2017 SGR 01551/2017 SGR 482 and by Agència de Gestió d'Ajuts Universitaris i de Recerca.

References

1. Adrot, O., Flaus, J.M.: Fault detection based on uncertain models with bounded parameters and bounded parameter variations. In: Proceedings of 17th IFAC World Congress, Seoul, Korea (2008)
2. Armengol, J., Vehí, J., Travé-Massuyès, L., Sainz, M.Á.: Interval model-based fault detection using multiple sliding windows. In: 4th IFAC Symposium on Fault Detection, Supervision and Safety for Technical Processes SAFEPROCESS, pp. 168–173 (2000)
3. Armengol, J., Vehí, J., Sainz, M.Á., Herrero, P., Gelso, E.: Squaltrack: a tool for robust fault detection. IEEE Trans. Syst. Man Cybern. Part B **39**(2), 475–488 (2008)
4. Blanke, M., Kinnaert, M., Lunze, J., Staroswiecki, M.: Diagnosis and Fault-Tolerant Control, 2nd edn. Springer, Berlin (2006)
5. Chen, J., Patton, R.: Robust Model-Based Fault Diagnosis for Dynamic Systems. Kluwer Academic Publishers, Dordrecht (1999)
6. Chow, E., Willsky, A.: Analytical redundancy and the design of robust failure detection systems. IEEE Trans. Autom. Control **AC-29**, 603–614 (1984)
7. Clark, R.N., Fosth, D., Walton, V.M.: Detection instrument malfunctions in control systems. IEEE Trans. Aerosp. Electron. Syst. **11**, 465–473 (1975)

8. Emami-Naeini, A., Akhter, M., Rock, S.: Effect of model uncertainty on failure detection: the threshold selector. IEEE Trans. Autom. Control **AC-33**, 1106–1115 (1988)
9. Escobet, T., Travé-Massuyès, L., Tornil, S., Quevedo, J.: Fault detection of a gas turbine fuel actuator based on qualitative causal models. In: Proceedings of European Control Conference (ECC'01), Porto, Portugal (2001)
10. Fagarasan, I., Ploix, S., Gentil, S.: Causal fault detection and isolation based on a set-membership approach. Automatica **40**, 2099–2110 (2004)
11. Fuente, M.J., Vega, P.: Neural networks applied to fault detection of a biotechnological process. Eng. Appl. Artificial Intell. **12**, 569–584 (1999)
12. Fuente, M.J., Vega, P., Zarrop, M., Poch, M.: Fault detection in a real wastewater plant using parameter-estimation techniques. Control Eng. Pract. **4**(8), 1089–1098 (1996)
13. Gertler, J.: Fault Detection and Diagnosis in Engineering Systems. Marcel Dekker, New York (1998)
14. Hamelin, F., Sauter, D.: Robust fault detection in uncertain dynamic systems. Automatica **36**(11), 1747–1754 (2000)
15. Hansen, E.: Global Optimization Using Interval Analysis. Marcel Dekker, New York (1992)
16. Horak, D.T.: Failure detection in dynamic systems with modelling errors. AIAA J. Guid. Control Dyn. **11**(6), 508–516 (1988)
17. Isermann, R.: Process fault detection based on modeling and estimation methods—a survey. Automatica **20**(4), 387–404 (1984)
18. Isermann, R.: Fault diagnosis of machines via parameter estimation and knowledge processing. Automatica **29**, 815–836 (1993)
19. Isermann, R.: Fault-Diagnosis Systems: An Introduction from Fault Detection to Fault Tolerance. Springer, Berlin (2006)
20. Kearfott, R.B.: Rigorous Global Search: Continuous Problems. Kluwer Academic Publishers, Dordrecht (1996)
21. Milanese, M., Norton, J., Piet-Lahanier, H., Walter, E.: Bounding Approaches to System Identification. Plenum Press, New York (1996)
22. Patton, R.J., Chen, J.: Observer-based fault detection and isolation: robustness and applications. Control Eng. Pract. **5**(5), 671–682 (1997)
23. Ploix, S., Adrot, O.: Parity relations for linear uncertain dynamic systems. Automatica **42**(6) (2006)
24. Ploix, S., Follot, C.: Fault diagnosis reasoning for set-membership approaches and application. In: IEEE International Symposium on Intelligent Control (2001)
25. Puig, V., Quevedo, J., Escobet, T., de las Heras, S.: Robust fault detection approaches using interval models. In: 16th IFAC World Congress (2002)
26. Puig, V., Saludes, J., Quevedo, J.: Worst-case simulation of discrete linear time-invariant interval dynamic systems. Reliab. Comput. **9**(4), 251–290 (2003)
27. Puig, V., Stancu, A., Escobet, T., Nejjari, F., Quevedo, J., Patton, R.: Passive robust fault detection using interval observers: application to the DAMADICS benchmark problem. Control Eng. Pract. **14**(6), 621–633 (2006)
28. Puig, V., Quevedo, J., Escobet, T., Nejjari, F., de las Heras, S.: Passive robust fault detection of dynamic processes using interval models. IEEE Trans. Control Syst. Technol. **16**(5), 1083–1089 (2008)
29. Rambeaux, F., Hamelin, F., Sauter, D.: Optimal thresholding for robust fault detection of uncertain systems. Int. J. Robust Nonlinear Control **10**(14), 1155–1173 (2000)
30. Rinner, B., Weiss, U.: Online monitoring by dynamically refining imprecise models. IEEE Trans. Syst. Man Cybern. **34**, 1811–1822 (2004)
31. Sainz, M.Á., Armengol, J., Vehí, J.: Fault detection and isolation of the three-tank system using the modal interval analysis. J. Process Control **12**(2), 325–338 (2002)
32. Tornil, S., Escobet, T., Travé-Massuyès, L.: Robust fault detection using interval models. In: Proceedings of European Control Conference (ECC'03), Cambridge, UK (2003)
33. Travé-Massuyès, L., Escobet, T., Pons, R., Tornil, S.: The Ca∼En diagnosis system and its automatic modelling method. Computación y Sistemas **5**(2), 648–658 (2001)

Chapter 5
Model-Based Diagnosis by the Artificial Intelligence Community: The DX Approach

Carlos J. Alonso-González and Belarmino Pulido

5.1 Introduction

The Artificial Intelligence community has shown interest in the diagnosis task for more than 40 years. The first approach to automated diagnosis systems was through rule-based expert systems. Mycin [17], which was capable to diagnose infectious diseases, is the paradigmatic example from that period. The diagnosis knowledge was stated as relationships between diseases and their symptoms, and they were modeled by means of rules. The diagnosis process used backward chaining reasoning to link available symptoms, to produce a set of feasible diagnoses, together with their certainty factors, somehow resembling a probability.

Diagnosis systems were not only applied in medicine, but also to perform electronic systems troubleshooting and diagnosis, as in the different SOPHIE works [1]. Device-dependency, maintenance and reusability problems, and the lack of answer for unexpected faults were known drawbacks of the Expert System approach to diagnosis. It is the work by Randall Davis [3] that provides the ground for what it is known as "reasoning from first principles" that proposes to use models, obtained from Physics First Principles, to reason about system behavior. Moreover, he also proposed to clearly separate the structure and the behavior of the system, working at the component level. These are the main features of the model-based reasoning approach provided from the Artificial Intelligence community, also known as the DX approach, which is in clear contrast with the system model view presented in the Control Theory approach.

C. J. Alonso-González · B. Pulido (✉)
Departamento de Informática, Universidad de Valladolid, Valladolid, Spain
e-mail: belar@infor.uva.es

C. J. Alonso-González
e-mail: calonso@infor.uva.es

T. Escobet et al. (eds.), *Fault Diagnosis of Dynamic Systems*,
https://doi.org/10.1007/978-3-030-17728-7_5

The DX approach also provides a definition of the diagnosis process, which is clearly linked to other high-level tasks such as repair and configuration [2]:

Definition 5.1 (*Diagnosis*) Diagnosis is the task, that given a system and a set of observations from a faulty situation, which is able to determine what is wrong in the system (and what would be necessary to repair in order to bring back the system to a normal state).

In this chapter, we provide the logical and computational foundations of early works in the DX approach, together with the early extensions to include fault modes in the models, leading to a more precise identification.

5.2 Reiter's Theory: A Logical Formalization of Model-Based Diagnosis

5.2.1 Consistency-Based Diagnosis

Consistency-based diagnosis, CBD, is the main approach to model-based diagnosis in the DX community. Originally, it was devised to diagnose component oriented, static and time-invarying systems. Later on, it has been extended to cope with dynamic and different kind of systems—process, biologic, socioeconomic, etc. However, the foundations of the method remain the same and are better introduced using its initial form.

In its original formulation, CBD can be characterized by:

- It is a *component-oriented approach*: it works on physical systems that can be described by a finite set of interconnected components that only interacts throughout their terminals.
- It only uses knowledge of the *systems structure*—i.e., components interconnections[1]—and the *components behavior*.
- It only uses local models of *correct behavior*.
- *Exoneration is not allowed* in the reasoning process: observations are used to reject models of correct behavior but never to state that a component is correct.

CBD is a domain-independent theory that can be applied to any static, time-invariant system as long as it can be properly described as a component-oriented system. Hence, it provides the foundation to develop a general diagnosis theory that can be used to engineer autonomous agents with built-in diagnosis capabilities.

In this chapter, we present first the ideas underlying this approach to diagnosis, followed by Reiter's logical theory of CBD, [16]. Reiter's theory provides a formal account of the main concepts introduced by the General Diagnostic Engine, GDE, an

[1]Sometimes, the component interconnections are also called the *topology* of the system.

efficient approach to compute diagnosis under the CBD framework, developed by de Kleer and Williams [6]. The origins of CBD can be traced, at least, to the constraints suspension technique proposed by [4].

5.2.1.1 Models and Prediction of Behavior

Consistency-based diagnosis requires that the system to be diagnosed is made up of individual components that interact through connections between their terminals. The *components* are usually physical devices (transistors, resistances, valves, etc.) that only interact with the environment through their terminals. *Terminals* have associated values of magnitudes, like voltage, current, or pressure. *Connections* are considered ideal and just propagate values between terminals. Real connections can be modeled as components plus ideal connections.

CBD relies on the *no function in structure principle*. This principle requires that the behavior of the each component is represented by a *local model* that doesn't make any assumption about its working environment: the presence of another component, state of another component, operation point, etc. Local models can only include relations among the input and output variables at the component terminals and perhaps, the internal parameters of the component. This requirement is needed for the *logical* approach that the model-based reasoning methodology applies to diagnosis: if the models are not local and rely, although implicitly, on assumptions about their working environment, any change in the environment invalidates the model. Local models provide an additional advantage: they foster model reuse and development of model libraries that can be used to generate models for component-oriented systems.

Certainly, the no function in structure principle is a very demanding requirement. But it is necessary to preserve the soundness of the logical reasoning approach to diagnosis. When real-world systems make impractical to apply this modeling principle, the components and interconnection that violates the locality requirement should be lumped and managed as a single component. The diagnosis will lose some localization capabilities but still will be sound.

Independently of the nature of the models, CBD exploits all the restrictions that the model imposes on the input and output variables of a component. In its logical approach to diagnosis, CBD does not assign causality to models: they simply restrict the set of values that a non-faulty component may exhibit at their terminals. Consequently, a model can be used to infer the value of any variable from the available data.

Using local models ensures that *the validity of a prediction* performed by a component model, or inference step, *only depends* on an additional assumption, that few model-based methodologies make explicit: *the assumption of correct behavior* of the component that supplies the model to perform the inference.

To make explicit the assumption of correct behavior of a component, CBD employs the literal $AB(.)$ to denote that a component is faulty—alternatively, the literal $OK(.)$ to refer to non-faulty components. Hence, if we want to model the correct behavior of a component named A, the literal $\neg AB(A)$ is included in the logical model, usually in the left side of a logical implication, while the right side describes the relations among variables and parameters. With this convention, the model of an adder of integers with inputs in_1 and in_2 and output out is represented by the universal quantified sentence:

$$ADD(x) \wedge \neg AB(x) \Rightarrow out(x) = in_1(x) + in_2(x)$$

.

It should be stressed that in Reiter's theory, the relation between the $\neg AB(x)$ literal and the model of correct behavior—in our example: $out(x) = in_1(x) + in_2(x)$ – should be articulated with the conditional relation, $\Rightarrow$, and not with the biconditional, $\Leftrightarrow$. This excludes exoneration from the reasoning process with any sound inference rule.

From a computational point of view, it is useful to keep track of the assumptions of correct behavior involved in a chain of inferences. This task is usually called *dependency recording* and it is performed by a Dependency-Recorder Engine, DRE. In the scope of CBD, a DRE keeps track of the $\neg AB(.)$ literals that support each prediction and it is able to provide the set(s) of assumptions that justify any inference. This is a very efficient approach to compute diagnosis in the CBD framework. It will be further explained in the section dedicated to the General Diagnostic Engine, GDE. However, DREs are not part of the formal theory of CBD. DREs are computational devices and Reiter's theory is a logical theory that is not bound to any computational approach, as long as it only employs a sound deduction method.

5.2.2 A Formal Model of Consistency-Based Diagnosis

Reiter, in his seminal paper *A Theory of diagnosis from First Principles*, [16], proposed a general formal theory for CBD. This formalization uses the language of First-Order Logic (FOL) and it is still the conceptual framework for CBD. It defines precisely what is a diagnosis problem and its set of solutions: a system and its diagnoses. We proceed to present the definitions and main results of this theory, using the polybox system to illustrate it. The interested reader should carefully read the original Reiter's paper. In this chapter, we are going to introduce the formulation of de Kleer, Mackworth and Reiter, [5] that subsumes the original Reiter's work and allows for further extensions of the basic theory. This theory works for static problems. There is still not an accepted formal theory of CBD for dynamic systems, as it will be explained in the next Chapter.

Definition 5.2 (*System*) A system is a triplet $(SD, COMPS, and OBS)$ where:

- SD, the system description is a finite set of first-order sentences.
- $COMPS$, the system components is a finite set of symbols of constants.
- OBS, the system observations is a finite set of first-order sentences.

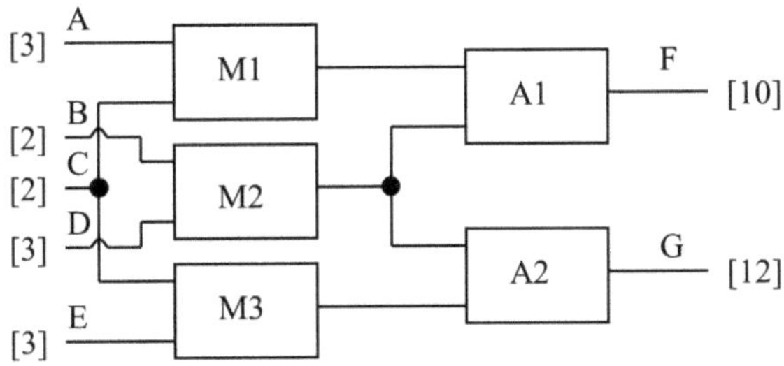

Fig. 5.1 The polybox system

In Reiter's theory, a *System* specifies a diagnosis problem.

SD defines the system structure and provides the models of correct behavior of its components. Connections are modeled with equality relations over the values of the signals of their terminals. The correct behavior assumption is made explicit with the $\neg AB(x)$ literal.

COMPS provides the names of the system components.

OBS states the observations of the system for the posed diagnosis problem. They are modeled with equality relations between constant values—the observed values—and the values of the signals at the terminals where observations take place. In other model-based approaches, it is called the observational model.

With the aforementioned conventions, the formal definition of the polybox system, whose behavior and model was introduced in Chap. 2, and with the observations shown in brackets in Fig. 5.1, is presented again for the sake of readability:

- $SD : \{MULT(M1), MULT(M2), MULT(M3), ADD(A1), ADD(A2),$ $in_2(M1) = in_1(M3), out(M1) = in_1(A1), out$ $(M2) = in_2(A1), out(M2) = in_1(A2), out(M3) = in_2(A2),$ $MULT(x) \wedge \neg AB(x) \Rightarrow out(x) = in_1(x) \times in_2(x),$ $ADD(x) \wedge \neg AB(x) \Rightarrow out(x) = in_1(x) + in_2(x).$
- $COMPS : \{M1, M2, M3, A1, A2\}.$
- $OBS = \{\ in_1(M1) = 3,\ \ in_2(M1) = 2,\ \ in_1(M2) = 2,\ \ in_2(M2) = 3,\ \ in_1$ $(M3) = 2, in_2(M3) = 3, out(A1) = 10, out(A2) = 12\}.$

Definition 5.3 ($D(\mathcal{C}_p, \mathcal{C}_n)$) Given two sets of components, $\mathcal{C}_p$ and $\mathcal{C}_n$, $D(\mathcal{C}_p, \mathcal{C}_n)$ is defined to be the conjunction: $\left[\bigwedge_{c \in \mathcal{C}_p} AB(c)\right] \bigwedge \left[\bigwedge_{c \in \mathcal{C}_n} \neg AB(c)\right]$.

$D(\mathcal{C}_p, \mathcal{C}_n)$ is a conjunctive clause of positive and negative literals. It declares that the components in $\mathcal{C}_p$ are faulty and that the components in $\mathcal{C}_n$ are non-faulty. For instance:

$$D(\{A2, M2\}, \{A1, M1, M3\}) = \\ AB(A2) \wedge AB(M2) \wedge \neg AB(A1) \wedge \neg AB(M1) \wedge \neg AB(M3).$$

Definition 5.4 (*Diagnosis*) Let $\Delta \subseteq COMPS$. $D(\Delta, COMPS - \Delta)$ is a diagnosis for $(SD, COMPS, OBS)$ if and only if $SD \cup OBS \cup \{D(\Delta, COMPS - \Delta)\}$ is satisfiable.

A *Diagnosis* is a solution to the diagnosis problem specified by *System*.

A *Diagnosis* is defined as a particular assignment of faulty behavior to the components of Δ together with an assignment of correct behavior to the other components of the system, those in $COMPS - \Delta$, such that the *System* with these assignments of behavior is consistent.

From the definition of diagnosis, the assumption of faulty behavior of the components in Δ plus the assumption of correct behavior of the components in $COMPS - \Delta$ removes any possible inconsistency from the *System*. "For a well-designed system," inconsistencies only occur when after injecting observations in the model, different values are assigned to the signal of a terminal. This is commonly called a symptom. Again, for a well-designed system, symptoms only occur if a component fails—assuming non- faulty observations, another potential source of inconsistencies. Consequently, we can interpret a diagnosis as a set of faulty/non-faulty assignments that removes all the symptoms from a *System*.

Remark 5.1 A diagnosis exists for $SD \cup COMPS \cup OBS$ if an only if $SD \cup OBS$ is consistent.

The consistency of SD is a necessary condition for a well-designed systems. If OBS is also consistent—and it has to be consistent if we assume that the sensors do not fail—$SD \cup OBS$ consistency is a necessary condition for a well-designed system.

The next proposition formalizes the intuition that if every component is correct, then no abnormal behavior—i.e., symptom—should be observed.

Proposition 5.1 *$D(\emptyset, COMPS - \emptyset)$ is a diagnosis for $(SD, COMPS, OBS)$ if and only if $SD \cup OBS \cup \{\neg AB(c) / c \in COMPS\}$ is consistent.*

Now, we introduce the key concept of minimal diagnosis.

Definition 5.5 (*Minimal diagnosis*) Let $\Delta \subseteq COMPS$. A diagnosis $D(\Delta, COMPS - \Delta)$ is a minimal diagnosis for $(SD, COMPS, OBS)$ if and only if for no proper subset Δ' of Δ, $D\left(\Delta', COMPS - \Delta'\right)$ is a diagnosis.

In a sense, a minimal diagnosis is the desired solution to a diagnosis problem because it involves a minimum number of faulty assignments to recover the consistency. Other criteria are possible, but without further domain knowledge, blaming a minimum number of components for the abnormal behavior of the systems is a very practical strategy. In the original Reiter formulation, diagnosis was defined to be always minimal.

Remark 5.2 If $D(\Delta, COMPS - \Delta)$ is a diagnosis, then there is a *minimal diagnosis* $D\left(\Delta', COMPS - \Delta'\right)$ such that $\Delta' \subseteq \Delta$.

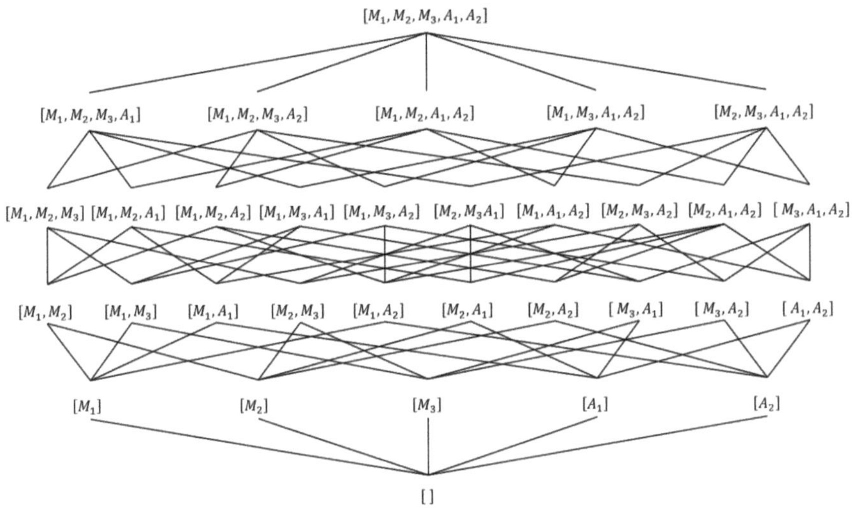

Fig. 5.2 Lattice structure of the candidate's space

This is a consequence of the previous definitions.

To fully characterize the set of diagnoses for a system, we need the following hypothesis, which holds if we only employ models of correct behavior and no exoneration:

Hypothesis 1 (*Minimal diagnosis hypothesis*) If $D(\Delta, COMPS - \Delta)$ is a minimal diagnosis, then for every Δ' superset of Δ, $D\left(\Delta', COMPS - \Delta'\right)$ is also a diagnosis.

Keep in mind that if we extend the basic Reiter's theory with models of faulty behavior or if we use observations to exonerate components, the *Minimal diagnosis hypothesis* may no longer hold.

From Remark 5.2 and Hypothesis 1, we can conclude that the set of minimal diagnoses efficiently represents all the *diagnoses* of a *system*.

Reiter's definitions of *diagnosis* and *minimal diagnosis* formalize the concept of *candidate* and *minimal candidate*, respectively, introduced by de Kleer and Williams [6]. If $D(\Delta, COMPS - \Delta)$ is a diagnosis, the set of components Δ is called a candidate. Candidates are usually represented with square brackets. If $[A2, M2]$ is a candidate, then both $A2$ and $M2$ are faulty while $A1$, $A3$, and $M2$ are correct.

In terms of candidates, for a system with n total number of components, we can potentially generate 2^n sets of different candidates. However, not all these potential candidates are valid diagnoses for a given set OBS. The space of diagnoses that solves a diagnosis problem can be efficiently represented by minimal candidates because of the lattice structure (see Fig. 5.2) that the subset relation induces on that space.

Finding the set of minimal diagnoses, or minimal candidates, we obtain a parsimonious representation of the solution space. In this sense, we can find all the solutions to a diagnosis problem if we find all the minimal candidates.

The following definitions formalize the concept of conflict, also introduced by de Kleer and Williams [6].

Definition 5.6 (*AB-literal*) An AB-literal is one of $AB(c)$ or $\neg AB(c)$ for some $c \in COMPS$.

Definition 5.7 (*AB-clause*) An AB-clause is a disjunction of AB-literals containing no complementary pair of AB-literals.

Definition 5.8 (*Conflict*) A conflict for $(SD, COMPS, OBS)$ is an AB-clause entailed by $SD \cup OBS$. A positive conflict is a conflict containing only positive AB-literals.

Conflicts are a key concept in Reiter's theory—and in any other model-based diagnosis approach. The original Reiter's theory only defines positive conflicts. Non-positive conflicts are important when using fault models or performing exoneration. From now on, we limit our discussion to positive conflicts, or conflicts for short.

Defining a positive conflict as an AB-clause provides a very intuitive interpretation of a conflict: A conflict identifies a set of components such that at least one of them is faulty—a necessary condition to make the corresponding AB-clause true when every AB-literal is positive.

An alternative interpretation is also possible. By definition, if the clause $AB(c_1) \vee AB(c_2) \vee \ldots \vee AB(c_k)$ is a conflict, we have:

$$SD \cup OBS \models AB(c_1) \vee AB(c_2) \vee \ldots \vee AB(c_k)$$

That can be rewritten as

$$SD \cup OBS \cup \{\neg AB(c_1) \wedge \neg AB(c_2) \wedge \ldots \wedge \neg AB(c_k)\}\, is\ inconsistent$$

Therefore, a conflict can also be interpreted as a set of assumptions of correct behavior that support a symptom leading to an inconsistency. This is the interpretation provided in [6] in the context of GDE—and the main computation performed by GDE.

From the formal definition of the polybox system shown in Fig. 5.1, it is easy to show that there are at least two conflicts:

$$\begin{aligned} C_1 &: \quad AB(M1) \vee AB(M2) \vee AB(A1). \\ C_2 &: AB(M1) \vee AB(M3) \vee AB(A1) \vee AB(A2). \end{aligned}$$

The conflict $C_1 : AB(M1) \vee AB(M2) \vee AB(A1)$ is a conflict because assuming correct behavior of $M1$, $M2$ and $M3$ we can infer a value of 12 at terminal F, while the observed value is 10. This correspond to a casual propagation of the input observations through the model, as shown in Fig. 5.3; propagation that leads to a symptom.

$C_2 : AB(M1) \vee AB(M3) \vee AB(A1) \vee AB(A2)$ is a conflict because assuming correct behavior of the components $M1$, $M3$, $A1$, and $A2$ allows to infer a value

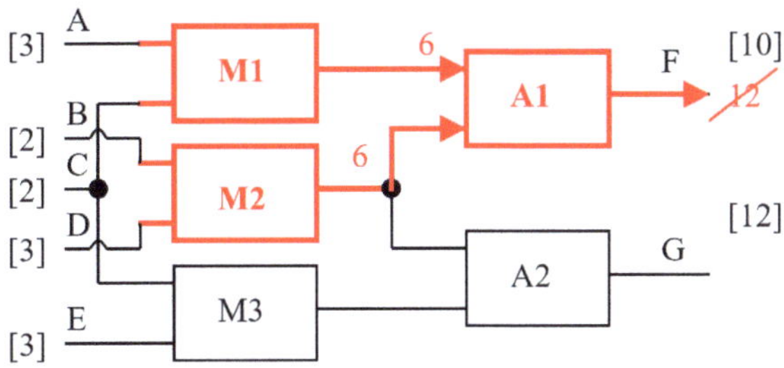

Fig. 5.3 The polybox system: first (minimal) conflict, $C_1 = \{M1, M2, A1\}$

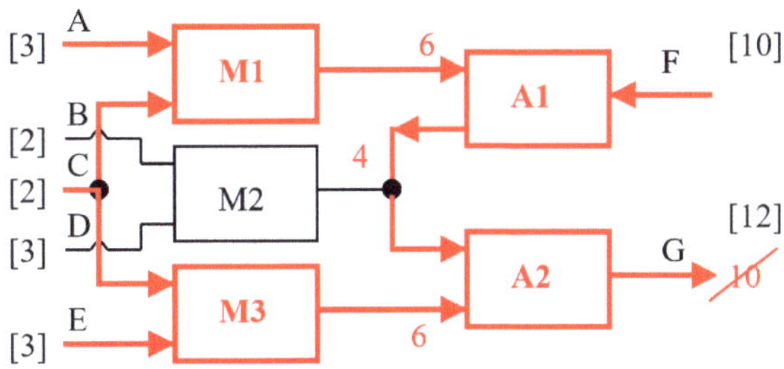

Fig. 5.4 The polybox system: second (minimal) conflict, $C_2 = \{M1, A1, A2, M3\}$

of 10 at terminal G, while the observed value is 12. This correspond to a noncausal propagation of values through the system, as shown in Fig. 5.4.

Positive conflicts are usually represented by the set of components referenced in the AB-clause. Hence, conflict C_1 is represented by the set $\{M1, M2, A1\}$.

Remark 5.3 If π is a conflict and $\pi \subset \pi' \subset COMPS$, then π' is also a conflict.

This is a consequence of the definition of *conflict* and limiting to positive conflicts. We now introduce the key concept of minimal conflict.

Definition 5.9 (*Minimal conflict*) A conflict π for $(SD, COMPS, OBS)$ is a minimal conflict if no proper subset of π is also a conflict.

It is easy to check that conflicts C_1 and C_2 are minimal conflicts.

For a system with n total number of components, we can generate 2^n sets of potential conflicts. However, not all these potential conflicts are valid conflicts for a given set OBS. The space of conflicts for a diagnosis problem can be efficiently represented by the set of minimal conflicts, in a similar way as the set of minimal diagnosis represents all the diagnosis for a system.

Conflicts and diagnosis are different concepts. However, they are intimately related. Assume that $D\,(\Delta, COMPS - \Delta)$ is a diagnosis for $(SD, COMPS, OBS)$. Remember that a diagnosis is an assignment of faulty behavior for the components in Δ that recovers the consistency of a system: $SD \cup OBS \cup D\,(\Delta, COMPS - \Delta)$ has to be consistent. If $\pi : AB\,(c_1) \vee AB\,(c_2) \vee \ldots \vee AB\,(c_k)$ is a conflict then, necessarily, some element of Δ must also be in π. Otherwise, the set $SD \cup OBS \cup \{\neg AB\,(c_1) \wedge \neg AB\,(c_2) \wedge \ldots \wedge \neg AB\,(c_k)\}$ would be consistent and π would not be a conflict.

The aforementioned argument is formalized in Reiter's theory and it's the foundation of Reiter's characterization of diagnoses. First, we refresh the concept of hitting set.

Definition 5.10 (*Hitting set*) Let $\mathcal{C}$ be a collection of sets. A hitting set for $\mathcal{C}$ is a set $\mathcal{H}$ such that $\mathcal{H} \subseteq \bigcup_{S \in \mathcal{C}} S$ and $\forall S \in \mathcal{C}, \mathcal{H} \cap S \neq \emptyset$. $\mathcal{H}$ is a *minimal hitting set* for $\mathcal{C}$ if no proper subset of $\mathcal{H}$ is a hitting set for $\mathcal{C}$.

Theorem 5.1 (Reiter characterization of diagnosis) *Let* $\Delta \subseteq COMPS$*,* $D(\Delta, COMPS - \Delta)$ *is a minimal diagnosis for* $(SD, COMPS, OBS)$ *if and only if* Δ *is a minimal hitting set for the collection of minimal conflicts of* $(SD, COMPS, OBS)$

This result is not only of theoretical importance. Its practical consequences permeate the entire field. Essentially, this theorem transforms the problem of computing diagnoses into the problem of computing minimal conflicts—although computing minimal hitting sets still remains a complex computational problem.

The theorem provides a systematic method to characterize the diagnoses of a system, generating the minimal diagnoses in two steps:

1. Find the minimal conflicts of the system.
2. Generate the minimal diagnoses as the minimal hitting set of the collection of minimal conflicts.

The majority of the CBD systems use some variant of this method, which can be applied incrementally whenever a new minimal conflict is found, because the minimal hitting sets can be computed incrementally.

We proceed to illustrate the method with the polybox system. For the given OBS, this system has only two minimal conflicts, $C_1 : \{M1, M2, A1\}$ and $C_2 : \{M1, M3, A1, A2\}$. From these minimal conflicts, we obtain the following minimal hitting sets: $\Delta_1 = \{M1\}$, $\Delta_2 = \{A1\}$, $\Delta_3 = \{M2, M3\}$, and $\Delta_4 = \{M2, A2\}$. The Fig. 5.5 shows schematically how those minimal hitting sets are built. For each Δ_i, we obtain a minimal diagnosis D_i. D_1 and D_2 are single-fault diagnosis. D_3 and D_4 are double-fault diagnosis. D_3, for instance, says that a simultaneous fault at $M2$ and $M3$ is consistent with SD and OBS, while a single fault at $M2$ or $M3$ is not. In the CBD framework, these four minimal diagnoses have the same status. To further discriminate among them, we have to inject additional information like new observations or more domain knowledge.

Algorithm 5.1: Conflicts guide candidates generation.

```
Inputs: MinimalConflicts
CandidatesCollection ← {∅}
for each Conflict ∈ MinimalConflics do
  CurrentCandidates ← CandidatesCollection
  for each Candidate ∈ CurrentCandidates do
    if Candidate ∩ Conflict = ∅ then
      CandidatesCollection ← UpdateCandidates(Candidate, CandidatesCollection, Conflict)
Return CandidatesCollection
```

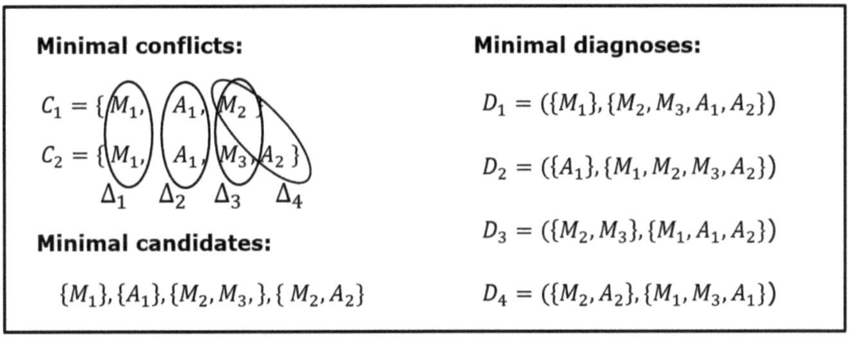

Fig. 5.5 Minimal diagnoses as minimal hitting sets of minimal conflicts

---

Algorithm 5.2: UpdateCandidates.

Inputs: Candidate, CandidatesCollection, Conflict
CandidatesCollection ← CandidatesCollection - Candidate
for each Component ∈ Conflict **do**
  NewCandidate ← Candidate ∪ { Component }
  CandidatesCollection ← CandidatesCollection ∪ NewCandidate
Remove duplicates and non-minimal elements from CandidatesCollection
Return CandidatesCollection

Algorithm 5.1 describes a procedure to compute the minimal diagnoses from the collection of conflicts. The algorithm assumes that every conflict has been computed before generating the diagnoses. Nevertheless, given that the algorithm works incrementally, processing one conflict at a time, it can be very easily adapted to update the set of minimal diagnoses as soon as a new minimal conflict is detected. The algorithm starts with a set of minimal diagnoses, $CandidatesCollection$ with a single element: the empty set. If there is no conflict, this is the unique minimal diagnosis. Consequently, in the first pass over the second **for** loop, the $Candidate$ has no element of the first conflict and the algorithm $UpdateCandidates$ is invoked. This algorithm replaces the $Candidate$ by all its superset constructed adding a single element of the conflict. Figure 5.6 shows the candidate space for the polybox example after the first conflict has been processed. The tentative minimal diagnoses are enclosed with dotted circles. Every potential candidate under the continuous line is discarded by the conflict. From now on, the algorithm refines the minimal diagnoses processing the following conflicts. Figure 5.7 shows an intermediate step, while Algorithm 5.2 is processing the second minimal conflict, replacing $[M2]$ by its minimal supersets but before removing duplicates and non-minimal candidates. Figure 5.8 shows the final effect of processing the second minimal conflict and the final result, with four minimal diagnoses.

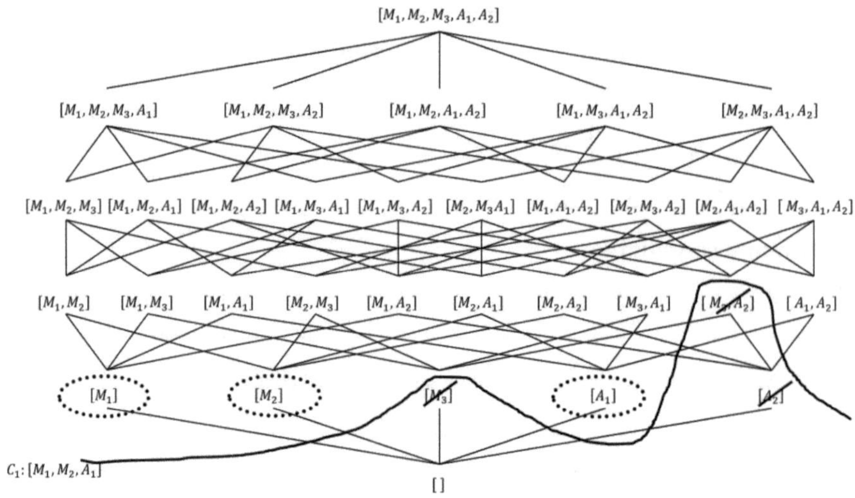

Fig. 5.6 Candidates space after processing first conflict

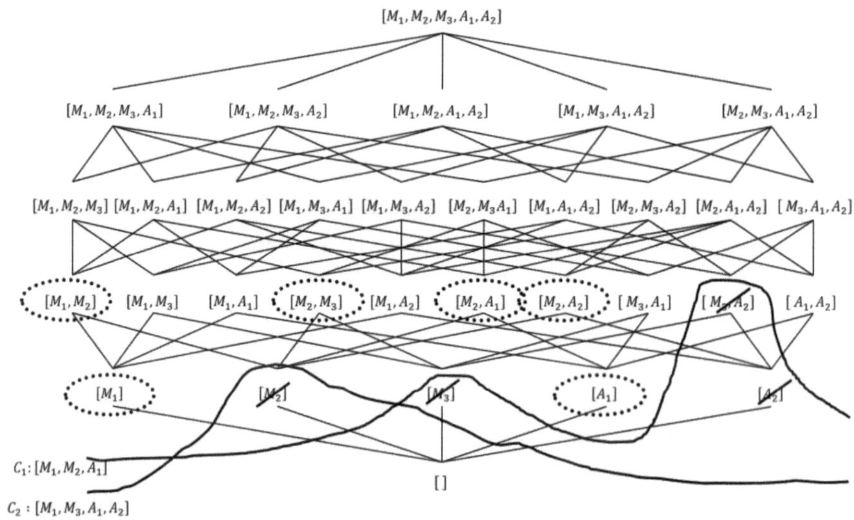

Fig. 5.7 Candidates space processing second conflict before removing duplicates and non-minimal

Reiter's theory of CBD had the merit of setting consistency-based diagnosis on a firm ground. It also offers a simple, elegant and general framework to diagnosis. Nevertheless, CBD has his own caveats and limitations, some of them due to the general approach taken and some of them due to the basic formulation of Reiter. We will discuss some of these issues in the following sections.

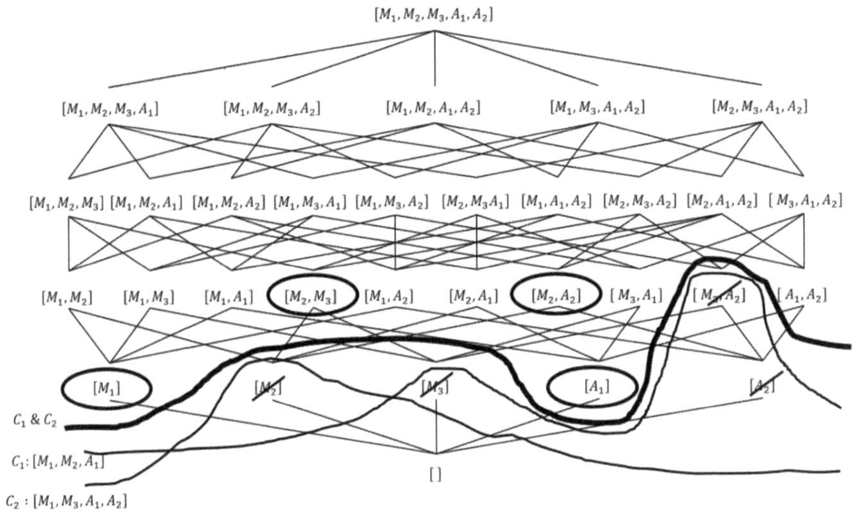

Fig. 5.8 Candidates space after processing second conflict

5.3 GDE: The Computational Approach

Reiter's theory provides a formal framework for consistency-based diagnosis. It also provides an important guideline to compute minimal diagnoses: compute minimal conflicts first, and then minimal diagnoses as their minimal hitting sets. However, it is a pure logical theory and no computational method is provided to compute minimal conflicts in an efficient manner. It only requires a solid and complete deduction procedure. However, deduction in First-Order Logic is an NP semi decidable problem. Actually, it is not practical except for simple systems with a small number of components.

Nevertheless, Reiter's theory was developed to formalize the already- existing computational methods that were indeed computing diagnosis *a la* consistency-based manner. One of these methods is of particular importance: The General Diagnosis Engine, GDE, proposed by de Kleer and Williams in their seminal "Diagnosing Multiple Faults" article [6].

5.3.1 GDE Fundamentals

GDE was the first model-based computational diagnosis system able to cope efficiently with multiple faults. It is still the main computational paradigm for consistency-based diagnosis and a reference to compare model-based proposals on the DX community. GDE performs consistency-based diagnosis. Hence, the general characterization of CBD applies to GDE. We recalled it here for the reader's convenience:

- It is a *component-oriented approach.*
- It only uses knowledge of the *systems structure* and the *components behavior.*
- It only uses local models of *correct behavior.*
- *Exoneration is not allowed* in the reasoning process.

According to de Kleer and Kurien [12], GDE makes the following basic assumptions:

- GDE works on physical systems that:
  - Are made of sets of interconnected components.
  - Its desired functionality is known.
  - The design of the system achieves the desired functionality.
  - The system is a correct instance of the design.
- All malfunctions are caused by faulty component(s).
- The behavioral information:
  - Only provides indirect evidence.
  - That can be used to reject potential behaviors, but not to confirm them.

The last assumption avoids exoneration. The second assumption requires considering any potential fault source as a component—for instance, a connection or a terminal may be modeled as a component if required. The first assumption declares that GDE works with component-oriented systems and makes explicit something that is usually taken for granted: that the device is a correct instance of the design.

5.3.2 How GDE Works

GDE works incrementally in the following manner:

- *Generating* every possible *prediction* from the available observations. This process obtains the global behavior of the system.
- *Detecting* every *symptom*—i.e., a discrepancy between a prediction and an observation or another prediction for the same variable.
- *Identifying* every (minimal) *conflict.*
- *Generating* every (minimal) fault *candidate.*

GDE provides an intuitive description of the concept of conflicts and candidates. A GDE "conflict is a set of assumptions that support a conflict, and thus leads to an inconsistency". "A candidate is a particular hypothesis of how the actual device differs from the model. Candidates are represented by a set of components. The components explicitly mentioned are faulty while the ones not mention are non-faulty. A candidate must explain all the current symptoms of the physical system given the available observations". These informal descriptions of the concept of conflict and candidate, which are what GDE computes, happen to be the conflicts and candidates that Reiter formally defines.

GDE generality derives from the fact that it works on a single class of models: for any device and any desired level of abstraction, the models provide the structure of the system and the local behavioral model of the components. Components communicate through their terminals according to the system structure. The behavioral models have to be local and must never reference any entity outside the terminal components. The models impose restrictions on the values that the terminal variables may take, but do not impose causality. Finally, connections only propagate values among terminals. These are the common request for the consistency-based approach to diagnosis.

GDE generates local predictions, causal and noncausal, using the behavioral models of the components.

GDE exploits the structure of the system to propagate local predictions in every possible direction allowed by the structure description. This allows GDE to obtain the global behavior of the system and to find out every symptom given the current observations.

Once the minimal candidates for the current set of observations are computed, GDE proceeds to refine the set of candidates proposing a new observation. This observation is selected among the non-measured variables according to an information theoretic loss function, the entropy of the diagnosis candidates. Hence, the diagnosis process in GDE is a cycle of finding minimal candidates for the current set of observations and requesting new measurements, until the diagnosis candidates can no longer be refined.

5.3.3 Efficiently Computing Minimal Conflicts with a Dependency-Recording Engine

The computational efficiency of GDE derives, first, of the concepts of minimal conflict and minimal candidates, introduced intuitively by de Kleer and Williams [6] and formalized by Reiter [16]. Nevertheless, care must be taken when generating the global behavior of the system to detect symptoms and compute conflicts from them.

The efficient computation of conflicts is accomplished with the help of a *Dependency-Recording Engine*, or DRE for short. Particularly, GDE employs a kind of DRE named *Assumption-based Truth Maintenance System*, ATMS, described in the following papers: [9–11].

GDE inference system consists of two independent components: a problem solver and an ATMS. The problem solver is in charge of making local inferences. It is also responsible for notifying the ATMS of the inference performed, providing the additional information that the ATMS needs. The communication from the problem solver to the ATMS must be performed after each local inference step.

An ATMS keeps track of all the inferences performed during the diagnosis process. This is achieved building a graph whose nodes correspond to a single problem datum. For the scope of this discussion, we are interested in three kinds of nodes: facts, assumed, and derived nodes. Facts are considered true, without any further assumption or justification. An example of a fact may be an observed datum. A derived node keeps the result of a local inference step and it is also a datum.

An assumed node represents a datum whose truth value is not known, but that must be true to perform a local inference. Assumed nodes allow managing the default assumption underlying every local inference step: that the component behaves correctly and then we can use the model of correct behavior to perform an inference.

When we have a chain of local inferences, then we have a set of assumptions that must hold for the last inference to be valid. This set of assumptions is called an *ATMS environment*.

Actually, each ATMS node has three components: the datum, its justification(s), and its label. A justification is the set of data -nodes- used to perform the local inference step. An *ATMS label* is the set of environments in which the datum is true. ATMS labels precise definition is provided in [9]. One of its key properties is that all of its environments are minimal.

Remember that the GDE inference system consists of two independent components: a problem solver and an ATMS. The problem solver performs local inferences and informs the ATMS. It also provides its justification, i.e., the data used, and an assumption, i.e., the component used.

The ATMS is responsible for registering the inference. This may require including a new node (in the case of a new datum), its justification (in case it is a new one), and computing its label (for a new node or new justification). It also adds an assumption node if it is a new assumption. Label computation is the main task of the ATMS.

Symptoms or discrepancies arise whenever a variable is assigned two different values. It is the responsibility of the problem solver to detect discrepancies and to inform the ATMS. In this case, a special falsity node is created. Its justification is made of the nodes that set the inconsistency. And its label is the empty set to maintain the consistency of the system. Note however, that a label can be (and actually it is) computed for a falsity node using the standard ATMS label computation procedure. The environments of the label computed for a falsity node are called NOGOOD for a very good reason: these are the environments that yield a symptom. Remember that an environment is a set of assumptions. If all the assumptions of a NOGOOD environment hold, then there is a discrepancy. As the reader may already suspect, NOGOODs are the (minimal) conflicts underlying the detected symptom. NOGOOD environments are registered by the ATMS, although the label assigned to the falsity node is the empty set.

The previous paragraphs try to describe the essentials of the GDE inference engine, although almost every technical detail has been omitted. The interested reader should consult [9–11]. We'll try to illustrate its working principles using the polybox example and the first symptom detected propagating in the causal direction the value of the observed inputs to the output terminal F.

Remember that $C_1 : AB\,(M1) \vee AB\,(M2) \vee AB\,(A1)$ is a conflict because assuming correct behavior of $M1$, $M2$, and $A1$, we can infer a value of 12 at terminal F, while the observed value is 10. It is a minimal conflict because if we remove any component from C_1, the resulting set is no longer a conflict. In GDE notation, C_1 is the set $\langle M1, M2, A1\rangle$.

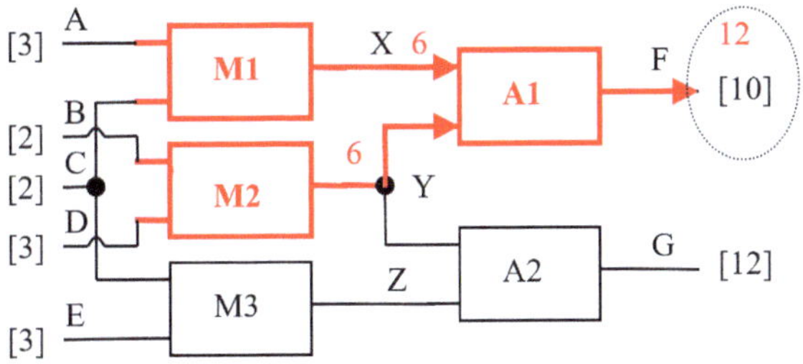

Fig. 5.9 The polybox system: first (minimal) conflict

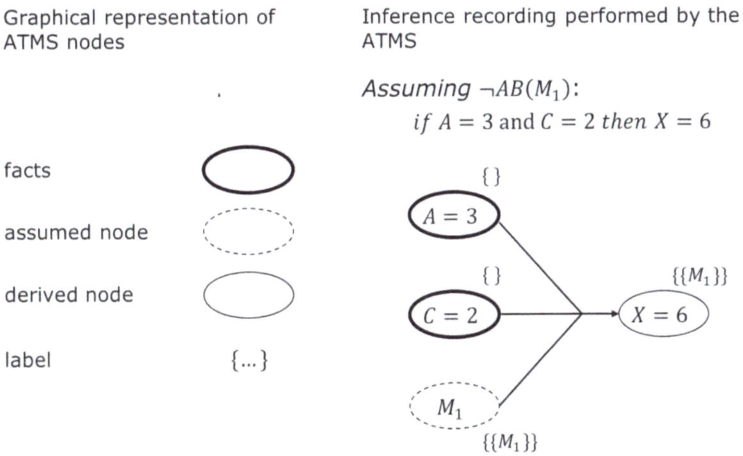

Fig. 5.10 Graphical representation of ATMS first inference step recording

GDE performs three local inference steps to derive the conflict C_1, as shown in Fig. 5.9.

Figure 5.10 shows a graphical representation, taken from [9], of the graph built by the ATMS if only the observations of $A = 3$ and $B = 2$ were available and the problem solver assumes that $M1$ is working properly. Different nodes of the graph are represented by different kind of circles that enclose the datum. The labels, computed by the ATMS, are shown in braces next to their datum. $A = 3$ and $B = 2$, the observations, are considered true, hence their labels are empty and have no justification. The problem solver, to perform the local inference, assumes that $M1$ is not faulty. This is represented by an assumed node, which has no justification and its label contains a single environment that simply describes the assumption. The result of the inference is $X = 6$. This is a derived node. Its justification is described by its parents in the net. Its label contains the only environment that it depends on, that is $\{M1\}$.

If we add observations $B = 2$ and $D = 3$, the problem solver can perform two additional inferences. The first one predicts $Y = 6$ and it's registered in a similar

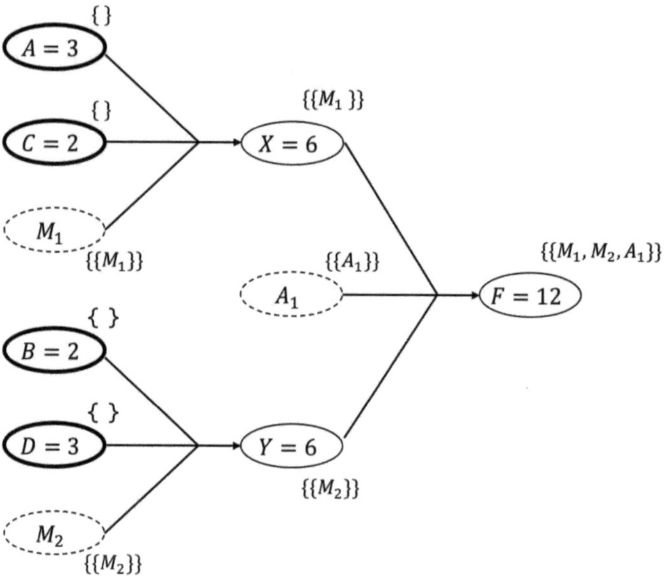

Fig. 5.11 ATMS recording after the third inference step

way to the previous one, but with the label $\{M2\}$. More interesting is the third one that concludes $F = 12$. The resulting graph is shown in Fig. 5.11. Notice that, in this simple example, the label of datum $F = 12$ contains a single environment, consisting of the union of the environments of its parents in the net.

If now we add the observation $F = 10$, the ATMS adds another fact node. The problem solver then has to inform the ATMS of the inconsistency that is justified by $F = 10$ and $F = 12$. The ATMS includes a faulty node, see Fig. 5.12, and computes the corresponding NOGOOD: $\{M1, M2, A1\}$.

It is also the responsibility of the problem solver to find every symptom or inconsistency. This is guaranteed by an exhaustive search that propagates every available value in every direction. Figure 5.13 shows some symptoms of the polybox. The ATMS also helps in this search, because it also works as a cache of the local inferences. Hence, no inference is duplicated. This is particularly useful when working with qualitative models and qualitative variables, which take values on a small domain.

From those symptoms, only two minimal conflicts are found, shown in Figs. 5.14 and 5.15, that also shows the label of the predicted datum. The reader can easily check that any other symptom shown in Fig. 5.13 yields one of those conflicts.

Once GDE has found the set of minimal conflicts, it can compute the set of minimal candidates as the minimal hitting sets of the minimal conflicts. This computation can also be made incrementally, as soon as a new minimal conflict is found.

For the polybox example, there are four minimal candidates: $[M1]$, $[A1]$, $[M2, M3]$, and $[M2, A2]$. Any of those candidates is able to explain the system behavior

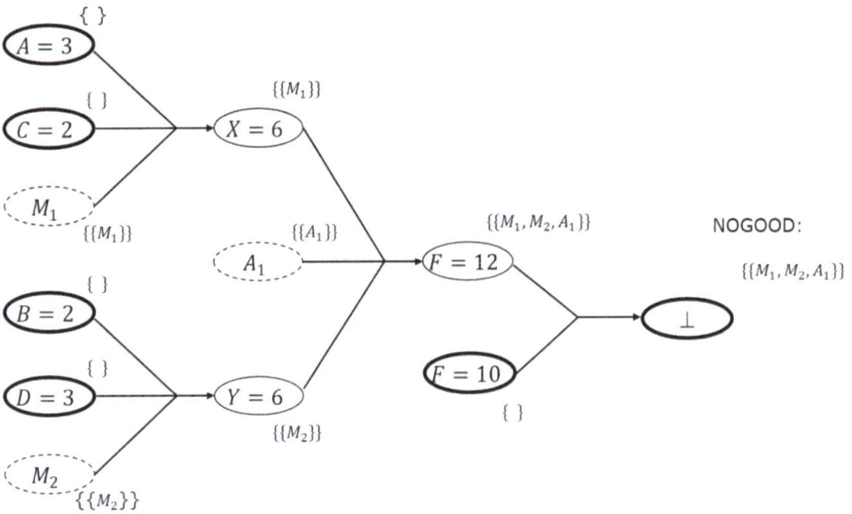

Fig. 5.12 ATMS recording after the third inference step

Fig. 5.13 Some symptoms for the polybox

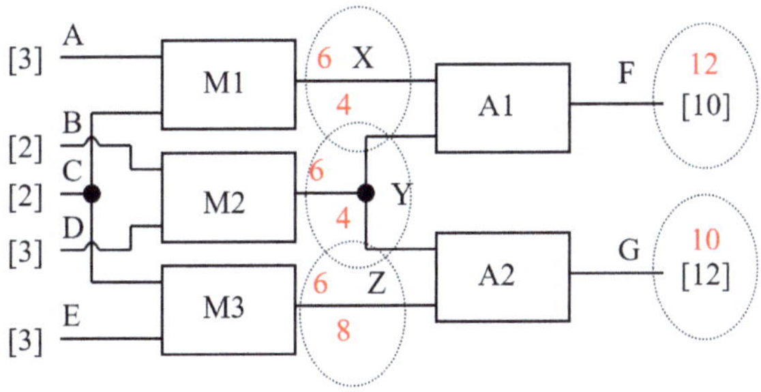

Fig. 5.14 First minimal conflict and ATMS label computation

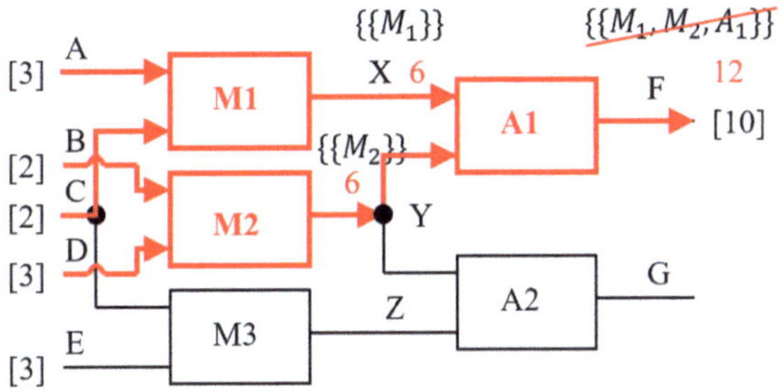

and to recover the *consistency* of the system model. Figure 5.16 shows the explanation provided by the single-fault candidates. For instance, for the candidate [$M1$], shown in red in the figure of the left, this component is faulty and obtains a value of 4 when multiplying 3×2, the prediction at terminal F is 10 and no symptom remains in the system. In a similar way if only $A1$ is faulty, we can blame it for computing the value

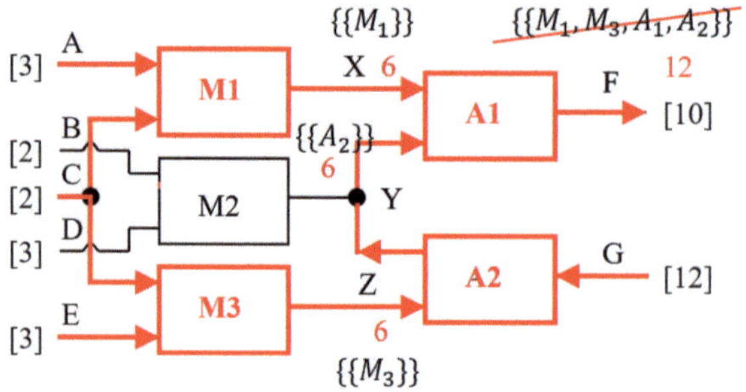

Fig. 5.15 Second minimal conflict and ATMS label computation

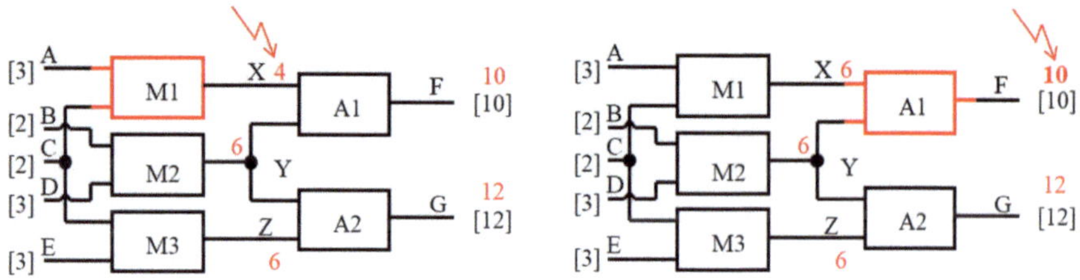

Fig. 5.16 Single minimal candidates explanation of symptoms

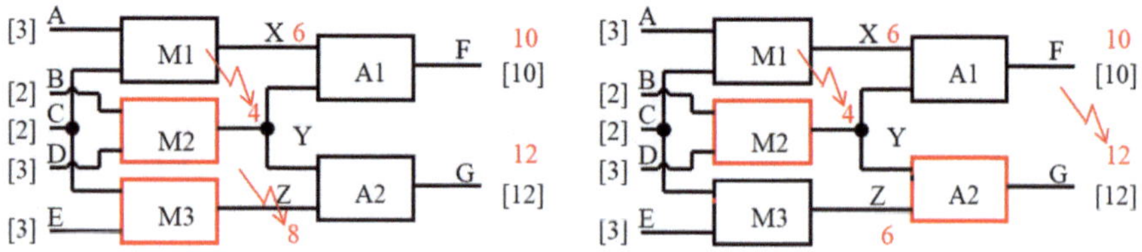

Fig. 5.17 Double minimal candidate explanation of symptoms

10 instead of 12, and again, it explains the behavior of the system. Figure 5.17 shows the less obvious explanations provided by the double-fault candidate. For instance, for candidate [$M2$, $M3$], a plausible explanation is that being $M2$ faulty it obtains the value of 4 at its output. This behavior removes the symptom at terminal F. But then a new symptom would appear at terminal G. Consequently, at least one of $M3$ or $A2$ must also be faulty. The fault at $M3$ removes the symptom from the system.

5.3.4 Selecting Observations to Refine the Set of Candidates

Given that the set of minimal candidates can be potentially large when multiple faults are considered, GDE includes a one look-ahead strategy to refine the set of minimal candidates. This strategy proposes to take the additional measure that, in average, minimize the entropy of the resulting candidates set. The strategy assumes that all measures have equal cost and that the components fail independently. This independency assumption is not critical for the strategy, but reduces the number of probabilities to manage.

Minimizing the average entropy minimizes, in average, the number of measurements needed to reduce the candidate set as much as possible. Under the hypothesis that all measures have equal cost, it also minimizes the cost of the measurements.

Complete details of the one look-ahead strategy used by GDE can be found in [6]. It is important to note, however, that computing the expected entropy of the candidate set requires knowing the new conflicts and candidates after a measure is taken. This has to be done for every measure and possible output of the observation. GDE is able to make this computation efficiently because the ATMS keeps the environment of every prediction for every variable. Hence, for each possible measurement outcome, it can easily determine the new conflicts and candidates. This property is crucial for the effective computation of the average entropy.

For the polybox example with the provided observations, we have that $X = 4$ in environments $\{A1, M2\}$ and $\{A1, A2, M3\}$, as shown in Figs. 5.18 and 5.19. And $X = 6$ in environment $\{M1\}$, see Fig. 5.14, for instance.

Given that there is no other registered value for the variable X, we have three significant outputs when measuring it:

- If we measure $X = 4$, then $\langle M1 \rangle$ is the unique minimal conflict and the new minimal candidate is $[M1]$.

Fig. 5.18 Environment of $X = 4$: $A1, M2$

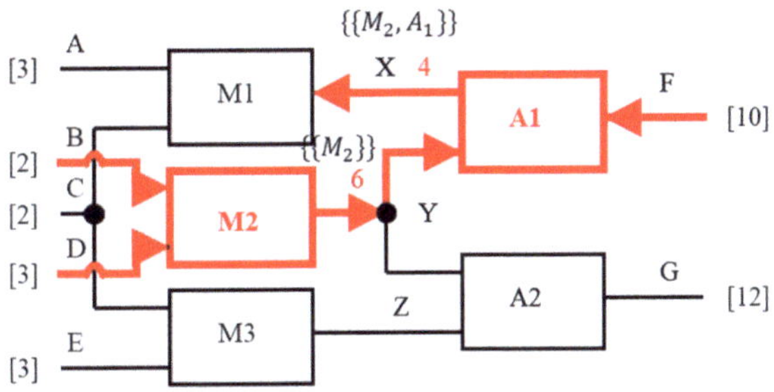

Fig. 5.19 Environment of $X = 4$: $A1, A2, M3$

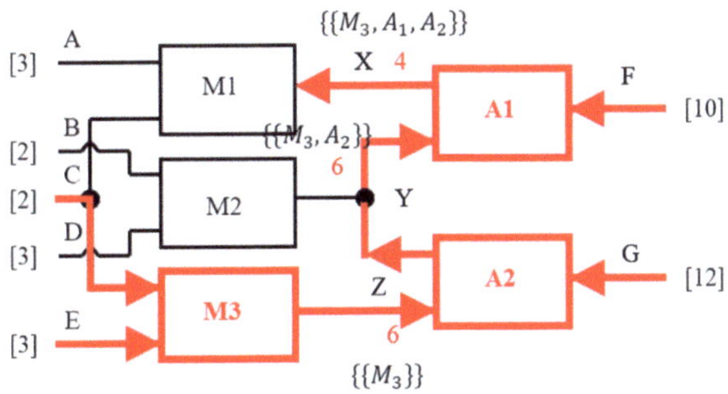

- If $X = 6$, $\langle A1, M2\rangle$ and $\langle A1, A2, M3\rangle$ are the new conflicts and the new minimal candidates are $[A1]$, $[M2, M3]$, and $[A2, M2]$.
- If $X \neq 4$ and $X \neq 6$, then $\langle M1\rangle$, $\langle A1, M2\rangle$ and $\langle A1, A2, M3\rangle$ are conflicts and $[A1, M1]$, $[M1, M2, M3]$, and $[A2, M1, M2]$ are minimal candidates.

In this way, GDE can immediately obtain the new minimal candidates for any measured output of any variable. Afterward, GDE has to compute the candidates expected entropy for every measure. In general, for each still not measured variable, GDE has to compute:

$$H_e(x_i) = \sum_{j=1,\ldots,k} p(x_i = v_{i,j}) H(x_i = v_{i,j}),$$

where $v_{i,1}, \ldots, v_{i,k}$ are all the possible values of x_i and $H(x_i = v_{i,j})$ is the entropy if the observation of x_i is $v_{i,j}$.

Fortunately, the computation of $p(x_i = v_{i,j})$ and $H(x_i = v_{i,j})$ can be performed using the current candidates probability distribution and counting over the set of candidates of the predicted outcomes, see [6] for a formal proof.

Summarizing, GDE is able to propose a measuring strategy that minimizes the cost of the observation to refine the set of candidates. And it can tackle this task because the ATMS keeps track of every inference, including the assumptions that support them.

5.3.5 *Advantages and Limitations of GDE*

GDE approach to model-based diagnosis is a key landmark of the Artificial Intelligence community to the diagnosis of physical devices. The basic ideas deployed in GDE have established firmly in the DX community and had influenced many posterior advances in the field. Clearly, GDE has also its own limitations, but more often than not they have opened new research lines that have expanded the application of diagnosis to real-world problems.

The main contributions of GDE are the following:

- GDE is a domain-independent diagnosis engine. All the knowledge that it uses of the system is provided by the models and the structure.
- GDE only uses models of correct behavior. It doesn't require knowledge of how a component may fail. As a consequence, GDE is able to diagnose new, unpredicted faults.
- GDE is able to cope with multiple faults in an efficient and automatic way.
- GDE includes a next measurement policy that in average minimizes the number of measures needed to isolate the fault as much as possible.

Among the main limitations of GDE, we should mention the following:

- As it is, GDE can only be applied to component-oriented systems. However, it can be extended to other kinds of systems. For instance, it can be applied to processes, as long as they are modeled as interconnected components and the no function in structure principle is followed.
- It is particularly apt to work with qualitative models and qualitative variables, but has some difficulties with models and variables in the real domain. For more on that topic, see the next section.
- Computational complexity is an issue; although it only computes minimal conflicts and minimal candidates, the number of minimal candidates can grow exponentially in the worst case. So difficulties arise for systems with a large set of components. Consider combining it with a hierarchical modeling approach, where the system is view as a small set of large components. Once a small number of components are identified as candidates, proceed to start a diagnosis process on those components.
- Limitations of local models and stepwise inference. Some global properties of a system may not be capture propagating local inferences. A global analysis of the model may be required. Consider, for instance, a multiple-input and multiple-output system where one of the outputs is independent of one of the inputs. Say, for the sake of ilustration, that the value at output terminal O_1 is indepentend of the value at input terminal I_1. Imagine also that there is a path between those independent input–output terminals. If a symptom appears at O_1, it is possible that some of the components in the path between I_1 and O_1 were included in some minimal conflict only because there is a path between I_1 and O_1. In this scenario, those components should not be part of a minimal conflict. The independence between connected terminals requires a global analysis of the system.
- Limitations of the knowledge. A logical approach to diagnosis using only models of correct behavior may generate candidates that, although logically sound, cannot be found in a real device. Imagine, for instance, that a candidate to explain the symptoms requires a pump operating with no power supply. This can be a formal logic explanation, but it is not acceptable because a pump cannot behave in this economic way. For an example of this kind of candidates, see the system discussed at the beginning of the following section.

5.4 Consistency-Based Diagnosis with Fault Modes

The main advantage of Consistency-based Diagnosis is that we only need correct behavior models to perform fault detection and isolation. In fact, it can automatically handle multiple faults [6]. However, using only correct behavior models also provides less discriminative power to perform fault identification, and it can additionally slow down dynamic systems diagnosis. But probably the most important drawback comes from the chance to obtain logically correct, but physically impossible diagnosis results. Let's use the example in Fig. 5.20 from Struss and Dressler [18] to illustrate

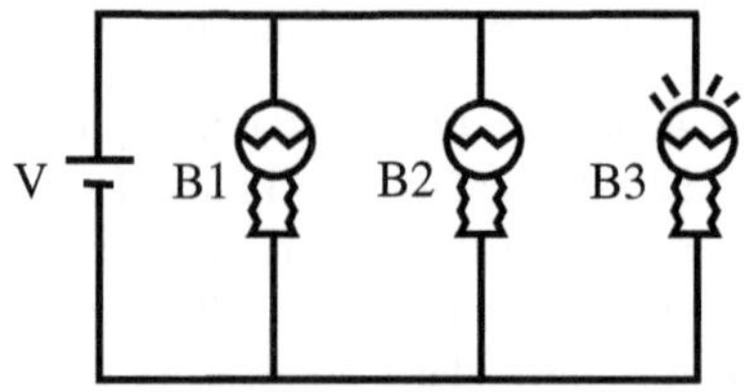

Fig. 5.20 Circuit with one battery and three bulbs

the point. The system is made up of one battery, V, which supplies power to three bulbs: $\{B1, B2, B3\}$.

A possible logical model for such system, including only correct behavior models, could be:

$$BULB(X) \wedge OK(X) \wedge VAL(PORT(X), +) \Rightarrow VAL(LIGHT(X), \text{ON}). \tag{5.1}$$

$$BULB(X) \wedge OK(X) \wedge VAL(PORT(X), 0) \Rightarrow VAL(LIGHT(X), \text{OFF}). \tag{5.2}$$

$$BULB(X) \wedge OK(X) \wedge VAL(LIGHT(X), \text{OFF}) \Rightarrow VAL(PORT(X), 0). \tag{5.3}$$

$$BULB(X) \wedge OK(X) \wedge VAL(LIGHT(X), \text{ON}) \Rightarrow VAL(PORT(X), +) \tag{5.4}$$

$$BATTERY(X) \wedge OK(X) \Rightarrow VAL(PORT(X), +). \tag{5.5}$$

The model must be supplemented with the set of components, $COMPS = \{V, B1, B2, B3\}$, the parallel connection between the ports in those components (connected components must share the same port values; we assume that the connections are not components and they do not fail, just to simplify the model):

$$BULB(X) \wedge OK(X) \wedge BULB(Y) \wedge OK(Y) \wedge$$
$$CONNECTED(X, Y) \wedge VAL(PORT(X), Z) \Rightarrow VAL(PORT(Y), Z). \tag{5.6}$$

$$BATTERY(X) \wedge OK(X) \wedge BULB(Y) \wedge OK(Y) \wedge$$
$$CONNECTED(X, Y) \wedge VAL(PORT(X), Z) \Rightarrow VAL(PORT(Y), Z). \tag{5.7}$$

$$CONNECTED(V, B1). \tag{5.8}$$

$$CONNECTED(V, B2). \tag{5.9}$$

$$CONNECTED(V, B3). \tag{5.10}$$

$$CONNECTED(B1, B2). \tag{5.11}$$

$$CONNECTED(B1, B3). \tag{5.12}$$

$$CONNECTED(B2, B3). \tag{5.13}$$

and the observational model: $\{V = \text{ON}, B1 = \text{OFF}, B2 = \text{OFF}, B3 = \text{ON}\}$.

It is clear from the observations that they are not consistent with the set of correctness assumptions. GDE would provide several diagnoses. For instance, $[B1, B2]$ would be a minimal diagnosis, which is clearly consistent with the observations: V is ON, providing power to $B3$, while $B1$ and $B2$ are faulty. However, we can also find the following minimal diagnosis: $[V, B3]$, that would mean that $B3$ is faulty producing light while V is off, which is physically impossible, but logically consistent with the model.

Dressler and Struss [7] clearly explained that "if GDE logically negates the correct mode of a component, anything is possible". But in the real world, a component usually does not fail in any possible way, but in a limited number of ways. Sometimes, we can even model the behavior of a faulty component. The idea is to add those possible ways a component can break as potential behavioral models for the component. This is the last and more informed way to include knowledge about faulty behavior in a diagnosis system. But there are others.

1. One of the first attempts to use knowledge about failure was the *no intermitency* approach [15]: components behave non-intermittently even when fault are present, as a consequence components must behave coherently between multiple observation tests. The approach was not valid for dynamic systems, and assumed, obviously, non-intermittent faults.
2. The *physical impossibility* axioms [8] were introduced to help the diagnosis process. We don't know precisely how a component fails, but we do know what is physically impossible. For instance, in the battery and bulbs example, we can state that a bulb cannot be lit if there is no power. This could be added as an additional axiom:

$$\neg(BATTERY(X) \wedge VAL(LIGHT(X), ON) \wedge VAL(PORT(X), 0)).$$

3. Finally, two different works proposed to use predictive fault models [13, 18]. In this case, each component will have multiple behavior modes, either nominal or fault. Each mode must have a predictive model stating what should be its intended behavior in that mode. Now, instead of having a binary assignment for each component (either $OK(c)$ or $\neg OK(C)$), we can have N different assignments. The diagnosis process changes and now it is defined as a Mode Assignment [7]:

Definition 5.11 (*Mode Assignment*) A (complete) mode assignment, D, is a conjunctive clause

$$D = \wedge_{c \in COMPS}\, m_{ji}(C_i) \text{where } m_{ji} \in modes(C_i)$$

Definition 5.12 (*Diagnosis*) A diagnosis for a system description SD, a set of components $COMPS$, and observations OBS, is a complete mode assignment, D, such that $SD \cup OBS \cup D$ is satisfiable.

Each fault mode will require a new predicate in the logical model:

$$ADDER(X) \wedge StuckAtZero(X) \Rightarrow Value(Output(X), 0).$$

Since the faulty behavior is explicitly modeled, it can also be refuted given the observations. Such is the idea behind Consistency-based Diagnosis with Fault Models, that provides not only Fault Detection and Isolation, but Fault Identification [13, 18]. Different behavioral models are checked against current observations. Faulty candidates are made up of those modes that are consistent with available observation, i.e., those that cannot be rejected.

At the same time, we can also cope with unknown faults by providing the **unknown** fault mode, whose behavior cannot be rejected.

The problem with introducing Fault Modes is that we increase exponentially the complexity of the Fault Diagnosis task. If we have a system with N components, and the average number of behavior modes is M, we must discriminate among M^N modes, which is much worse than the binary decision which needs to discriminate among 2^N potential mode assignments.

Consistency-based diagnosis is not the only way to use logical models to perform automated diagnosis. Abductive diagnosis [14] is another way to model the relation between the model, the components and the observations. Abductive diagnosis goes a step further in the fault identification process trying to solve the following problem:

Definition 5.13 (*Abductive Diagnosis*) A diagnosis D for a system description SD, a set of components $COMPS$, and the set observations OBS, that can be split between inputs and outputs, $OBS = \mathcal{I} \cup \mathcal{O}, \mathcal{I} \cap \mathcal{O} = \emptyset$, is a complete mode assignment, D, such that $SD \cup D \cup \mathcal{I} \models \mathcal{O}$.

5.5 Conclusions

Consistency-based diagnosis provides a formal and sound method to perform model-based diagnosis using just models of nominal behavior as a set of First-Order Logic formula.

Providing a formal theory of diagnosis was an important milestone in the field. Moreover, it had, at least, two direct benefits. From a theoretical point of view, it established the basic for a systematic research in the area, which in the view of the authors had a strong influence on the advances of the field. From a practical point of view, it made clear the assumptions that were implicitly exploited by many other diagnosis approaches, allowing for a systematic comparison of different diagnostic methods and a sound explanation of the different results that each method can obtain.

Reiter's characterization [16] allows to compute the complete set of minimal diagnoses from the set of minimal conflicts, if they are positive (meaning that there is neither fault modes nor exoneration), providing a domain-independent method that relied upon just a theorem prover. This was the first logical characterization of diagnosis, and the first step toward other forms of diagnosis, such as abductive diagnosis, default diagnosis, or diagnosis with fault modes.

Almost simultaneously, de Kleer and Williams [6] proposed the computational alternative for consistency-based diagnosis: the General Diagnosis Engine. GDE relied also upon models made up of FOL formula, together with an inference engine capable to do estimations based on the models, and an online Dependency-Recording

Engine (ATMS) that recorded nominal behavior assumptions. GDE is still the *de facto* paradigm for consistency-based diagnosis and it was the first diagnosis engine to handle automatically and in a domain-independent way multiple faults. GDE was also able to diagnose minimal diagnoses from the set of minimal conflicts.

The basic consistency-based diagnosis approach, using only correct behavior models, caused the lack of discriminative power in some diagnosis results. In this chapter we have revised existing proposals, which depends on the amount of knowledge provided about fault models.

Finally, both seminal works used FOL models that worked well for static systems or discrete-valued domains. Actually, those works are still the basic references for static systems. However, there is no general extension of these works for continuous or dynamic systems. The next chapter will deal with this issue.

References

1. Brown, J., Burton, R., de Kleer, J.: Pedagogical and knowledge engineering techniques in SOPHIE I, II and III. In: Sleeman, D., Brown, J.S. (eds.) Intelligent Tutoring Systems (1982)
2. Console, L.: Model-based diagnosis. MONET Summer School, Lecture A3 (2000). http://www.qrg.northwestern.edu/resources/monet_summer_school_2000/monet-summer-school-announcement.htm
3. Davis, R.: Expert systems: where are we? And where do we go from here? Artif. Intell. **3**, 3–22 (1982)
4. Davis, R.: Diagnostic reasoning based on structure and behavior. In: Bobrow, D.G. (ed.) Qualitative Reasoning About Physical Systems, pp. 347–410. Elsevier, Amsterdam (1984). https://doi.org/10.1016/B978-0-444-87670-6.50010-8
5. De Kleer, J., Mackworth, A.K., Reiter, R.: Characterizing diagnoses and systems. Artif. Intell. **56**(2–3), 197–222 (1992)
6. De Kleer, J., Williams, B.C.: Diagnosing multiple faults. Artif. Intell. **32**(1), 97–130 (1987)
7. Dressler, O.: On-line diagnosis and monitoring of dynamic systems based on qualitative models and dependency-recording diagnosis engines. In: Proceedings of the Twelfth European Conference on Artificial Intelligence (ECAI-96), pp. 461–465 (1996)
8. Friedrich, G., Gottlob, G., Nejdl, W.: Physical impossibility instead of fault models. In: Proceeding of the American Asociation of Artificial Intelligence, AAAI, vol. 90, pp. 331–336 (1990)
9. de Kleer, J.: An assumption-based TMS. Artif. Intell. **28**, 127–162 (1986)
10. de Kleer, J.: Extending the ATMS. Artif. Intell. **28**, 163–196 (1986)
11. de Kleer, J.: Problem solving with the ATMS. Artif. Intell. **28**, 197–224 (1986)
12. de Kleer, J., Kurien, J.: Fundamentals of model-based diagnosis. In: Proceedings of the 5th IFAC Symposium on Fault Detection, Supervision and Safety of Technical Processes, pp. 25–36. Safeprocess, Elsevier, Washington, DC. http://dekleer.org/Publications/DXSafeProcessv7_files/frame.htm (2003)
13. de Kleer, J., Williams, B.: Diagnosing with behavioral modes. In: Proceedings of the Eleventh International Joint Conference on Artificial Intelligence (IJCAI-89). Detroit, Michigan, USA (1989)
14. Poole, D.: Explanation and prediction: an architecture for default and abductive reasoning. Comput. Intell. **5**(2), 97–110 (1989)
15. Raiman, O., de Kleer, J., Saraswat, V.A., Shirley, M.: Characterizing non-intermittent faults. In: AAAI, vol. 91, pp. 849–854 (1991)
16. Reiter, R.: A theory of diagnosis from first principles. Artif. Intell. **32**(1), 57–95 (1987)

17. Shortliffe, E.H.: MYCIN: Computer-Based Medical Consultations (1976)
18. Struss, P., Dressler, O.: Physical negation: introducing fault models into the general diagnostic engine. In: Proceedings of the Eleventh International Joint Conference on Artifical Intelligence (IJCAI-89), pp. 1318–1323. Detroit, Michigan, USA (1989)

Chapter 6
Model-Based Diagnosis by the Artificial Intelligence Community: Alternatives to GDE and Diagnosis of Dynamic Systems

Belarmino Pulido and Carlos J. Alonso-González

Abstract In this chapter, we analyze main problems found by the Artificial Intelligence approach to Model-based diagnosis (DX): the online computation of minimal conflicts by means of an ATMS-like dependency-recording engine, and the need for an extension to deal with dynamic systems diagnosis. To cope with the first problem we will see different options: from extensions to the original GDE to the description of several topological methods, explaining deeply one of them: the Possible Conflict (PC) approach, and its relation with minimal conflicts and ARRs. To cope with the second problem, dynamics, we review the whole set of proposals made to extend Reiter's formalization and the GDE to dynamic systems: from GDE extensions to the natural extension of topological methods to include temporal information. In this chapter we provide the complete extension of the PCs approach to diagnose dynamic systems, and their relation not only with ARRs, but with another FDI proposals for systems tracking: state-observers.

6.1 Computational Alternatives to GDE

Even though Reiter's theory [38] provides a formal framework for consistency-based diagnosis, and the GDE [26] provides a domain- independent tool to compute those diagnoses, it was obvious from the early works in the DX community that some extensions were needed to diagnose real-life systems.

One of its main drawbacks, as stated in the previous chapter, comes from using an Assumption-based Truth Maintenance System (or ATMS), or any other *Dependency-Recording Engine* (DRE), to record online dependencies about correct behavior in

B. Pulido (✉) · C. J. Alonso-González
Departamento de Informática, Universidad de Valladolid, Valladolid, Spain
e-mail: belar@infor.uva.es

C. J. Alonso-González
e-mail: calonso@infor.uva.es

T. Escobet et al. (eds.), *Fault Diagnosis of Dynamic Systems*,
https://doi.org/10.1007/978-3-030-17728-7_6

order to compute minimal conflicts [23–25]. Using any DRE introduces a computational overload in the system, because correctness labels need to be computed for every new set of observations. There is a similar problem regarding memory if we need to keep track of previous computations and we have real-valued domains.

Additionally, when the original First-Order Logic (FOL) models were extended to deal with dynamic systems, there were problems because there was no direct extension for Reiter's theory, and also due to the use of qualitative modeling for the simulation of dynamic systems behavior, as reported by several authors [16, 20].

As a consequence of these drawbacks, several proposals have been made in the past 30 years. To cope with the first issue, there are two main trends: to provide extensions to the GDE proposal or to avoid using a DRE online. Regarding the second issue, there are two main trends: first, those systems that rely upon an alternative to the simulation of qualitative models—known as state-based diagnosis, and based on works by Struss and co-workers [41]—, and second those approaches that do not use an online DRE. These later techniques are also known as topological methods, because they use information provided by the model/system topology (derived from connections between components, and represented by shared variables among models) in order to find the set of minimal conflicts avoiding to use an online DRE.

Several works from the DX community and the QR (Qualitative Reasoning) communities have used topological approaches to avoid online dependency-recording in Consistency-based Diagnosis. Most of them have proposed to use graphs to represent the topological or causal relation among system variables, and to explore those graphs to find the relations or dependencies among system variables [31, 33, 42]. Other authors proposed to use different kinds of models at the same time in a hierarchical way [11, 12].

The exploration of the system-related graph can be done online as in Ca∼En [42], TRANSCEND [31], and DYNAMIS [11], or can be done offline (this approach is also known as pre-compilation methods, because they compile off-line the set of potential dependencies, i.e., they perform off-line dependency-recording). Both Ca∼En and TRANSCEND combine two different kinds of models: quantitative models to perform behavior estimation and discrepancy detection, and graphs to explore the dependencies. Once a discrepancy is found between the observed and the estimated value for a variable—which is a nontrivial task as explained in the previous chapters—these techniques propose to explore the graphs backward from the node related to the discrepancy. Each edge in the graph represent a constraint that is used to estimate the variable, hence if the constraint does not hold, it is a potential source of discrepancy, and can be part of a conflict. The exploration finishes when it reaches observed variables. Such exploration is performed every time a discrepancy is found. Pre-compilation methods perform a similar search, but the search is carried only once, and it is done offline, before the diagnosis process begins.

While we will go back to these methods for dynamic systems diagnosis in the next section, we will focus on one pre-compilation method known as Possible Conflicts to explain offline dependency recording as an alternative to the GDE, because it provides a natural extension to diagnose dynamic systems, while retaining main concepts in Reiter's theory and the GDE.

6.1.1 Possible Conflicts

The main idea behind the Possible Conflict (PC) approach is that every interaction between system variables is constrained by topological restrictions in the models, i.e., propagation paths are limited. Moreover, the system observability is defined by a set of sensors, that in most systems remain fixed. With those two ideas in mind, PCs propose to study offline the system model looking for every potential path to the set of observations, which should be the source of potential discrepancies and therefore the base for conflicts. At the same time, since the main idea is to provide the same results as the GDE, propagation of information in the model must be done solving just one variable in one equation in each step, mimicking the GDE inference step. Consequently, the main goal using PCs is to search for sets of equations with minimal redundancy, because minimal conflicts in GDE represent a subsystem with minimal redundancy, as pointed out by Pulido and Alonso-González [36].

PCs computation [35, 36] is considered as a pre-compilation method from the DX community, that searches for subsets of equations having the minimal analytical redundancy necessary to perform fault diagnosis. PCs are computed offline based on the analysis of the available system model, and each PC is represented as a set of equations, together with its known and unknown variables.[1]

Looking for an easy comparison between PCs computation and minimal conflict calculation as in the GDE framework, and given that PCs exploit both the structural and behavioral information in the System Description, *SD*, provided as a set of First-Order Logic formula: $\{f_1, \ldots, f_n\}$, we need to provide a mechanism to transform *SD* into the new model $\mathcal{M}$:

Definition 6.1 *(Model)* The system model can be defined as $\mathcal{M}(\Sigma, \mathcal{U}, \mathcal{Y}, \mathcal{X}, \Theta)$, where $\Sigma = \bigcup_{c_i \in COMPS} r_i$ is a set of differential-algebraic equations (DAEs), defined over a collection of known and unknown variables: $\mathcal{U}$ is a set of inputs, $\mathcal{Y}$ a set of outputs, $\mathcal{X}$ a set of state and/or intermediate, i.e., unknown variables, and Θ is the set of model parameters.

- Without loss of generality, we assume that each component model in *SD* will have exactly one formula f_i to model its behavior.
- Each formula f_i in the behavioral model in *SD* provides exactly one equation or relation in $\mathcal{M} : r_i$.
- The explicit assumption of correct behavior for a component $c_i \in COMPS$ in *SD*, provided by the presence of the predicate $\neg AB(c_i)$, is stated in $\mathcal{M}$ as a label related to the relation r_i, obtained from f_i. Additionally, for each $\neg AB(c_i)$ we will include a parameter with the same name, c_i, to the set of parameters Θ: using the relation r_i to model the behavior of c_i implies that we assume that the predicate $AB(c_i)$ is false, i.e., component c_i is behaving properly.
- Faults related to connections between components are not allowed. The topological information provided by FOL formulas such as $input_a(c_i) = output_b(c_j)$,

[1] Software for PCs computation is available at: http://www.infor.uva.es/~belar/SoftwareCPCs/.

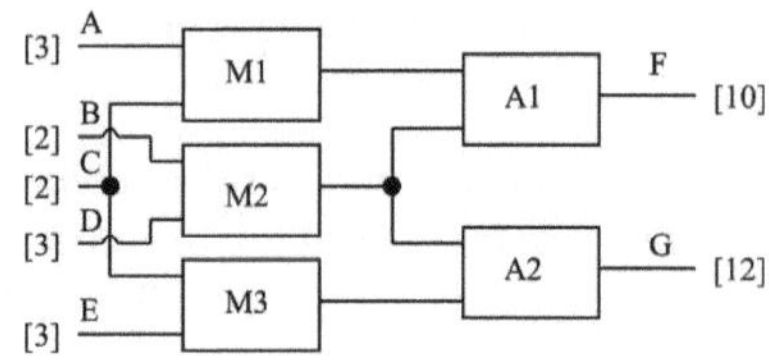

Fig. 6.1 The polybox system

is replaced by using the same variable x_k for both terminals: $input_a(c_i)$ and $output_b(c_j)$. If we want to model faults in a connection, we should introduce the connection as another component $c \in COMPS$.

- In *SD*, the observational model is provided as a collection of FOL formula obs_i. In $\mathcal{M}$ each observed variable, acting as input or output (represented by variables in $\mathcal{U} \cup \mathcal{Y}$), is related to exactly one variable in obs_i. For the sake of simplicity we do not consider sensor faults at this stage. If we want to model sensor faults, we should introduce the sensor as another component $c \in COMPS$.

Because we will later extend our system model definition to cope with continuous dynamic behavior, we define our model as a collection of Differential Algebraic Relations, DAEs. It is clear that for static systems we only need algebraic relations.

For instance, in Chaps. 2 and 5 the classical polybox example, that can be seen in Fig. 6.1, was introduced; the reader can find in Chap. 2 that the System Description $SD_{polybox}$ has an equivalent of System Model $\mathcal{M}_{polybox}$.

To reason about the potential propagation paths of information in the model, we can abstract the information in $\mathcal{M}$ even more. In fact, offline PCs computation requires three steps:

1. First step consists of abstracting the representation of the system as a hypergraph: $H_M = \{\mathcal{V}_\mathcal{M}, \mathcal{R}_\mathcal{M}\}$. The nodes of the hypergraph $\mathcal{V}_\mathcal{M}$ are system variables in $\mathcal{U} \cup \mathcal{Y} \cup \mathcal{X}$, while the set of hyperarcs $\mathcal{R}_\mathcal{M}$ represent an abstraction of the constraints between these variables in the equations $r_i \in \Sigma$. Thus, each equation $r_i \in \Sigma$ will provide one structural constraint: $(r_i : \mathcal{S}_i, \mathcal{X}_i)$, where $\mathcal{S}_i \subseteq \mathcal{U} \cup \mathcal{Y}$ accounts for the measured input or output variables, and $\mathcal{X}_i \subseteq \mathcal{X}$ accounts for the unknown (state or intermediate variables) in r_i.
   For instance in the polybox example, equation $M1$ will be abstracted to $(M1 : \{a, c\}, \{x\})$, and $M1$ is a part of $\Theta_{polybox}$.
   For the polybox example, $H_M = \{\mathcal{V}_\mathcal{M}, \mathcal{R}_\mathcal{M}\}$:

$$\mathcal{V}_\mathcal{M} = \{\{a, b, c, d, e\} \cup \{f, g\} \cup \{x, y, z\}\}.$$
$$\mathcal{R}_\mathcal{M} = \{(M1 : \{a, c\}, \{x\}), (M2 : \{b, d\}, \{y\}), (M3 : \{c, e\}, \{z\}), (A1 : \{f\}, \{x, y\}), (A2 : \{g\}, \{y, z\})\}.$$

2. Second step is to derive *Minimal Evaluation Chains* (MECs), which are minimal overconstrained subsystems: $H_{mec} \subseteq H_M$. These MECs are computed using depth-first search, starting from each variable in H_M, and looking for those subsystems with exactly n unknown variables and $n + 1$ equations.

The existence of a MEC is a necessary condition for analytical redundancy to exist. MECs have the potential to be solved using local propagation (solving one equation in one unknown) from the set of available measurements.
Each MEC will define a model for a subsystem: $MEC_m = (\Sigma_m, \mathcal{S}_m, \mathcal{Y}_m, \mathcal{X}_m, \Theta_m) \subseteq M$ where $\Sigma_m \subseteq \Sigma$, $\mathcal{S}_m \subseteq (\mathcal{U} \cup \mathcal{Y})$, $\mathcal{Y}_m \subseteq \mathcal{Y}$, $\mathcal{X}_m \subseteq \mathcal{X}$, $\Theta_m \subseteq \Theta$, $\mathcal{S}_m$ has at least one element, and $\mathcal{Y}_m$ has exactly one element.
For instance, in the polybox example equations $\{M1, M2, A1\}$ defines MEC_1 for variable f. We have three equations, with two unknown variables $\{x, y\}$, and five observed variables $\{a, c, b, d, f\}$. In a similar way MEC_2 have also three equations and two unknowns, and MEC_3 has four equations and three unknowns.
MECs for the polybox example are

$$
\begin{aligned}
H_{MEC_1} &= \{\{a, b, c, d, f\} \cup \{x, y\}\}, \{(M1 : \{a, c\}, \{x\}), (M2 : \{b, d\}, \{y\}), \\
&\quad (A1 : \{f\}, \{x, y\})\} \\
H_{MEC_2} &= \{\{\{b, c, d, e, g\}, \{y, z\}\}, \{(M2 : \{b, d\}, \{y\}), (M3 : \{c, e\}, \{z\}), \\
&\quad (A2 : \{g\}, \{y, z\})\}\} \\
H_{MEC_3} &= \{\{a, c, e, f, g\} \cup \{x, y, z\}\}, \{(M1 : \{a, c\}, \{x\}), (A1 : \{f\}, \{x, y\}), \\
&\quad (A2 : \{g\}, \{y, z\}), (M3 : \{c, e\}, \{z\})\}\}
\end{aligned}
$$

Figure 6.2 shows the hypergraph for the polybox *SD*, and the three sub-hypergraphs defined by the MECs.

3. The third step is to generate *Minimal Evaluation Models* (MEMs) by assigning causality[2] to the constraints in the set of MECs, if possible.
Each constraint $(r_i : \mathcal{S}_i, \mathcal{X}_i) \in \Sigma_m$ can be solved in one or more causal ways. Hence, for each r_i there are $(r_{i_j} : v_{i_j}, \mathcal{S}_{i_j} \cup \mathcal{X}_{i_j} \setminus \{v_{i_j}\})$, where $j \geq 1$. In this way, we assume that variable v_{i_j} appearing in r_{i_j} is computed given $\mathcal{S}_{i_j} \cup \mathcal{X}_{i_j} \setminus \{v_{i_j}\}$.
For instance, in the polybox example it is clear that $A1$ can be solved in three different ways:
$A1_1 : f := x + y$, but also
$A1_2 : x := f - y$, and
$A1_3 : y := f - x$.
The same happens for $A2$. $M1$, $M2$, and $M3$ can be solved also in three different ways, but we need to assume that there is no division by zero. For the sake of simplicity, we will assume that $M1$, $M2$, and $M3$ can only be solved in the causal form:

$$M_i : out_i := in1_i \times in2_i$$

Each MEM is a directed hypergraph derived from MEC_m, $H_{mem} = \{\mathcal{V}_{mem}, \mathcal{R}_{mem}\}$, where $\mathcal{V}_{mem} = \mathcal{U}_m \cup \mathcal{Y}_m \cup \mathcal{X}_m$ and $\mathcal{R}_{mem} \subset \mathcal{V}_{mem} \times \cdots \times \mathcal{V}_{mem}$. The leaves in the hypergraph are the variables in $\mathcal{S}_m$, the intermediate nodes are the variables in $\mathcal{X}_m$, and the root is $\hat{y}_m \in \mathcal{Y}_m$, which is the estimated variable and the potential source

[2] In this context, by causality assignment we mean every possible way one variable in one equation can be solved assuming the remaining variables are known.

of a discrepancy. Each hyperarc $r_{i_j} \in \mathcal{R}_{mem}$ in the MEM represents one causal assignment for each $(c_{i_j} : v_{i_j}, \mathcal{S}_{i_j} \cup \mathcal{X}_{i_j} \setminus v_{i_j}) \in \Sigma_m$. That is, the MEM specifies the order equations in a MEC should be locally solved starting from the leaves (input measurements $\mathcal{S}_m$) to generate the subsystem output, i.e., the root: $\hat{y}_m \in \mathcal{Y}_m$. There will be zero or more MEMs for each MEC. If there is zero, it means that the MEC cannot be solved using local propagation. Otherwise, there is at least one MEM that represents an ordering about how the set of equations can be solved to estimate the variable $\hat{y}_m$, and such ordering can be used to build an executable model. For the polybox example, each MEC has at least one MEM. Next, we show one of the MEMs for each MEC in the previous step:

$$\begin{aligned}
H_{MEM_1} &= \{\{a, b, c, d, f\} \cup \{x, y\}\}, \{(M1 : \{x\}, \{a, c\}), (M2 : \{y\}, \{b, d\}),\\
&\quad (A1 : \{f\}, \{x, y\})\}.\\
H_{MEM_2} &= \{\{\{b, c, d, e, g\}, \{y, z\}\}, \{(M2 : \{y\}, \{b, d\}), (M3 : \{z\}, \{c, e\}),\\
&\quad (A2 : \{g\}, \{y, z\})\}.\\
H_{MEM_3} &= \{\{a, c, e, f, g\} \cup \{x, y, z\}\}, \{(M1 : \{x\}, \{a, c\}), (M3 : \{z\}, \{c, e\}),\\
&\quad (A2 : \{y\}, \{g, z\}), (A1 : \{f\}, \{x, y\})\}.
\end{aligned}$$

Figure 6.3 shows three directed hypergraphs representing one possible MEM for each MEC in the polybox example. Each directed hyperarc represents one possible causal interpretation for one constraint in the original model, and must be interpreted as the variables in the tail of the hyperarc must be known before computing the head of the hyperarc.
There are additional MEMS for each MEC, but all of them share the same structural information.

Although these steps are done offline, the executable model provided by the MEM (in the form of simulation or state observer, as it will be explained later) can be used afterward online to estimate nominal behavior, thus allowing to track the system and computing a residual for each MEM, being the difference between the estimated and the measured variables: $\hat{y}_m - y_m$.

In consistency-based diagnosis [26, 38], a conflict arises given a discrepancy between observed and predicted values for a variable.[3] Hence, conflicts are the result of the fault detection stage. Algorithms used to compute PCs were designed to explore the system model and to find offline those subsystems capable to become minimal conflicts online. That notion leads to the definition of a Possible Conflict:

Definition 6.2 (*Possible Conflict (PC)*) The set of constraints in a MEC that give rise to at least one MEM.

Different works have demonstrated the relation between the DX and FDI approaches [13], between PCs/MECs and minimal conflicts [36], Analytical Redundancy Relations (ARRs), and other structural model decomposition methods for

[3] In FDI terminology, a conflict arises when the residual deviates significantly from zero.

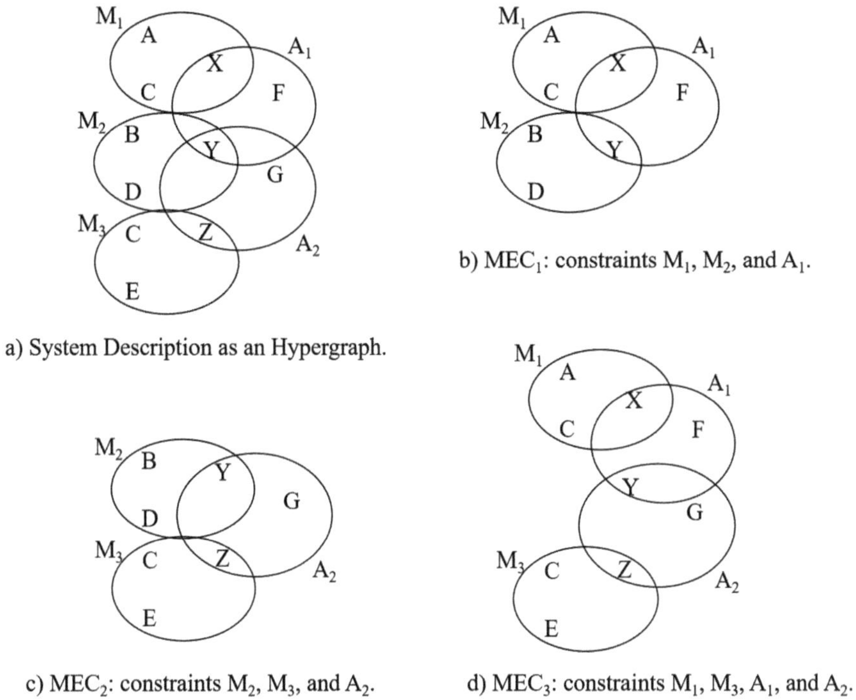

Fig. 6.2 **a** Polybox SD as an hypergraph. **b**, **c**, and **d** show three MECs for the polybox example

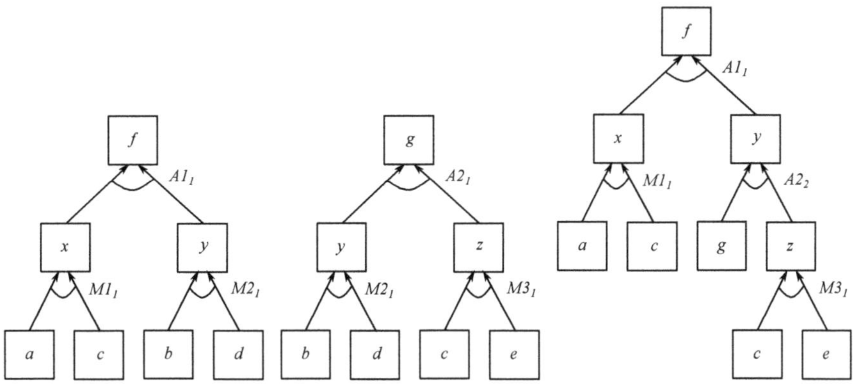

Fig. 6.3 Three MEMS, one for each MEC. All the leaves and the root for each MEC are observations: $\{a, b, c, d, f\}$ for MEC_1, $\{b, c, d, e, g\}$ for MEC_2, and $\{a, c, e, g, f\}$ for MEC_3. For the sake of readability, we show multiple times the same leaf variable

static systems [1], and linear dynamic systems [7]. Because we are interested in diagnosing continuous dynamic systems, we need to include additional knowledge to make explicit this similarity.

The software required to find the complete set of PCs in a system model can be found at https://www.infor.uva.es/~belar/SoftwareCPCs/PCC-setup.exe.

6.1.2 PCs and Their Relation to Minimal Conflicts

There is a strong relationship between PCs and minimal conflicts as computed by the GDE. It has been demonstrated by Pulido and Alonso-González [36] that assuming that both GDE and the algorithms to find the set of PCs use the same structural and behavioral models:

- Each PC is related to any strictly overdetermined set of equations/constraints capable to perform a double estimation for a variable in the set, known as the discrepancy node in the associated hypergraph.
- The algorithms used to compute PCs are able to find any strictly overdetermined set of constraints, known as MECs, in the system model. Additionally, for each MEC, the algorithms are able to find any possible way to solve the set of equations, known as a MEM, using local propagation alone, as it is done by the GDE.
- Since the GDE solves one equation in one unknown in its inference process, whenever GDE finds a minimal conflict, MC, related to one variable, there will be a MEC containing the same set of constraints as those of MC.
- If MC is a minimal conflict found by the GDE, and MC is related to a discrepancy variable $v \in V_{SD}$, then there will be a MEM with v as a discrepancy node and capable to obtain such discrepancy.
- Whenever there is a discrepancy in a Possible Conflict, PC, related to a discrepancy node v, GDE will always find a minimal conflict, MC, with the same set of constraints as in PC. But, unfortunately, the contrary does not always hold.
  The number of MEMs for each MEC is exponential in the average number of interpretations for each hyperarc in the MEC. If we do not use the complete set of MEMs for a MEC, we cannot guarantee that the set of PCs will find the complete set of minimal conflicts, as the GDE does, unless the system is made up of linear constraints. This is because usually only one MEM is selected, and only one executable model is built and used for fault detection; and it is known from algebra and linear analysis that a set of nonlinear equations can provide different solutions if solved by numerical methods.
  Nevertheless, it is still possible to claim that the set of PCs can provide in practice the same results as the GDE. Pulido and Alonso-González [36] introduced *the Equivalence Assumption*: if all the MEMs for a MEC provide the same solution, i.e., all the ways to solve the set of equations is unique, then we can guarantee that the set of PCs and the GDE provide the same set of minimal conflicts.

It must be also noticed that in the GDE each correctness assumption is related to exactly one equation, and each model is made up of one equation. In that sense, in GDE each minimal conflict represents exactly a minimal set of components. Whenever a component model is made up of several equations, this is not straightforwardly applicable. But it is just a problem of working at different levels of abstraction. It does not introduce any novelty in the reasoning process.

6.2 Consistency-Based Diagnosis of Continuous Systems

Dynamic systems diagnosis has been a major research topic since the beginning of the DX approach. Even before Reiter's theory and the GDE proposal, in 1984 the work by Hamscher and Davis [21] proposed to extend constraint suspension techniques to deal with dynamic models, and also pointed out the need to observe the system state variables to be able to provide deterministic results. This early work allowed to reason with different levels of temporal abstraction, together with multiple levels of model abstraction.

A major problem in the DX community to reason about system dynamics is the clear absence of a formal theory about temporal information modeling. This problem is clearly stated in the work by Brusoni et al. [9] that analyzed the different alternatives for modeling temporal behavior in the late 90s. Therefore, and in clear contrast to the FDI community, there is no general theory to extend Reiter's proposal to dynamic systems, but a collection of works that have established themselves as references for the community.

Additionally, in the late 80s and early 90s, there was a strong cooperation between the DX and the QR research communities. QR had additional ontologies and modeling paradigms [18], making even more difficult to find a general framework. Since early works by Williams [45, 46] reasoning about temporal evolution with qualitative dynamic models—introducing the notions of history and episodes—was also a major issue in QR research.

The main issue in modeling dynamics comes from the representation of the system state variables, and the influence in the current state from previous states (assuming that the diagnosis of combinatorial circuits, i.e., with no state, can be seen as a sequence of static diagnosis processes). In quantitative terms, as expressed in Chap. 2, we should be able to somehow represent the relation in discrete time as

$$x(t) = x(t-1) + \Delta t \frac{dx(t-1)}{dt}$$

as pointed out by Chantler et al. [10]. This kind of relation is usually named a *differential* constraint, while algebraic or functional relations in the model are usually named *instantaneous* constraints.

If we can compute $\frac{dx(t-1)}{dt}$ then we can perform integration and obtain $x(t)$ by simulation. If we can observe the variables $x(t)$, we can compute $\frac{dx(t)}{dt}$ and later estimate the next state $x(t+1) = x(t) + \Delta t \frac{dx(t)}{dt}$.

The QR community was also aware of this dependency (about how past and present values influence the new system states) in qualitative simulation, where it was named the "feedback effect".

6.2.1 Direct Extensions to GDE

There have been several proposals in the DX and QR communities to perform consistency-based diagnosis of dynamic systems, some of them directly extending GDE or Reiter's theory.

- SIDIA [20] extended GDE for dynamic systems. SIDIA had the ability to reason at different levels of abstraction to cope with complexity. To reason about time, SIDIA extended Williams' notion of episodes (single fault, single context) to Extended Episode. A special extension of episode propagation was developed, and evaluation of probes had to be adapted. According to Gückenbiehl and Schäfer-Richter [20]:

  > All of these extensions to GDE proved easily integratable without touching the basic mechanisms of prediction-based diagnosis. In particular, GDE's clear separation between diagnosis and behavior prediction allowed the straightforward integration of the new predictive engine. However, the work also showed that the application of GDE to complex and dynamic models requires further elaboration of some of its basic features. In particular, GDE's information theoretic probe selection procedure and the predictive engine should be supported by stronger heuristic knowledge.

- Early works in the community were led by Philippe Dague and Patrick Taillibert in France. DEDALE [14, 15] introduced the notion of time index for each state variable together with qualitative order of magnitude reasoning to diagnose dynamic systems in the steady state. Later, DEDALE was extended to reason about transient behavior using an array of numeric intervals in DIANA/CATS [14]. Numeric intervals were used to model uncertainty in both the model parameters and the measurements. Arrays of such estimations were used to represent reasoning in a given time interval. In DIANA, CATS were responsible for minimal conflict detection, as in the GDE, by means of a dependency recording similar to the ATMS [23, 24]. In CATS models, correct behavior is represented as functional relations between physical quantities. Finally, the notion of Temporal Band Sequences (TBS) were developed to reason about dynamic behavior and uncertainty in DOGS [28]. Using quantitative models, we have some problems to estimate derivatives. DOGS, which follows Reiter's approach to diagnosis, avoided such computation by representing the evolution of the system as a sequence of temporal bands. Each temporal band corresponds to a time interval where the behavior is bounded by two polynomial functions.

However, the problem of introducing dependencies with previous states persisted, because tracking dependencies from one state leads to the dependency on the same state in the previous time step. Different solutions were provided. Dressler et al. [17] proposal made explicit the difference between inter- and intra-state constraints in the models. They solved the issue by reasoning first at the intra-state and later using inter-state constraints to move to the next state. This two-step reasoning has been tackled in two different ways, leading to simulation-based diagnosis—which is the most used approach—and to state-based diagnosis [41]. The main idea in state-based diagnosis is that the next state depends only on the current state and the qualitative deviations observed in the state variables. This solution is only possible using a special kind of qualitative models.

Simulation-based approaches assume that finding an inter-state constraint (a relation between a state variable and its derivative) halts the propagation process in dependency recording, because there is no new dependency in terms of variables, but a dependency on previous values for the same set of variables.

6.2.2 Topological Methods

An alternative to the previous types of analysis is to use graphical models that resemble the topology of the system model including temporal arcs to reason about time. We next discuss the Ca∼En and TRANSCEND approaches.

6.2.2.1 Ca∼En

The Ca∼En [4] proposal was developed by the LAAS team led by Louise Travé-Massuyès, and was tested in several industrial scenarios, including the successful TIGER project [30], where it was applied to diagnose a gas turbine.

As an alternative to online dependency recording as performed by the GDE, Ca∼En can be classified as a topological method. It used the topology of the system model, i.e., the structural model provided by the equations, to find the set of conflicts.

Ca∼En formalism provided a system model at two different levels of abstraction: first, the *local level*, using causal models, where variables were related through links called *influences*, providing qualitative causality relationships. Second, the *global level*, was able to manipulate DAEs representing numerical relations among variables.

At the local level, causal influences can be modeled at five different levels, including temporal and instantaneous constraints. According to Travé-Massuyès and Pons [44]: temporal or differential influences are implemented as a predicate $I+(X, Y, c, K, Td, Tr)$ or $I-()$, where Y is the variable influenced by X, c is the activation condition, K is the gain of the influence (the ratio between the variation of amplitude for Y with respect to X), Td is the temporal delay (time taken by Y to react to changes in X), and Tr is the response time (time needed by Y to reach a new stationary state after being altered).

In Fig. 6.4, we can see both levels of abstraction for Ca∼En [42]: the global level, and the local level, together with examples of their respective operators.

Ca∼En has fault detection and isolation capabilities. Fault detection was able to detect discrepancies between predicted and observed variables by means of interval

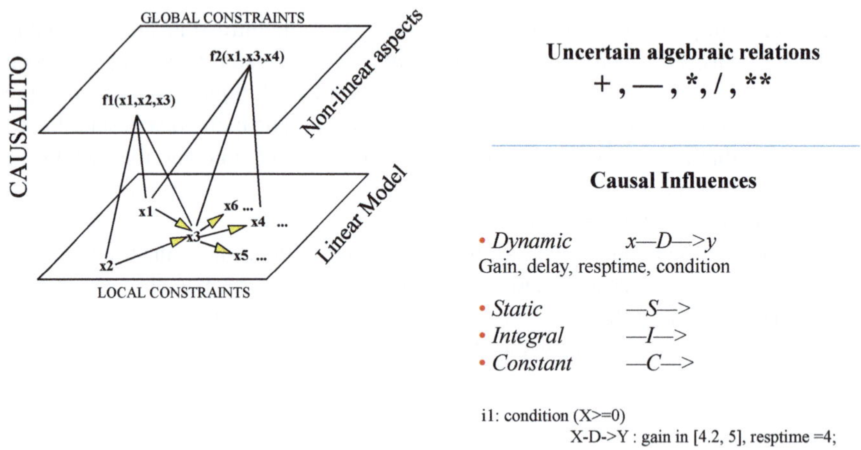

Fig. 6.4 Ca∼En modeling formalism (figure by courtesy of Louise Travé-Massuyès)

Fig. 6.5 Ca∼En causal model for the three-tank system in Chap. 2 (figure provided by Louise Travé-Massuyès and Renaud Pons), including the link between causal relations and global equations as provided in Chap. 2

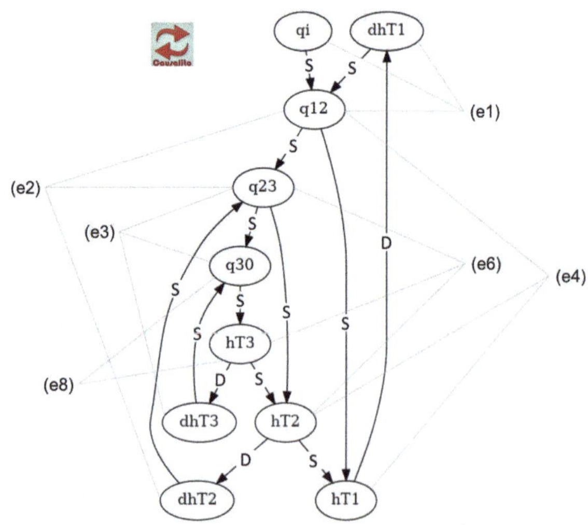

models, that used adaptive thresholds to deal with parameter and noise uncertainty. Fault isolation analyzed these discrepancies, propagating them through the temporal causal model, checking qualitative consistency and providing the set of fault candidates (the causal paths leading to inconsistencies provided the set of conflicts).

For instance, for the three-tank system in Chap. 2, the causal graph generated by Ca∼En using Causalito [44] is shown in Fig. 6.5.

Further information about the Ca∼En diagnosis system and its automatic modeling method can be found in Travé-Massuyès et al. [43].

6.2.2.2 TRANSCEND

TRANSCEND [31] proposes another alternative to online dependency recording using a topological approach. The approach departs from a Bond Graph (BG) model [8], which is a general, domain- independent modeling approach, as introduced in Sect. 1.3.3. From the BG system model, two different types of models are generated. First, a numerical, real-valued state-space model, which is used to design an Extended Kalman Filter (EKF) (for additional information on EKFs, see Chap. 4) for tracking the nominal system behavior with noisy measurements. Second, a Temporal Causal Graph (TCG) is obtained, whose nodes are the system variables, and whose labeled edges are related to parameters in the system model.

The TCG is a qualitative abstraction of the state-space model and it is used to reason about the consistency of the qualitative values in the model. TCGs, defined by Mosterman and Biswas [31] for diagnosis tasks, are an extended form of signal flow graphs for dynamic systems. They capture the causal and temporal relations between process parameters and the measurement variables in the system. TCGs can be directly derived from the bond graph model of the system [31] with effort and flow variables represented as vertices, and relations between the variables represented as directed edges. Formally, a TCG can be defined as [39]:

Definition 6.3 (*Temporal Causal Graph*) A TCG is a directed graph $< \mathcal{V}, \mathcal{L}, \mathcal{D} >$. $\mathcal{V} = \mathcal{E} \cup \mathcal{F}$, where $\mathcal{V}$ is a set of vertices, $\mathcal{E}$ is a set of effort variables, and $\mathcal{F}$ is a set of flow variables in the bond graph system model. $\mathcal{L}$ is the label set $\{=, 1, -1, p, p^{-1}, p \cdot dt, p^{-1} \cdot dt\}$ (p is a parameter name of the physical system model). The dt specifier indicates a temporal edge relation, which implies that a vertex affects the derivative of its successor vertex across the temporal edge. $\mathcal{D} \subseteq \mathcal{V} \times \mathcal{L} \times \mathcal{V}$ is a set of edges [32].

The edges in the TCG are labeled $(=, 1, -1, p, p^{-1}, p \cdot dt, p^{-1} \cdot dt)$ according to the temporal and the causal relation between the variables connected by the edge. $=$ denotes equality between the variables, 1 denotes direct proportionality, -1 denotes inverse proportionality, p and p^{-1} denote a parametric relation, and $p \cdot dt$ and $p^{-1} \cdot dt$ denote integration.

TCGs can be easily derived from the bond graph model of a system by a three-step process [31]:

1. Assign causality in the bond graph model using the Sequential Causality Assignment Procedure (SCAP) algorithm [22].
2. Generate a representation of the system as a directed graph encapsulating relations among power variables in the bond graph. Each bond provides two system variables (one effort and one flow), and the BG elements provide edges connecting the system variables.
3. Add temporal information and component parameters to the individual causal edges connecting the system variables to form the TCG.

Regarding components in the bond graph model, junctions and resistors define instantaneous magnitude relations. On the other hand, capacitors and inductors define

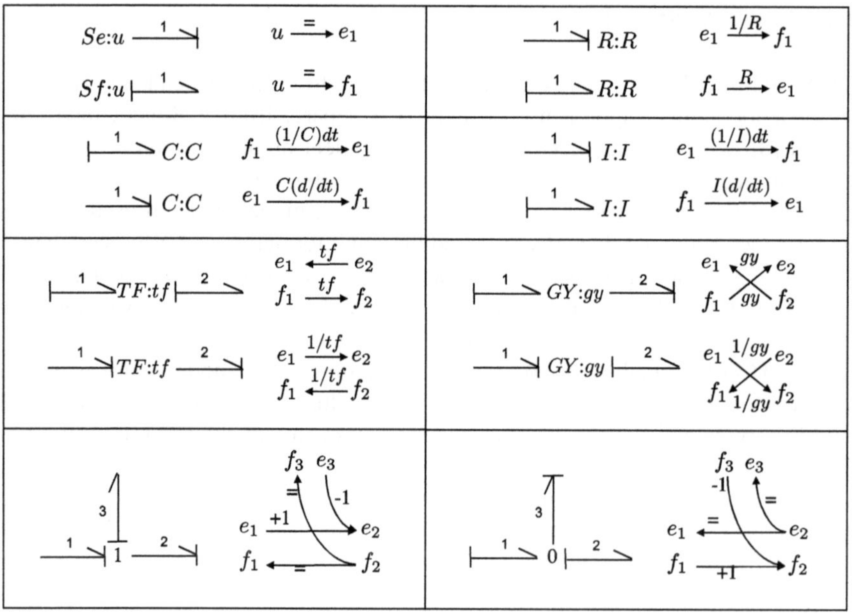

Fig. 6.6 Temporal causal graph transformations for each bond graph component

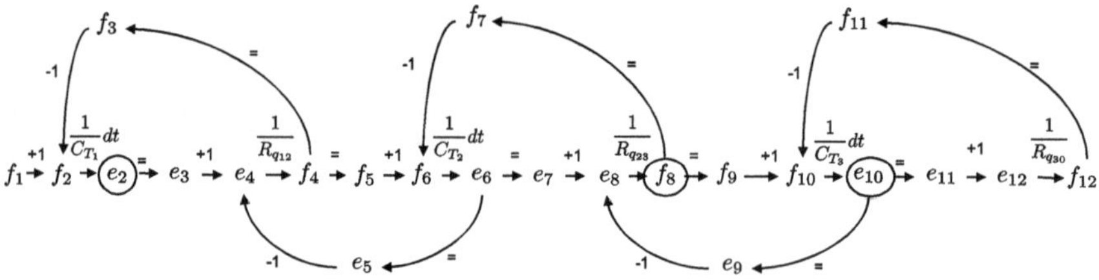

Fig. 6.7 Temporal causal graph for the three-tank system

both magnitude and temporal effects on causal edges. Figure 6.6 shows the TCG transformations for each one of the components in a bond graph model.

Figure 6.7 shows the Temporal Causal Graph of the bond graph model of the three-tank system (Fig. 2.9). In the example, bond 1 imposes flow from the source $Sf : q_i$ to the 0-junction, so in this case bond 1 only becomes flow f_1, and the edge $f_1 \xrightarrow{+1} f_2$ is drawn in the TCG. Bond 2 becomes effort e_2 and flow f_2, and, as integral causality is considered, the edge $f_2 \xrightarrow{dt/C_{T_1}} e_2$ is included in the TCG. At the 0-junction, we have $e_2 = e_3$ (in this case e_1 is not drawn because sources do not get any input), and $f_2 = f_1 - f_3$, hence we draw the edges $e_2 \xrightarrow{=} e_3$, $f_3 \xrightarrow{-1} f_2$ in the TCG. Following these ideas, the complete TCG of the three-tank system is obtained, where measured variables are marked with a circle.

It is clear that TCGs represent a particular case of directed hypergraphs, where all the edges in the TCG can be represented by either instantaneous ($=, 1, -1, p, p^{-1}$) or differential ($p \cdot dt, p^{-1} \cdot dt$) constraints.

The TCG approach to online fault detection and isolation is implemented as the TRANSCEND system [2, 31]. Unlike the ARR and PC schemes, the fault detection approach is implemented as an independent process with an observer and a statistical fault detector. The fault isolation scheme is implemented by a three-step process: (1) hypothesis generation after fault detection, (2) fault signature generation for all hypothesized faults, and (3) hypothesis refinement through progressive monitoring.

The residual generation scheme used in TRANSCEND models the deviations in measurements and the fault effects using a qualitative reasoning framework. Deviation in individual measurements is represented as a $\pm$ change in the magnitude and slope of a measured signal residual (i.e., $y(t) - \hat{y}(t)$), where a $\pm$ value indicates a change above (below) normal for a measurement residual (or a positive (negative) residual slope). A 0 implies no change in the measurement value (or a 0 (flat) residual slope). Note that for dynamic systems, both the measurement deviation and slope can change with time, i.e., a measurement representation can go from $(+, +)$ to $(+, 0)$. Fault parameter changes are also represented as $\pm$ values. For example, R^+_{Pipe} implies a fault where the pipe resistance increases above normal, and R^-_{Pipe} implies a fault where the pipe resistance decreases to below its nominal value.

Fault signatures, i.e., the effect of a fault on a measurement, are also expressed in the qualitative framework defined above. More formally, a qualitative fault signature is expressed as [29]:

Definition 6.4 (*Qualitative fault signature*) Given a fault, f, the time of fault occurrence, t_f, and a measurement, m, the qualitative fault signature, $QFS(f, m)$, of order k, is an ordered $(k + 1)$-tuple consisting of the predicted magnitude, and $1st$ through kth-order time-derivative effects of a residual signal of measurement m, at the point of failure of fault t_f, expressed as qualitative values: below normal $(-)$, normal (0), and above normal $(+)$. Typically, k is chosen to be the order of the system.

Given a set of possible fault parameters and a set of measurements associated with a system, the qualitative fault signatures can be derived from the TCG model of the system by forward propagation from the fault parameter along the edges of the TCG to the measurement nodes [31]. All deviation propagations start off as zeroth-order effects (magnitude changes). When an integrating edge in the TCG is traversed, the magnitude change becomes a first-order change, i.e., the first derivative of the affected quantity changes. Each subsequent traversal of an integral edge increases the order of the fault signature by 1. As an example, in the nonlinear three-tank system, the fault signature for $C^-_{T_1}$ on pressure measurement $h_{T_{3,obs}}$ is $(0, +)$, indicating a decrease in the tank capacity will cause no change in tank 3 pressure at the point of fault occurrence, and a gradual increase in pressure after the occurrence of the fault.

Table 6.1 lists the qualitative fault signatures derived for all possible single faults on the three measurements $h_{T_{1,obs}}(e_2)$, $h_{T_{3,obs}}(e_{10})$, and $q_{23,obs}(f_8)$ for the three-tank system. An additional symbol used in the fault signatures could be $*$, which captures indeterminate effects due to the qualitative framework used for signatures

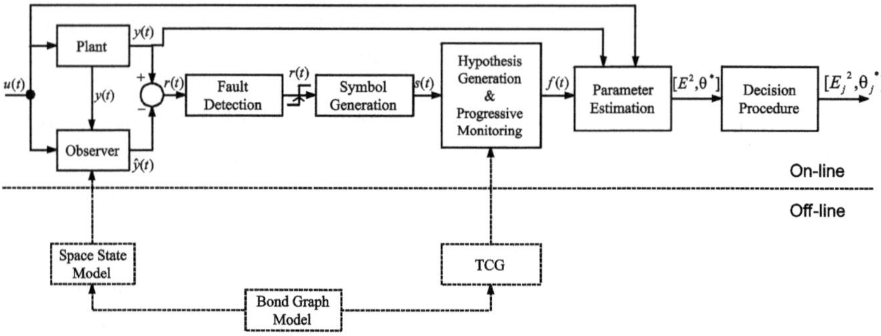

Fig. 6.8 Computational scheme for the TRANSCEND diagnosis approach, extended with fault identification capabilities (© 2012 IEEE. Reprinted, with permission, from [6])

as discussed above. An indeterminate effect occurs when there are at least two paths of the same order in the TCG that propagate + and − effects, so the resultant effect is unknown.

Summarizing, fault candidates whose signatures remain consistent with the observed measurements are considered fault candidates, $f(t)$ in Fig. 6.8, others are dropped.

TRANSCEND avoided some of the computational difficulties associated with numerical schemes, but its qualitative framework lacked discriminative power. The work by Manders et al. [29] extended the initial TRANSCEND approach with a parameter estimation method (using a hill climbing method for error minimization applied to the state-space model; for additional information see Chap. 4), which can be seen on the right side in Fig. 6.8. The parameter estimation step was employed on the reduced candidate set to provide the most likely fault candidate and the fault parameter value changes. A separate parameter estimator is initiated for each one of the hypothesized faults. The fault parameter estimate with the smallest least squares error is labeled the most likely candidate.

Later, the work by Bregon et al. [6] improved the fault identification stage by coupling TRANSCEND with a system decomposition model and using smaller models for the fault identification step.

Table 6.1 Qualitative fault signature matrix for the three-tank system

	$h_{T_{1,obs}}(e_2)$	$h_{T_{3,obs}}(e_{10})$	$q_{23,obs}(f_8)$
$C_{T_1}^-$	+−	0+	0+
$C_{T_2}^-$	0+	0+	+−
$C_{T_3}^-$	0+	+−	−+
$R_{q_{12}}^+$	0+	0−	0−
$R_{q_{23}}^+$	0+	0−	−+
$R_{q_{30}}^+$	0+	0+	0−

The combined qualitative/quantitative fault isolation scheme provides some advantages against traditional numeric schemes, but it also experiences computational problems when applied to online fault isolation/identification for complex, nonlinear systems.

Both Ca∼En and TRANSCEND combined online behavior estimation coupled with online dependency recording by traversing some kind of graphical structure representing the dependencies. Whenever a discrepancy is found, just tracking back the path from the discrepancy to observed variables provide an automatic way of computing conflicts online. But there is also the option to avoid online traversing of the graphical structures using offline dependency recording as in the Possible Conflict approach or as it will be explained in the next chapter. PCs can be easily extended to include dynamic system models, providing another alternative for Consistency-based Diagnosis for dynamic systems [35, 36].

6.3 Offline Dependency Recording for CBD of Dynamic Systems

As previously mentioned, the main difference between static and dynamic system models is the presence of equations or constraints linking state variables and their derivatives [16, 19, 37]. Following Dressler's proposal [16], we can make a difference of two kinds of constraints in the model: *differential* constraints, those used to model dynamic behavior—*inter-state constraints* in Dressler's terminology—and *instantaneous* constraints, those used to model static or instantaneous relations between system variables—*intra-state constraints* in Dressler's terminology. As shown by Frisk et al. [19], the fact that differential constraints are explicitly represented or not does not introduce genuinely new structural information. In the PCs approach, this representation is always explicit.

As we have seen in the previous section, differential constraints can be interpreted in two ways, depending on the selected causality assignment. In integral causality,

Fig. 6.9 Schematic representation of three-tank system

constraint is solved by computing $x(t)$ using $x(t-1)$ and $\dot{x}(t-1)$. In derivative causality, it is assumed that the derivative $\dot{x}(t)$ can be computed based on present and past samples, being x known. Integral causality usually implies using simulation, and it is the preferred approach in the DX field. Derivative causality is the preferred approach in the FDI approach. Both have been demonstrated to be equivalent for numerical models, assuming adequate sampling rates and precise approximations for derivative computation are available, and assuming initial conditions for simulation are known [10]. PCs can easily handle both types of causality, since they only represent a different causal assignment while building MEMs [37].

Special attention must be paid to loops in the MEM (set of equations that must be solved simultaneously). Loops containing differential constraints in integral causality are allowed, because under integral causality the time indices are different to both sides of the differential constraint [16, 17]. It is generally accepted that loops containing differential constraints in derivative causality cannot be solved [3]. We will now illustrate the three steps required to compute PCs in the continuous three-tank case study in Fig. 6.9 that was introduced in Chap. 2.

As mentioned above, we need to introduce in the system model $\mathcal{M}$ three differential constraints: $\{e13, e14, e15\}$ to make explicit the relation between the three state variables $\mathcal{X} = \{h_{T_1}, h_{T_2}, h_{T_3}\}$, and their derivatives $\{\dot{h}_{T_1}, \dot{h}_{T_2}, \dot{h}_{T_3}\}$.

1. To obtain the hypergraph related to the system model, we just need to add one hyperarc for each DAE in the model. The set Σ_{3Ts} provides the set of hyperarcs $\mathcal{R}_{3Ts}$:

$$
\begin{gathered}
(e1 : \{\dot{h}_{T_1}, q_i, q_{12}, q_{f_1}\}) \\
(e2 : \{\dot{h}_{T_2}, q_{12}, q_{23}, q_{f_2}\}) \\
(e3 : \{\dot{h}_{T_3}, q_{23}, q_{30}, q_{f_3}\}) \\
(e4 : \{q_{12}, h_{T_1}, h_{T_2}\}) \\
(e5 : \{q_{f_1}, h_{T_1}\}) \\
(e6 : \{q_{23}, h_{T_2}, h_{T_3}\}) \\
(e7 : \{q_{f_2}, h_{T_2}\}) \\
(e8 : \{q_{30}, h_{T_3}\}) \\
(e9 : \{q_{f_3}, h_{T_3}\}) \\
(e13 : \{h_{T_1}, \dot{h}_{T_1}\}) \\
(e14 : \{h_{T_2}, \dot{h}_{T_2}\}) \\
(e15 : \{h_{T_3}, \dot{h}_{T_3}\}).
\end{gathered}
$$

The reader should notice that in the behavioral model we only have non-observed variables in the constraints. Hence, the set of nodes in the hypergraph is obtained from variables in $\mathcal{X}$ alone. In Chap. 2, we decided not to include sensor faults. Hence, it makes no sense to make a difference between an observed variable (in $\mathcal{U} \cup \mathcal{Y}$) and its corresponding internal variables in $\mathcal{X}$. For that reason we have decided against including equations $\{e10, e11, e12\}$ that represent the observation model.[4] Instead, to make clear which are the observed variables, we have

---

[4]Equations $\{e_{10}, e_{11}, e_{12}\}$ define the observational model, just linking each internal variable in $\mathcal{X}$ with its sensor in $\mathcal{U} = \{q_i\}$ or $\mathcal{Y} = \{h_{T1,obs}, h_{T3,obs}, q_{23,obs}\}$.

Table 6.2 Set of minimal evaluable chains for the three-tank system. Only three of them have a Minimal Evaluable Model and identify a Possible Conflict

MEC$_m$	$\mathcal{R}_{MEC_m}$	$\mathcal{V}_{MEC_m} = \mathcal{Y}_m \cup \mathcal{X}_m \cup \mathcal{S}_m$
MEC_1	$\{e3, e8, e9, e15\}$	$\{h^*_{T_3}\} \cup \{\dot{h}_{T_3}, q_{30}, q_{f_3}\} \cup \{q^*_{23}\}$
MEC_2	$\{e2, e4, e6, e7, e14\}$	$\{q^*_{23}\} \cup \{h_{T_2}, \dot{h}_{T_2}, q_{12}, q_{f_2}\} \cup \{h^*_{T_3}, h^*_{T_1}\}$
MEC_3	$\{e1, e2, e4, e5, e6, e7, e13, e14\}$	$\{h^*_{T_1}\} \cup \{\dot{h}_{T_1}, q_{12}, h_{T_2}, \dot{h}_{T_2}, q_{f_1}, q_{f_2}\} \cup \{q^*_i, q^*_{23}\}$
MEC_4	$\{e1, e4, e5, e6, e13\}$	$\{\} \cup \{h_{T_2}, \dot{h}_{T_1}, q_{12}, q_{23}, q_{f_1}\} \cup \{q^*_i, h^*_{T_1}, h^*_{T_3}\}$
MEC_5	$\{e1, e2, e5, e6, e7, e13, e14\}$	$\{\} \cup \{\dot{h}_{T_1}, \dot{h}_{T_2}, q_{12}, h_{T_2}, q_{f_1}, q_{f_2}\} \cup \{h^*_{T_1}, q^*_{23}, h^*_{T_3}\}$

added an asterisk to each one of them: $\{q^*_i, q^*_{23}, h^*_{T_1}, h^*_{T_3}\}$.
The hypergraph: $H_{3Ts} = \{\mathcal{V}_{3Ts}, \mathcal{R}_{3Ts}\}$, being $\mathcal{V}_{3Ts} = \{q^*_i, h^*_{T_1}, h_{T_2}, h^*_{T_3}, \dot{h}_{T_1}, \dot{h}_{T_2}, \dot{h}_{T_3}, q_{12}, q^*_{23}, q_{30}, q_{f_1}, q_{f_2}, q_{f_3}\}$, and $\mathcal{R}_{3Ts} = \{e1, e2, e3, e4, e5, e6, e7, e8, e9, e13, e14, e15\}$.

2. In this work, we have used the algorithms described in Pulido and Alonso-González [36] to derive the set of MECs: $H_{MEC_m} = (\mathcal{V}_{MEC_m}, \mathcal{R}_{MEC_m})$, for $m = 1$ to 5 that can be seen in Table 6.2.
In the case of MEC_4 and MEC_5, we cannot determine if there is a designated output $\mathcal{Y}_m$. This fact will be made clear next.
3. In the three-tank system, we have provided for each constraint exactly one potential causal assignment, and it is the one shown in equations eq_1 to eq_9. Regarding differential constraints, we have allowed only integral causality, as shown in equations eq_{13} to eq_{15}.
Using those causality assignments, we have obtained exactly one MEM for MEC_1, MEC_2, and MEC_3; MEC_4 and MEC_5 have no global valid causal assignment, hence they are not Possible Conflicts. On the other hand, each MEC with one MEM defines a Possible Conflict: $\{PC_1, PC_2, PC_3\}$, capable to estimate variables $\{h^*_{T3}, q^*_{23}, h^*_{T1}\}$, respectively. The graphical description of these MEMs for the PCs can be seen in Figs. 6.10, 6.11, and 6.12.
The reader should notice that leakage flows $\{q_{f_1}, q_{f_2}, q_{f_3}\}$ have zero value, when we use only models of correct behavior (i.e., for fault detection). They could be used in case that predictive fault models are used.

Summarizing, each MEM for a PC represents how to build an executable model to monitor the behavior of the subsystem defined by the PC, just traversing the MEM structure from leaves to the root (which is $\hat{y}_m$). Such executable model can be implemented as a simulation model or as a state observer [5, 37]. However, building such model for complex nonlinear systems is not a trivial task.

The software required to perform a simple dynamic system diagnosis given a system model specified as a set of DAEs and the set of PCs, named *DBCwithPCs*, can be found at https://www.infor.uva.es/~belar/SoftwareCPCs/.

Fig. 6.10 MEM_1 for PC_1. Estimation of h_{T_3} via $\dot{h}_{T_3}$, given q^*_{23}, q_{30}, and q_{f_3}. Both q_{30} and q_{f_3} are obtained from the previous value of h_{T_3}

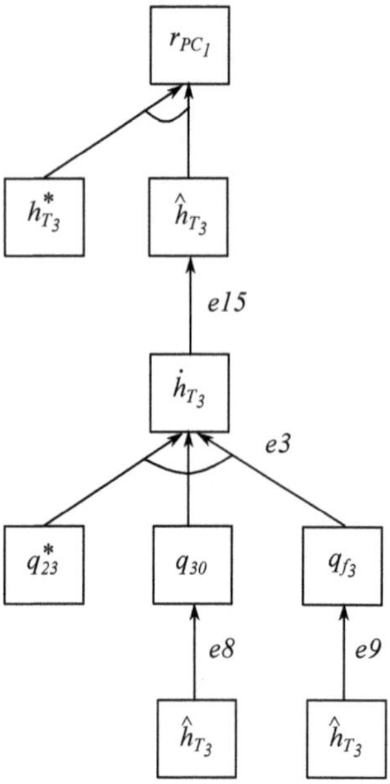

Fig. 6.11 MEM_2 for PC_2. Estimation of q_{23} given h_{T_2} and $h^*_{T_3}$. h_{T_2} is obtained via $\dot{h}_{T_2}$, obtained from q_{12}, q_{f_2}, and the previous value of q_{23}. q_{12} is computed given state variables $h^*_{T_1}$ and the estimation of h_{T_2}. q_{f_2} only needs the value of h_{T_2}

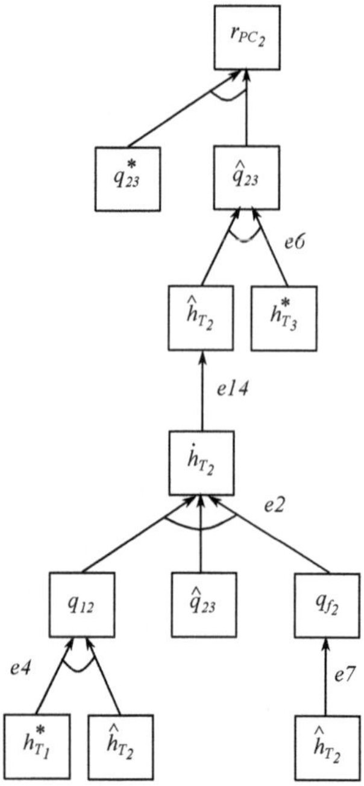

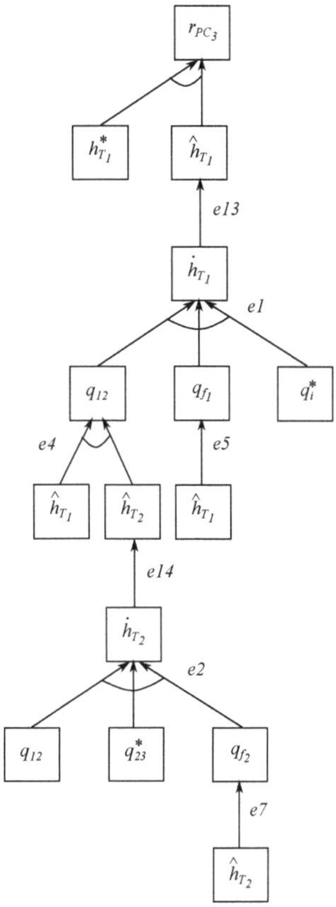

Fig. 6.12 *MEM_3* for PC_3. Estimation of h_{T_1} via $\dot{h}_{T_1}$, given q_{12}, q_{f_1} and q^*_i; q_{12} is computed given state variables $h^*_{T_1}$ and the estimation of h_{T_2} via $\dot{h}_{T_2}$, which is computed given q_{12}, q_{f_2}, and q^*_{23}. q_{f_1} and q_{f_2} are computed as in the two previous PCs

6.3.1 From DAEs to State-Space Representation in PCs

Our PCs are made up of a collection of DAEs: $\Sigma_m \in \Sigma$, and when we include differential constraints they contain state variables. Can we obtain a state-space representation equivalent to the DAE form? We will explain that in this subsection.

Let us consider a dynamic nonlinear system as described in Chap. 2:

$$\dot{\mathbf{x}}(t) = \mathbf{f}(\mathbf{x}(t), \mathbf{u}(t), \mathbf{v}(t)).$$
$$\mathbf{y}(t) = \mathbf{g}(\mathbf{x}(t), \mathbf{u}(t), \mathbf{w}(t)).$$

As mentioned in Chap. 2, it can be implemented as a simulation or a state observer model as follows [34], depending on the values of k:

$$\dot{\hat{\mathbf{x}}}(t) = \mathbf{f}(\hat{\mathbf{x}}(t), \mathbf{u}(t), \mathbf{v}(t)) + \mathbf{k}(\mathbf{y}(t) - \hat{\mathbf{y}}(t)).$$
$$\hat{\mathbf{y}}(t) = \mathbf{g}(\hat{\mathbf{x}}(t), \mathbf{u}(t), \mathbf{w}(t)).$$

A MEM is built in such a way that guarantees structural observability, according to Staroswiecki [40]:

Definition 6.5 (*Structural observability*) A system in integral causality is structurally observable if there exists a matching[5] that is complete on the unknown variables.

As mentioned before in this chapter we have imposed integral causality on Possible Conflicts calculation. Moreover, each PC defines a strictly overdetermined set of constraints, i.e., a MEC. By construction each MEC defines a bipartite graph, made up of the constraints and the unknown variables, with a complete matching in the unknowns.[6] Consequently, once differential equations are introduced, each MEM defines a subsystem that is structurally observable [5] if it contains at least one state variable and its differential constraint.

Theorem 6.1 *Given a PC, if its MEM describes a dynamic subsystem, this subsystem is structurally observable.*

Given a Minimal Evaluation Model, MEM_i, $H_{mem_i} = \{\mathcal{V}_{mem_i}, \mathcal{R}_{mem_i}\}$, its state-space representation can be expressed in a general way by the tuple $(\mathbf{x_i}, \mathbf{u_i}, y_i, \mathbf{f_i}, \mathbf{g_i})$, where

- $\mathbf{x_i} =< x_{i_1}, x_{i_2}, \ldots, x_{i_n} >$ is the state vector of the system described by MEM_i,
- $\mathbf{u_i} =< u_{i_1}, u_{i_2}, \ldots, u_{i_m} >$ is the input vector of the system described by MEM_i,
- y_i is the output of the system described by MEM_i
- $\mathbf{f_i}$ is the state function of the system described by MEM_i,
- $\mathbf{g_i}$ is the output function of the system described by MEM_i.

with $\{x_{i_1}, x_{i_2}, \ldots, x_{i_n}, u_{i_1}, u_{i_2}, \ldots, u_{i_m}, y_i\} \subseteq \mathcal{V}_{mem_i}$, $\{\dot{x}_{i_1}, \dot{x}_{i_2}, \ldots, \dot{x}_{i_n}\} \subseteq \mathcal{V}_{mem_i}$; y_i is the discrepancy node of MEM_i, and $\{u_{i_1}, u_{i_2}, \ldots, u_{i_m}, y_i\}$ are the only observed variables of MEM_i and

$$\dot{\mathbf{x}}_\mathbf{i}(t) = \mathbf{f_i}(\mathbf{x_i}(t), \mathbf{u_i}(t)). \tag{6.1}$$

$$y_i(t) = \mathbf{g_i}(\mathbf{x_i}(t), \mathbf{u_i}(t)). \tag{6.2}$$

When MEM_i has no algebraic loops, each component of the state function, f_{i_j}, is obtained from H_{mem_i} by the following procedure:

- Build subgraph $H_{f_{i_j}} \subseteq H_{mem_i}$, traversing downward H_{mem_i} from the occurrence of $\dot{x}_j$ to the first occurrence of either an input or a state variable.
- Compose the equations that label the arcs of $H_{f_{i_j}}$ from inputs and state variables to $\dot{x}_j$.

[5]In the structural approach defined by Staroswiecki, the structural model defines a bipartite graph for the constraints and the unknown variables in the system. The matching in the definition refers to a matching in that bipartite graph. The reader can find additional information in those structural issues in the work by Blanke et al. [3] and in Chap. 3 in this book.

[6]This is easy to understand because MECs are built adding one constraint in each step to determine an unknown variable, mimicking how the CBD computational paradigm, GDE, works online [26, 36].

Similarly, the output function g_i is obtained from H_{mem_i} by the procedure:

- Built subgraph $H_{g_i} \subseteq H_{mem_i}$, traversing H_{mem_i} from the output y_i to the first occurrence of either an input or state variable.
- Compose the equations that label the arcs of H_{g_i} from inputs and state variables to y_i.

By construction, the causal matching in each MEM guarantees that $\forall i, j, H_{f_{i_j}}$ and H_{g_i} can be built for any MEM_i and state variable x_{i_j} in MEM_i. Consequently, when MEM_i has no algebraic loops, the analytical expression of f_{i_j} and g_i can always be obtained from MEM_i.

If MEM_i has an algebraic loop, we cannot obtain the analytical expression of f_{i_j} and/or g_i. Nevertheless, we still can build $H_{f_{i_j}}$ and H_{g_i}, which provide the structural description of f_{i_j} and g_i, respectively. From these structural descriptions, an external solver can compute the value of all the unknown variables in state-space formulation.

In the three-tank system example, we can obtain a state-space representation for each PC following these instructions and just by composing functions. Looking at Figs. 6.10, 6.11, 6.12, we only need to traverse the graph and compose the functions until we just use observed variables. For instance, to estimate $h^*_{T_3}$ in PC_1 in Fig. 6.10 it is straightforward from $e15$, once we obtain h_{T_3} from $\dot{h}_{T_3}$ once we perform integration. In this case, this would provide the observational model for PC_1: $g_1()$, which assumes that the sensor model is the identity operator for the sake of simplicity. This assumption will be made in all the remaining observational models.

$$h_{T_3} = g_1(\hat{h}_{T_3}) = \hat{h}_{T_3}. \tag{6.3}$$

It should be clear that this estimation of h_{T_3} is completely local to PC_1 observational model, and this variable must be different from the one in the main model. For that reason we do not add an additional sub-index to the estimated variables $\hat{h}_{T_i}$.

To obtain $f_1()$ for the transitional model for PC_1 we just need to compose $e3$ with $e8$ and $e9$. Again, all the internal variables in the equations belong to PC_1.

$$\begin{aligned} \dot{\hat{h}}_{T_3} &= e3(q^*_{23}, q_{30}, q_{f_3}) \\ &= e3(q^*_{23}, e8(\hat{h}_{T_3}), e9(\hat{h}_{T_3})) \\ &= f_1(\hat{h}_{T_3}, q^*_{23}). \end{aligned} \tag{6.4}$$

Following the same reasoning we can obtain the state-space representation for Possible Conflicts PC_2, building the transitional $f_2()$, and the observational model $g_2()$ to estimate q_{23}:

$$\begin{aligned} \dot{\hat{h}}_{T_2} &= e2(q_{12}, q_{f_2}, q^*_{23}) \\ &= e2(e4(h^*_{T_1}, \hat{h}_{T_2}), e7(\hat{h}_{T_2}), q^*_{23}) \\ &= f_2(\hat{h}_{T_2}, h^*_{T_1}, q^*_{23}). \end{aligned} \tag{6.5}$$

Table 6.3 Relations between equations and fault parameters

Equation	$Leakage_{T_1}$	$Leakage_{T_2}$	$Leakage_{T_3}$	$Stuck_{q12}$	$Stuck_{q23}$	$Stuck_{q30}$
$e1$						
$e2$						
$e3$						
$e4$				1		
$e5$	1					
$e6$					1	
$e7$		1				
$e8$						1
$e9$			1			

$$
\begin{aligned}
q_{23}^* &= e6(\hat{h}_{T_2}, h_{T_3}^*) \\
&= g_2(\hat{h}_{T_2}, h_{T_3}^*).
\end{aligned}
\tag{6.6}
$$

and in a similar way we build $f_3()$ and $g_3()$ for PC_3:

$$
\begin{aligned}
\dot{\hat{h}}_{T_1} &= e1(q_{12}, q_i^*, q_{f_1}) \\
&= e1(e4(\hat{h}_{T_1}, \hat{h}_{T_2}), q_i^*, e5(\hat{h}_{T_1})) \\
&= f_{3_1}(\hat{h}_{T_1}, \hat{h}_{T_2}, q_i^*).
\end{aligned}
\tag{6.7}
$$

$$
\begin{aligned}
\dot{\hat{h}}_{T_2} &= e2(q_{12}, q_{23}^*, q_{f_2}) \\
&= e2(e4(\hat{h}_{T_1}, \hat{h}_{T_2}), q_{23}^*, e7(\hat{h}_{T_2})) \\
&= f_{3_2}(\hat{h}_{T_1}, \hat{h}_{T_2}, q_{23}^*).
\end{aligned}
\tag{6.8}
$$

$$
h_{T_1} = g_3(\hat{h}_{T_1}) = \hat{h}_{T_1}
\tag{6.9}
$$

As stated before, using the state-space representation or the collection of DAEs, we can build an executable model for each PC and use them to estimate the evolution of the system variables.

Moreover, again the PCs provide offline the relation between the estimations and the potential faults in the system. The reader should notice that we just need to annotate the relation between the faulty parameters $\{Leakage_{T_i}, Stuck_{q_j}\}$, with $i = 1..3$, and $j = 12, 23, 30$, and the equations ($\{e5, e7, e9\}$ for leakages, and $\{e4, e6, e8\}$ for stuck pipes), using Table 6.3.

The relation between faults and PCs can be summarized using the notation of the Theoretical Fault Signature Matrix as explained in Chap. 3. Table 6.4 provides the Theoretical Fault Signature Matrix the three-tank example: PC_1 will be related to a leakage in tank T_3 or a blockage at the T_3 output; PC_2 will be related to leakages

Table 6.4 Relations between Possible Conflicts and fault parameters

Equation	$Leakage_{T_1}$	$Leakage_{T_2}$	$Leakage_{T_3}$	$Stuck_{q12}$	$Stuck_{q23}$	$Stuck_{q30}$
PC_1	0	0	1	0	0	1
PC_2	0	1	1	1	1	0
PC_3	1	1	0	1	0	0

in tanks T_2 or T_3, or blockages between tanks T_1 and T_2, or between tanks T_2 and T_3; finally, PC_3 will be related to leakages in tanks T_1 or T_2, or a blockage between tanks T_1 and T_2.

The simple way to perform consistency-based diagnosis is to simulate all the *PC* models online and to generate the set of fault candidates as minimal hitting sets of those Possible Conflicts that are confirmed as real conflicts. Whenever a new conflict is confirmed, we recalculate the set of minimal fault candidates in an incremental way.

In that sense, once we have built an executable model for each *PC* we can start tracking system behavior online. Each PC_i will provide a signal, related to the discrepancy node, that can be used to generate a residual r_i. Let's consider the following diagnosis scenarios:

- A fault by a leakage in tank T_1 will be detected only by PC_3. After a period of time, the residual r_3 will be activated and following the consistency-based approach we will determine that the set of fault candidates consistent with the fault are[7]: $\{[Leakage_{T_1}], [Leakage_{T_2}], [Stuck_{q_{12}}]\}$. All of them are singleton. Without additional information about fault modes, we cannot discriminate further.
- A fault due to a pipe stuck closed between tanks T_1 and T_2 will activate two residuals: r_2 and r_3. Let's assume that r_2 is activated first. Immediately after the detection the set of potential fault candidates is made up of the set of singletons: $\{[Leakage_{T_2}], [Leakage_{T_3}], [Stuck_{q_{12}}], [Stuck_{q_{23}}]\}$. Afterward, when r_3 is activated, we compute the set of fault candidates as the minimal hitting set of the faults from the two PCs: $\{[Leakage_{T_2}], [Stuck_{q_{12}}], [Leakage_{T_3}, Leakage_{T_1}], [Stuck_{q_{23}}, Leakage_{T_1}]\}$. We have two singleton and two double faults. The real fault is included among these candidates. Without additional information, we cannot further discriminate.

6.4 Conclusions

Initial drawbacks for consistency-based diagnosis of dynamics systems have been addressed in this chapter. First, online dependency recording for non-discrete domains can be solved using topological methods, similar to those used in the FDI

[7]Following the convention in [26], fault candidates are presented in brackets.

field and known as structural methods (as described in Chap. 3). Such similarity is deeply discussed in the next chapter.

We have presented several topological methods, and we have introduced the Possible Conflict approach [36], that avoids using a Dependency-Recording Engine by computing offline those sets of equations or subsystems capable to become minimal conflicts once the observations are known. We have summarized how they can be computed, paving the way to the comparison [1] with minimal conflicts, conflicts, ARRs [3], and MSOs [27].

Topological methods can be easily extended to diagnose also dynamic systems, by merely including explicitly differential constraints (which model the relation between state variables and their derivatives) in the model. Because there is no general extension to Reiter's characterization for dynamic systems, we have presented three topological methods: Ca$\sim$En [4, 42], TRANSCEND [31], and Possible Conflicts [35]. We have illustrated the PCs approach for the three-tank system case study, providing a complete CBD approach for dynamic systems, and further integration with FDI techniques such as state observers [5, 7].

The main conclusion is that topological approaches such as PCs allow to perform CBD of dynamic systems without DREs, and they ease the integration with FDI techniques. This issue will become clear in the next chapter.

Acknowledgements The authors would like to thank the valuable contributions of Anibal Bregon, Teresa Escobet, Louise Travè-Massuyés, and Renaud Pons for the material related to Ca$\sim$En, TRANSCEND, and the BRIDGE references.

References

1. Armengol, J., Bregon, A., Escobet, T., Gelso, E., Krysander, M., Nyberg, M., Olive, X., Pulido, B., Travé-Massuyès, L.: Minimal structurally overdetermined sets for residual generation: a comparison of alternative approaches. In: Proceedings of the IFAC-Safeprocess 2009. Barcelona, Spain (2009)
2. Biswas, G., Simon, G., Mahadevan, N., Nararsimhan, S., Ramirez, J., Karsai, G.: A robust method for hybrid diagnosis of complex systems. In: Proceeding of the 5th IFAC Symposium on Fault Detection. Supervision and Safety of Technical Processes, SAFEPROCESS03, pp. 1125–1130. Washington D.C, USA (2003)
3. Blanke, M., Kinnaert, M., Lunze, J., Staroswiecki, M.: Diagnosis and Fault-Tolerant Control. Springer, Berlin (2006)
4. Bousson, K., Travé-Massuyès, L., Zimmer, L.: Causal model-based diagnosis of dynamic systems. LAAS Report No 94231 (1994)
5. Bregon, A., Alonso-González, C.J., Pulido, B.: Integration of simulation and state observers for online fault detection of nonlinear continuous systems. IEEE Trans. Syst. Man Cybern.: Syst. **44**(12), 1553–1568 (2014)
6. Bregon, A., Biswas, G., Pulido, B.: A decomposition method for nonlinear parameter estimation in TRANSCEND. IEEE Trans. Syst. Man Cybern.-Part A: Syst. Humans **42**(3), 751–763 (2012)
7. Bregon, A., Biswas, G., Pulido, B., Alonso-González, C., Khorasgani, H.: A common framework for compilation techniques applied to diagnosis of linear dynamic systems. IEEE Trans. Syst. Man. Cybern.: Syst. **44**(7), 863–876 (2014). https://doi.org/10.1109/TSMC.2013.2284577

8. Broenink, J.: Introduction to Physical Systems Modelling with Bond Graphs. SiE Whitebook on Simulation Methodologies (1999)
9. Brusoni, V., Console, L., Terenziani, P., Dupré, D.T.: A spectrum of definitions for temporal model-based diagnosis. Artif. Intell. **102**(1), 39–79 (1998)
10. Chantler, M., Daus S. Vikatos, T., Coghill, G.: The use of quantitative dynamic models and dependency recording engines. In: Proceedings of the Seventh International Workshop on Principles of Diagnosis (DX-96), pp. 59–68. Val Morin, Quebec, Canada (1996)
11. Chittaro, L., Guida, G., Tasso, C., Toppano, E.: Functional and teleological knowledge in the multimodeling approach for reasoning about physical systems: a case study in diagnosis. IEEE Trans. Syst. Man Cybern. **23**(6), 1718–1751 (1993)
12. Chittaro, L., Ranon, R.: Hierarchical diagnosis guided by observations. In: Proceedings of the 17th International Joint Conference on Artificial Intelligence-Volume 1, pp. 573–578. Morgan Kaufmann Publishers Inc., Burlington (2001)
13. Cordier, M.O., Dague, P., Lévy, F., Montmain, J., Staroswiecki, M., Travé-Massuyès, L.: Conflicts versus analytical redundancy relations: a comparative analysis of the model based diagnosis approach from the Artificial Intelligence and Automatic Control perspectives. IEEE Trans. Syst. Man Cybern. Part B (Cybern.) **34**(5), 2163–2177 (2004)
14. Dague, P.: Model-based diagnosis of analog electronic circuits. Ann. Math. Artif. Intell. **11**(1), 439–492 (1994). https://doi.org/10.1007/BF01530755
15. Dague, P., Deves, P., Luciani, P., Taillibert, P.: Analog systems diagnosis. In: Proceedings of European Conference on Artificial Intelligence, ECAI, pp. 173–178 (1990)
16. Dressler, O.: On-line diagnosis and monitoring of dynamic systems based on qualitative models and dependency-recording diagnosis engines. In: Proceedings of the Twelfth European Conference on Artificial Intelligence (ECAI-96), pp. 461–465 (1996)
17. Dressler, O., Freitag, H.: Prediction sharing across time and contexts. In: Proceedings of the Eighth International Workshop on Qualitative Reasoning about Physical Systems (QR-94), pp. 63–68. Nara, Japan (1994)
18. Forbus, K.: Qualitative reasoning about physical processes. In: Proceedings of the Seventh International Joint Conference on Artificial Intelligence (IJCAI-81). Vancouver, Canada (1981)
19. Frisk, E., Dustegor, D., Krysander, M., Cocquempot, V.: Improving fault isolability properties by structural analysis of faulty behavior models: application to the DAMADICS benchmark problem. In: Proceedings of SAFEPROCESS-2003. Washington, DC, USA (2003)
20. Guckenbiehl, T., Schäfer-Richter, G.: SIDIA: extending prediction based diagnosis to dynamic models. In: Expert Systems in Engineering Principles and Applications, pp. 53–68. Springer (1990)
21. Hamscher, W., Davis, R.: Diagnosing circuits with state: An inherently underconstrained problem. In: Proceedings of AAAI, pp. 142–147 (1984)
22. Karnopp, D., Rosenberg, R., Margolis, D.: System Dynamics, A Unified Approach, 3rd edn. Wiley, New York (2000)
23. de Kleer, J.: An assumption-based TMS. Artif. Intell. **28**, 127–162 (1986)
24. de Kleer, J.: Extending the ATMS. Artif. Intell. **28**, 163–196 (1986)
25. de Kleer, J.: Problem solving with the ATMS. Artif. Intell. **28**, 197–224 (1986)
26. de Kleer, J., Williams, B.: Diagnosing multiple faults. Artif. Intell. **32**(1), 97–130 (1987)
27. Krysander, M., Åslund, J., Nyberg, M.: An efficient algorithm for finding minimal overconstrained subsystems for model-based diagnosis. IEEE Trans. Syst. Man Cybern.-Part A: Syst. Humans **38**(1), 197–206 (2008)
28. Loiez, E., Taillibert, P.: Polynomial temporal band sequences for analog diagnosis. In: Proceedings of the Fifteenth International Joint Conference on Artificial Intelligence (IJCAI-97), pp. 474–479. Nagoya, Japan (1997)
29. Manders, E.J., Narasimhan, S., Biswas, G., Mosterman, P.J.: A combined qualitative/quantitative approach for fault isolation in continuous dynamic systems. SafeProcess, Budapest, Hungary (2000)
30. Milne, R., Travé-Massuyès, L.: Model based aspects of the TIGER gas turbine condition monitoring system. IFAC Proc. Vol. **30**(18), 405–410 (1997)

31. Mosterman, P., Biswas, G.: Diagnosis of continuous valued systems in transient operating regions. IEEE Trans. Syst. Man Cybern. **29**(6), 554–565 (1999)
32. Narasimhan, S., Biswas, G.: Model-based diagnosis of hybrid systems. IEEE Trans. Syst. Man Cybern. Part A **37**(3), 348–361 (2007). https://doi.org/10.1109/TSMCA.2007.893487
33. Oyeleye, O., Finch, F., Kramer, M.: Qualitative modeling and fault diagnosis of dynamic processes by MIDAS. Chem. Eng. Commun. **96**, 205–228 (1990)
34. Puig, V., Quevedo, J., Escobet, T., Meseguer, J.: Toward a better integration of passive robust interval-based FDI algorithms. In: Proceedings of the 6th IFAC Symposium on Fault Detection, Supervision and Safety of Technical Processes, SAFEPROCESS06. Beijing, China (2006)
35. Pulido, B., Alonso, C., Acebes, F.: Consistency-based diagnosis of dynamic systems using quantitative models and off-line dependency-recording. In: 12th International Workshop on Principles of Diagnosis (DX-01), pp. 175–182. Sansicario, Italy (2001)
36. Pulido, B., Alonso-González, C.: Possible Conflicts: a compilation technique for Consistency-based diagnosis. IEEE Trans. Syst. Man Cybern. Part B: Cybern. **34**(5), 2192–2206 (2004)
37. Pulido, B., Bregon, A., Alonso-González, C.: Analyzing the influence of differential constraints in possible conflicts and ARR computation. In: Meseguer, P., Mandow, L., Gasca, R.M. (eds.) Current Topics in Artficial Intelligence, CAEPIA 2009 Selected Papers. Springer, Berlin (2010)
38. Reiter, R.: A theory of diagnosis from first principles. Artif. Intell. **32**(1), 57–95 (1987)
39. Roychoudhury, I., Biswas, G., Koutsoukos, X.: Designing distributed diagnosers for complex continuous systems. IEEE Trans. Autom. Sci. Eng. **6**(2), 277–290 (2009)
40. Staroswiecki, M.: A structural view of fault-tolerant estimation. In: Proceedings of IMechE. Part I: J. Systems and Control Engineering, vol. 221 (2007)
41. Struss, P.: Fundamentals of model-based diagnosis of dynamic systems. In: Proceedings of the Fifteenth International Joint Conference on Artifical Intelligence (IJCAI-97), pp. 480–485. Nagoya, Japan (1997)
42. Travé-Massuyès, L.: Model based thoughts Ca-En and TIGER then and now. In: Bundy, A., Wilson, S. (eds.) ROB MILNE: A Tribute to a Pioneering AI Scientist, Entrepreneur and Mountaineer, pp. 1–28. IOS Press (2006). ISBN 1-58603-639-4
43. Travé-Massuyès, L., Escobet, T., Pons, R., Tornil, S.: The Ca-En diagnosis system and its automatic modelling method. Computación y Sistemas **5**(2), 128–143 (2001)
44. Travé-Massuyès, L., Pons, R.: Causal ordering for multiple mode systems. In: Proceedings of the Eleventh International Workshop on Qualitative Reasoning-QR97, pp. 203–214 (1997)
45. Williams, B.: Doing time: putting qualitative reasoning on firmer ground. In: Proceedings of the Fifth AAAI National Conference on Artificial Intelligence (AAAI-86), pp. 105–112. Philadelphia, Pennsylvania, USA (1986)
46. Williams, B.: Temporal qualitative analysis: explaining how physical systems work. In: Readings in Qualitative Reasoning about Physical Systems, pp. 133–177. Morgan-Kaufmann Pub., San Mateo (1990). Revised version of Qualitative analysis of MOS circuits, in Artificial Intelligence, vol. 24, No. 1-3 pp. 281–346 (1984)

Chapter 7
BRIDGE: Matching Model-Based Diagnosis from FDI and DX Perspectives

Louise Travé-Massuyès and Teresa Escobet

7.1 Introduction

As introduced in Chap. 1, the goal of diagnosis is to identify the possible causes of explaining a set of observed symptoms. The following three tasks are commonly identified:

- *fault detection* discriminates normal system states from faulty states,
- *fault isolation*, also called *fault localization*, points at the faulty components, and
- *fault identification*, identifies the type of fault.

Several scientific communities have addressed these tasks and contributed with a large spectrum of methods, in particular, the signal processing, control, and Artificial Intelligence (AI) communities. Diagnosis spreads from the signal acquisition level up to levels in which relevant abstractions are used to interpret the available signals qualitatively.

Qualitative interpretations of the signals exist in terms of symbols or events. To do that, discrete formalisms borrowed from AI find a natural link with continuous models from the control community. Different facets of diagnosis investigated in the control or the AI fields have been discussed in the literature. References [24–26] provide three interesting surveys of the different approaches that exist in these fields. These two communities have their own model-based diagnosis track:

- the FDI (Fault Detection and Isolation) track, whose foundations are based on engineering disciplines, such as control theory and statistical decision-making,

---

L. Travé-Massuyès (✉)
LAAS-CNRS, University of Toulouse, CNRS, Toulouse, France
e-mail: louise@laas.fr

T. Escobet
Reseach Center for Supervision, Safety and Automatic Control (CS2AC), Universitat Politècnica de Catalunya, Terrassa, Spain
e-mail: teresa.escobet@upc.edu

T. Escobet et al. (eds.), *Fault Diagnosis of Dynamic Systems*,
https://doi.org/10.1007/978-3-030-17728-7_7

- the DX (Diagnosis) track, whose foundations are derived from the fields of logic, combinatorial optimization, search, and complexity analyses.

There have been a growing number of researchers in both communities who have tried to understand and bridge FDI and DX approaches to build better, more robust, and effective diagnostic systems. In particular, the concepts and results of the FDI and DX tracks have been put in correspondence and the lessons learned from this comparative analysis pointed out.

The FDI and DX streams both consider the diagnosis problem from a *system* point of view, which results in significant overlaps. Even the names of the two tracks are the same: *Model-Based Diagnosis* (MBD). This chapter presents and examines this "bridge".

The diagnosis principles are the same, although each community has developed its own concepts and methods, guided by different modeling paradigms and solvers. FDI relies on analytical models, linear algebra, and nonlinear system theory, whereas DX takes its foundations in logic.

In the 2000s, although the common goals were quite clear, the underlying concepts and the procedures of the two fields would remain mutually obscure. There were more and more researchers who tried to understand and synergistically integrate methods from the two tracks to propose more efficient diagnostic solutions.

The chapter is organized as follows. After the introduction section, Sect. 7.2 first presents a brief overview of the approaches proposed by the FDI and DX model-based diagnosis communities. Although quite commonplace, this overview is necessary because it provides the basic concepts and principles that form the foundations of our comparative analysis. It is followed by the comparison of the mathematical objects used as input of the diagnosis procedures. Section 7.3 then establishes the correspondences of concepts on both sides and compares the techniques used by the two communities. Interestingly, the results obtained by the two approaches are shown to be the same under some assumptions that are exhibited. Finally, Sect. 7.4 illustrates the DX-FDI MBD bridge with the classical example of the polybox.

7.2 DX and FDI MBD Approaches

Both the FDI and DX communities have a Model-Based Diagnosis (MBD) track which can be put in correspondence. After a brief reminder of the concepts of the two tracks, this section compares the models used on each side and sets the assumptions that are adopted to favor the comparative analysis.

7.2.1 Brief Overview of the FDI Approach

This section briefly summarizes the concepts presented in Chap. 4. The FDI community generally deals with dynamic systems represented by behavioral models that relate system inputs $u \in \mathcal{U}$ and outputs $y \in \mathcal{Y}$, gathered in the set of measurable variables $\mathcal{Z}$, and system internal states defining the set of unknown variables $\mathcal{X}$. The variables $z \in \mathcal{Z}$ and $x \in \mathcal{X}$ are functions of time. The typical model can be formulated in the temporal domain, then known as a *state-space model*:

$$\begin{aligned} BM &: d\mathbf{x}/dt = f(\mathbf{x}(t), \mathbf{u}(t), \theta), \\ OM &: \mathbf{y}(t) = g(\mathbf{x}(t), \mathbf{u}(t), \theta), \end{aligned} \tag{7.1}$$

where $\mathbf{x}(t) \in \Re^{n_x}$ is the state vector, $\mathbf{u}(t) \in \Re^{n_u}$ is the input vector, and $\mathbf{y}(t) \in \Re^{n_y}$ is the output vector. Then, $\mathbf{z}(t) = (\mathbf{u}(t), \mathbf{y}(t))^T$. $\theta \in \Re^{n_\theta}$ is a constant parameter vector. The components of f and g are real functions over $\Re$. BM is the behavioral model and OM is the observation model. The whole system model is noted $SM(\mathbf{z}, \mathbf{x})$, like in [13], and assumed noise-free. The equations of $SM(\mathbf{z}, \mathbf{x})$ may be associated to components but this information is not represented explicitly. The models can also be formulated in the frequency domain, for instance, in the form of transfer functions in the linear case.

Model (7.1) can be illustrated by the example of two coupled water tanks T_t and T_b. T_t is the top tank and its output fills the bottom tank T_b:

$$BM : \begin{cases} \dot{x}_1(t) = a_1\, u(t) - a_2 \sqrt{x_1(t)}, \\ \dot{x}_2(t) = a_3 \sqrt{x_1(t)} - a_4 \sqrt{x_2(t)}, \end{cases} \tag{7.2}$$

$$OM : \begin{cases} y_1(t) = \sqrt{x_1(t)}, \\ y_2(t) = \sqrt{x_2(t)}, \end{cases} \tag{7.3}$$

where $a_i, i = 1, \ldots, 4$, $a_i \neq 0$, are model parameters. $\mathbf{x}(t) = (x_1(t), x_2(t))^T$ represents the state vector and corresponds to the level in each tank, $u(t) \not\equiv 0$ is the input vector, and $\mathbf{y}(t) = (y_1(t), y_2(t))^T$ is the output vector. Measurable variables are given by the vector $\mathbf{z}(t) = (u(t), y_1(t), y_2(t))^T$.

The books [4, 8, 10, 16] provide excellent surveys, which cite the original papers that the reader is encouraged to consult. The paper [26] also provides a quite comprehensive survey. The equivalence between observers, parity space, and parameter estimation has been proved in the linear case [15].

The concept central to FDI methods is the concept of *residual*, and one of the main problems is to *generate residuals*. Residual generators are defined in Definition 3.15 of Chap. 3 and this definition is recalled below.

Definition 7.1 (*Residual generator for $SM(z, x)$*) A system that takes as input a subset of measured variables $\tilde{\mathcal{Z}} \subseteq \mathbf{z}$ and generates as output a scalar r is a residual generator for the model $SM(\mathbf{z}, \mathbf{x})$ if for all z consistent with $SM(\mathbf{z}, \mathbf{x})$, $\lim_{t \to +\infty} r(t) = 0$.

Let us consider the model $SM(\mathbf{z}, \mathbf{x})$ given by (7.1), then $SM(\mathbf{z}, \mathbf{x})$ is said to be consistent with an observed trajectory $\mathbf{z}$, or simply *consistent with measurements* $\mathbf{z}$, if there exists a trajectory of $\mathbf{x}$ such that the equations of $SM(\mathbf{z}, \mathbf{x})$ are satisfied. The residuals tend to zero as t tends to infinity when the system model is consistent with measurements; otherwise, some residuals may be different from zero. In practice, noises may affect the residuals that are never exactly *zero*. Indeed, the noise-free assumption adopted for (7.1) is never met. Statistical tests that account for the statistical characteristics of noise [3, 8] are used to evaluate the residuals as a Boolean value 0 or 1. The residuals are often optimized to be robust to disturbances [18] and to take into account uncertainties [1]. The reader can refer to Chaps. 4 and 10 for more details about how to deal with uncertainty, using decoupling methods or with interval methods, respectively.

Among the three standard FDI approaches, the bridge comparison is carried out based on the so-called parity space approach [5]. In this approach, residuals are generated from relations that are inferred from the system model by eliminating unknown variables, i.e., state variables. These relations, called *Analytical Redundancy Relations* (ARR), are determined offline. ARRs are constraints that only involve measured input and output variables and their derivatives. For linear systems, ARRs are obtained eliminating unknown state variables by linear projection on a particular space, called the *parity pace* [5]. An extension to nonlinear systems is proposed in [21]. On the other hand, structural analysis [2, 22] is an interesting approach because it allows one to obtain, for linear or nonlinear systems, the just overdeterminated sets of equations from which ARRs can be derived (see Chap. 3).

Every ARR can be put in the form $r(t) = 0$, where $r(t)$ is the *residual.*

For the two-tank system (7.2)–(7.3), the two following residuals can be obtained from the Rosenfeld–Groebner algorithm as explained in [7]:

$$\begin{aligned} r_1(t) &= -a_1\,u + a_2\,y_1 + 2\,y_1\,\dot{y}_1, \\ r_2(t) &= a_4\,y_2 - a_3\,y_1 + 2\,y_2\,\dot{y}_2. \end{aligned} \tag{7.4}$$

If the behavior of the system satisfies the model constraints, then the residuals are zero because the ARRs are satisfied. Otherwise, some of them may be different from zero when the corresponding ARRs are violated. Given a set of n residuals, a *theoretical fault signature* $FS_j = [s_{1j}, s_{2j}, \ldots, s_{nj}]$ given by the Boolean evaluation of each residual is associated with each fault F_j. Note that F_j may be a simple or multiple faults. The *signature matrix* is then defined as follows.

Definition 7.2 (*Signature Matrix*) Given a set of n ARRs, the signature matrix FS associated with a set of n_f faults $\mathcal{F} = [F_1, F_2, \ldots, F_{n_f}]$ is the matrix that crosses corresponding residuals as rows and faults as columns, and whose columns are given by the theoretical signatures of the faults, i.e., $FS = [FS_1, FS_2, \ldots, FS_{n_f}]$.

In the two-tank example system (7.2)–(7.3), let us consider a fault f_{T_b} on the bottom tank, i.e., a leak l_b, and a fault f_{T_t} on the top tank, i.e., a leak l_t. The leak l_b of the bottom tank impacts residual $r_1(t)$, whereas the leak l_t of the top tank impacts

Table 7.1 Fault signature of the two-tank system

	f_{T_b}	f_{T_t}	$f_{T_b T_t}$
$r_1(t)$	1	1	1
$r_2(t)$	0	1	1

both residuals $r_1(t)$ and $r_2(t)$. Multiple faults composed of the two leaks obviously affect the two residuals as well. The signature matrix is, hence, given in Table 7.1.

Diagnosis is achieved by matching the observed signature, i.e., the Boolean residual values obtained from the actual measurements to one of the theoretical signatures of the n_f faults.

7.2.2 Brief Overview of the DX Logical Diagnosis Theory

In the model-based logical diagnosis theory of DX as proposed by [12, 19] and presented in depth in Chap. 5, the description of the system is driven by components and relies, in its original version, on first-order logic. A system is given by a tuple $(SD, COMPS, OBS)$, where

- SD is the *system description* in the form of a set of first-order logic formulas with equality,
- $COMPS$ represents the *set of components* of the system given by a finite set of constants, and
- OBS is a set of first-order formulas, which represent the *observations*.

SD uses the specific predicate AB, meaning *abnormal*. Applied to a component c of $COMPS$, $\neg AB(c)$ means that c is normal and $AB(c)$ that c is faulty. For instance, the model of a two inputs adder would be given by

$$\neg AB(x) \wedge ADD(x) \Rightarrow out(x) := in_1(x) + in_2(x). \tag{7.5}$$

Definition 7.3 (*Diagnosis*) A diagnosis for the system $(SD, COMPS, OBS)$ is a set $\Delta \subseteq COMPS$ such that $SD \cup OBS \cup \{AB(c) \mid c \in \Delta\} \cup \{\neg AB(c) \mid c \in OBS - \Delta\}$ is satisfiable.

The above definition means that the assumption stating that the components of Δ are faulty and all the others are normal is consistent with the observations OBS and the system description SD. A diagnosis, hence, consists in the assignment of a mode, normal or faulty,[1] to each component of the system, which is consistent with the model and the observations.

Definition 7.4 (*Minimal diagnosis*) A *minimal diagnosis* is a diagnosis Δ such that $\forall \Delta' \subset \Delta$, Δ' is not a diagnosis.

[1]This framework has been extended to fault models in [12].

To obtain the set of diagnoses, it is usual to proceed in two steps, basing the first step on the concept of conflict introduced in [19] and later extended in [12]. The original definition, which we call *R-conflict*, i.e., conflict in the sense of Reiter, is the following.

Definition 7.5 (*R-conflict and minimal R-conflict*) An *R-conflict* is a set $\mathcal{C} \subseteq COMPS$ such that the assumption that all the components of $\mathcal{C}$ are normal is not consistent with SD and OBS. A *minimal R-conflict* is an R-conflict that does not contain any other conflict.

The set of diagnoses can be generated from the set of conflicts. Reference [19] proved that minimal diagnoses are given by the *hitting sets*[2] of the set of minimal R-conflicts. An algorithm based on the construction of a tree, known as the HS-tree, was originally proposed in [19].

The parsimony principle indicates that preference should be given to minimal diagnoses. Another reason why minimal diagnoses are important is because in many cases, they characterize the whole set of diagnoses. In other words, all the supersets of minimal diagnoses are diagnoses. The conditions for this to be true were provided in [12] by extending the definition of an R-conflict to a disjunction of AB-literals, $AB(c)$ or $\neg AB(c)$, containing no complementary pair, entailed by $SD \cup OBS$. Then, a *positive conflict* is a conflict for which all of its literals are positive and one can identify a positive conflict with an R-conflict [19] as defined above.

Diagnoses are characterized by minimal diagnoses if and only if all minimal conflicts are positive [12]. Unfortunately, only sufficient conditions exist on the syntactic form of SD and OBS. One of those is that the clause form of $SD \cup OBS$ only contains positive AB-literals. This is verified, for instance, if all sentences of SD are of the same form as (7.5), which means that only necessary conditions of correct behavior are expressed.

7.2.3 Modeling Comparison

Given the frameworks defined in Sects. 7.2.1 and 7.2.2, it is important to compare the models that are used on both sides to represent the knowledge useful to diagnosis. Three dimensions can be analyzed:

- the system representation,
- observations, and
- faults.

[2]The hitting sets of a collection of sets are given by the sets that intersect every set of the collection.

7.2.3.1 System Representation

The modeling paradigm of FDI does not make explicit use of the concept of component, and the system model SM is composed of the behavior model BM and the observation model OM of the non-faulty system. The behavioral model (7.1) describes the system as a whole. On the contrary, the DX approach models every component independently, and specifies the structure of the system, i.e., how the different components are connected. Another important difference is that the assumption of correct behavior is represented explicitly in SD, thanks to the predicate AB. If $\mathcal{F}$ is a formula describing the normal behavior of a component, SM only contains $\mathcal{F}$, whereas SD contains the formula $\neg AB(c) \Rightarrow \mathcal{F}$.

The comparison of the two approaches is only possible if the models on both sides represent the same system and the observations/measurements capture the same reality. This is formalized by the *System Representation Equivalence* (SRE) property introduced in [6], which requires that SM is obtained from SD by setting to *false* all the occurrences of the predicate AB. It is also assumed that the same observation language is used, i.e., OBS is a conjunction of equality relations, which assign a value to every measured variable.

7.2.3.2 Observations

In DX, the set of observations expresses as a set of first-order formulas. It is, hence, possible to express disjunctions of observations, which provides a powerful language. However, very often, only conjunctions of atomic formulas are used. In FDI, the observations are always conjunctions of equalities assigning a real value and/or possibly an interval value to an observed variable. In the following, to favor the comparative analysis, we do assume that we have the same observation language. In both FDI and DX approaches, OBS is identical and made up of relations $OBS = \mathbf{z}$.

7.2.3.3 Faults

DX adopts a component-centered modeling approach and defines a diagnosis as a set of (faulty) components. In FDI, the concept of component is not central. FDI represents faults as variables that are explicitly involved in the equations of BM and/or OM [9]. For deterministic models, fault variables can be associated:

- to parameters, indicating that the parameter changes value when the fault is present, in which case they are referred as *multiplicative faults*, and
- to input and/or output variables, indicating actuators and sensors faults, in which case they are referred as *additive faults*.

FDI faults rather correspond to the DX concept of fault mode. In general, several parameters can be associated with a given component, giving rise to different fault modes. FDI faults are viewed as deviations with respect to the models of normal behavior, whereas in the DX's logical view the faulty behavior cannot be predicted from the normal model.

The parameters of FDI models may not have straightforward physical semantics. The model developer must be able to link model parameters to physical parameters to perform fault isolation.

Note that the DX approach can account for parametric faults by expressing the model at a finer level. For instance, considering a single-input single-output (static) component c whose behavior depends on two parameters θ_1 and θ_2, the standard DX model given by

$$COMPS(x) \wedge \neg AB(x) \Rightarrow out(x) = f(in(x), \theta_1, \theta_2) \tag{7.6}$$

could be replaced by

$$\begin{aligned} COMPS(x) \wedge PARM1(y) \wedge PARM2(z) \wedge \neg AB(x) \wedge \neg AB(y) \wedge \neg AB(z) \Rightarrow \\ out(x) = f(in(x), y, z) \\ PARM1(\theta_1), PARM2(\theta_2), COMPS(c). \end{aligned} \tag{7.7}$$

The component-based DX approach can, hence, be generalized by allowing the set $COMPS$ to include not only components (including sensors and actuators) but also parameters.

7.3 DX and FDI Model-Based Diagnosis Bridge

This section provides the theoretical links and an analysis of the diagnosis results of the DX approach and the parity space FDI approach as presented in [6, 23]. Practical comparison and potential synergies are also discussed.

7.3.1 ARR Versus R-Conflict

In the two approaches, diagnosis is triggered when discrepancies occur between the modeled (correct) behavior and the observations (OBS). As shown in Sect. 7.2, the detection of discrepancies corresponds to

- R-conflicts in DX and
- ARRs that are not satisfied by OBS in FDI.

The fault signature matrix FS, as defined in Definition 7.2, can be used to explain the relation between R-conflicts and ARRs. FS crosses ARRs in rows and faults/components in columns (here faults are univocally associated to components). The concept of *ARR Support* is also necessary.

Definition 7.6 (*ARR Support*) Consider ARR_i to be an ARR for $SM(\mathbf{z}, \mathbf{x})$, then the *support* of ARR_i, noted $supp(ARR_i)$, is the set of components $\{c_j\}$ (columns of the signature matrix FS) whose corresponding matrix cells FS_{ij} are nonzero on the ARR_i line.

The support of an ARR of the form $r(\mathbf{z}, \dot{\mathbf{z}}, \ddot{\mathbf{z}}, \ldots) = 0$ indicates the set of components whose models, or submodels, are involved in the obtention of the relation $r(\mathbf{z}, \dot{\mathbf{z}}, \ddot{\mathbf{z}}, \ldots) = 0$. The equations of the model $SM(\mathbf{z}, \mathbf{x})$ can indeed be partitioned in component models and every equation of $SM(\mathbf{z}, \mathbf{x})$ can be labelled as being part of the model of some component. Let $SM(C)$ denote the subset of equations defining the model of a component $c \in COMPS$ and $SM(C) = \bigcup_{c \in C} SM(c)$ the subset of equations corresponding to $C \subseteq COMPS$.

Let us now introduce two completeness properties, which refer to *detectability* indicated by a *d*, and to *isolability* indicated by an *i*.

Property 7.1 (ARR–d–completeness) *A set E of ARRs is said to be* d-complete *if*

- *E is finite;*
- $\forall OBS$, *if* $SM \cup OBS \models \perp$, *then* $\exists ARR_i \in E$ *such that* $\{ARR_i\} \cup OBS \models \perp$.

Property 7.2 (ARR–i–completeness) *A set E of ARRs is said to be* i-complete *if*

- *E is finite;*
- $\forall C$, *set of components such that* $C \subseteq COMPS$, *and* $\forall OBS$, *if* $SM(C) \cup OBS \models \perp$, *then* $\exists ARR_i \in E$ *such that* $supp(ARR_i)$ *is included in C and* $\{ARR_i\} \cup OBS \models \perp$.

ARR–d–completeness and ARR–i–completeness express the theoretical capability of a set of ARRs to be sensitive, hence, to detect any inconsistency between the corresponding sub-model of SM and observations OBS.

Example 7.1 Consider the small polybox example represented in Fig. 7.1.

The elementary components are one multiplier M, two adders A1 and A2 together with a set of sensors. The behavioral model BM is the following:

$$\begin{aligned} \mathrm{M} &: d = a \times b, \\ \mathrm{A1} &: f = c + d, \\ \mathrm{A2} &: g = e + d. \end{aligned} \tag{7.8}$$

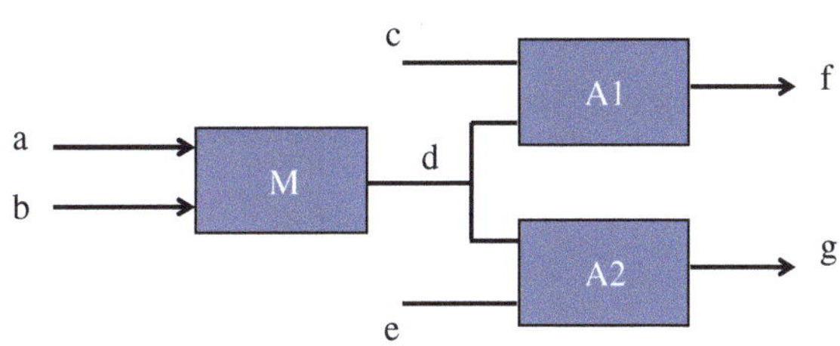

Fig. 7.1 Small polybox example

All the variables are censored but d. For the sake of simplicity, let us assume that sensor models are identity operators, then the observation model OM is the following:

$$\begin{aligned} \text{Sa} &: a = a_{obs}, \\ \text{Sb} &: b = b_{obs}, \\ \text{Sc} &: c = c_{obs} \\ \text{Se} &: e = e_{obs}, \\ \text{Sf} &: f = f_{obs}, \\ \text{Sg} &: g = g_{obs}. \end{aligned} \tag{7.9}$$

We can easily obtain three ARRs for this simple system by following the paths between inputs and/or outputs.

$$\begin{aligned} ARR1 &: r_1 = 0 \text{ where } r_1 \equiv f_{obs} - a_{obs} \cdot b_{obs} - c_{obs}, \\ ARR2 &: r_2 = 0 \text{ where } r_2 \equiv g_{obs} - a_{obs} \cdot b_{obs} - e_{obs}, \\ ARR3 &: r_3 = 0 \text{ where } r_3 \equiv f_{obs} - g_{obs} - c_{obs} + e_{obs}. \end{aligned} \tag{7.10}$$

Their supports are

$$\begin{aligned} supp(ARR1) &= \{A1, M\}, \\ supp(ARR2) &= \{A2, M\}, \\ supp(ARR3) &= \{A1, A2\}. \end{aligned} \tag{7.11}$$

Hence, the signature matrix for the set of single fault corresponding to components A1, A2, and M is given by Table 7.2.

We can notice that the set of ARRs composed of $ARR1$ and $ARR2$ seems to be sufficient to guaranty detectability and isolability of the three possible faults. As a matter of fact, $ARR3$ can be obtained by combining $ARR1$ and $ARR2$ (more precisely subtracting $ARR1$ and $ARR2$). The detectability power of $\{ARR1, ARR2\}$ is confirmed by the fact that this set is d-complete according to Property 1 defined above. However, let us now consider the following set of observations $OBS = \{a = 2, b = 3, c = 2, e = 2, f = 12, g = 10\}$ and the set of components $C = \{A1, A2\}$. Then, we have $SM(C) \cup OBS \models \perp$ but:

Table 7.2 Small polybox single-fault signature matrix

	f_{A1}	f_{A2}	f_M
$ARR1$	1	0	1
$ARR2$	0	1	1
$ARR3$	1	1	0

Table 7.3 Small polybox double-fault signature matrix for $c_i, c_j \in \{A1, A2, M\}, c_i \neq c_j$

	f_{A1}	f_{A2}	f_M	$f_{c_i c_j}$
$ARR1$	1	0	1	1
$ARR2$	0	1	1	1
$ARR3$	1	1	0	1

$$\begin{aligned} supp(ARR1) &= \{A1, M\} \nsubseteq C, \\ supp(ARR2) &= \{A2, M\} \nsubseteq C. \end{aligned} \tag{7.12}$$

Hence, the set of ARRs $\{ARR1, ARR2\}$ is not i-complete. This means that the isolability power of this set of ARRs is not maximal. For this simple example, this is easy to show by considering double faults (see Table 7.3).

Obviously, the set of ARRS $\{ARR1, ARR2\}$ does not differentiate any double faults from f_M and $ARR3$ turns to be useful for isolating the faults.

ARR–d–completeness and ARR–i–completeness are keys to the comparison of the FDI and DX approaches. The main results can be summarized by the following proposition [6].

Proposition 7.1 *Assuming the SRE property and that OBS is the set of observations for the system given by SM (or SD), then*

1. *If ARR_i is violated by OBS, then $supp(ARR_i)$ is an R-conflict;*
2. *If E is a d-complete set of ARRs, and if C is an R-conflict for $(SD, COMPS, OBS)$, then there exists $ARR_i \in E$ that is violated by OBS;*
3. *If E is an i-complete set of ARRs, then given an R-conflict C for $(SD, COMPS, OBS)$, there exists $ARR_i \in E$ that is violated by OBS and $supp(ARR_i)$ is included in C.*

The result 1 of Proposition 7.1 is intuitive and can be explained by the fact that the inconsistencies between the model and observations are captured by R-conflicts in the DX approach and by ARRs violated by OBS in the FDI approach. Consequently, the support of an ARR can be defined as a *potential R-conflict*. This concept is also called *possible conflict* in [17].

The results 2 and 3 of Proposition 7.1 refer to fault detectability and fault isolability. The result 2 outlines the ARR–d–completeness property as the condition for *fault detectability*. From the result 3, the ARR–i–completeness property appears as the condition under which a formal equivalence between R-conflicts and ARR supports holds, as stated by the following corollary.

Corollary 7.1 *If both the SRE and the ARR–i–completeness properties hold, the set of minimal R-conflicts for OBS and the set of minimal supports of ARRs (taken in any i-complete set of ARRs) violated by OBS are identical.*

The detailed proofs of Proposition 7.1 and Corollary 7.1 can be found in [6].

7.3.2 *Redundant ARRs*

An important result coming from ARR–i–completeness refers to redundant ARRs. In FDI, it is generally accepted that if ARR_j is obtained from a linear combination of two other ARRs, ARR_{i_1} and ARR_{i_2}, then ARR_j is redundant (unless some considerations about noises and sensitivity to faults come into play). Nevertheless, the i-completeness property states that not only the analytical expression of ARR_j must be taken into account but also its support to conclude about the fact that it is redundant. The formal conditions are stated in the proposition below from [6].

Proposition 7.2 *A given ARR_j is redundant with respect to a set of ARR_is, $i \in I$, $j \notin I$, where I is a set of integer indexes such that $card(I) \geq 2$, if and only if $\exists I' \subseteq I$ such that*

(1) $\forall OBS$, if all ARR_is, $i \in I'$, are satisfied by OBS, then ARR_j is satisfied by OBS,
(2) $supp(ARR_j) \supseteq supp(ARR_i)$, $\forall i \in I'$.

The above proposition can be explained by the fact that if $supp(ARR_j)$ does not satisfy condition 2, then it captures an inconsistency that is not captured by the initial ARR_is, $i \in I$. Added to the initial ARR_is, it, hence, contributes to the achievement of ARR–i–completeness.

Example 7.2 Let us consider the small polybox of Example 7.1, then $ARR3$ can be obtained as a linear combination of $ARR1$ and $ARR2$; however, it is not redundant. This has already been shown by noticing that it is necessary to discriminate any double faults from f_M. This can also be confirmed because it does not satisfy condition 2 of Proposition 7.2. Indeed, $supp(ARR3) = \{A1, A2\} \not\subseteq supp(ARR1) = \{A1, M\}$ and $supp(ARR3) = \{A1, A2\} \not\subseteq supp(ARR2) = \{A2, M\}$.

7.3.3 Exoneration Assumptions

The exoneration assumptions, *ARR-exoneration* and *component-exoneration*, used by DX and FDI, respectively, are different.

Definition 7.7 (*ARR-exoneration*) Given OBS, any component in the support of an ARR satisfied by OBS is exonerated, i.e., considered as normal.

Definition 7.8 (*Component-exoneration*) Given OBS and $c \in COMP$, if $SM(c) \cup OBS$ is consistent, then c is exonerated, i.e., considered as normal.

The FDI approach generally uses the ARR-exoneration assumption without formulating it explicitly. On the other hand, the DX approach generally proceeds with no exoneration assumption at all. When this is not the case, it uses the component-exoneration assumption and represents it explicitly. If a component c is exonerated, its model is written as

$$COMP(c) \wedge \neg AB(c) \Longleftrightarrow SM(c),$$

where the simple logical implication, found in (7.5), for instance, is replaced by a double implication. Explicit assumptions guarantee logical correctness of the DX diagnoses obtained by the DX method. Interestingly, *ARR-exoneration* cannot be expressed in the DX formalism and conversely, *component-exoneration* cannot be expressed in the FDI formalism.

It has been shown that under the same assumptions, in particular, in the case of no exoneration, the diagnoses that are obtained by the DX and the FDI approach are the same [6].

Theorem 7.1 *Under the i-completeness and no exoneration assumptions, the diagnoses obtained by the FDI approach are identical to the (non-empty) diagnoses obtained by the DX approach.*

7.3.4 Comparison in Practice

Although diagnosis results have been shown to be the same with the DX and the FDI approach, the frameworks and procedures adopted by the two approaches have practical impacts. In particular, we can note the following two points:

- *Handling single and multiple faults*: in the FDI approach, because the fault signatures are determined offline for every fault, the number of considered faults is generally limited. Most of the time, only single fault are considered. On the contrary, the DX approach naturally deals with multiple faults. A consequence is that the number of diagnoses is exponential and this is why it is common to introduce preference criteria, like fault probabilities, to order the diagnoses. Several search methods have been proposed to find the *preferred diagnoses* or to retrieve the diagnoses in preference order (see, for instance, [20, 29]).
- *Offline versus online processing*: in the FDI approach, ARRs are determined offline and only a simple consistency check is performed online. This may be quite relevant for real-time applications with hard temporal constraints. Inversely, in the DX approach, the whole diagnosis process is online, the advantage being that only the models need to be updated in case of any evolution of the system. The two approaches have been integrated to obtain the advantages of both: some DX works have used the idea of the FDI community to construct ARRs offline [11, 14, 17, 28] and some FDI works have proposed to base the fault isolation phase on the conflicts derived from violated ARRs [27].

7.4 Case Studies

7.4.1 Polybox Case Study

The comparison of the DX and FDI approach is first performed on the well-known polybox example.

7.4.1.1 FDI Approach

The elementary components of the polybox example are the adders A1 and A2, multipliers M1, M2, and M3 together with the set of sensors. The behavioral model BM is the following:

$$\begin{aligned} \text{M1} &: x = a \times c, \\ \text{M2} &: y = b \times d, \\ \text{M3} &: z = c \times e, \\ \text{A1} &: f = x + y, \\ \text{A2} &: g = y + z, \end{aligned} \tag{7.13}$$

and the observation model OM assumes that sensor models are identity operators for the sake of simplicity:

$$\begin{aligned} \text{Sa} &: a = a_{obs}, \\ \text{Sb} &: b = b_{obs}, \\ \text{Sc} &: c = c_{obs}, \\ \text{Sd} &: d = d_{obs}, \\ \text{Se} &: e = e_{obs}, \\ \text{Sf} &: f = f_{obs}, \\ \text{Sg} &: g = g_{obs}. \end{aligned} \tag{7.14}$$

The set of observations is, for example, $OBS = \{a_{obs} = 2, b_{obs} = 2, c_{obs} = 3, d_{obs} = 3, e_{obs} = 2, f_{obs} = 10, g_{obs} = 12\}$.

Three redundancy relations can be found:

$$\begin{aligned} ARR1 &: r_1 = 0 \text{ where } r_1 \equiv f_{obs} - a_{obs} \cdot c_{obs} - b_{obs} \cdot d_{obs}, \\ ARR2 &: r_2 = 0 \text{ where } r_2 \equiv g_{obs} - d_{obs} \cdot d_{obs} - c_{obs} \cdot e_{obs}, \\ ARR3 &: r_3 = 0 \text{ where } r_3 \equiv f_{obs} - g_{obs} - a_{obs} \cdot c_{obs} - c_{obs} \cdot e_{obs}. \end{aligned} \tag{7.15}$$

$ARR1$, $ARR2$, and $ARR3$ are obtained from the models of {M1, M2, A1}, {M2, M3, A2}, and {M1, M3, A1, A2}, respectively. If we assume that the sensors are not faulty, the ARRs can be written as

Table 7.4 Polybox single-fault signature matrix

	f_{A1}	f_{A2}	f_{M1}	f_{M2}	f_{M3}
$ARR1$	1	0	1	1	0
$ARR2$	0	1	0	1	1
$ARR3$	1	1	1	0	1

$$
\begin{aligned}
&ARR1 : f - (a \cdot c + b \cdot d) = 0,\\
&ARR2 : g - (b \cdot d + c \cdot e) = 0,\\
&ARR3 : f - g - a \cdot c + c \cdot e = 0.
\end{aligned}
\tag{7.16}
$$

The signature matrix for the set of single fault corresponding to components A1, A2, M1, M2, and M3 in the case of component-exoneration assumption defined in Definition 7.8 is given in Table 7.4.

The case of multiple faults can be dealt with by expanding the number of columns of the signature matrix, leading to a total number of $2^m - 1$ columns if all the possible multiple faults are considered.

The interpretation of multiple-fault signature entries is the same as for single fault. Given the way multiple-fault signatures are derived from single-fault signatures, this interpretation implies that the simultaneous occurrence of several faults is not expected to lead to situations in which the faults compensate, resulting in the non-observation of the multiple faults. As it will be stated later more formally, this is known as the *multiple-fault exoneration assumption*, which is a generalization of the exoneration assumption defined for single fault.

For different *Observed Signatures* (OSs) formed by the observed residual vector $(r_1, r_2, r_3)^T$, the diagnosis results are summarized in Table 7.6 that resumes single- and multiple-fault signatures from Tables 7.4 and 7.5.

Another interesting point to note is that, in the polybox example, the same diagnosis results would be obtained using the partial signature corresponding to ARR1 and ARR2 only in these three cases:

- $(r_1, r_2) = (0, 0)$: no fault,
- $(r_1, r_2) = (0, 1)$: A2 or M3 faulty,
- $(r_1, r_2) = (1, 0)$: A1 or M1 faulty.

In these three cases, the use of ARR3, associated with r_3, does not provide any more localization power. This is obviously not the case for the two last observed signatures (columns 5 and 6 of Table 7.6) for which r_3 is needed to disambiguate the signature ($r_1 = 1, r_2 = 1$). It can be noticed that ARR3 was obtained from the combination of ARR1 and ARR2.

7.4.1.2 DX Logical Diagnosis Approach

The system description is

Table 7.5 Polybox double-fault signature matrix

	f_{A1}	f_{A2}	f_{M1}	f_{M2}	f_{M3}	f_{A1A2}	f_{A1M1}	f_{A1M2}	f_{A1M3}	f_{A2M1}	f_{A2M2}	f_{A2M3}	f_{M1M2}	f_{M1M3}	f_{M2M3}
$ARR1$	1	0	1	1	0	1	1	1	1	1	1	0	1	1	1
$ARR2$	0	1	0	1	1	1	0	1	1	1	1	1	1	1	1
$ARR3$	1	1	1	0	1	1	1	1	1	1	1	1	1	1	1

Table 7.6 Polybox FDI diagnosis results for different observation signatures

	OS				
$ARR1$	0	0	1	1	1
$ARR2$	0	1	0	1	1
$ARR3$	0	1	1	0	1
Single-fault diagnoses	None	{A2}, {M3}	{A1}, {M1}	{M2}	None
Multiple-fault diagnoses	None	{A2, M3}	{A1, M1}	None	All double faults but {A2, M3} and {A1, M1}

$$\begin{aligned}
COMPS &= \{\text{M1, M2, M3, A1, A2}\} \\
SD &= \{ADD(\text{c}) \wedge \neg AB(\text{c}) \Rightarrow Output(\text{c}) = Input1(\text{c}) + Input2(\text{c}), \\
&\quad MULT(\text{c}) \wedge \neg AB(\text{c}) \Rightarrow Output(\text{c}) = Input1(\text{c}) \times Input2(\text{c}), \\
&\quad MULT(\text{M1}), MULT(\text{M2}), MULT(\text{M3}), ADD(\text{A1}), ADD(\text{A2}), \\
&\quad Output(\text{M1}) = Input1(\text{A1}), Output(\text{M2}) = Input2(\text{A1}), \\
&\quad Output(\text{M2}) = Input1(\text{A2}), Output(\text{M3}) = Input2(\text{A2})\}, \\
&\quad \text{for c} \in COMPS \\
OBS &= \{Input1(\text{M1}), Input2(\text{M1}), Input1(\text{M2}), Input2(\text{M2}), \\
&\quad Input1(\text{M3}), Input2(\text{M3}), Output(\text{A1}), Output(\text{A2})\}.
\end{aligned}$$

Suppose the polybox is given the inputs $a = 2$, $b = 2$, $c = 3$, $d = 3$, $e = 2$, and it outputs $f = 10$, $g = 12$ in response. The set of observations is represented by

$$\begin{aligned}
OBS = \{&Input1(M1) = 2, Input2(M1) = 3, Input1(M2) = 2, Input2(M2) = 3, \\
&Input2(M3) = 2, Output(A1) = 10, Output(A2) = 12\}.
\end{aligned}$$

The polybox with the observations as seen above ($f = 10$, $g = 12$) has the following minimal R-conflicts: $\{A1, M1, M2\}$ and $\{A1, A2, M1, M3\}$ due to the abnormal value of 10 for f. Symmetrically, $f = 12$ and $g = 10$ yield $\{A2, M2, M3\}$ and $\{A1, A2, M1, M3\}$. In the case $f = 10$ and $g = 10$, the two minimal R-conflicts are $\{A1, M1, M2\}$ and $\{A2, M2, M3\}$. In the case $f = 10$ and $g = 14$, the three minimal R-conflicts are $\{A2, M2, M3\}$, $\{A1, M1, M2\}$, and $\{A1, A2, M1, M3\}$.

The corresponding minimal diagnoses are presented in Table 7.7.

7.4.1.3 Bridge

Releasing the exoneration assumption in the polybox example leads to the single-fault signature matrix shown in Table 7.8 and to the extended-fault signature matrix presented in Table 7.9. These are obtained from the standard ones (see Table 7.4 and 7.5) by replacing 1's by $\times$'s, which allows these entries to be matched with 0 or 1 in the observed signature. Note that all signatures of triple faults and more are equal to $(\times, \times, \times)^T$).

Table 7.7 Polybox DX diagnosis results for different observation signatures

	OBS				
$f =$	12	12	10	10	10
$g =$	12	10	12	10	14
Minimal R-conflicts	None	{A2, M2, M3} {A1, A2, M1, M3}	{A1, M1, M2} {A1, A2, M1, M3}	{A1, M1, M2} {A2, M2, M3}	{A1, M1, M2} {A1, A2, M1, M3} {A2, M2, M3}
Minimal diagnoses	{}	{A2} {M3} {A1, A2} {M1, M2}	{A1} {M1} {A2, M2} {M2, M3}	{M2} {A1, A2} {A1, M3} {M1, M3} {M1, M3}	All double faults but {A2, M3} and {A1, M1}

Table 7.8 Polybox single fault signature matrix without exoneration

	f_{A1}	f_{A2}	f_{M1}	f_{M2}	f_{M3}
$ARR1$	×	0	×	×	0
$ARR2$	0	×	0	×	×
$ARR3$	×	×	×	0	×

The following results are then obtained:

- With outputs $f = 12$ and $g = 10$, i.e., observed signature (0,1,1), there are four minimal diagnoses: two single-fault diagnoses {A2} and {M3} and two double-fault diagnoses {A1, A2} and {M1, M2}, and 22 superset diagnoses.
- With outputs $f = 10$ and $g = 12$, i.e., observed signature (1,0,1), there are four minimal diagnoses: two single-fault diagnoses {A1} and {M1} and two double-fault diagnoses {A2, M2} and {M2, M3}, and 22 superset diagnoses.
- With outputs $f = 10$ and $g = 10$, i.e., observed signature (1,1,0), there are five minimal diagnoses: one single-fault diagnosis {M2} and four double-fault diagnoses {A1, A2}, {A1, M3}, {M1, A2} and {M1, M3}, and 20 superset diagnoses.
- With outputs $f = 10$ and $g = 14$, i.e., observed signature (1,1,1), there are eight minimal double-fault diagnoses: {A1, A2}, {A1, M2}, {A1, M3}, {A2, M1}, {A2, M2}, {M1, M2}, {M1, M3}, and {M2, M3}, and 16 superset diagnoses.

These results obtained by FDI are identical to those obtained by DX (see Table 7.7). In the case where $f = 12$ and $g = 12$, i.e., observed signature (0,0,0), the empty subset is a minimal diagnosis according to DX and any non-empty subset of components is a diagnosis according to both approaches: there are 5 minimal single-fault diagnoses and 26 superset diagnoses. The only difference between FDI and DX is that the "no-fault" column of signature (0,0,0) which would correspond to the empty diagnosis subset is left implicit in the signature matrix.

It can be noticed that, except in the $f = 10$ and $g = 14$ case (where anyhow, no exoneration can apply as no ARR is satisfied), the results are different from those obtained under the default exoneration assumption (see Table 7.6).

Table 7.9 Polybox double-fault signature matrix without exoneration

	f_{A1}	f_{A2}	f_{M1}	f_{M2}	f_{M3}	f_{A1A2}	f_{A1M1}	f_{A1M2}	f_{A1M3}	f_{A2M1}	f_{A2M2}	f_{A2M3}	f_{M1M2}	f_{M1M3}	f_{M2M3}
$ARR1$	×	0	×	×	0	×	×	×	×	×	×	0	×	×	×
$ARR2$	0	×	0	×	×	×	0	×	×	×	×	×	×	×	×
$ARR3$	×	×	×	0	×	×	×	×	×	×	×	×	×	×	×

7.4.2 Three-Tank Case Study

For this case study, we use the FDI approach to derive ARRs, then apply the bridge result to obtain the DX counterpart results.

The system (cf. Sect. 2.2) is made up of three identical tanks T_1, T_2, T_3. All three tanks have the same physical features such as height and cross-sectional area, A. There is a measured input flow q_i for tank T_1, which is drained into T_2 via a pipe q_{12}. A similar process gets the flow from T_2 to T_3 via pipe q_{23}. Finally, there is an output flow q_{30} from T_3. The system has three sensors measuring the level in tanks T_1 and T_3 (level transducers $LT1$ and $LT2$, respectively), and another sensor measuring the flow through pipe q_{23} (flow transducer $FT1$).

Adopting the FDI approach, the following three dynamic system equations model the normal behavior of the system. The change in the level in each tank, $\dot{h}_{T_i}$, is computed according to mass balances:

$$
\begin{aligned}
e1_n &: \dot{h}_{T_1} = \frac{q_i - q_{12}}{A},\\
e2_n &: \dot{h}_{T_2} = \frac{q_{12} - q_{23}}{A},\\
e3_n &: \dot{h}_{T_3} = \frac{q_{23} - q_{30}}{A}.
\end{aligned}
$$

Flows between tanks q_{12}, q_{23}, q_{30} are modeled as

$$
\begin{aligned}
e4_n &: q_{12} = S_{p_1} \cdot \mathrm{sign}(h_{T_1} - h_{T_2}) \cdot \sqrt{2g \mid h_{T_1} - h_{T_2} \mid},\\
e6_n &: q_{23} = S_{p_2} \cdot \mathrm{sign}(h_{T_2} - h_{T_3}) \cdot \sqrt{2g \mid h_{T_2} - h_{T_3} \mid},\\
e8_n &: q_{30} = S_{p_3} \cdot \sqrt{2g \cdot h_{T_3}},
\end{aligned}
$$

being S_{p_i}, $i = 1, \ldots, 3$, the cross-sectional area of the pipes. The relation between the state variables, h_{T_i}, and their derivatives $\dot{h}_{T_i}$ is given by

$$
\begin{aligned}
e13 &: h_{T_1} = \int \dot{h}_{T_1} \cdot dt,\\
e14 &: h_{T_2} = \int \dot{h}_{T_2} \cdot dt,\\
e15 &: h_{T_3} = \int \dot{h}_{T_3} \cdot dt.
\end{aligned}
$$

The above nine equations form the behavioral model BM.

The observational model OM is given by the following equations:

$$
\begin{aligned}
e10 &: h_{T_1,obs} = h_{T_1},\\
e11 &: h_{T_3,obs} = h_{T_3},\\
e12 &: q_{23,obs} = q_{23}.
\end{aligned}
$$

Six possible faults are considered: three-tank leakages and three pipe blockages as presented in Sect. 2.2.

To obtain ARRs, we can use the structural analysis approach as presented in Sect. 3.4.3 or the possible conflicts approach as presented in Sect. 6.1.1. Three ARRs are obtained from the three equation sets below:

$$\{(e3_n)(e8_n)(e11)(e12)(e15)\},$$
$$\{(e2_n)(e4_n)(e6_n)(e10)(e11)(e12)(e14)\},$$
$$\{(e1_n)(e4_n)(e6_n)(e10)(e11)(e12)(e13)\}.$$

The fault signature matrix with and without exoneration are given in Table 7.10 and in Table 7.11, respectively, where f_{T_i} represents a leakage at the bottom of tank T_i, $i = 1, \ldots, 3$, and $f_{P_{jk}}$ represents a stuck closed fault on the pipe connecting tank T_j and tank T_k or the atmosphere, $i = 1, \ldots, 3$, $k \in \{2, 3, 0\}$.

Assume that the observed signature is $(0, 0, 1)^T$ then by item 1 of Proposition 7.1, we know that $\{T_1, P_{12}, P_{30}\}$ is an R-conflict.

If we use the standard FDI approach and the fault signature matrix of Table 7.10, f_{T_1} is the only fault signature that can be matched. So we get one unique diagnosis $\Delta = \{T_1\}$.

On the other hand, the DX approach can obtain the diagnoses as the hitting sets of the R-conflicts [19]. In this case, we have one single R-conflict $\{T_1, P_{12}, P_{30}\}$, which indicates that three minimal diagnoses are $\Delta_1 = \{T_1\}$, $\Delta_2 = \{P_{12}\}$, and $\Delta_3 = \{P_{30}\}$. This result seems in contradiction with the result obtained with the standard FDI approach. However, let us now apply the FDI approach by relaxing the ARR-exoneration assumption.

If we now use the FDI approach with no ARR-exoneration, the fault signature matrix of Table 7.11 must be considered. Given that a "×" entry can be matched to "0" or "1" in the observed signature, not only f_{T_1} but also $f_{P_{12}}$ and $f_{P_{30}}$ can be matched. The diagnosis results are, hence, the same as for the DX approach, exemplifying Theorem 7.1.

Table 7.10 Three-tank fault signature matrix

	f_{T_1}	f_{T_2}	f_{T_3}	$f_{P_{12}}$	$f_{P_{23}}$	$f_{P_{30}}$
$ARR1$	0	0	1	0	0	1
$ARR2$	0	1	0	1	1	0
$ARR3$	1	0	0	1	1	0

Table 7.11 Three-tank fault signature matrix without exoneration

	f_{T_1}	f_{T_2}	f_{T_3}	$f_{P_{12}}$	$f_{P_{23}}$	$f_{P_{30}}$
$ARR1$	0	0	×	0	0	×
$ARR2$	0	×	0	×	×	0
$ARR3$	×	0	0	×	×	0

Assume now that the observed signature is $(1, 1, 0)^T$; then by item 1 of Proposition 7.1, we know that $\{T_3, P_{30}\}$ and $\{T_2, P_{12}, P_{23}\}$ are R-conflict.

If we use the FDI approach with (cf. Table 7.10) or without ARR-exoneration (cf. Table 7.11), no-fault signature matches the observed signature.

On the other hand, if we obtain the diagnoses from the two R-conflicts following the DX approach, we obtain six minimal diagnoses $\Delta_1 = \{T_3, T_2\}$, $\Delta_2 = \{T_3, P_{12}\}$, $\Delta_3 = \{T_3, P_{23}\}$, $\Delta_4 = \{T_2, P_{30}\}$, $\Delta_5 = \{P_{30}, P_{12}\}$, and $\Delta_6 = \{P_{30}, P_{23}\}$. As a matter of fact, all of them are double-fault diagnoses. This is the reason why they could not be found with the single-fault signature matrix. Interestingly, the DX approach handles single and multiple faults in the same framework, while the FDI approach requires to generate the extended-fault signature matrix.

7.5 Conclusions

In this chapter, the FDI approach based on ARRs and the DX logical approach have been compared and the hypotheses underlying the two approaches have been clearly stated. The concepts on both sides have been bridged, and it has been shown that the two approaches provide the same diagnosis results under some assumptions referring to exoneration. The classical polybox and the three-tank case studies have been used to illustrate the MBD bridge from which potential synergies are possible.

The links and the understanding of the MBD bridge provide sound ground for merging the best of both worlds and producing ever better diagnosers for real complex systems.

References

1. Adrot, O., Maquin, D., Ragot, J.: Fault detection with model parameter structured uncertainties. In: Proceedings of the European Control Conference, ECC'99. Karlsruhe (1999)
2. Armengol, J., Bregon, A., Escobet, T., Gelso, E., Krysander, M., Nyberg, M., Olive, X., Pulido, B., Travé-Massuyès, L.: Minimal Structurally Overdetermined sets for residual generation: a comparison of alternative approaches. In: Proceedings of the 7th IFAC Symposium on Fault Detection, Supervision and Safety of Technical Processes, Safeprocess'09 (2009)
3. Basseville, M., Nikiforov, I.: Detection of Abrupt Changes: Theory and Application. Citeseer (1993)
4. Blanke, M., Kinnaert, M., Lunze, J., Staroswiecki, M.: Diagnosis and Fault-Tolerant Control. Springer, Berlin (2003)
5. Chow, E., Willsky, A.: Analytical redundancy and the design of robust failure detection systems. IEEE Trans. Autom. Control **29**(7), 603–614 (1984)
6. Cordier, M., Dague, P., Lévy, F., Montmain, J., Staroswiecki, M., Travé-Massuyès, L.: Conflicts versus analytical redundancy relations: a comparative analysis of the model based diagnosis approach from the artificial intelligence and automatic control perspectives. IEEE Trans. Syst. Man Cybern. Part B **34**(5), 2163–2177 (2004)
7. Denis-Vidal, L., Joly-Blanchard, G., Noiret, C.: Some effective approaches to check the identifiability of uncontrolled nonlinear systems. Math. Comput. Simul. **57**(1–2), 35–44 (2001)

8. Dubuisson, B.: Automatique et statistiques pour le diagnostic. Hermes Science Europe Ltd (2001)
9. Gertler, J.: Analytical redundancy methods in failure detection and isolation. In: Preprints of the IFAC SAFEPROCESS Symposium, pp. 9–21 (1991)
10. Gertler, J.: Fault Detection and Diagnosis in Engineering Systems. Marcel Deker, New York City (1998)
11. Katsillis, G., Chantler, M.: Can dependency-based diagnosis cope with simultaneous equations. In: Proceedings of the 8th International Workshop on Principles of Diagnosis DX-97, pp. 51–59 (1997)
12. Kleer, J., Mackworth, A., Reiter, R.: Characterizing diagnoses and systems. Artif. Intell. **56**(2–3), 197–222 (1992)
13. Krysander, M., Aslund, J., Nyberg, M.: An efficient algorithm for finding minimal overconstrained subsystems for model-based diagnosis. IEEE Trans. Syst. Man Cybern. Part A: Syst. Hum. **38**(1), 197–206 (2008)
14. Loiez, E., Taillibert, P.: Polynomial temporal band sequences for analog diagnosis. In: Proceedings of the Fifteenth International Joint Conference on Artificial Intelligence IJCAI-97, Nagoya, Japan, 23–29 Aug 1997, p. 474 (1997)
15. Patton, R., Chen, J.: A re-examination of the relationship between parity space and observer-based approaches in fault diagnosis. Eur. J. Diagn. Saf. Autom. **1**(2), 183–200 (1991)
16. Patton, R., Frank, P., Clark, R.: Fault diagnosis in dynamic systems. Theory and Applications (1989)
17. Pulido, B., Gonzalez, C.: Possible conflicts: a compilation technique for consistency-based diagnosis. IEEE Trans. Syst. Man Cybern. - Part B: Cybern. **34**(5), 2192–2206 (2004)
18. Qiu, Z., Gertler, J.: Robust FDI and H_{inf} optimization. In: Proceedings of the 32nd IEEE Conference on Control and Decision CDC'93. San Antonio, Texas (1993)
19. Reiter, R.: A theory of diagnosis from first principles. Artif. Intell. **32**(1), 57–95 (1987)
20. Sachenbacher, M., Williams, B.: Diagnosis as semiring-based constraint optimization. In: Proceedings of the European Conference on Artificial Intelligence ECAI'04, vol. 16, p. 873 (2004)
21. Staroswiecki, M., Comtet-Varga, G.: Analytical redundancy relations for fault detection and isolation in algebraic dynamic systems. Automatica **37**(5), 687–699 (2001)
22. Staroswiecki, M., Declerck, P.: Analytical redundancy in non linear interconnected systems by means of structural analysis. In: Proceedings of the IFAC Symposium on Advanced Information Processing in Automatic Control, pp. 51–55 (1989)
23. Travé-Massuyès, L.: Bridging control and artificial intelligence theories for diagnosis: a survey. Eng. Appl. Artif. Intell. **27**, 1–16 (2014)
24. Venkatasubramanian, V., Rengaswamy, R., Kavuri, S.N.: A review of process fault detection and diagnosis part ii: qualitative models and search strategies. Comput. Chem. Eng. **27**(3), 313–326 (2003)
25. Venkatasubramanian, V., Rengaswamy, R., Kavuri, S.N., Yin, K.: A review of process fault detection and diagnosis part iii: process history based methods. Comput. Chem. Eng. **27**(3), 327–346 (2003). https://doi.org/10.1016/S0098-1354(02)00162-X-a3
26. Venkatasubramanian, V., Rengaswamy, R., Yin, K., Kavuri, S.N.: A review of process fault detection and diagnosis part i: quantitative model-based methods. Comput. Chem. Eng. **27**(3), 293–311 (2003)
27. Vento, J., Puig, V., Sarrate, R., Travé-Massuyès, L.: Fault detection and isolation of hybrid systems using diagnosers that reason on components. IFAC Proc. Vol. **45**(20), 1250–1255 (2012)
28. Washio, T., Motoda, H., Niwa, Y., INSS, I.: Discovering admissible model equations from observed data. In: Proceedings of the 16th International Joint Conference on Artificial Intelligence IJCAI'99, vol. 2, pp. 772–779. Citeseer (1999)
29. Williams, B., Ragno, R.: Conflict-directed A* and its role in model-based embedded systems. J. Discret. Appl. Math. Citeseer (2003)

Chapter 8
Data-Driven Fault Diagnosis: Multivariate Statistical Approach

Joaquim Melendez i Frigola

8.1 Introduction

Generally speaking, process monitoring is a continuous real-time task of determining the state of a physical system, by gathering information, and from it, recognize and indicate any anomalies in its behavior [3]. The objective is to discriminate between two principal states (normal and abnormal operation) in order to detect possible faults with reduced uncertainty. With this aim, process monitoring methods consist in analyzing the consistency of process variables (either measured or estimated) with respect to what is considered the normal operation conditions (NOC) of the system. Whenever this consistency does not hold, the process is considered to operate under Abnormal Operating Conditions (AOC); and this is what is expected to happen in presence of a fault (fault detection). Thus, model-based approaches require an accurate description of the system dynamics (the model), represented in Fig. 8.1 by the functions $\mathbf{f}(\mathbf{u}(\mathbf{k}), \mathbf{s}(\mathbf{k}))$ and $\mathbf{g}(\mathbf{s}(\mathbf{k}), \mathbf{u}(\mathbf{k}))$. In those approaches, usually $\mathbf{f}()$ and $\mathbf{g}()$ are described by analytical equations obtained from first principles. The output of the system ($\mathbf{y}(k)$) at a given time instant, k, is being determined by the input vector ($\mathbf{u}(k)$) and the current state of the process ($\mathbf{s}(k)$) (Fig. 8.1):

$$\dot{\mathbf{s}}(k) = \mathbf{f}\left(\mathbf{u}(k), \mathbf{s}(k)\right), \tag{8.1}$$

$$\mathbf{y}(k) = \mathbf{g}(\mathbf{s}(k), \mathbf{u}(k)). \tag{8.2}$$

Thus, model-based approaches take advantage of analytical redundancy, given by the model and measurements, to check consistency of acquired data coming from actuators and sensors - $\mathbf{y}(k)$, $\mathbf{u}(k)$ -. A convenient reorganization of the model equations allows computing the *analytical redundancy equations* used for that purpose.

J. Melendez i Frigola (✉)
Institut d'Informatica i Aplicacions, Universitat de Girona, EPS-P4,
c/Maria Aurelia Capmany, 61, 17003 Girona, Spain
e-mail: joaquim.melendez@udg.edu

T. Escobet et al. (eds.), *Fault Diagnosis of Dynamic Systems*,
https://doi.org/10.1007/978-3-030-17728-7_8

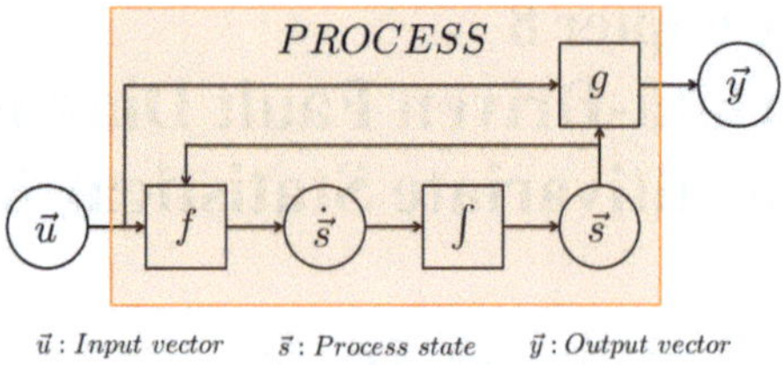

Fig. 8.1 Example of a general process relating the input vector (**u**), the state of the process (**s**), and the output of the system (**y**)

Once an inconsistency is detected, it results in an alarm generation (fault detection) and a diagnosis procedure can follow. Diagnosis consists in a logical search aiming to identify and isolate the variables, or parameters, in the model that are responsible for this inconsistency between acquired data and the model.

Main concern on these methods relies on the quality of the model (modeling errors, completeness, identifiable, parameter adjusting, drifts, etc.) and their manageability (dimensionality, nonlinearity in the model, adaptation to working point operation, etc.) in real scenarios. Moreover, despite the quality of the model, there are several properties that these models should fulfill, w.r.t. available data, in order to guaranty the detectability and isolability of faults.

On the other hand, data-driven methods do not assume the pre-existence of an analytical model -**f**() and **g**() are not required—and focus modeling efforts in discovering possible relationships and constraints in the process variables using a data mining approach. Thus, data-driven methods for fault detection/diagnosis imply two differentiate stages: first, data-driven modeling (learning from data), and second, model exploitation for monitoring tasks, i.e., fault detection and isolation. Usually, the modeling stage is performed with data acquired during Normal Operation Conditions (NOC) of the process, whereas fault detection is performed continuously by checking consistency of new data (coming from process) with the learned model. Although some data-driven methods can focus on fault modeling (*supervised learning*), when enough representative data is available for every fault, this approach will not be considered in this chapter.

Data-driven approaches reduce substantially the effort and time required to build a reference model representing normal operating conditions. These methods assume that variations present in data, gathered from process instruments, obey to physical constraints and principles governing the process (affected by some measurement noise, of course). The performance of data-driven methods depends on both data quality (noise, blanks, errors, outliers, coverage, etc.) and performance (in terms of accuracy, precision, recall, or type I and type II errors, for example) of the method used in the learning and decision procedures. Thus, it is very important to assure the representativeness of historical data being used during the learning stage. It is important to notice that usually process are operated around a working point and consequently available data usually only represent partially the dynamics of the process. A similar consideration should be done w.r.t to available data. Decision of which sensors should be installed usually responds to control and supervision requirements, so they usually represent the system governing variables; however, some dynamics could not be observable with those sensors. Thus, data-driven modes can result in a partial representation of the system and/or could require re-computation when normal operating conditions change (change to a different working point, for example) to some non-previously observed state (in this historical data set used for building the model).

Despite the previous considerations, it is important to note that the objective is not to discover equations describing physical behavior of the system, but simply discovering dependencies among variables useful for fault detection and diagnosis. Thus, many data-driven models do not preserve causality among variables and usually models are reduced to either black-box models or simple algebraic functions that are enough to check consistency of data required for fault detection.

Since data-driven methods take advantage of well-known data mining methods (machine learning, statistics), data process is usually organized in a matrix structure, $\mathbf{X}$, defined by the m variables (combination of input, $\mathbf{u}(k)$, and output, $\mathbf{y}(k)$, vectors) distributed in columns, where every sampling time corresponds to a row and defines an observation of the process (observation given by the m variables) at the time instant k:

$$\mathbf{x}(k) = \left[\mathbf{u}(k)\ \mathbf{y}(k) \right] = \left[x_{k,1}\ x_{k,2}\ \dots\ x_{k,i}\ \dots\ x_{k,m} \right]. \tag{8.3}$$

8.1.1 Scope

In this chapter, a specific data-driven method for fault detection and isolation that uses multivariate statistical principles is presented. Interest of multivariate statistical methods is mainly because the number of sensors being installed in any system is drastically increasing, increasing the redundancy of information being captured. In such situations, the use of multivariate statistical techniques, based on the existence of correlations among measured variables, provides a solid framework for fault detection based on well-proved theoretical background. Principal Component Analysis (PCA) or Partial Least Squares (PLS), also known as projection to latent structures, are fundamental techniques that support fault detection and isolation strategies commonly used in chemical and other process industries, and also admit extensions to deal with particularities of batch processes. These methods are also described in the literature as multivariate process control techniques, because they extend traditional (uni)variate statistical approaches for process control based on the $\pm 3\sigma$ criteria for steady-state single-variable monitoring.

The method described in the following sections relies on PCA principles and extends it toward a complete methodology for fault detection and isolation. PCA is a well-known multivariate statistical technique that exploits the correlation analysis, among all the variables being studied of a process, to identify existing linear relationships and represent them in a lower dimension space (projection technique) with the ability to decouple them from noise. Based on this technique, it is possible to model process behavior during Normal Operating Conditions (NOC) and use this model as reference for monitoring. Projection of new data into this model allows to detect when an observation is inconsistent with the learned reference model (representing normal operating conditions) and consequently alert of a possible fault. In such case, a detailed analysis of the fault magnitude w.r.t the contributions of involved variables allows also to identify the variable, or a set of them, most probably causing the misbehavior (fault isolation).

In the following section, the background of principal component analysis is introduced with a practical perspective and oriented to the use of this technique to model the normal operation of the process. Then, in Sect. 8.3, this model is used as a reference to perform the fault detection tasks by evaluating the adequacy of a new observation to this model, in terms of distance; and Sect. 8.4 tackles the diagnosis procedure from the perspective of identifying the variables that most probably cause the faulty situation.

8.2 Background on Principal Component Analysis (PCA)

Principal Component Analysis (PCA) is a technique for data compression and information extraction. PCA is used to find combinations of variables, or factors, that describe major trends in a data set. PCA is concerned to explain the variance–covariance structure through a few linear combinations of the original variables. Processes involving a large number of variables can be monitored using this technique as a fundamental principle. Thus, observations (historic data) collected during Normal Operation Conditions (NOC) are used to build a data-driven model. Then, this NOC statistical model will be used to assess the behavior of the process by checking consistency of new observations with respect to this reference model (fault detection).

Thus, multivariate data (NOC observations) are expected to be organized in a matrix structure, $\mathbf{X}^{n\times m}$, with m variables (columns) and n observations (rows). Data contained in columns (variables) are assumed to be centered (zero mean) and standardized (unit variance).

$$\mathbf{X} = \begin{bmatrix} x_{1,1} & x_{1,2} & \cdots & x_{1,m} \\ x_{2,1} & x_{2,2} & \dots & x_{2,m} \\ \vdots & \vdots & \ddots & \vdots \\ x_{k,1} & x_{k,2} & \dots & x_{k,m} \\ \vdots & \vdots & \ddots & \vdots \\ x_{n,1} & x_{n,2} & \cdots & x_{n,m} \end{bmatrix}. \tag{8.4}$$

From $\mathbf{X}$, the sample covariance matrix ($\mathbf{S}$) can easily be computed with the following expression:

$$\mathbf{S} = \frac{1}{n-1}\mathbf{X}^T\mathbf{X} \tag{8.5}$$

And using an eigenvalue decomposition of the sample covariance matrix $\mathbf{S}$ $(m \times m)$, the two matrices $\mathbf{V}$ and Λ are obtained -both $m \times m$-.

$$\mathbf{S} = \frac{1}{n-1}\mathbf{X}^T\mathbf{X} = \mathbf{V}\Lambda\mathbf{V}^\mathbf{T}. \tag{8.6}$$

Columns in matrix $\mathbf{V}$ are known as *eigenvectors*, or *loadings*, and contain orthonormal vectors representing directions where data express the major variability (*principal components*). Since $\mathbf{S}$ is the variance–covariance matrix, variability in the direction of the ith *eigenvector*, or principal component (ith column in $\mathbf{V}$), is expressed by the associated variance (σ_i^2) and is represented by λ_i ($\lambda_i = \sigma_i^2$), the ith element of the diagonal matrix $\mathbf{\Lambda}$, also known as *eigenvalues*.

$$\Lambda = \begin{bmatrix} \lambda_1 & 0 & \cdots & 0 \\ 0 & \lambda_2 & \cdots & 0 \\ \vdots & \vdots & \ddots & \vdots \\ 0 & \cdots & \lambda_i & \cdots \\ \vdots & \vdots & \ddots & \vdots \\ 0 & 0 & \cdots & \lambda_m \end{bmatrix}. \tag{8.7}$$

Eigenvectors are unitary vectors, and every component in the ith vector represents the weights (*loadings*) of the original variables in that principal component and λ_i is the variance of data moving in that direction. Thus, matrix $\mathbf{V}$ allows transforming original data, $\mathbf{X}$, into the new representation space (principal components space) without loosing information. Transformation can be applied to an single observation ($\mathbf{t} = \mathbf{xV}$) or to the whole data set contained in the original matrix $\mathbf{X}$ resulting in a

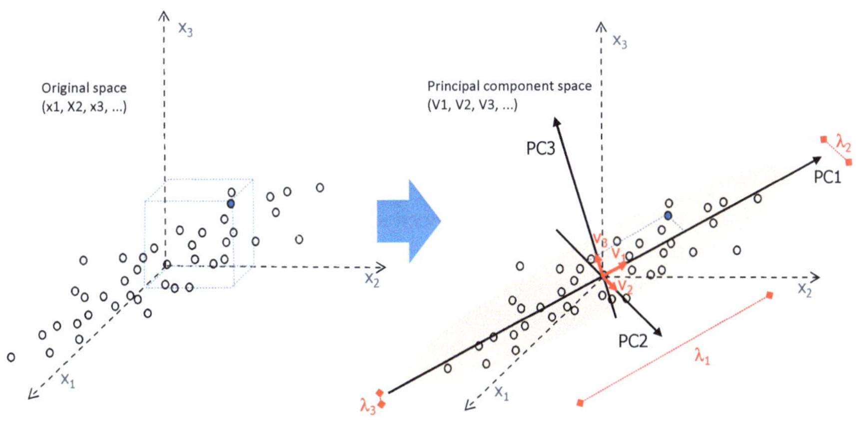

Fig. 8.2 Principal component space defined by ($\mathbf{V}_1$,$\mathbf{V}_2$,$\mathbf{V}_3$) and associated eigenvalues (λ_1, λ_2, λ_3)

new matrix of transformed data **T** (Fig. 8.2), also known as *scores*.

$$\mathbf{T} = \mathbf{X}\mathbf{V}. \tag{8.8}$$

Additional benefit of the unitary matrix **V** is that the inverse operation is carried out with the transpose ($\mathbf{V}\mathbf{V}^T = \mathbf{I}$). Thus, transformation of *scores*, **T**, into the original data is given by

$$\mathbf{X} = \mathbf{T}\mathbf{V}^T. \tag{8.9}$$

8.2.1 Dimension Reduction

One of the most important characteristics when applying PCA is the dimensionality reduction in the number of variables. This reduction is attained by retaining only the r principal components with larger *eigenvalues*. Assuming that eigenvalues in $\mathbf{\Lambda}$ are ordered according to their decreasing magnitude ($|\lambda_1| > |\lambda_2| > \cdots > |\lambda_r| > \cdots > |\lambda_m| \geq 0$) and that each one corresponds to the associated ith *eigenvector* in **V**, the total variance captured with the first r principal components can be computed as follows:

$$\Sigma_r = \sum_{i=1}^{r} \lambda_i. \tag{8.10}$$

Thus, when only r *eigenvectors* of **V** are considered, the $m \times m$ matrix **V** becomes $m \times r$ and is called **P**, the projection matrix that defines a lower dimension subspace, represented by r principal components.

Although there are many criteria to select the appropriate r (Scree test, cross-validation, minimum variance reconstruction [9], etc.), one of the most commonly used consists in selecting the r principal components that represent a percentage of the total variance of the data under study, as represented in (8.11)

$$r : \frac{1}{m} \sum_{i=1}^{r} \lambda_i \times 100 \geq Var(\%). \tag{8.11}$$

Observe that when original data has been standardized (unit variance), the total variance coincides with the number of variables (m) and this is the total accumulated variance in the original data. Another simple criterion consists in selecting those components with $\lambda_i > 1$, aiming to gather in every principal component, at least as many information (in terms of variance) as in a single-original variable (unit variance).

Following the previous example (Fig. 8.2), where initial space was described by three variables, the selection of $r = 2$ principal components will result in a projection $m \times 2$ matrix with two *eigenvectors*, $\mathbf{P} = [\mathbf{V}_1\mathbf{V}_2]$, that projects original data onto the subspace defined by $\mathbf{V}_1$ and $\mathbf{V}_2$ (Fig. 8.3).

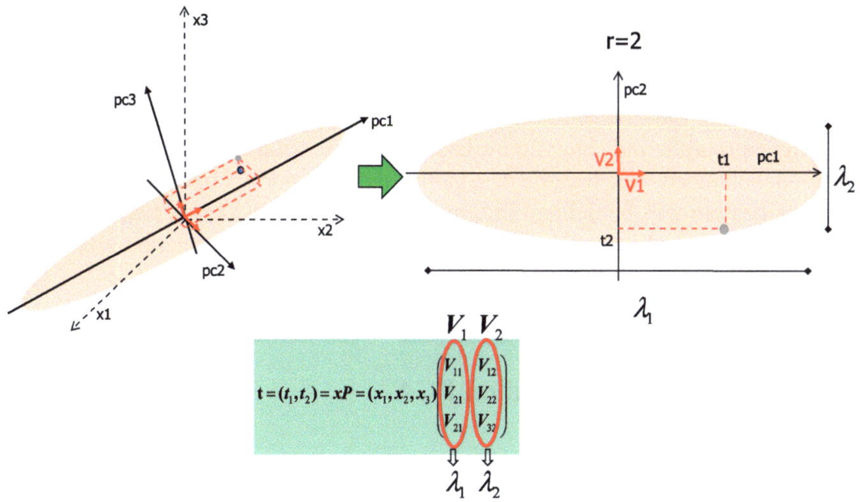

Fig. 8.3 Dimension reduction: Example from 3D to 2D

8.2.2 *The PCA Model: Projection and Residual Spaces*

Defining **P** as a $m \times r$ matrix, containing only the first r *eigenvectors* with larger *eigenvalues*; it is used, instead of the original **V**, to transform data from original space to the principal component space, resulting in a projection onto a lower dimensional space. The following equations represent this transformation for an array (set of observations) and a vector (single observation), respectively.

$$\mathbf{T} = \mathbf{XP} \tag{8.12}$$

$$\mathbf{t} = \mathbf{xP}. \tag{8.13}$$

The application of this projection operation (**P**) implies that some information contained in the original data is lost (corresponding to the discarded principal components). Therefore, when projecting back the scores, **T**, into the original m-dimensional space the data will not be exactly the same. This is represented by $\hat{\mathbf{X}}$ and $\hat{\mathbf{x}}$ in (8.14) and (8.15) as follows:

$$\hat{\mathbf{X}} = \mathbf{TP}^T \tag{8.14}$$

$$\hat{\mathbf{x}} = \mathbf{tP}^T. \tag{8.15}$$

The difference between **X** and $\hat{\mathbf{X}}$ is known as residual, $\tilde{\mathbf{X}}$, or error, **E**, matrix, $\mathbf{E} = \tilde{\mathbf{X}} = \mathbf{X} - \hat{\mathbf{X}}$ [7, 10]. It resumes the information of the $m - r$ not retained components (residual space). Thus, the complete PCA model can be described as

$$\mathbf{X} = \hat{\mathbf{X}} + \tilde{\mathbf{X}} = \hat{\mathbf{X}} + \mathbf{E} = \mathbf{TP}^T + \mathbf{E}. \tag{8.16}$$

E is a $n \times m$ matrix that contains a row vector for each observation, orthogonal to the *scores*, with the information not captured for the r components selected in the projection subspace. Decomposition described in (8.16) defines the PCA model that will be used for fault detection and isolation. Thus, fault detection will consist in defining appropriate indicators to evaluate the quality of a new observation w.r.t to this model obtained from historic data gathered during normal operation conditions. These indicators will be defined in terms of distance, considering both projection and residual of the observation after applying the model transformation given by **P**. Next subsections introduce these distance criteria.

8.3 Fault Detection Based on PCA

The purpose of defining a fault detection mechanism is to assess new observations w.r.t. a reference model, which defines the normal operation conditions, and gives a fitness measure. In this section, we assume that this model is given by (8.16), and in particular matrix **P** has been obtained with NOC data. Two complementary indicators are usually proposed for fault detection in multivariate process monitoring using PCA: The Hotteling's T^2 and the square prediction error (SPE) statistics. The first is a kind of square distance of the projected data to the center of the model, whereas the second measures the distance (square distance) of the observation to the projection subspace. The benefit of being defined as distances instead of vectors is that it allows a simple visualization of these indicators. Since new observations in a process are usually gathered periodically, this visualization results in simple *control charts* (this is a common terminology in process control) that are updated every time a new observation is obtained. This section details how these two indicators are defined and what information they provide (Sect. 8.3.1 for the T^2 index and Sect. 8.3.3 for the SPE index), as well as the detectability criteria for both indices (Sect. 8.3.2 for the T^2 index and Sect. 8.3.4 for the SPE index).

8.3.1 *Fault Detection in the Projection Subspace: The Hotteling's T^2 Index*

A simple measure of how an observation fits the PCA model is to evaluate how far is the projection of the observation from the center of the model (mean value). This can be computed with the Hotteling's T^2, that is, a weighted distance that considers the importance of every component using λ_i as weight.

$$T_{\mathbf{x}}^2 = \sum_{i=1}^{r} \frac{{t_i}^2}{\lambda_i}. \tag{8.17}$$

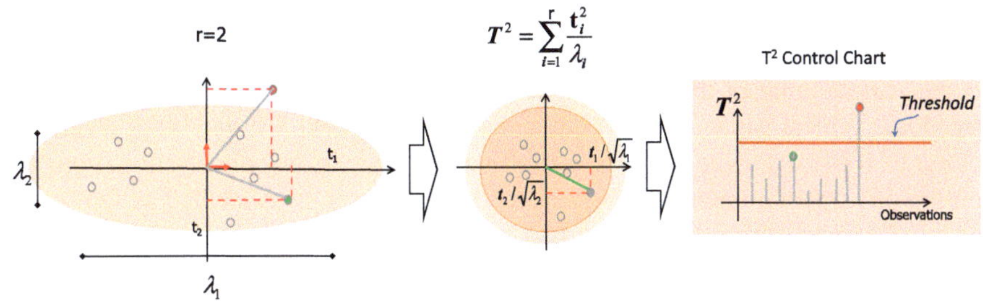

Fig. 8.4 T^2: Graphical representation

Since the objective is to evaluate only the projection, the indicator considers only the retained first r principal components. Figure 8.4 offers a graphical representation ($r = 2$).

$T_{\mathbf{x}}^2$ stands for the T^2 indicator of an observation $\mathbf{x}$, and t_i is the projection of this single observation $\mathbf{x}$ into the ith principal component (ith *score*) computed using (8.13). This expression can be rewritten using the *score* vector $\mathbf{t}$, which is a row vector containing the *scores* for the r retained principal components of $\mathbf{x}$ in this way:

$$T_{\mathbf{x}}^2 = \mathbf{t}\Lambda^{-1}\mathbf{t}^T, \qquad (8.18)$$

Λ^{-1} being a diagonal matrix defined as

$$\Lambda^{-1} = \begin{bmatrix} \frac{1}{\lambda_1} & 0 & \cdots & 0 \\ 0 & \frac{1}{\lambda_2} & \cdots & 0 \\ \vdots & \vdots & \ddots & \vdots \\ 0 & 0 & \cdots & \frac{1}{\lambda_r} \end{bmatrix}. \qquad (8.19)$$

Substituting (8.13) in (8.18), the relation of the original (standardized) observation with the T^2 index can be identified in this way:

$$T_{\mathbf{x}}^2 = \mathbf{x}\mathbf{P}\Lambda^{-1}\mathbf{P}^T\mathbf{x}^T = \mathbf{x}\mathbf{D}\mathbf{x}^T, \qquad (8.20)$$

with $\mathbf{D} = \mathbf{P}\Lambda^{-1}\mathbf{P}^T$. If data have been auto-scaled, this is a Mahalanobis distance measure of each observation to the center of the model. A graph, or control chart, built with such data is useful for detecting variations in the projection space, defined by the r principal components, greater (with respect to the centered data) than common-cause variations, while preserving the structure correlation gathered by the PCA model.

The control limit for the T^2 statistic for a given significance level α (τ_α) can be estimated using the following expression [1, 5]:

$$\tau_\alpha = \frac{(n^2 - 1)r}{n(n - r)} F_\alpha(r, n - r), \qquad (8.21)$$

where n is the number of observations/rows of the data matrix $\mathbf{X}$ (in this case filled with NOC data); r is the number of principal components retained; and $F_\alpha(r, n-r)$ is the critical point of the Fisher–Snedecor distribution for r and $n-r$ degrees of freedom. Typical values for the confidence level α are 90, 95, and 99%, where the closer to 100% the value is, the fewer false alarm rate. So, any observation projected on the principal component subspace, $\hat{\mathbf{x}}$, that surpasses τ_α is labeled as faulty due to the T^2 index. Of course, there is a trade-off between false alarms and missed detection, and consequently the selection of such confidence level is a decision to be made.

8.3.2 Fault Detectability in the Projection Subspace

In order to determine the detectability based on T^2 for a given PCA model, let us define a faulty observation ($\mathbf{x}_f$) as an additive simple fault represented by an unitary direction vector ($\boldsymbol{\xi}_i$) and with a fault magnitude f_i:

$$\mathbf{x}_f = \mathbf{x}_o + f_i \boldsymbol{\xi}_i. \tag{8.22}$$

The fault-free measurement of the observation is represented by $\mathbf{x}_o$. As stated in the previous subsection, $T^2_{\mathbf{x}} > \tau_\alpha$ in order to be detectable by means of the T^2 index, so if we substitute the previous equation into (8.20):

$$\left(\mathbf{x}_o\mathbf{P} + f_i\boldsymbol{\xi}_i\mathbf{P}\right)\Lambda^{-1}\left(\mathbf{P}^T\boldsymbol{\xi}_i^T f_i + \mathbf{P}^T\mathbf{x}_o^T\right) > \tau_\alpha, \tag{8.23}$$

$$\left(\hat{\mathbf{x}}_o + f_i\hat{\boldsymbol{\xi}}_i\right)\Lambda^{-1}\left(\hat{\boldsymbol{\xi}}_i^T f_i + \hat{\mathbf{x}}_o^T\right) > \tau_\alpha. \tag{8.24}$$

Vectors $\hat{\boldsymbol{\xi}}_i = \boldsymbol{\xi}_i\mathbf{P}$ and $\hat{\mathbf{x}}_o = \mathbf{x}_o\mathbf{P}$ are the projections, onto the projection space, of the fault direction and fault-free values, respectively. Assuming that the expected value of $\mathbf{x}_o$ is an exemplar value that falls close to the center of the model ($T^2_{\mathbf{x}_o} \approx 0$), then detectability on the projection space depends on the following relation:

$$\left\| f_i\boldsymbol{\xi}_i\mathbf{P}\Lambda^{\frac{-1}{2}} \right\|^2 > \tau_\alpha. \tag{8.25}$$

$\Lambda^{\frac{-1}{2}}$ is a diagonal matrix containing the root square of the eigenvalues associated with the r retained principal components in $\mathbf{P}$ defined as follows:

$$\Lambda^{\frac{-1}{2}} = \begin{bmatrix} \frac{1}{\sqrt{\lambda_1}} & 0 & \cdots & 0 \\ 0 & \frac{1}{\sqrt{\lambda_2}} & \cdots & 0 \\ \vdots & \vdots & \ddots & \vdots \\ 0 & 0 & \cdots & \frac{1}{\sqrt{\lambda_r}} \end{bmatrix}. \tag{8.26}$$

Thus, faults occurring in a direction orthogonal to the projection hyperplane ($\left\| \hat{\xi}_i \right\| = 0$) will never be detected with this criterion. This is the case of faults affecting variables with a low representation in the loading matrix, **P**, since their variations mainly will appear in the residual space.

8.3.3 Fault Detection in the Residual Subspace: The SPE Index

When an observation, **x**, produces a large variation out of the projection hyperplane described by the r principal components, this means that the data structure has been broken and the expected correlation condition among some variables does not fulfill. Consequently, there is a fault in the system. Thus, in order to detect this type of faults, it is necessary to observe the residual subspace and analyze the error ($\tilde{\mathbf{x}}$) component of **x**. Usually, the squared prediction error (SPE) of the observation is used for that purpose [8]:

$$SPE_{\mathbf{x}} = \sum_{i=1}^{m} \left(x_i - \hat{x}_i\right)^2, \tag{8.27}$$

$$SPE_{\mathbf{x}} = \left(\mathbf{x} - \hat{\mathbf{x}}\right)\left(\mathbf{x} - \hat{\mathbf{x}}\right)^T. \tag{8.28}$$

Computation of $\hat{\mathbf{x}} = (\hat{x}_1, \ldots, \hat{x}_i, \ldots, \hat{x}_m)$ is done using (8.15). It corresponds to the modulus of the residual vector, so it can also be expressed as

$$SPE_{\mathbf{x}} = \tilde{\mathbf{x}}\tilde{\mathbf{x}}^T = \|\tilde{\mathbf{x}}\|^2. \tag{8.29}$$

Again, $\tilde{\mathbf{x}} = \mathbf{x}\left(\mathbf{I} - \mathbf{P}\mathbf{P}^T\right) = \mathbf{x}\tilde{\mathbf{C}}$ is the projection of **x** into the residual subspace given by matrix $\tilde{\mathbf{C}} = \left(\mathbf{I} - \mathbf{P}\mathbf{P}^T\right)$.

Assuming that after preprocessing the data (centering, scaling, etc.), NOC observations can be modeled using a multivariate normal distribution, the SPE will present very small values (typically associated with noise) close to zero, while T^2 presents a larger variation (associated with the variability of the normal operating conditions). Consequently, the SPE will be much more sensitive than T^2 to variations in the process structure (faults), since they alter the correlation structure of the observed data and it is expected that these changes will be observed in the residual vector and consequently emphasized by the SPE square index. Usually, the index is represented in a control chart (Fig. 8.5). The SPE threshold can be computed using the approximation presented in [4] for a given confidence level α (assuming that observations were auto-scaled and the correlation matrix has full rank):

$$\delta_\alpha = \theta_1 \left[\frac{h_0 c_\alpha \sqrt{2\theta_2}}{\theta_1} + 1 + \frac{\theta_2 h_0 (h_0 - 1)}{\theta_1^2}\right]^{\frac{1}{h_0}}, \tag{8.30}$$

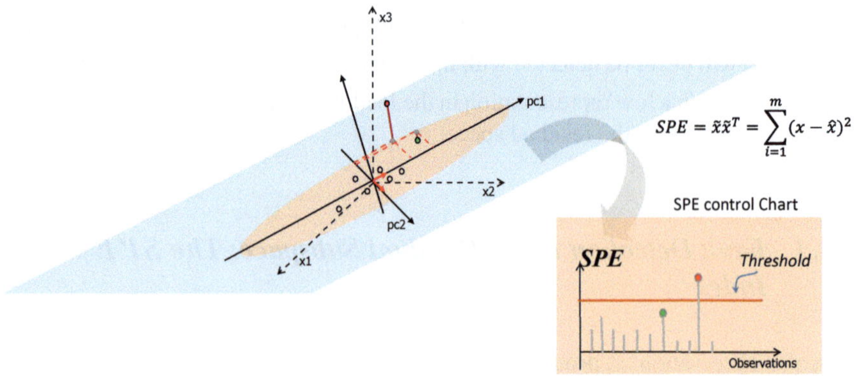

Fig. 8.5 SPE: Graphical representation

with

$$\theta_j = \sum_{i=r+1}^{m} \lambda_i^j \quad h_0 = 1 - \frac{2\theta_1\theta_3}{3\theta_2^2}, \tag{8.31}$$

where c_α is the value of the normal distribution for a significance level α and m is the number of variables in the original space. Whenever an observation has a value greater than the SPE and/or T^2 control limit, it is labeled as faulty (fault detection).

8.3.4 Fault Detectability in the Residual Subspace

In order to detect a fault in the residual subspace, its projection in this subspace must be greater than δ_α. Using the faulty observation defined in (8.22) and substituting it in (8.29), a detectable fault must hold that

$$\begin{aligned} \left\|\tilde{\mathbf{x}}_f\right\|^2 &> \delta_\alpha \\ \left\|\mathbf{x}_o\tilde{\mathbf{C}} + f_i\boldsymbol{\xi}_i\tilde{\mathbf{C}}\right\|^2 &> \delta_\alpha. \end{aligned} \tag{8.32}$$

As stated in Sect. 8.3.2, fault-free situations mostly project into the projection space ($\hat{\mathbf{x}}_o$). Consequently, their presence into the residual subspace ($\tilde{\mathbf{x}}_o$) is approximately null, and $SPE_{\mathbf{x}_o} \approx 0$. As a result, the fault appearance is the one responsible for surpassing the SPE limit, which allows to simplify the previous expression in this way:

$$\|f_i\tilde{\boldsymbol{\xi}}_i\|^2 > \delta_\alpha. \tag{8.33}$$

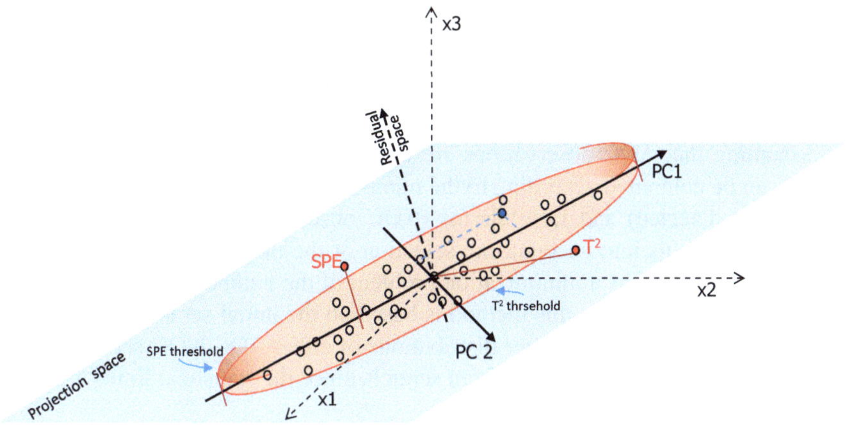

Fig. 8.6 Graphical representation of the PCA space decomposition

Furthermore, from a geometrical point of view, if we admit that normal conditions are below this threshold, in the worst situation, the fault requires a magnitude double to move outside the threshold when moving in the opposite direction. Thus, the detectability conditions are guaranteed for faults twofold larger than the interval confidence used to compute the statistical limits, as follows:

$$\|f_i\tilde{\boldsymbol{\xi}}_i\|^2 > 2\delta_\alpha. \tag{8.34}$$

Figure 8.6 presents a graphical representation of the original space and the space decomposition given by PCA, indicating the control limits for $T^2(\tau_\alpha)$ and the $SPE(\delta_\alpha)$ indices and the PCS for all NOC observations (points inside the ellipsoid delimited by τ_α). Two abnormal observations (red dots) illustrate a faulty observation (red dot) whose SPE value surpasses δ_α and an unusual (or extreme) observation whose T^2 value surpasses τ_α.

8.3.5 *False Alarms and Missed Detections When Evaluating the Model*

As stated in Sect. 8.2, the statistical model of a process is built up from a series of observations of the process under Normal Operating Conditions (NOC) (denoted as training set). Taking this model as a base, the acceptance limit of all the observations is defined on the basis of the variability observed in the data used to build the model. Sections 8.3.1 and 8.3.3 presented the corresponding statistical limits for T^2 and the SPE, respectively, based on a certain confidence level α. Based on these distributions, the number of false alarms and missed detections found in the training set can be determined. In case that either the number of false alarms or missed detections within the training set is significantly different from the expected values, then the statistical limits have to be computed based on the SPE and T^2 values within the training set. Therefore, the first step when monitoring a process is to check the validity of the statistical limits used for fault detection.

Assuming that AOC observations are available, the detection capability of the model can be computed according to the number of missed detections (AOC observations not detected) and false alarms (NOC observations incorrectly labeled as faulty). Taking this into account, the selection of the number of principal components can be defined as a minimization problem of the number of missed detection and/or false alarms. Since this method is based on the initial set of NOC and AOC observations, whenever the inner distributions of either AOC/NOC set change, the statistical model has to be rebuilt from scratch, since the statistical limits are based on the number of principal components.

8.4 Fault Isolation

Fault isolation on PCA models is usually conducted through the *contribution analysis* of the statistical indices T^2 and SPE. The idea is to compute the influence of each original variable in the total magnitude of SPE or T^2. Usually, a graphical representation is used to help identifying the larger values and consequently the variables that more contribute to the faulty, or *out of control* situation. Note that contribution plots only indicate which variables are related to the fault, but they do not reveal the primary cause of it [6]. Finally, note that the information contained in the contributions can be used to determine both the fault direction (which variables are involved) and magnitude (contribution of each variable) affecting an observation.

8.4.1 Fault Isolation in the Projection Subspace: T^2 Contributions

When a fault is detected, because the index T^2 takes an excessively large value (over the threshold), it is necessary to analyze which variables, x_j, from the original space, $\mathbf{x} = (x_1, x_2, \dots, x_j, \dots, x_m)$ are responsible for this situation. To deal with this challenge, consider (8.17) and observe that ${T_x}^2$ for a given observation, $\mathbf{x}$, is obtained as the addition of as many terms $\frac{{t_i}^2}{\lambda_i}$ as r components retained in the model. From (8.13), we can deduce that every component, t_i of the *scores* vector is obtained as a contribution of original variables weighted by the corresponding components of the ith loading vector

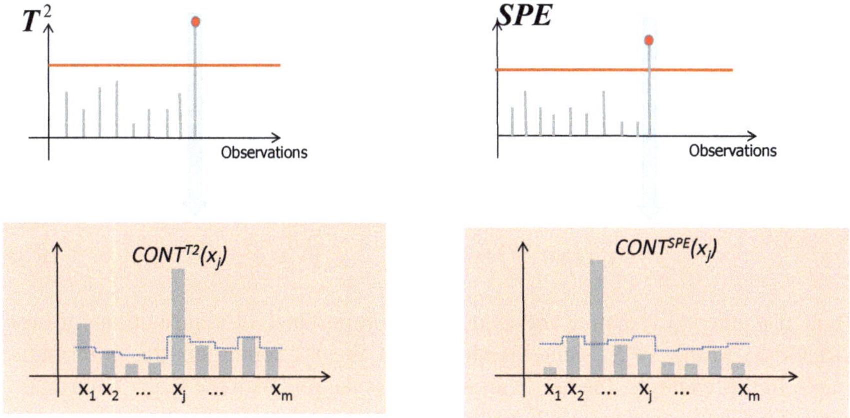

Fig. 8.7 Graphical representation of contributions for T^2 (left) and SPE (right)

$$t_i = \left[x_1, x_2, \ldots, x_j, \ldots, x_m\right] \begin{bmatrix} P_{1,i} \\ P_{2,i} \\ \cdots \\ P_{j,i} \\ \cdots \\ P_{m,i} \end{bmatrix}. \tag{8.35}$$

Thus, contribution of every variable x_j to the t_i component of ${T_x}^2$ can be expressed as in (8.36) [11].

$$cont(t_i, x_j) = \frac{t_i x_j P_{j,i}}{\lambda_i}. \tag{8.36}$$

Finally, the total contribution of variable x_j to ${T_x}^2$ can be computed as the addition of the individual contributions of x_j to every t_i as

$$CONT^{{T_x}^2}(x_j) = \sum_{i=1}^{r} cont(t_i, x_j) = \sum_{i=1}^{r} \frac{t_i x_j P_{j,i}}{\lambda_i}. \tag{8.37}$$

Contributions of each variable to the index T^2 in (8.37) are magnitudes, easily representable in a graph as depicted in Fig. 8.7 (left). Thus, it is easy to identify those variables that most contribute to the a large T^2. Statistical thresholds, based on expected values of these contributions obtained during normal operation conditions, can also be used to evaluate he importance of these contributions. In the example of Fig. 8.7, variables x_1 and x_j present the largest contributions and consequently are identified as the most probable location of the fault, in terms of involved variables.

8.4.2 Diagnosis in the Residual Subspace: SPE Contributions

Likewise T^2, considering that the SPE_x for a given observation, **x**, is given by (8.27), it can be stated that every variable, x_j, has an additive contribution to the SPE_x given by the following expression:

$$CONT^{SPE}(x_j) = \left(x_j - \hat{x}_j\right)^2 = \tilde{x}_j^2. \tag{8.38}$$

Thus, it is easy to visualize and evaluate the importance of contributions of every variable x_j to the SPE_x value. Right side of Fig. 8.7 represents this contribution graph. In this example, the third variable (x_3) is the one with the larger contribution to the SPE and consequently this allows identifying the possible fault cause. Also, statistical thresholds can be obtained for these contributions from the distributions of observations used to build the model in normal operation conditions (represented in the same picture by the threshold dotted line).

8.5 Illustrative Example

This section illustrates the data-driven approach presented in this chapter using a previous work where data from ARIANE missions were analyzed [2]. The example considers a reduced subset of 5 variables governing the Vulcain engine measured as average values for 15 different missions (name of variables has been omitted for confidentiality reasons). Table 8.1 represents the original data (before centering and standardization) for these 15 missions. From data in the table, matrix **X** is filled after centering (zero mean) and standardizing (unit variance). The covariance matrix (8.5) and the eigenvalue decomposition (8.6) have been computed. From the decomposition, the eigenvalues and associated percentage of total variance are $\lambda_1 = 2.68(53.7\%)$, $\lambda_2 = 1.15(22.9\%)$, $\lambda_3 = 0.67(53.3\%)$, $\lambda_4 = 0.46(9.3\%)$, and $\lambda_5 = 0.04(0.8\%)$. After applying the criteria of retaining the 99% of the total variance (8.11), a matrix **P** with four *eigenvectors* has been obtained as the projection model. Figure 8.8 represents these loadings.

From a simple observation of Fig. 8.8, we can corroborate that variables x_1 and x_2 have very similar behavior (similar components in all the loadings). We also observe that $\boldsymbol{P}_1$ represents mainly variability in variable x_1 and x_4, whereas $\boldsymbol{P}_2$ is contributed mainly by x_2 and x_4, and $\boldsymbol{P}_4$ is directly related to x_5. On the other hand, $\boldsymbol{P}_3$ captures in similar way the behavior of x_2 and x_5 and presents a negative (opposite sign) influence of x_3.

Although this is a small dataset (only 5 variables and 15 observations), some interesting results can be derived from a multivariate statistical monitoring. In order to evaluate the data, the matrix $\boldsymbol{P}$ has been used to project (Eq. 8.15) the data corresponding to the 15 missions and after computing T^2 ($\alpha = 90\%$) and SPE ($\alpha = 95\%$)

Table 8.1 Original data: 5 variables and 15 missions

x_1	x_2	x_3	x_4	x_5
121.9	91.137	108.096	126.2	24.2
120.9	94.685	111.838	124.3	19.8
124.1	90.322	107.391	126.9	19.2
122.2	90.025	106.646	125.9	20.06
119.8	86.933	103.966	123.8	19.5
125.745	89.343	105.897	129.197	21.75
123.43	89.125	107.203	129.044	21.79
123.883	94.878	107.497	126.674	20.06
125.353	89.303	107.976	129.283	22.43
126.0153	93.7609	109.373	128.8173	22.52
125.3861	92.4001	111.612	128.9062	23.32
124.4702	90.0061	109.7599	128.2996	24.32
121.3318	87.8128	105.9164	125.1422	17.68
123.4016	88.6698	110.0603	126.7221	17.38
121.654	85.324	110.419	124.701	17.71

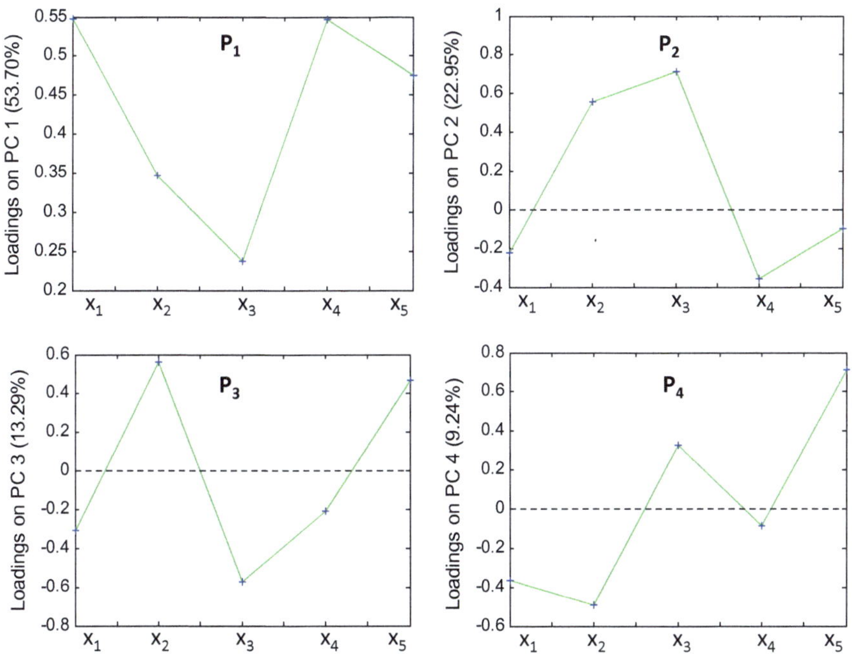

Fig. 8.8 Loadings of the first four principal components

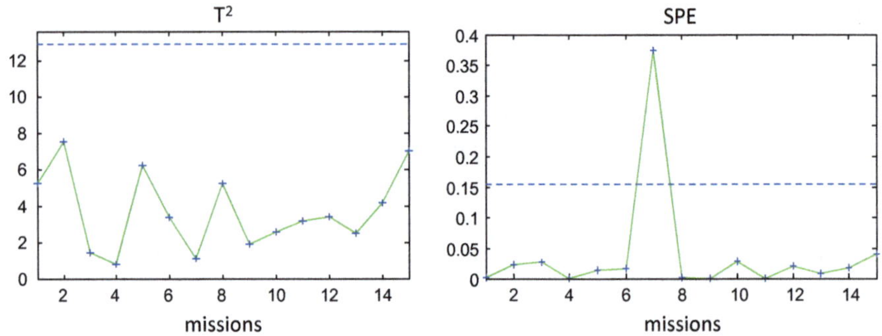

Fig. 8.9 T^2 and SPE for the 15 missions

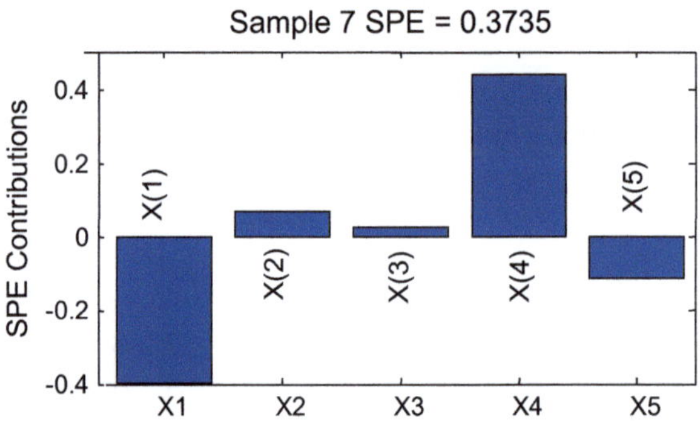

Fig. 8.10 SPE Contributions for mission number 7

for all of them using (8.17 and (8.27), respectively, resulting in the values depicted in Fig. 8.9.

From the analysis of T^2, we observe that all the missions are under the threshold. Probably, the short number of data used in the construction of the model evidences the differences in the values of this index but there are no significant variations. Since T^2 represents the main variability of data, only major changes will be detected with it. However, from the analysis of SPE (95% confidence limit correspond to dashed line in Fig. 8.9), we observe that mission 7 presents a very high value, over the threshold. This indicates that data from this flight do not preserve the same data structure that the others have. Then, a singularity is suspected in mission (observation) number 7. A fault was detected.

Once a fault is detected in an observation, the contribution analysis for fault isolation can be activated. With this aim, the contribution analysis for mission number 7 is performed using decomposition of SPE given by (8.38). These contributions are depicted in Fig. 8.10. In this case, the sign of contributions has been preserved for a better understanding. From the quick analysis of SPE contributions of mission 7, we can observe that variables x_1 and x_4 present the larger contributions; consequently, these are isolated as the responsible for the faulty situations of mission 7.

Acknowledgements This work has been developed within the eXiT (https://exit.udg.edu) research group (2017 SGR 1551) and supported by the CROWDSAVING project (Ref. TIN2016-79726-C2-2-R), funded by the Spanish Ministerio de Industria y Competitividad within the Research, Development and Innovation Program oriented toward the Societal Challenges.

References

1. Anderson, T.W.: An Introduction to Multivariate Statistical Analysis. Wiley, New York (1984)
2. Barta, C., Meléndez, J., Colomer, J.: Off line Diagnosis of Ariane Flights Using PCA, pp. 704–708. Elsevier, New York (2007)
3. Isermann, R., Ballé, P.: Trends in the application of model-based fault detection and diagnosis of technical processes. Control Eng. Pract. **5**(5), 709–719 (1997). http://www.sciencedirect.com/science/article/pii/S0967066197000531
4. Jackson, J.E., Mudholkar, G.S.: Control procedures for residuals associated with principal component analysis. Technometrics **21**(3), 341–349 (1979). https://doi.org/10.2307/1267757
5. Johnson, R.A., Wichern, D.W.: Applied Multivariate Statistical Analysis. Prentice-Hall, Upper Saddle River (1988)
6. Kourti, T.: Application of latent variable methods to process control and multivariate statistical process control in industry. Int. J. Adapt. Control Signal Process. **19**(4), 213–246 (2005). https://doi.org/10.1002/acs.859
7. MacGregor, J.F.: Multivariate statistical approaches to fault detection and isolation. In: SAFEPROCESS (2003)
8. Nomikos, P., MacGregor, J.F.: Monitoring Batch Process Using Multiway Principal Componet Analysis. AIChE J. **40**(8), 1361–1375 (1994). http://www3.interscience.wiley.com/journal/109063733/abstract
9. Qin, S.J., Dunia, R.: Determining the number of principal components for best reconstruction. J. Process Control **10**, 245–250 (2000)
10. Russell, E.L., Chiang, L.H., Braatz, R.D.: Data-driven Methods for Fault Detection and Diagnosis in Chemical Processes (Advances in Industrial Control), 1st edn. Springer, Berlin (2000)
11. Wise, B.M., Gallagher, N.B., Bro, R., Shaver, J.M., Windig, W., Koch, R.S.: Chemometrics Tutorial for PLS _ Toolbox and Solo. Eigenvector Research Incorporated (2006)

Chapter 9
Discrete-Event Systems Fault Diagnosis

Alban Grastien and Marina Zanella

Keywords Discrete-event systems · Automata theory · Events · Diagnosis · Diagnosability

9.1 Introduction

In this chapter, we discuss the problem of diagnosis of discrete-event systems. A discrete-event system, DES for short, is a model for a dynamic system where the state evolves through discrete (rather than continuous) occurrences, called events.

DESs are typically used to model systems that are discrete in nature, for instance, mechanical systems with on/off positions or processes such as controllers. It is natural to represent the evolution of such systems through events that, for diagnosis purposes, can be seen as instantaneous: the switch opens, the controller moves to the next state in its specification, etc.

DESs can also be used to reason about systems for which one would be more inclined to use continuous variables. In the case of a power system, for instance, one might want to discuss voltage, which is typically a continuous variable; however, for diagnosis purposes it may be sufficient to partition the voltage domain (i.e., to discretize said domain) into three discrete values: *null*, *low*, and *normal*. Similarly, in order to reason about the state of a circuit breaker based on its input and output voltage, a precise power equation might be available; however, it may be sufficient to look at the discrete question of whether the voltage at both ends is equal.

A. Grastien (✉)
Data61, Canberra, Australia
e-mail: alban.grastien@data61.csiro.au

M. Zanella
University of Brescia, Brescia, Italy
e-mail: marina.zanella@unibs.it

T. Escobet et al. (eds.), *Fault Diagnosis of Dynamic Systems*,
https://doi.org/10.1007/978-3-030-17728-7_9

The problem of building the DES model from a system is a hot research topic and will not be discussed in this chapter. We will assume that the DES is an input to the diagnosis problem.

The seminal work on diagnosis of DESs is the 1995 paper in Transactions on Automated Control by Meera Sampath and her co-authors [31]. This paper is still relevant and identifies the main ideas of the problem.

Reasoning about DESs is computationally hard. In systems of equations with unknown continuous variables, one can either compute analytically the value of these variables or estimate these values, and then iteratively improve these estimates until one gets arbitrarily close to the real value. This is not possible in discrete settings, where the best option is generally to perform a *search* (i.e., enumerate the possible solutions and verify their validity). Consequently, many problems in discrete domains are NP-hard. When considering DESs, such problems can be PSPACE-hard, which means that DES algorithms are often much slower than their counterparts in other domains.

Another characteristic of research on DESs is the emphasis on diagnosability. This property is particularly useful to determine the abstraction level at which you want to model the system.

When should one consider to use a DES to reason about his own system? One of the important features of DESs is the fact that they handle dynamic systems, i.e., systems for which it is important to reason not only about what has been observed but also about the *order* in which these observations were made. Second, DESs are good at handling discrete aspects of the system behavior. Many real-world systems cannot be modeled only with continuous variables and their derivatives; such systems have several operational *modes* that feature different derivatives: hysteresis behaviors, saturations, switches, etc. There remains obviously the question of whether the model should be a DES or a hybrid model (i.e., a model that includes both discrete and continuous variables). Finally, DESs offer a very clean formalism that can make it easier both building and providing explanations to a user than using continuous variables or continuous variables and their derivatives.

Section 9.2 presents the notation that we use in this chapter and defines the notion of a DES diagnosis problem and its solution. In Sect. 9.3, we discuss diagnosis algorithms. Section 9.4 is dedicated to DES diagnosability. Finally, in Sect. 9.5, we provide some pointers and short descriptions of extensions relevant to the previous sections.

9.2 Problem Definition

9.2.1 Discrete-Event Systems

A DES is essentially an automaton where transitions are labeled by events. Those events are partitioned into different categories, according to their purpose for diagnosis: observable, unobservable, faulty. There exist different definitions of a DES,

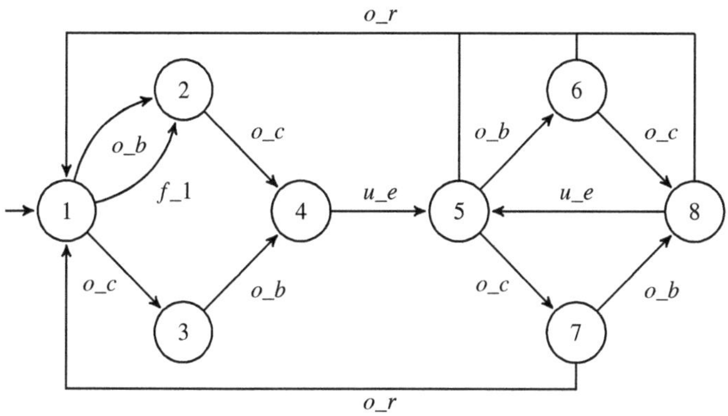

Fig. 9.1 A DES with eight states, three observable events, one unobservable event, and one faulty event

for instance, with several events on transitions or "unstable states" where the system cannot stay indefinitely; we settle here for a simple definition.

Definition 9.1 A *discrete-event system* is a finite automaton $A = \langle \mathcal{Q}, \mathcal{T}, \mathcal{E}, \mathcal{E}_o, \mathcal{E}_f, q_{init} \rangle$ where

- $\mathcal{Q}$ is a finite set of *states* and $q_{init} \in \mathcal{Q}$ is the initial state;
- $\mathcal{E}$ is a finite set of *events*; $\mathcal{E}_o \subseteq \mathcal{E}$ is the set of *observable events*; $\mathcal{E}_f \subseteq \mathcal{E} \setminus \mathcal{E}_o$ is the set of *faulty events* (or simply *faults*); and
- $\mathcal{T} \subseteq \mathcal{Q} \times \mathcal{E} \times \mathcal{Q}$ is a finite set of *transitions* that satisfy the following condition (for determinism):

$$\langle q, e, q_1' \rangle \in \mathcal{T} \ \wedge \ \langle q, e, q_2' \rangle \in \mathcal{T} \ \Rightarrow \ q_1' = q_2'.$$

Automata are represented graphically as in Fig. 9.1. States are displayed as circles, and the initial state (here: 1) is indicated by an incoming edge with no source. The transition $\langle q, e, q' \rangle$ is represented by an edge from q to q' labeled with e. In our examples, observable events will be prefixed with $o\_$ and faulty events with $f\_$; other events are prefixed with $u\_$ (for unobservable).

To simplify the notations, without loss of generality, we assume that every DES is deterministic, i.e., that (i) the initial state is known precisely (instead of ranging over a set of initial states), and (ii) given a state and an event, the next state is known. Transforming a nondeterministic DES into a deterministic one is trivial: if the DES includes a number of initial states $\{q_1, \ldots, q_k\}$, we can create a single dummy state q_{init} and dummy transitions from this state to the states $q_i \in \{q_1, \ldots, q_k\}$ (cf. Fig. 9.2). Similarly if there are several transitions leaving the same state labeled with the same event, it is possible to split each of these transitions into two transitions, one of which represents the nondeterministic effect of the event (cf. Fig. 9.3). Such changes do not modify the semantics of the DES significantly.

The DES represents how the system state changes over time.

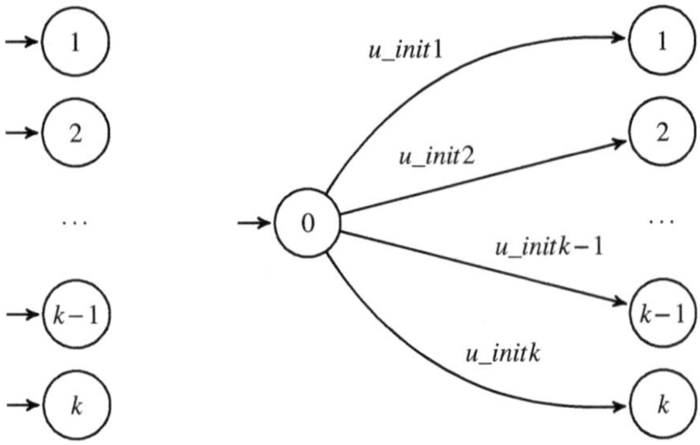

Fig. 9.2 Transforming a DES with several initial states into a DES with one initial state

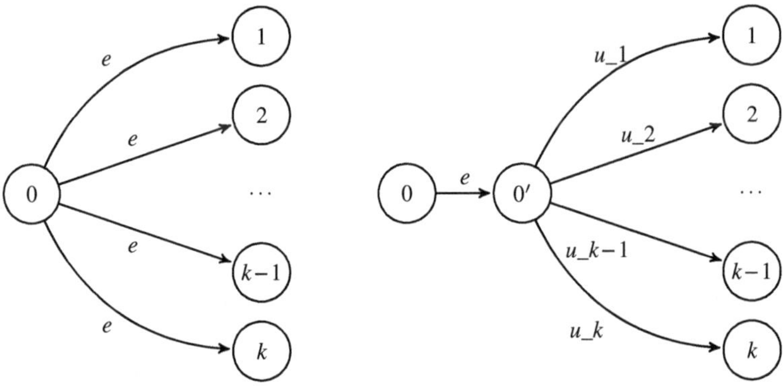

Fig. 9.3 Transforming a DES with nondeterministic transitions into a DES with deterministic transitions

Definition 9.2 (*DES Semantics*) Given the DES $A = \langle \mathcal{Q}, \mathcal{T}, \mathcal{E}, \mathcal{E}_o, \mathcal{E}_f, q_{\text{init}} \rangle$, a *path* is a double sequence $\rho = \langle [q_0, \ldots, q_k], [e_1, \ldots, e_k] \rangle$ of states and events linked through transitions, i.e., a sequence that satisfies the condition $\forall i \in \{1, \ldots, k\}$. $\langle q_{i-1}, e_i, q_i \rangle \in \mathcal{T}$. The path is also denoted $q_0 \xrightarrow{e_1} q_1 \xrightarrow{e_2} \ldots \xrightarrow{e_k} q_k$ or concisely $q_0 \xrightarrow{e_1 e_2 \ldots e_k} q_k$. An *execution* is a path that starts in the initial state ($q_0 = q_{\text{init}}$); if ρ is an execution, we write $\rho \in A$. A *trace* σ is the label (i.e., the sequence of events) $e_1 e_2 \ldots e_k$ of some path. The *language* of the DES is the set $\mathcal{L}(A) \subseteq \mathcal{E}^\star$ of all traces that label some execution. Since the DES is deterministic, a trace σ and a state q define uniquely the state reached from q through σ: $q[\sigma]$ is defined as the state q' such that $q \xrightarrow{\sigma} q'$.

It is assumed that the model represents all possible system behaviors, i.e., for every way that the system may behave, there is at least one execution in the model that represents this behavior. This is different from other modeling frameworks where the faulty behavior is often left unspecified. However, this is not restrictive as it is possible to model the fact that the system behavior is unknown after a fault occurrence. This

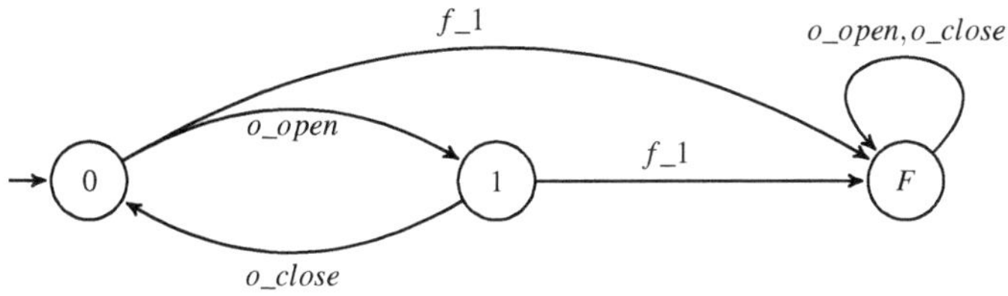

Fig. 9.4 A DES with no knowledge of the faulty behavior

is illustrated in Fig. 9.4: in nominal conditions, the system alternates between states 0 and 1 and the observable events o_open and the o_close are interleaved. The lack of knowledge is about the behavior of the system when faulty is modeled by means of a single state, F: since we do not know the order of occurrence of the observable events, according to the model these events may occur in any order. Obviously, more precise models lead to more precise diagnoses.

When the system performs an execution $q \xrightarrow{\sigma} q'$, it generates an observation that is precisely the sequence of observable events in the trace.

Definition 9.3 The *(certain) observation* $obs(\sigma)$ of the trace $\sigma \in \mathcal{E}^\star$ is the projection of σ on the observable events, formally:

$$obs(\sigma) = \begin{cases} \varepsilon & \text{if } \sigma = \varepsilon, \\ o\ obs(\sigma') & \text{if } \sigma = o\sigma' \text{ and } o \in \mathcal{E}_o, \\ obs(\sigma') & \text{if } \sigma = e\sigma' \text{ and } e \in \mathcal{E} \setminus \mathcal{E}_o, \end{cases}$$

where ε is the empty sequence. We write $\mathcal{L}_{obs}$ the language of observations that can be generated by the system, i.e., $\mathcal{L}_{obs} = \{obs(\sigma) \mid \sigma \in \mathcal{L}(A)\}$.

Notice that the assumption that the DES is deterministic is not restrictive from the point of view of the diagnosis task as such a task does not take as input a trace σ, which, if A is deterministic, corresponds to just one path. Instead, the diagnosis task takes as input the observation $obs(\sigma)$, that is, what has been perceived of the occurred path (typically through sensors). Since several distinct deterministic traces can have the same projection on $\mathcal{E}_o$, the same as when traces are nondeterministic, assuming that A is deterministic does not simplify the diagnosis task.

9.2.2 Diagnosis

The last ingredient needed to define the diagnostic problem is the fault hypothesis.

Definition 9.4 A *fault hypothesis* $\delta \subseteq \mathcal{E}_f$ is a set of faulty events. The *hypothesis* of the trace σ is the set of faulty events that occur in the trace, formally:

$$h(e_1 \dots e_k) = \{e_1, \dots, e_k\} \cap \mathcal{E}_f.$$

We also use the notation $2^{\mathcal{E}_f} = \{\delta \subseteq \mathcal{E}_f\}$ to represent the powerset of $\mathcal{E}_f$, i.e., the collection of subsets of $\mathcal{E}_f$. Consequently, we use the notation $\delta \in 2^{\mathcal{E}_f}$.

We are now in a position to define the problem of diagnosis, whose solution is a set of fault hypotheses.

Definition 9.5 Given the DES $A = \langle \mathcal{Q}, \mathcal{T}, \mathcal{E}, \mathcal{E}_o, \mathcal{E}_f, q_{init} \rangle$, given an observation θ relevant to it, a diagnosis problem instance is the pair (A, θ). The solution of such an instance, which is called *diagnosis* and denoted $\Delta(A, \theta)$, is the set of fault hypotheses of the traces that produce the observation:

$$\Delta(A, \theta) = \{\delta \in 2^{\mathcal{E}_f} \mid \exists \sigma \in \mathcal{L}(A) \cdot obs(\sigma) = \theta \wedge h(\sigma) = \delta\}. \tag{9.1}$$

Each trace σ in the formula and the corresponding execution are called *explanations* of the observation.

The parameters A or θ will be dropped from the expression $\Delta(A, \theta)$ when it is assumed that they are obvious from the context.

Example Let us consider the system modeled in Fig. 9.1. Assume that the sequence of events $\theta_1 = o\_b, o\_c, o\_c, o\_b, o\_b$ has been observed while the system was running. The only explanation for θ_1 is the following: $\rho_1 = 1 \xrightarrow{o\_b} 2 \xrightarrow{o\_c} 4 \xrightarrow{u\_e} 5 \xrightarrow{o\_c} 7 \xrightarrow{o\_b} 8 \xrightarrow{u\_e} 5 \xrightarrow{o\_b} 6$ with trace $\sigma_1 = o\_b, o\_c, u\_e, o\_c, o\_b, u\_e, o\_b$. The hypothesis associated with ρ_1 is $h(\sigma_1) = \emptyset$. Therefore, the diagnosis is $\Delta(\theta_1) = \{\emptyset\}$, which means that the faulty event definitely did not occur.

Assume now that the observation is $\theta_2 = o\_c, o\_c, o\_b, o\_b$. The only explanation for this observation is $\rho_2 = 1 \xrightarrow{f\_1} 2 \xrightarrow{o\_c} 4 \xrightarrow{u\_e} 5 \xrightarrow{o\_c} 7 \xrightarrow{o\_b} 8 \xrightarrow{u\_e} 5 \xrightarrow{o\_b} 6$. The diagnosis is therefore $\Delta(\theta_2) = \{\{f\_1\}\}$, which means that the faulty event definitely did occur.

Finally, assume that the observation is $\theta_3 = o\_c, o\_b, o\_c$. There are this time several explanations: $\rho_3 = 1 \xrightarrow{o\_c} 3 \xrightarrow{o\_b} 4 \xrightarrow{u\_e} 5 \xrightarrow{o\_c} 7$; $\rho_3' = 1 \xrightarrow{f\_1} 2 \xrightarrow{o\_c} 4 \xrightarrow{u\_e} 5 \xrightarrow{o\_b} 6 \xrightarrow{o\_c} 8$; and $\rho_3'' = 1 \xrightarrow{f\_1} 2 \xrightarrow{o\_c} 4 \xrightarrow{u\_e} 5 \xrightarrow{o\_b} 6 \xrightarrow{o\_c} 8 \xrightarrow{u\_e} 5$. In general, we will ignore explanations that include non-observable suffixes, i.e., we are only interested in diagnosing the system after the last observable event occurred, as it is assumed that there is no sufficient information about the state since this last observed event. The diagnosis is therefore $\Delta(\theta_3) = \{\emptyset, \{f\_1\}\}$, which means that, based on the observation gathered so far, it is impossible to determine whether the faulty event f_1 occurred. ◊

If the explanations that include non-observable suffixes are ignored, Eq. (9.1) has to be updated as follows:

$$\Delta(A, \theta) = \{\delta \in 2^{\mathcal{E}_f} \mid \exists \sigma \in \mathcal{L}(A) \cdot \sigma \in (\mathcal{E}^\star \mathcal{E}_o) \cup \{\varepsilon\} \cdot obs(\sigma) = \theta \wedge h(\sigma) = \delta\}.$$

9.3 How to Perform Diagnosis

Most complete methods for computing diagnoses of DES follow, in one way or another, an approach that consists in tracking the belief state.[1] This approach can be summarized as follows. First, reformulate the model so that each state includes the diagnostic information, i.e., the set of faults that affected the system on the way to the current state. Second, compute the set of (reformulated) states that the system could be in at the end of the observation that has been gathered thus far; this set is called the *belief state*. Third, extract from the belief state the diagnostic information.

The approach, as presented here, is fairly naïve and in many instances impractical. There are a number of techniques that can be used to simplify the task. Diagnosability analysis can be utilized to identify how much the observability of the system can be reduced without the precision of the diagnosis being compromised. The operations described further on can also be reordered. For instance, the global model of the system is defined as the synchronous product of its component models. Consequently, this global model is generally exponentially large with respect to the number of components. It is possible instead to use the observation to decide which part of the global model needs to be generated [2, 29]. Finally, symbolic methods described at the end of this section are also a good way to address this problem.

9.3.1 Incorporating the Diagnostic Information into the States

In our definition of a diagnostic problem, we assume that faults are modeled by the set of events $\mathcal{E}_f$. The belief state approach requires the faults to be embedded in the states of the model. In this subsection, we formalize this notion and show how the model can be reformulated so that this property is guaranteed.

Definition 9.6 A DES model $A = \langle \mathcal{Q}, \mathcal{T}, \mathcal{E}, \mathcal{E}_o, \mathcal{E}_f, q_{init} \rangle$ is *state-fault* if there exists a *fault labeling* function $D : \mathcal{Q} \to 2^{\mathcal{E}_f}$ such that, for all traces $\sigma \in \mathcal{L}(A)$, $D(q_{init}[\sigma]) = h(\sigma)$.

The state-fault property indicates that, if the system is known to be in state q, then the set of faulty events that occurred on the system is known to be $D(q)$. This is a very powerful property, because it indicates that if we know that the execution $q \xrightarrow{\sigma} q'$ occurred on the system, then the diagnostic information is included in the state q' and, therefore, the specific trace σ can be ignored. Notice, however, if the system is only known to be in either q_1, or q_2, …, or q_k, then the set of faulty events is $D(q_1)$, or $D(q_2)$, …, or $D(q_k)$.

Models are sometimes built as state-fault. For instance, if a fault corresponds to a component being physically broken, then the behavior of this component will

[1]Notable exceptions include methods based on model checking [10] or SAT problem solving [13, 16].

be significantly different once the fault has occurred, and it is very likely that the states reachable through traces with a fault will be disjoint from those without. As, however, models are not always state-fault, we need to introduce the notion of state-fault reformulation.

Definition 9.7 Given a non-state-fault DES model A, a *state-fault reformulation* of A is a state-fault DES A' with the same sets of events and the same language: $\mathcal{L}(A') = \mathcal{L}(A)$.

Lemma 9.1 indicates that, for diagnosis purposes, a non-state-fault DES can be replaced by its reformulation.

Lemma 9.1 *Let A be a DES model and let A' be a state-fault reformulation of A. For any observation θ the following property holds:*

$$\Delta(A, \theta) = \Delta(A', \theta).$$

A model can be reformulated automatically, as described in the following.

Definition 9.8 Given a DES model $A = \langle \mathcal{Q}, \mathcal{T}, \mathcal{E}, \mathcal{E}_o, \mathcal{E}_f, q_{\text{init}} \rangle$ the *automatic reformulation* of A is the model $A' = \langle \mathcal{Q}', \mathcal{T}', \mathcal{E}', \mathcal{E}_o', \mathcal{E}_f', q'_{\text{init}} \rangle$ where

- $\mathcal{Q}' = \mathcal{Q} \times 2^{\mathcal{E}_f}$, and $q'_{\text{init}} = \langle q_{\text{init}}, \emptyset \rangle$,
- $\mathcal{E}' = \mathcal{E}$, $\mathcal{E}_o' = \mathcal{E}_o$, $\mathcal{E}_f' = \mathcal{E}_f$, and
- $\mathcal{T}' = \{\langle \langle q, \mathcal{F} \rangle, e, \langle q', \mathcal{F}' \rangle \rangle \in \mathcal{Q}' \times \mathcal{E}' \times \mathcal{Q}' \mid \langle q, e, q' \rangle \in \mathcal{T} \ \wedge \ \mathcal{F}' = \mathcal{F} \cup (\{e\} \cap \mathcal{E}_f)\}$.

Notice that, since A is assumed to be deterministic, also A' is deterministic. Hence, we can denote the only state in A' that can be reached starting from a state $\langle q, \mathcal{F} \rangle$ when a string of events $\sigma \in \mathcal{E}^\star$ has occurred as $\langle q, \mathcal{F} \rangle[\sigma]$.

Lemma 9.2 *The automatic reformulation as presented in Definition 9.8 is state-fault.*

Proof Consider a path on either the model A or the automatic reformulation A'. It is quite easy to see that there is an equivalent path in the other finite automaton (for any $q \xrightarrow{e} q'$ transition, there is a $\langle q, \mathcal{F} \rangle \xrightarrow{e} \langle q', \mathcal{F}' \rangle$ transition, and conversely). This implies that the model and its automatic reformulation share the same language, that is, $\mathcal{L}(A') = \mathcal{L}(A)$.

We now show that the automatic reformulation is state-fault. Consider a path $\langle q_0, \mathcal{F}_0 \rangle \xrightarrow{e_1} \dots \xrightarrow{e_k} \langle q_k, \mathcal{F}_k \rangle$ on the automatic reformulation A' starting with $q_0 = q_{\text{init}}$ and $\mathcal{F}_0 = \emptyset$. We show that for all i, $\mathcal{F}_i$ is precisely $\mathcal{F}_i = \{e_1, \dots, e_i\} \cap \mathcal{E}_f$. We prove it by induction:

- This is true for $i = 0$ since the unique initial state $\langle q_0, \mathcal{F}_0 \rangle$ is such that $\mathcal{F}_0 = \emptyset$.
- Assume that, for some i, $\mathcal{F}_i = \{e_1, \dots, e_i\} \cap \mathcal{E}_f$ does hold. By definition of the automatic reformulation, $\mathcal{F}_{i+1} = \mathcal{F}_i \cup (\{e_{i+1}\} \cap \mathcal{E}_f) = \{e_1, \dots, e_i, e_{i+1}\} \cap \mathcal{E}_f$.

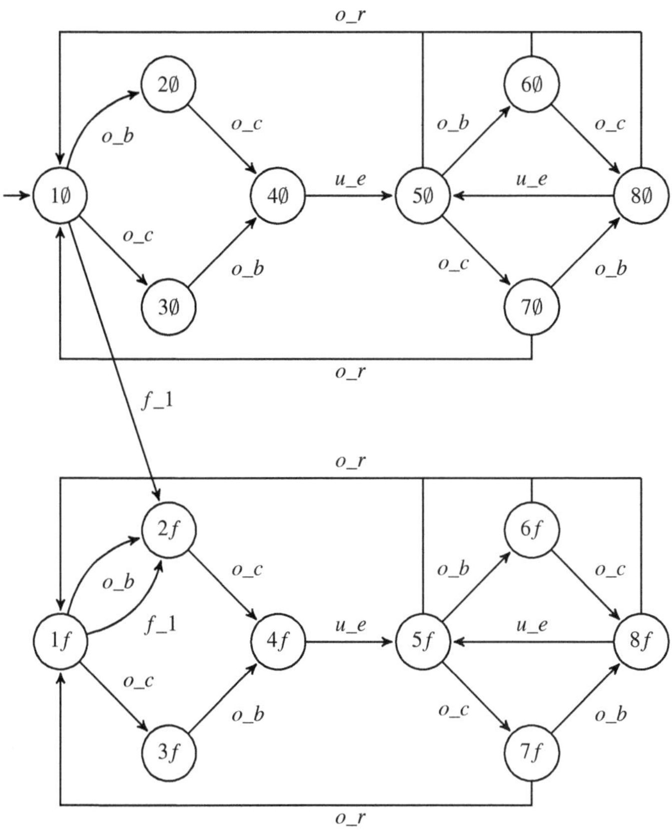

Fig. 9.5 Automatic reformulation of Fig. 9.1: $x\emptyset$ represents the state $\langle x, \emptyset\rangle$; yf represents the state $\langle y, \{f\_1\}\rangle$

Let $D : \mathcal{Q} \times 2^{\mathcal{E}_f} \to 2^{\mathcal{E}_f}$ be the function that associates each state $\langle q, \mathcal{F}\rangle$ of A' with the second argument $\mathcal{F}$ of the state itself. We can prove that D is the function that associates each state with the set of faults that occur on all paths to the state. Indeed, let $\langle q_0, \mathcal{F}_0\rangle \xrightarrow{e_1} \ldots \xrightarrow{e_k} \langle q_k, \mathcal{F}_k\rangle$ be a path on the automatic reformulation. By the result above, $D(\langle q_0, \mathcal{F}_0\rangle[e_1 \ldots e_k]) = D(\langle q_k, \mathcal{F}_k\rangle) = \mathcal{F}_k = \{e_1, \ldots, e_k\} \cap \mathcal{E}_f = h(e_1 \ldots e_k)$. □

The automatic reformulation of the model from Fig. 9.1 is provided in Fig. 9.5.

The automatic reformulation transforms a model with $|\mathcal{Q}|$ states in a model with up to $|\mathcal{Q}| \times 2^{|\mathcal{E}_f|}$ states. This means that, for a nontrivial number of faults, the automatic reformulation will become intractable. There are several ways to avoid this issue. First, this transformation is generally not performed explicitly [2]. Second, if each fault is diagnosed separately, then a reformulation is produced for each fault, which is at most twice as large as the original model. Finally, notice that the automatic reformulation of a state-fault model has the same size as the original model (since there will be only one state $\langle q, D(q)\rangle$ for each q in the original model).

9.3.2 Diagnosis by Belief State

The "belief state" refers to the knowledge (or "belief") about the current state of the system. This knowledge can be made explicit by enumerating the states that are logically consistent with the knowledge itself. This is how we represent the belief state: as the smallest set of states that we know for sure includes the current state.

Definition 9.9 Let $A = \langle \mathcal{Q}, \mathcal{T}, \mathcal{E}, \mathcal{E}_o, \mathcal{E}_f, q_{\text{init}} \rangle$ be a model, and $\mathcal{S} \subseteq \mathcal{Q}$ be a subset of states. Let $\theta \in \mathcal{E}_o{}^\star$ be an observation. The *belief state* of θ from $\mathcal{S}$ in model A is the set of states $\mathcal{B}_A(\mathcal{S}, \theta) \subseteq \mathcal{Q}$ that end any path that starts in a state in $\mathcal{S}$ and generates the observation θ:

$$\mathcal{B}_A(\mathcal{S}, \theta) = \{q' \in \mathcal{Q} \mid \exists q \in \mathcal{S} \cdot \exists \sigma \in \mathcal{E}^\star \cup \{\varepsilon\} \cdot obs(\sigma) = \theta \ \wedge \ (q \xrightarrow{\sigma} q') \in A\}. \tag{9.2}$$

If we assume that the sequence of events σ ends with an observable event (unless it is empty), Eq. (9.2) has to be changed as follows:

$$\mathcal{B}_A(\mathcal{S}, \theta) = \{q' \in \mathcal{Q} \mid \exists q \in \mathcal{S} \cdot \exists \sigma \in (\mathcal{E}^\star \mathcal{E}_o) \cup \{\varepsilon\} \cdot obs(\sigma) = \theta \ \wedge \ (q \xrightarrow{\sigma} q') \in A\}.$$

When the DES model is obvious from the context, we write $\mathcal{B}(\mathcal{S}, \theta)$. When $\mathcal{S}$ is the singleton $\{q_{\text{init}}\}$, we omit it.

We extend the definition of the fault labeling function D to a set of states this way: $D(\mathcal{S}) = \{D(q) \mid q \in \mathcal{S}\}$. Diagnosis by belief state relies on Theorem 9.1.

Theorem 9.1 *Let A be a state-fault DES model with function D. Let θ be an observation. The following result holds:*

$$\Delta(A, \theta) = D(\mathcal{B}(\theta)).$$

Proof This proof is based on the most general definitions of diagnosis and belief state, as of Eqs. (9.1) and (9.2), respectively. The proof of the theorem relevant to the case when only traces ending with an observable event are taken into account is left to the reader.

Let us consider the RHS of the equality in the theorem. Based on Definition 9.9,

$$\mathcal{B}(\theta) = \{q' \in \mathcal{Q} \mid \ \exists \sigma \in \mathcal{L}(A) \cdot obs(\sigma) = \theta \ \wedge \ (q_{init} \xrightarrow{\sigma} q') \in A\}.$$

Since each $q' \in \mathcal{B}(\theta)$ can be represented as $q_{init}[\sigma]$,

$$D(\mathcal{B}(\theta)) = \{D(q_{init}[\sigma]) \mid \exists \sigma \in \mathcal{L}(A) \cdot obs(\sigma) = \theta\}. \tag{9.3}$$

According to Definition 9.6, $D(q_{init}[\sigma]) = h(\sigma)$. By replacing $D(q_{init}[\sigma])$ with $h(\sigma)$ in (9.3), we obtain

$$D(\mathcal{B}(\theta)) = \{h(\sigma) \mid \exists \sigma \in \mathcal{L}(A) \cdot obs(\sigma) = \theta\},$$

which equals $\Delta(A, \theta)$ as given in Definition 9.5. □

Example We illustrate Theorem 9.1 with the system of Fig. 9.1 reformulated in Fig. 9.5. Consider the observation $\theta_1 = o\_b, o\_c, o\_c, o\_b, o\_b$. The belief state in the state-fault model is $\mathcal{B}(\theta_1) = \{6\emptyset\}$. Therefore, the diagnosis is $\Delta(\theta_1) = \emptyset$.

Similarly for observation $\theta_2 = o\_c, o\_c, o\_b, o\_b$, the belief state is $\mathcal{B}(\theta_2) = \{6f\}$ and the diagnosis is $\Delta(\theta_2) = \{\{f\_1\}\}$.

Finally, for observation $\theta_3 = o\_c.o\_b, o\_c$, the belief state is $\mathcal{B}(\theta_3) = \{7\emptyset, 8f\}$ and the diagnosis is therefore $\Delta(\theta_3) = \{\emptyset, \{f\_1\}\}$. ◊

9.3.3 Computing the Belief State

We provided a formal definition of the belief state, but not a procedure to compute it. This is the topic of this subsection. The main property that makes the computation of the belief state easy is the fact that it can be incremental. This is formally captured by the following lemma.

Lemma 9.3 *For any set of states $\mathcal{S}$ and any pair of sequences of observable events θ_1 and θ_2, the following holds:*

$$\mathcal{B}(\mathcal{S}, \theta_1\theta_2) = \mathcal{B}(\mathcal{B}(\mathcal{S}, \theta_1), \theta_2).$$

The proof essentially comes from the fact that any path that generates $\theta_1\theta_2$ can be split into two connected paths that generate θ_1 and θ_2, respectively, and conversely.

Proof This proof is based on the definitions of diagnosis and belief state specialized to the case when explanations do not include unobservable suffixes. The proof of the lemma in the most general case is left to the reader.

Let us consider the RHS of the equality in Lemma 9.3. Based on Definition 9.9,

$$\mathcal{B}(\mathcal{S}, \theta_1) = \{q_1 \in \mathcal{Q} \mid \exists q \in \mathcal{S} \cdot \exists \sigma_1 \in (\mathcal{E}^\star \mathcal{E}_o) \cup \{\varepsilon\} \cdot obs(\sigma_1) = \theta_1 \ \wedge \ (q \xrightarrow{\sigma_1} q_1) \in A\}.$$

By exploiting the same definition, we obtain

$$\mathcal{B}(\mathcal{B}(\mathcal{S}, \theta_1), \theta_2) = \{q_2 \in \mathcal{Q} \mid \exists q_1 \in \mathcal{B}(\mathcal{S}, \theta_1) \cdot \exists \sigma_2 \in (\mathcal{E}^\star \mathcal{E}_o) \cup \{\varepsilon\}.$$
$$obs(\sigma_2) = \theta_2 \ \wedge \ (q_1 \xrightarrow{\sigma_2} q_2) \in A\}.$$

By replacing in this equation the qualification of the states $q_1 \in \mathcal{B}(\mathcal{S}, \theta_1)$ as provided by the previous equation, we obtain

$$\mathcal{B}(\mathcal{B}(\mathcal{S}, \theta_1), \theta_2) = \{q_2 \in \mathcal{Q} \mid \exists q \in \mathcal{S} \cdot \exists \sigma_1 \in (\mathcal{E}^\star \mathcal{E}_o) \cup \{\varepsilon\}.$$
$$\exists \sigma_2 \in (\mathcal{E}^\star \mathcal{E}_o) \cup \{\varepsilon\} \cdot obs(\sigma_1) = \theta_1 \wedge obs(\sigma_2) = \theta_2 \wedge \ (q \xrightarrow{\sigma_1} q_1 \xrightarrow{\sigma_2} q_2) \in A\}.$$

Now let us consider the LHS of the equality in Lemma 9.3. Based on Definition 9.9,

$$\mathcal{B}(\mathcal{S}, \theta_1\theta_2) = \{q_2 \in \mathcal{Q} \mid \exists q \in \mathcal{S} \cdot \exists \sigma \in (\mathcal{E}^\star\mathcal{E}_o) \cup \{\varepsilon\} \cdot obs(\sigma) = \theta_1\theta_2 \wedge (q \xrightarrow{\sigma} q_2) \in A\}. \tag{9.4}$$

Since the sequence of events σ in (9.4) is projected in the concatenation of two sequences of observable events, $\theta_1\theta_2$, we can divide σ into two sequences, σ_1 and σ_2, where $obs(\sigma_1) = \theta_1$ and $obs(\sigma_2) = \theta_2$. Hence, $\sigma = \sigma_1\sigma_2$, where $\sigma_1 \in (\mathcal{E}^\star\mathcal{E}_o) \cup \{\varepsilon\}$ and $\exists\sigma_2 \in (\mathcal{E}^\star\mathcal{E}_o) \cup \{\varepsilon\}$. By replacing σ with the concatenation $\sigma_1\sigma_2$ in (9.4), we obtain

$$\mathcal{B}(\mathcal{S}, \theta_1\theta_2) = \{q_2 \in \mathcal{Q} \mid \exists q \in \mathcal{S} \cdot \exists \sigma_1 \in (\mathcal{E}^\star\mathcal{E}_o) \cup \{\varepsilon\}.$$

$$\exists \sigma_2 \in (\mathcal{E}^\star\mathcal{E}_o) \cup \{\varepsilon\} \cdot obs(\sigma_1) = \theta_1 \wedge obs(\sigma_2) = \theta_2 \wedge (q \xrightarrow{\sigma_1} q_1 \xrightarrow{\sigma_2} q_2) \in A\},$$

which proves the identity of Lemma 9.3. □

The incremental property of Lemma 9.3 is very powerful for two reasons. First, it shows that the belief state $\mathcal{B}(\theta_1)$ computed in order to produce the diagnosis $\Delta(\theta_1)$ can be leveraged to compute $\mathcal{B}(\theta_1\theta_2)$ when it comes to producing the diagnosis $\Delta(\theta_1\theta_2)$; we discuss this further in Sect. 9.3.4, which deals with the diagnoser. Second, it demonstrates that the computation of a belief state relevant to a sequence of events can be implemented as a sequence of belief state computations for single observable events.

More specifically, we define two operations that allow us to compute the belief state. First, given a belief state $\mathcal{B}$ and a set of events $\mathcal{E}'$, the *next* belief state is the set of states reached from some state in $\mathcal{B}$ through a single transition labeled with an event from $\mathcal{E}'$:

$$next_{\mathcal{E}'}(\mathcal{B}) = \{q' \in \mathcal{Q} \mid \exists q \in \mathcal{B} \cdot \exists e \in \mathcal{E}' \cdot q \xrightarrow{e} q'\}.$$

When the set $\mathcal{E}'$ contains a single event e, we write $next_e$ instead of $next_{\{e\}}$.

Second, given a belief state $\mathcal{B}$, the *silent-closure* of $\mathcal{B}$ is the set of states reached from some state in $\mathcal{B}$ only through unobservable transitions:

$$silent\text{-}closure(\mathcal{B}) = \{q' \in \mathcal{Q} \mid \exists q \in \mathcal{B} \cdot \exists \sigma \in (\mathcal{E} \setminus \mathcal{E}_o)^\star \cdot q \xrightarrow{\sigma} q'\}.$$

The silent-closure can be computed by repeatedly invoking the *next* function, which adds the states reached through unobservable events, until convergence is reached.

Lemma 9.4 *Let $\mathcal{B}$ be a belief state. Define $\mathcal{B}_i$ as follows:*

- $\mathcal{B}_0 := \mathcal{B}$,
- *for all index $i \in \mathbf{N}$, $\mathcal{B}_i := \mathcal{B}_{i-1} \cup next_{\mathcal{E}\setminus\mathcal{E}_o}(\mathcal{B}_{i-1})$,*
- *$\mathcal{B}_\infty$ is the set $\mathcal{B}_k$ such that $\mathcal{B}_k = \mathcal{B}_{k+1}$. ($\mathcal{B}_\infty$ is bound to exist since $\mathcal{B}_i$ grows monotonically and $\mathcal{Q}$ is finite.)*

Then $silent\text{-}closure(\mathcal{B}) = \mathcal{B}_\infty$.

The belief state $\mathcal{B}(e_1 \ldots e_k)$ after the observation $e_1 \ldots e_k$ can be computed as follows:

$$\mathcal{B}(e_1 \ldots e_k) = \textit{silent-closure} \circ \textit{next}_{e_1} \circ \cdots \circ \textit{silent-closure} \circ \textit{next}_{e_k}(\{q_{\text{init}}\}).$$

This leads us to the diagnosis algorithm of Fig. 9.6a. This algorithm assumes that the input is a state-fault model with fault labeling function D. It computes the belief state at the end of the observed sequence and extracts the diagnosis of the final belief state.

Let m be the number of transitions in the fault-state DES (i.e., up to $2^f \times m'$, where f is the number of faulty events and m' the number of transitions in the non-fault-state DES), and k be the number of observed events. For any given observed event, the silent-closure operation and the next operation must be performed once, which

a. Explicit implementation

- BELIEF-STATE
  Input: $A = \langle \mathcal{Q}, \mathcal{T}, \mathcal{E}, \mathcal{E}_o, \mathcal{E}_f, q_{\text{init}} \rangle$
  Input: $o_1, \ldots, o_k \in \mathcal{E}_o{}^\star$
  $\mathcal{S} := \{q_{\text{init}}\}$
  for all $i \in \{1, \ldots, k\}$
      $\mathcal{S} :=$ SILENT-CLOSURE$(A, \mathcal{S})$
      $\mathcal{S} :=$ NEXT$(A, \mathcal{S}, \{o_i\})$
  return $\{\delta \in 2^{\mathcal{E}_f} \mid \exists q \in \mathcal{S}.\ \delta = D(q)\}$

- SILENT-CLOSURE
  Input: $A = \langle \mathcal{Q}, \mathcal{T}, \mathcal{E}, \mathcal{E}_o, \mathcal{E}_f, q_{\text{init}} \rangle$
  Input: $\mathcal{S} \subseteq \mathcal{Q}$
  $\textit{Result} := \mathcal{S}$
  $\textit{New} := \mathcal{S}$
  while $\textit{New} \neq \emptyset$
      $\textit{NextStates} :=$ NEXT$(A, \textit{New}, \mathcal{E} \setminus \mathcal{E}_o)$
      $\textit{New} := \textit{NextStates} \setminus \textit{Result}$
      $\textit{Result} := \textit{Result} \cup \textit{NextStates}$
  return $\textit{Result}$

- NEXT
  Input: $A = \langle \mathcal{Q}, \mathcal{T}, \mathcal{E}, \mathcal{E}_o, \mathcal{E}_f, q_{\text{init}} \rangle$
  Input: $\mathcal{S} \subseteq \mathcal{Q}$
  Input: $\mathcal{E}' \subseteq \mathcal{E}$
  return $\{q' \mid \exists q \in \mathcal{S}.\ \exists e \in \mathcal{E}'.\ \langle q, e, q' \rangle \in \mathcal{T}\}$

b. Symbolic implementation

- BELIEF-STATE
  Input: $A = \langle \Phi_{\mathcal{Q}}, \Phi_{\mathcal{T}}, \mathcal{E}, \mathcal{E}_o, \mathcal{E}_f, q_{\text{init}} \rangle$
  Input: $o_1, \ldots, o_k \in \mathcal{E}_o{}^\star$
  $\varphi := \Phi_{q_{\text{init}}}$
  for all $i \in \{1, \ldots, k\}$
      $\varphi :=$ SILENT-CLOSURE(A, φ)
      $\varphi :=$ NEXT(A, φ, Φ_{o_i})
  return $\{\delta \in 2^{\mathcal{E}_f} \mid \varphi \wedge \Phi_\delta \not\equiv \bot\}$

- SILENT-CLOSURE
  Input: $A = \langle \Phi_{\mathcal{Q}}, \Phi_{\mathcal{T}}, \mathcal{E}, \mathcal{E}_o, \mathcal{E}_f, q_{\text{init}} \rangle$
  Input: $\varphi_S \in \textit{Forms}(V_{\mathcal{Q}})$
  $\varphi_R := \varphi_S$
  $\varphi_N := \varphi_S$
  while $\varphi_N \not\equiv \bot$
      $\varphi_{XS} :=$ NEXT$(A, \varphi_N, \Phi_{\mathcal{E}} \wedge \neg \Phi_{\mathcal{E}_o})$
      $\varphi_N := \varphi_{XS} \wedge \neg \varphi_R$
      $\varphi_R := \varphi_R \vee \varphi_{XS}$
  return φ_R

- NEXT
  Input: $A = \langle \Phi_{\mathcal{Q}}, \Phi_{\mathcal{T}}, \mathcal{E}, \mathcal{E}_o, \mathcal{E}_f, q_{\text{init}} \rangle$
  Input: $\varphi_S \in \textit{Forms}(V_{\mathcal{Q}})$
  Input: $\varphi_{\mathcal{E}'} \in \textit{Forms}(V_{\mathcal{Q}} \cup V_{\mathcal{E}} \cup V_{\mathcal{Q}'})$
  return $\textit{Ex}\ V_{\mathcal{E}} \cup V'_{\mathcal{Q}}.\ \big((\varphi_S \wedge \Phi_{\mathcal{T}} \wedge \varphi_{\mathcal{E}'})[V/V']\big)$

Fig. 9.6 Belief state-based diagnosis (left: explicit implementation; right: symbolic implementation)

requires $O(m)$ operations. Therefore, the complexity of the algorithm of Fig. 9.6.a is $O(k \times m)$.

9.3.4 Diagnoser

The diagnoser, proposed by Sampath et al. [31], is an extension of the belief state method that leverages from Lemma 9.3 to precompute all possible belief states in advance; this makes the diagnostic task trivial to perform once the observation is available.

Definition 9.10 Let $A = \langle \mathcal{Q}, \mathcal{T}, \mathcal{E}, \mathcal{E}_o, \mathcal{E}_f, q_{\text{init}} \rangle$ be a state-fault model with fault labeling function D. The *diagnoser* is the finite automaton $A' = \langle \mathcal{Q}', \mathcal{T}', \mathcal{E}', q'_{\text{init}} \rangle$ defined as follows:

- $\mathcal{Q}' = 2^{\mathcal{Q}}$ and $q'_{\text{init}} = \{q_{\text{init}}\}$,
- $\mathcal{E}' = \mathcal{E}_o$, and
- $\mathcal{T}' = \{\langle \mathcal{S}, e, \mathcal{S}' \rangle \in \mathcal{Q}' \times \mathcal{E}' \times \mathcal{Q}' \mid \mathcal{S}' = \textit{silent-closure} \circ \textit{next}_e(\mathcal{S})\}$.

Notice that the diagnoser is defined over the set of observable events. Notice also that the diagnoser is deterministic, i.e., given a sequence of observable events, there is only one path from the initial state q'_{init} labeled with this sequence. The diagnosis procedure based on the diagnoser relies on the result in Lemma 9.5.

Lemma 9.5 *Given an observation θ and the diagnoser $A' = \langle \mathcal{Q}', \mathcal{T}', \mathcal{E}', q'_{init} \rangle$, the following holds:*

$$\mathcal{B}(\theta) = q'_{init}[\theta].$$

Lemma 9.5 indicates that the belief state at the end of θ can be computed by following the unique path labeled by θ on the diagnoser.

Example The diagnoser of the DES in Fig. 9.5, which is the reformulation of the DES in Fig. 9.1, is given in Fig. 9.7. The states corresponding to each identifier are shown in Table 9.1.

Coming back to the examples provided before, consider the observation $\theta_1 = o\_b, o\_c, o\_c, o\_b, o\_b$. By construction, there is precisely one path on the diagnoser that corresponds to this sequence, here $01 \xrightarrow{o\_b} 12 \xrightarrow{o\_c} 13 \xrightarrow{o\_c} 14 \xrightarrow{o\_b} 16 \xrightarrow{o\_b} 15$, which ends up in a nominal state (i.e., a state denoting a normal behavior). Therefore, the diagnosis is $\Delta(\theta_1) = \{\emptyset\}$.

Consider the observation $\theta_2 = o\_c, o\_c, o\_b, o\_b$. This time, the path is $01 \xrightarrow{o\_c} 02 \xrightarrow{o\_c} 06 \xrightarrow{o\_b} 07 \xrightarrow{o\_b} 09$. This last state being faulty, the diagnosis is $\Delta(\theta_2) = \{\{f\_1\}\}$. Notice that, if the DES can be affected by just one fault, as in the current example, the faulty states (those with a black background in Fig. 9.7) could be merged, as this part of the diagnoser is essentially a sink.

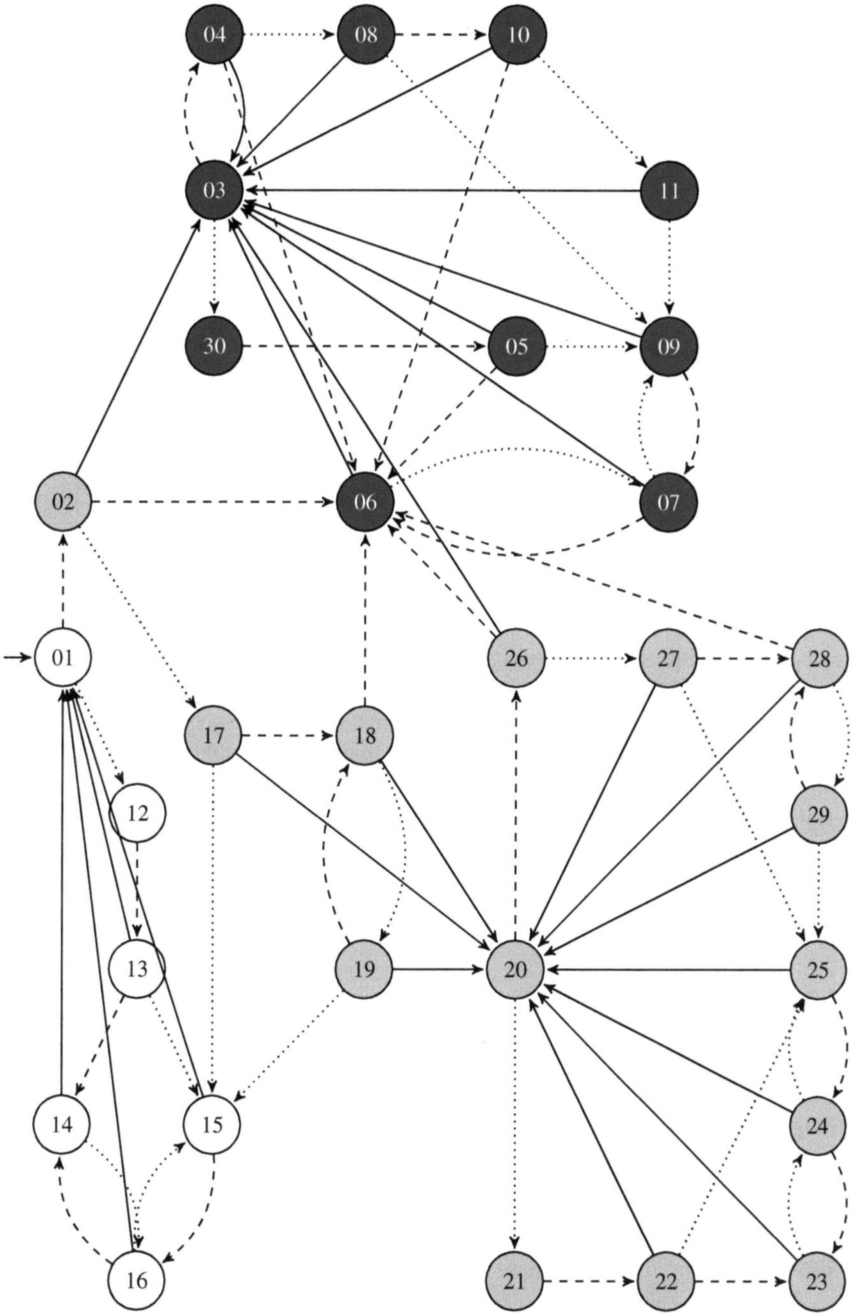

Fig. 9.7 Diagnoser of the DES in Fig. 9.5. Solid lines are labeled with *o_r*; dotted lines with *o_b*; dashed lines with *o_c*. A white background is associated with nominal states only; a gray background denotes ambiguous states; states with a black background are faulty

Table 9.1 Diagnoser states corresponding to the identifiers in Fig. 9.7

Id	State	Id	State	Id	State	Id	State	Id	State
01	$\{1\emptyset\}$	07	$\{8f\}$	13	$\{4\emptyset\}$	19	$\{8\emptyset, 6f\}$	25	$\{6\emptyset, 6f\}$
02	$\{3\emptyset, 4f\}$	08	$\{4f, 6f\}$	14	$\{7\emptyset\}$	20	$\{1\emptyset, 1f\}$	26	$\{3\emptyset, 3f, 4f\}$
03	$\{1f\}$	09	$\{6f\}$	15	$\{6\emptyset\}$	21	$\{2\emptyset, 2f\}$	27	$\{4\emptyset, 4f, 6f\}$
04	$\{3f, 4f\}$	10	$\{7f, 8f\}$	16	$\{8\emptyset\}$	22	$\{4\emptyset, 4f\}$	28	$\{7\emptyset, 7f, 8f\}$
05	$\{4f\}$	11	$\{6f, 8f\}$	17	$\{4\emptyset, 6f\}$	23	$\{7\emptyset, 7f\}$	29	$\{8\emptyset, 6f, 8f\}$
06	$\{7f\}$	12	$\{2\emptyset\}$	18	$\{7\emptyset, 8f\}$	24	$\{8\emptyset, 8f\}$	30	$\{2f\}$

Finally, consider the observation $\theta_3 = o\_c, o\_b, o\_c$. Here, the path is $01 \xrightarrow{o\_c} 02 \xrightarrow{o\_b} 17 \xrightarrow{o\_c} 18$, and the diagnosis is $\Delta(\theta_3) = \{\emptyset, \{f\_1\}\}$. State 18 is an ambiguous state of the diagnoser, since, when an observation reaches it, we do not know whether the DES is faulty or not. ◊

9.3.5 Symbolic Methods

Symbolic techniques for reasoning about DESs were originally developed in the context of model checking. The purpose of model checking is to verify whether an implementation represented by a state machine satisfies a specification described in some temporal logic [9]. Symbolic techniques, in particular, the development of binary decision diagrams (BDD) [5], made model checking applicable outside trivial examples although, as one can expect, they require careful utilization and they have shortcomings. We only provide a short introduction on this topic; for more information, we recommend the work from Schumann et al. [32] and the references in said work.

9.3.5.1 Propositional Logic

Propositional logic manipulates *propositions* (that are either *true* or *false*), which are either atomic propositions or composed as a collection of propositions linked through a logical connective. Formally, given a finite set of *propositional variables* $\mathcal{V}$, a *propositional formula* Φ is

- either a variable: v for some variable $v \in \mathcal{V}$;
- or a negation: $\neg\Phi'$ where Φ' is a propositional formula;
- or a disjunction: $\Phi_1 \vee \Phi_2$ where Φ_1 and Φ_2 are propositional formulas.

For convenience, redundant operators are defined:

- $\Phi_1 \wedge \Phi_2$ (conjunction) stands for $\neg((\neg\Phi_1) \vee (\neg\Phi_2))$.
- $\Phi_1 \rightarrow \Phi_2$ (implication) stands for $(\neg\Phi_1) \vee \Phi_2$.
- $\Phi_1 \leftrightarrow \Phi_2$ (equivalence) stands for $(\Phi_1 \rightarrow \Phi_2) \wedge (\Phi_2 \rightarrow \Phi_1)$.

- $\top$ (tautology) stands for $v \lor \neg v$ for some variable $v \in \mathcal{V}$.
- $\bot$ (contradiction) stands for $\neg\top$.

We write $Forms(\mathcal{V})$ the set of propositional formulas defined on the set of variables $\mathcal{V}$.

A *complete assignment* of $\mathcal{V}$ is a mapping $\alpha : \mathcal{V} \to \{T, F\}$ that associates each variable $v \in \mathcal{V}$ with either T (*true*) or F (*false*). An assignment α is *consistent* with propositional formula Φ, written $\alpha \models \Phi$, iff:

- either $\Phi = v$ is a variable and $\alpha(v) = T$;
- or $\Phi = \neg\Phi'$ is a negation and $\alpha \not\models \Phi'$;
- or $\Phi = \Phi_1 \lor \Phi_2$ is a disjunction and either $\alpha \models \Phi_1$ or $\alpha \models \Phi_2$.

Two propositional formulas that share the same set of consistent assignments are *equivalent*:

$$(\Phi_1 \equiv \Phi_2) \quad \Leftrightarrow \quad (\{\alpha : \mathcal{V} \to \{T, F\} \mid \alpha \models \Phi_1\} \;=\; \{\alpha : \mathcal{V} \to \{T, F\} \mid \alpha \models \Phi_2\}).$$

9.3.5.2 Symbolic Representation of Sets

The representation of a set is symbolic if its elements are not enumerated but are instead described as the property that only they share. This is a representation that we use on a daily basis, e.g., when we talk about the ripe tomatoes in a vegetable garden rather than enumerating every tomato that is ripe. We use this representation not only when we do not know all the elements in the set but also when it is convenient. For instance, a teacher would ask all the students on his left to work on some exercise while the other students would be asked to work on another exercise, rather than enumerating the students who are supposed to work on each exercise.

Let $\mathcal{O}$ be a finite set of objects. We want to be able to represent and manipulate subsets of $\mathcal{O}$. In our context, $\mathcal{O}$ will be either the set of states or the set of transitions of a DES.

We define a finite set of propositional variables $\mathcal{V}$ with the only requirement that their number should be at least the logarithmic number of the objects in $\mathcal{O}$: $|\mathcal{V}| \geq \log_2(|\mathcal{O}|)$.

It is then possible to use $Forms(\mathcal{V})$, the set of propositional formulas over $\mathcal{V}$, in order to represent every subset of $\mathcal{O}$. Formally, we define a function $\alpha : \mathcal{O} \to \mathcal{V} \to \{T, F\}$ that associates each object $s \in \mathcal{O}$ with a complete assignment $\alpha(s)$ of $\mathcal{V}$. This assignment is unique for each element (w.r.t. the equivalence function), i.e., $\forall\{s_1, s_2\} \subseteq \mathcal{O}.\ s_1 \neq s_2 \Rightarrow \alpha(s_1) \not\equiv \alpha(s_2)$. Notice that, because of the constraint on the size of $\mathcal{V}$, it is always possible to define such a function.

Example Consider the set of integers ranging between 0 and 20: $\mathcal{O} = \{0, \ldots, 20\}$. We define five variables $\mathcal{V} = \{v_0, \ldots, v_4\}$. The assignment function is as follows: consider the binary representation of the number s; the ith variable will evaluate to T iff the ith digit in the binary representation of s is 1. For instance,

$12_{10} = 01100_2$ (i.e., starting with the most significant bit), so the assignment $\alpha(12)$ is $\{v_0 \to F, v_1 \to T, v_2 \to T, v_3 \to F, v_4 \to F\}$. Similarly $\alpha(13) = \{v_0 \to F, v_1 \to T, v_2 \to T, v_3 \to F, v_4 \to T\}$. In this example, v_4 evaluates to F in $\alpha(s)$ iff s is an even number. ◊

Given a propositional formula $\Phi \in Forms(\mathcal{V})$ over $\mathcal{V}$, we now define the subset $Repr_\alpha(\Phi)$ of objects represented by Φ as the objects that match any consistent assignment of Φ:

$$Repr_\alpha(\Phi) \quad = \quad \{s \in \mathcal{O} \mid \alpha(s) \models \Phi\}.$$

If some assignment consistent with Φ matches no object of $\mathcal{O}$, then Φ is invalid. We can prove that equivalent propositional formulas represent the same subsets of objects: $(\Phi_1 \equiv \Phi_2) \Leftrightarrow Repr_\alpha(\Phi_1) = Repr_\alpha(\Phi_2)$. One of the goals of the tool that manipulates the propositional formulas is to find good (compact) representations of these formulas (this is briefly discussed in Sect. 9.3.5.5). For any subset $\mathcal{S}$ of objects, we write $\Phi_\mathcal{S}$ one of the formulas that satisfy $Repr_\alpha(\Phi_\mathcal{S}) = \mathcal{S}$.

Example According to the previous example, the following statements holds:

- $\Phi_{\{12\}} \equiv \neg v_0 \wedge v_1 \wedge v_2 \wedge \neg v_3 \wedge \neg v_4$.
- $\Phi_{\{13\}} \equiv \neg v_0 \wedge v_1 \wedge v_2 \wedge \neg v_3 \wedge v_4$.
- $\Phi_{\{12,13\}} \equiv \neg v_0 \wedge v_1 \wedge v_2 \wedge \neg v_3$.

It is worth noticing that the formula that represents the subset $\{12, 13\}$ is smaller than the formula for any of the subsets $\{12\}$ and $\{13\}$. This is precisely the goal of using symbolic representations: if the function α is defined in a "sensible" way, then many "sensible" sets of objects will be represented quite compactly. For instance, the set of even numbers of $\mathcal{S}$ in our example is represented by $\Phi_\mathcal{O} \wedge \neg v$ where v is the last variable regardless of the size of $\mathcal{S}$. ◊

In order to be able to manipulate subsets of objects through propositional formulas, we need to be able to translate the most common operators on subsets as propositional formula operators. Fortunately, this is very simple:

- $Repr_\alpha(\Phi_1) \cup Repr_\alpha(\Phi_2) = Repr_\alpha(\Phi_1 \vee \Phi_2)$.
- $Repr_\alpha(\Phi_1) \cap Repr_\alpha(\Phi_2) = Repr_\alpha(\Phi_1 \wedge \Phi_2)$.
- $\mathcal{S} \setminus Repr_\alpha(\Phi) = Repr_\alpha(\Phi_\mathcal{S} \wedge \neg\Phi)$.
- $(Repr_\alpha(\Phi) = \emptyset) \Leftrightarrow (\Phi \equiv \bot)$.

The proof of these statements is left as an exercise to the reader.

9.3.5.3 Symbolic Representation of Automata

The symbolic representation described in the previous subsection will be used to manipulate subsets of states and subsets of transitions.

Given the set of states $\mathcal{Q}$, we assume that a set of *state variables* $\mathcal{V}_\mathcal{Q}$ can be used to represent $\mathcal{Q}$. This set of variables needs to be carefully defined so that the formulas representing useful subsets of states have compact representations. Any subset $\mathcal{S} \subseteq \mathcal{Q}$ of states is represented by a formula $\Phi_{\mathcal{S}} \in Forms(\mathcal{V}_{\mathcal{Q}})$.

Example With reference to the automaton of Fig. 9.5, we define four variables A, B, C, and F. Given the state $q = \langle x, \delta \rangle$, we use the following assignment α:

- $\alpha(q)(A) = T$ iff $x \in \{5, 6, 7, 8\}$;
- $\alpha(q)(B) = T$ iff $x \in \{2, 4, 6, 8\}$;
- $\alpha(q)(C) = T$ iff $x \in \{3, 4, 7, 8\}$; and
- $\alpha(q)(F) = T$ iff $\delta = \{f\_1\}$.

For instance, the initial state is such that $\Phi_{q_{\text{init}}} = \neg A \wedge \neg B \wedge \neg C \wedge \neg F$.

While this assignment may seem arbitrary, it is actually convenient for many sets. For instance, the set $\mathcal{S}_{o\_b}$ of states reached after any transition labeled with o_b (which contains eight states) is such that $\Phi_{\mathcal{S}_{o\_b}} = B$. ♢

In order to represent sets of transitions, we need to define more variables. A transition is a tuple $\langle q, e, q' \rangle$ where q and q' are states and e is an event. While more sophisticated representations can be used, a disjunction over the set $\mathcal{V}_{\mathcal{E}'}$ of variables used to represent the events in $\mathcal{E}$ can be adopted, i.e., for each event $e \in \mathcal{E}$, a variable v_e is defined.

Furthermore, we need two sets of variables to represent the two states (the origin and the target of a transition); therefore, we define $\mathcal{V}'_{\mathcal{Q}}$ a copy of the variables $\mathcal{V}_{\mathcal{Q}}$: $\mathcal{V}'_{\mathcal{Q}} = \{v' \mid v \in \mathcal{V}_{\mathcal{Q}}\}$. A formula over $\mathcal{V}'_{\mathcal{Q}}$ represents a set of states after the next transition.

Given a formula Φ, we define $\Phi[\mathcal{V}_{\mathcal{Q}}/\mathcal{V}'_{\mathcal{Q}}]$ the formula Φ where every occurrence of $v \in \mathcal{V}_{\mathcal{Q}}$ is replaced by the corresponding variable $v' \in \mathcal{V}'_{\mathcal{Q}}$ and every occurrence of $v' \in \mathcal{V}'_{\mathcal{Q}}$ is replaced by the corresponding variable $v \in \mathcal{V}_{\mathcal{Q}}$. Notice that if the formula Φ was defined on $\mathcal{V}_{\mathcal{Q}}$, $\Phi[\mathcal{V}_{\mathcal{Q}}/\mathcal{V}'_{\mathcal{Q}}]$ is defined on $\mathcal{V}'_{\mathcal{Q}}$.

The set of variables for the transitions is the union $\mathcal{V}_{\mathcal{T}} = \mathcal{V}_{\mathcal{Q}} \cup \mathcal{V}_{\mathcal{E}'} \cup \mathcal{V}'_{\mathcal{Q}}$. A transition $\langle q, e, q' \rangle$ is then represented by the formula $\Phi_q \wedge \Phi_e \wedge (\Phi_{q'}[\mathcal{V}_{\mathcal{Q}}/\mathcal{V}'_{\mathcal{Q}}])$.

Example With reference to the automaton of Fig. 9.5, consider the transition $t = \langle 1, \emptyset \rangle \xrightarrow{o\_b} \langle 2, \emptyset \rangle$. This transition is represented by the following formula:

$$\Phi_t = \neg A \wedge \neg B \wedge \neg C \wedge \neg F \wedge o\_b \wedge \neg o\_c \wedge \neg u\_e \wedge \neg o_r \wedge \neg A' \wedge B' \wedge \neg C' \wedge \neg F'.$$

The set T_{o_b} of transitions labeled with o_b can be represented as follows:

$$\neg B \wedge B' \wedge o\_b \wedge \neg o\_c \wedge \neg u\_e \wedge \neg o_r \wedge (A \leftrightarrow A') \wedge (C \leftrightarrow C') \wedge (F \leftrightarrow F').$$

This example shows some patterns that we often find in the symbolic representation of automata: the subformula $(A \leftrightarrow A')$ indicates that the transitions do not change the value of variable A; $\neg B \wedge B'$ indicates that B should be false before the transition and that an effect of the transition is to make B true. Such representations are extremely compact, and for "well-behaved" systems, the size of these sets of transitions remains linear in the number of components in the model. ♢

Table 9.2 Common operations to manipulate states and transitions symbolically

Let $\mathcal{R}$ be a set of transitions; let $\mathcal{S} = \{q \in \mathcal{Q} \mid \exists e \in \mathcal{E}.\exists q' \in \mathcal{Q}.\ \langle q, r, q' \rangle \in \mathcal{R}\}$ be the set of origins of $\mathcal{R}$, then

$$\Phi_{\mathcal{S}} \equiv Ex\ \mathcal{V}_{\mathcal{E}} \cup \mathcal{V}'_{\mathcal{Q}}.\ \Phi_{\mathcal{R}}.$$

Let $\mathcal{R}$ be a set of transitions; let $\mathcal{R}' = \{\langle q', e, q \rangle \in \mathcal{Q} \times \mathcal{E} \times \mathcal{Q} \mid \langle q, e, q' \rangle \in \mathcal{R}\}$ be the flipping of the transitions head-to-tail, then

$$\Phi_{\mathcal{R}'} \equiv \Phi_{\mathcal{R}}[\mathcal{V}_{\mathcal{Q}}/\mathcal{V}'_{\mathcal{Q}}].$$

Let $\mathcal{R}$ be a set of transitions; let $\mathcal{S}$ be a subset of states; let $\mathcal{R}_{\mathcal{S}} = \{\langle q, e, q' \rangle \in \mathcal{R} \mid q \in \mathcal{S}\}$ be the subset of transitions of $\mathcal{R}$ that start from some state of $\mathcal{S}$, then

$$\Phi_{\mathcal{R}_{\mathcal{S}}} \equiv \Phi_{\mathcal{R}} \wedge \Phi_{\mathcal{S}}.$$

Let $\mathcal{R}$ be a set of transitions; let $\mathcal{E}' \subseteq \mathcal{E}$ be a subset of events; let $\mathcal{R}_{\mathcal{E}'} = \{\langle q, e, q' \rangle \in \mathcal{R} \mid e \in \mathcal{E}'\}$ be the subset of transitions of $\mathcal{R}$ that involve an event of $\mathcal{E}'$, then

$$\Phi_{\mathcal{R}_{\mathcal{E}'}} = \Phi_{\mathcal{R}} \wedge (\textstyle\bigvee_{e \in \mathcal{E}'} v_e).$$

In order to manipulate sets of states and transitions together, we need to present one last operator. Given a set of variables $\mathcal{V}$ and a variable $v \in \mathcal{V}$, given a proposition formula $\Phi \in Forms(\mathcal{V})$, we denote with $Ex\ v.\ \Phi$ the formula $\Phi' \in Forms(\mathcal{V} \setminus \{v\})$ defined by $\Phi_{v:=\top} \vee \Phi_{v:=\bot}$, where $\Phi_{v:=x}$ indicates that every occurrence of v is replaced with x.[2] For instance, if $\Phi = a \wedge b \wedge c$ then $Ex\ b.\ \Phi = (a \wedge \top \wedge c) \vee (a \wedge \bot \wedge c) \equiv (a \wedge c)$. Notice that the operation performed by Ex is more complex than just removing the variable from the formula. Indeed, if $\Phi = a \vee b \vee c$, then $Ex\ b.\ \Phi = (a \vee \top \vee c) \vee (a \vee \bot \vee c) \equiv \top$.

The operator Ex is extended to sets of variables: $Ex\ \{v_1, \ldots, v_k\}.\ \Phi = Ex\ v_1.\ \cdots Ex\ v_k.\ \Phi$. The formula $Ex\ W.\ \Phi$ has the property that it has the same consistent assignments as Φ, except that those assignments are projected on the variables of $\mathcal{V} \setminus W$. This is very valuable to represent the fact that some aspects about Φ can be ignored.

We are now in a position to show how to implement the basic operations to manipulate sets of transitions. These basic operations are described in Table 9.2. The first operation allows us to extract the origin states of a set of transitions. The second operation allows us to flip a set of transitions (notice that, combined with the previous operation, it can be used to compute the targets of a set of transitions). The third operation allows us to extract the transitions leaving a set of states. Finally, the fourth operation allows us to extract the transitions labeled with a given event.

We illustrate these concepts with a more complex query. Let $\mathcal{S}$ be a set of states, let e and f be two events. We want to find the set of states that can be reached from $\mathcal{S}$ through the occurrence of zero or one instance of e followed by f. This set can be defined as follows:

$$Ex\ \mathcal{V}_{\mathcal{E}} \cup \mathcal{V}'_{\mathcal{Q}} \cdot \left((\varphi \wedge \Phi_{\mathcal{T}} \wedge v_f)[\mathcal{V}/\mathcal{V}']\right),$$

[2]In general, the operator is written $\exists$ instead of Ex; we have introduced the second notation here to avoid confusion with the use of $\exists$ outside of symbolic contexts.

where

$$\varphi := \Phi_{\mathcal{S}} \ \vee \ Ex\ \mathcal{V}_{\mathcal{E}} \cup \mathcal{V}'_{\mathcal{Q}} \cdot \big((\Phi_{\mathcal{S}} \wedge \Phi_{\mathcal{T}} \wedge v_e)[\mathcal{V}/\mathcal{V}']\big)\,.$$

9.3.5.4 Symbolic Implementation of DES Diagnosis

The symbolic implementation of the diagnosis procedure simply consists in replacing the explicit (set-based) methods with the symbolic methods presented before. In Fig. 9.6b, the symbolic implementation is shown side by side with the explicit algorithm.

9.3.5.5 Other Symbolic Representations

We now discuss briefly the existing symbolic tools. We defined propositional formulas as the combination of disjunctions, negations, and variables. There exist other representations of these formulas, and finding good ones has been the focus of part of the knowledge representation community.

A good representation of propositional formulas should satisfy the following properties:

- It should be compact.
- Operations (e.g., the conjunction of two formulas) should be easy to perform (in the computational sense).
- Queries (e.g., are two given formulas equivalent?) should be easy to perform.

It is well known that there is no perfect solution: regardless of the implementation chosen, it will be unsatisfactory in at least one of the dimensions. Nevertheless, classes of implementations have been identified that either are good across all the dimensions of some specific problems (unfortunately, this is not the case with diagnosis of DESs) or perform well for many realistic domains. We refer to the work from Darwiche and co-authors for details on these approaches [11, 17, 39].

9.4 Diagnosability

Diagnosability is the problem of deciding whether there is enough observability on the system that the occurrence of a fault will always be detected and the fault will be identified precisely.

Diagnosability acts as a verification process, which provides the system designer with a guarantee that faults will be detected and isolated, thus enabling one to perform appropriate countermeasures, such as repair.

In this section, we first present the definition of diagnosability. We next discuss its link with the diagnoser proposed by Sampath et al. Then we deal with the twin plant and its use to verify diagnosability. Finally, we discuss some benefits of diagnosability in terms of system design.

9.4.1 Definition of Diagnosability

Diagnosability of DES is generally defined under the assumption that there is only one faulty event. We start by justifying this assumption.

Diagnosability is the property that the diagnosis state of the system (the set of faults that occurred on the system) can always be retrieved. If all the sets of faults can be identified accurately, then each fault can be identified individually and conversely. For this reason, while the set of faults $\mathcal{E}_f$ may be large, diagnosability is generally checked for each fault f separately. This implies that the set of faults is rewritten $\mathcal{E}_f = \{f\}$. Notice that this does not imply that the occurrence of other faults is deemed impossible (this would be the *single fault assumption*); instead, this rewriting means that the other faults are considered nominal events.

The definition of diagnosability of DES is not as straightforward as one would expect it to be. Ideally, diagnosability would be defined as the property that the diagnosis $\Delta(\theta)$ consists of a single set of faults for any observation θ that can be generated by the system, i.e., that there is no ambiguity as to what faults did or did not occur. Because the system is dynamic, however, we allow for a reasonable *delay* between the moment when any fault occurs and the moment when the fault is identified by the diagnostic engine. This delay is represented by k in the rest of this section.

Two definitions of diagnosability can be found in the literature. The first one, proposed by Sampath et al. [31], is more a *characterization* of a diagnosable system. We therefore concentrate on the second definition and discuss its implicit assumptions.

The second definition is based on that of a k-diagnoser *Dser* (Definition 9.11), which returns either N or F; F should be interpreted as *surely faulty* while N has the meaning *probably nominal*. The input of the k-diagnoser is a flow of observed events. A k-diagnoser must be correct and precise.

Correctness: the k-diagnoser must not return *faulty* if the behavior is normal (first item of the definition).

Precision: the k-diagnoser must return *faulty* if a fault occurred more than k events ago (second item of the definition).

Definition 9.11 Given a DES model $A = \langle \mathcal{Q}, \mathcal{T}, \mathcal{E}, \mathcal{E}_o, \mathcal{E}_f, q_{init} \rangle$ where $\mathcal{E}_f = \{f\}$, given a value $k \in \mathbf{N}$, a *k-diagnoser* of A is a classifier $Dser : \mathcal{E}_o{}^\star \to \{N, F\}$ that satisfies the following properties:

1. $\forall \sigma \in \mathcal{L}(A).\ \sigma \in (\mathcal{E} \setminus \mathcal{E}_f)^\star \Rightarrow Dser(obs(\sigma)) = N$;
2. $\forall \sigma \in \mathcal{L}(A).\ \sigma \in \mathcal{E}^\star f \mathcal{E}^k \mathcal{E}^\star \Rightarrow Dser(obs(\sigma)) = F$.

The two items of Definition 9.11 may conflict: there may be two traces σ and σ', one of which is nominal and the other faulty that generate the same observation ($\theta = obs(\sigma) = obs(\sigma')$). Hence, given the observation θ relevant to trace σ, Definition 9.11 requires both $Dser(\theta) = N$ and $Dser(\theta) = F$. In such situation, the system is not k-diagnosable because it is impossible to decide whether the observation is the result of a nominal or a faulty execution.

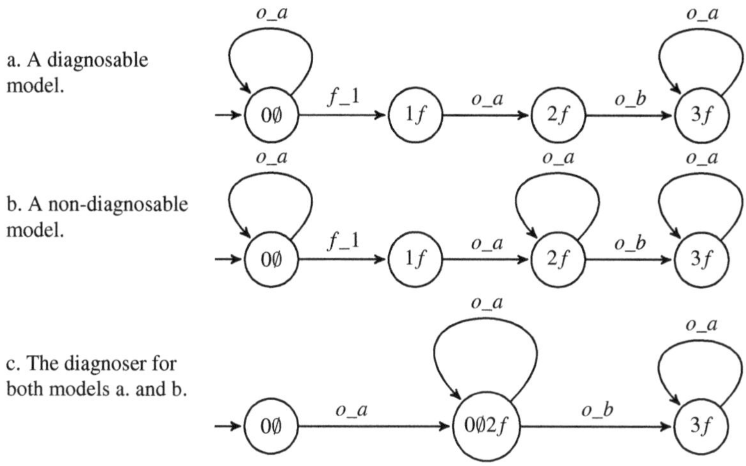

Fig. 9.8 Two models (with $\mathcal{E}_o = \{a, b\}$ and $\mathcal{E}_f = \{f\}$) having the same diagnoser but different diagnosability

Definition 9.12 A DES model is *k-diagnosable* if it admits a k-diagnoser. A DES model is *diagnosable* if it is k-diagnosable for some $k \in \mathbf{N}$.

We illustrate this definition with the two examples of Fig. 9.8. In the example a., we define *Dser* as follows:

$$Dser(\theta) = \begin{cases} N & \text{if } \theta = o\_a^{\star} \\ F & \text{otherwise,} \end{cases}$$

i.e., $Dser(\theta)$ returns N if and only if θ only contains observable event o_a (whereas it returns F if there is at least one o_b). It is easy to see that *Dser* is a 2-diagnoser, because as soon as a fault occurs, it is followed by the observation of an o_a event and then an o_b event, where the latter enables *Dser* to catch the fault.

Looking now at the example b, consider the following two traces: $\sigma_1 = o\_a^k$ and $\sigma_2 = f\_1\ o\_a^k$. From Definition 9.11 we infer $Dser(obs(\sigma_1)) = N$ and $Dser(obs(\sigma_2)) = F$. Since the two traces produce the same observation ($obs(\sigma_1) = obs(\sigma_2) = o\_a^k$), a k-diagnoser does not exist; hence, the model is not diagnosable.

Although we concentrate here on the notion of diagnosability, k-diagnosability is also important, in particular, when we want to fix an upper bound on the delay for fault identification.

The above definition of diagnosability and k-diagnosability makes a number of assumptions about the system.

- First, it assumes that the DES model is correct. However, it is very difficult to build a model that specifies precisely the possible sequences of events. In general, the model is an abstraction of the real behavior of the system, meaning that some executions may be added to the model even if they cannot occur. In particular, it is possible not to know what happens once a fault has occurred: the model allows any sequence of events after the occurrence of the fault; this implies that the diagnosis $\Delta(\theta)$ will include this fault for any observation θ, which means that such a model could still be diagnosable.

- Second, it assumes that the only purpose of diagnosis is to detect and identify the fault. Specifically, it does not consider that we would like to identify the fault before some undesirable consequence triggers. Faulty components can often be dangerous for other components or even people, and unfortunately sometimes it is impossible to detect the faults before the harm is done. More sophisticated definitions of diagnosability, such as that proposed by Cimatti et al. [8], allow one to express the fact that a fault is diagnosable if it can be identified before a critical state is reached.
- Value k for a k-diagnoser represents the maximum amount of time between the moment when the fault occurred and the moment when the fault is diagnosed precisely. Here k is expressed in terms of number of events in the system. There is an implicit assumption that the number of events is an acceptable proxy for the amount of time. This is not always realistic. On the one hand, a system can remain in the same state without any transition being triggered. On the other hand, in many systems faults can produce cascades of effects in a very short time span [25]. Notice that, according to the definition above, k relates to the overall number of events in σ, but there exist definitions where k refers to the number of observed events only.

Diagnosability, as presented in Definitions 9.11 and 9.12, can be formally translated as follows:

$$\forall \sigma \in (\mathcal{L}(A) \cap \mathcal{E}^{\star}\mathcal{E}_{\mathrm{f}}).\ \forall \sigma' \in \mathcal{E}^{k} \cdot \sigma\sigma' \in \mathcal{L}(A) \Rightarrow$$

$$\forall \sigma'' \in \mathcal{L}(A) \cdot (obs(\sigma\sigma') = obs(\sigma'')) \Rightarrow \sigma'' \in \mathcal{E}^{\star} f \mathcal{E}^{\star}.$$

This expression, which is used as the definition of diagnosability in a large part of the literature, can be interpreted as follows: given a trace σ that ends with a faulty event, given any extension σ' of σ with k events, any trace σ'' that generates the same observation as $\sigma\sigma'$ should include a fault.

The expression is indeed correct because its negation includes the existence of $\sigma\sigma'$, a faulty trace with k events since the occurrence of the fault, and σ'', a nominal trace, that produce the same observation $\theta = obs(\sigma\sigma') = obs(\sigma'')$. Therefore, the k-diagnoser would not be a classifier as it should return both $Dser(\theta) = F$ (because of $\sigma\sigma'$) and $Dser(\theta) = N$ (because of σ'').

9.4.2 Diagnosability and Diagnoser

There are obvious connections between diagnosability and the diagnoser defined by Sampath et al. [31] (Definition 9.10). In this subsection, we show that the diagnoser can be used to compute a k-diagnoser (Definition 9.11). We show, however, that the diagnoser is not sufficient to verify diagnosability.

Theorem 9.2 *Let $A = \langle \mathcal{Q}, \mathcal{T}, \mathcal{E}, \mathcal{E}_o, \mathcal{E}_f, q_{init} \rangle$ be a fault-state DES model with $\mathcal{E}_f = \{f\}$ and fault labeling function D. Let $A' = \langle \mathcal{Q}', \mathcal{T}', \mathcal{E}', q'_{init} \rangle$ be the diagnoser of A, as presented in Definition 9.10. Let $Dser : \mathcal{E}_o^\star \to \{N, F\}$ be defined as follows:*

$$Dser(\theta) = \begin{cases} F & \text{if } D(q'_{init}[\theta]) = \{\{f\}\} \\ N & \text{otherwise.} \end{cases}$$

Then Dser is a k-diagnoser of A iff A is k-diagnosable.

Remember that $\mathcal{E}_f$ is the singleton $\{f\}$; hence, the labels of the diagnoser states are $\{\emptyset\}$, $\{\{f\}\}$, and $\{\emptyset, \{f\}\}$. The construction of the k-diagnoser consists in taking the diagnoser of the DES model, labeling with F all states q' of the diagnoser such that $D(q') = \{\{f\}\}$ (i.e., those states for which we know for sure that the fault occurred), and labeling with N all the other states. The proof of Theorem 9.2 is left as an exercise to the reader. It is actually possible to prove that *Dser* is the most precise diagnoser.

The problem with Theorem 9.2 is that it does not tell us whether *Dser* is indeed a k-diagnoser. There exist some sufficient conditions to determine from the diagnoser that the DES is diagnosable, such as the one provided by Lemma 9.6.

Lemma 9.6 *Let $A = \langle \mathcal{Q}, \mathcal{T}, \mathcal{E}, \mathcal{E}_o, \mathcal{E}_f, q_{init} \rangle$ be a fault-state DES model with $\mathcal{E}_f = \{f\}$ and fault labeling function D. Let $A' = \langle \mathcal{Q}', \mathcal{T}', \mathcal{E}', q'_{init} \rangle$ be the diagnoser of A, as presented in Definition 9.10. If A' has no cycle of states labeled with $\{\emptyset, \{f\}\}$, then the model is diagnosable.*

The proof is left as an exercise for the reader.

Ironically, while the diagnoser may be the k-diagnoser, it is impossible in general to decide from the diagnoser whether the model is diagnosable. This is surprising because the diagnoser contains precisely all the information that is relevant for the diagnosis task, but not for determining whether diagnosability holds.

We illustrate this issue with the example of Fig. 9.8 that we have already introduced. Remember that the model in Fig. 9.8a is diagnosable, while the model in Fig. 9.8b is not. These two models, however, share the same diagnoser, which is presented in Fig. 9.8c. We see that knowing the diagnoser is not sufficient to determine diagnosability.

It is possible to leverage the computation of the diagnoser in order to test diagnosability. We briefly describe this procedure here. We call *ambiguous* any state q' of the diagnoser that includes at least one nominal state ($\exists q \in q'.\ D(q) = \emptyset$) and at least one faulty state ($\exists q \in q'.\ D(q) = \{f\}$); in the example of Fig. 9.8, one such state is $\{0\emptyset, 2f\}$. As hinted in Lemma 9.6, cycles over such ambiguous states in the diagnoser are potentially indicative of non-diagnosability. The procedure then verifies whether such cycles have a "realization" in the faulty part of original model. In our example, consider the cycle $\{0\emptyset, 2f\} \xrightarrow{o\_a} \{0\emptyset, 2f\}$; this cycle has a realization in the model b., namely, $2f \xrightarrow{o\_a} 2f$. This realization witnesses non-diagnosability. On the other hand, the model a. does not have any realization of the ambiguous cycle (the state $2f$ does not have an outgoing transition labeled with o_a). Therefore, model a. is diagnosable.

We do not recommend using the diagnoser to verify diagnosability in general. Indeed, the size of the diagnoser is exponential in the number of states of the original model. In comparison, the procedure presented next runs in polytime. A diagnoser-based procedure should be limited to situations where it is guaranteed that the diagnoser will remain small, or when the diagnoser will be computed for diagnosis purposes.

9.4.3 Twin Plant

The twin plant is a construction proposed by Jiang et al. [18] and independently by Yoo and Lafortune [42] under the name of *verifier*. The approach relies on the following property: a model is non-diagnosable if and only if there exist two infinite traces $\{\sigma, \sigma'\} \subseteq \mathcal{L}(A)$ that satisfy the following three conditions:

1. they produce the same observation ($obs(\sigma) = obs(\sigma')$),
2. one of them is nominal ($\sigma \in (\mathcal{E} \setminus \mathcal{E}_f)^\infty$), and
3. the other one is faulty ($\sigma' \in \mathcal{E}^\star \mathcal{E}_f \mathcal{E}^\infty$).

These two traces are called a *critical pair*.

It is possible to build a Büchi automaton whose infinite language consists precisely in all critical pairs. This automaton is obtained by *composing* the DES model with its copy, so that a path on the resulting automaton corresponds to two parallel paths on the original DES model. In practice, a state of the twin plant is a pair $\langle q_1, q_2 \rangle$, where both q_1 and q_2 are states of the model. In order to guarantee the first condition of a critical pair (specifically, the two traces produce the same observation), the observable transitions are *synchronized*, i.e., a transition $\langle q_1, q_2 \rangle \xrightarrow{o} \langle q_1', q_2' \rangle$ corresponds to two transitions $q_1 \xrightarrow{o} q_1'$ and $q_2 \xrightarrow{o} q_2'$. The second condition is guaranteed by forbidding in the first copy of the model the transitions associated with a fault. The third property is guaranteed by only considering traces that end up in a state $\langle q_1, q_2 \rangle$ where q_2 is faulty.

Definition 9.13 Let $A = \langle \mathcal{Q}, \mathcal{T}, \mathcal{E}, \mathcal{E}_o, \mathcal{E}_f, q_{init} \rangle$ be a fault-state DES model with $\mathcal{E}_f = \{f\}$ and fault labeling D. The *twin plant* of A is the Büchi automaton $\langle \mathcal{Q}', \mathcal{T}', \mathcal{E}', q_{init}', \mathcal{R}' \rangle$ defined as follows:

- $\mathcal{Q}' \subseteq \mathcal{Q} \times \mathcal{Q}$ is the set of states;
- $\mathcal{E}' = \mathcal{E}_o \cup (\mathcal{E} \times \{l, r\})$ is the set of events;

- $\mathcal{T}' = \mathcal{T}_o \cup \mathcal{T}_l \cup \mathcal{T}_r$ is the set of transitions, where
  - $\mathcal{T}_o = \{\langle q_1, q_2\rangle, e, \langle q_1', q_2'\rangle \mid e \in \mathcal{E}_o \wedge \langle q_1, e, q_1'\rangle \in \mathcal{T} \wedge \langle q_2, e, q_2'\rangle \in \mathcal{T}\}$,
  - $\mathcal{T}_l = \{\langle q_1, q_2\rangle, \langle e, l\rangle, \langle q_1', q_2'\rangle \mid e \in \mathcal{E} \setminus \mathcal{E}_o \setminus \mathcal{E}_f \wedge \langle q_1, e, q_1'\rangle \in \mathcal{T} \wedge q_2 = q_2'\}$, and
  - $\mathcal{T}_r = \{\langle q_1, q_2\rangle, \langle e, r\rangle, \langle q_1', q_2'\rangle \mid e \in \mathcal{E} \setminus \mathcal{E}_o \wedge q_1 = q_1' \wedge \langle q_2, e, q_2'\rangle \in \mathcal{T}\}$;
- $q_{\text{init}}' = \langle q_{\text{init}}, q_{\text{init}}\rangle$ is the initial state; and
- $\mathcal{R}' = \{\langle q_1, q_2\rangle \in \mathcal{Q}' \mid D(q_2) = \{f\}\}$ is the set of accepting states.

We define a fault labeling function L for the states of the twin plant as follows: $L(\langle q_1, q_2\rangle) = D(q_2)$. A state of the twin plant that satisfies $L(\langle q_1, q_2\rangle) = \{f\}$ is called *ambiguous* as it corresponds to two paths, one of which is nominal and the other one faulty.

We show how the twin plant can be used to test diagnosability of the class of DESs whose model is observably live (defined next). Notice, however, that this assumption can be relaxed and the condition of diagnosability is then slightly modified.

Definition 9.14 A model is *observably live* iff any cyclic path on the model includes at least one observable event.

Notice that if a model is *observably live*, then any infinite path on the model actually includes infinitely many observable events.

Theorem 9.3 *Let $A = \langle \mathcal{Q}, \mathcal{T}, \mathcal{E}, \mathcal{E}_o, \mathcal{E}_f, q_{init}\rangle$ be a state-fault, observably live DES model with a single fault $f \in \mathcal{E}_f$. Fault f is diagnosable iff the twin plant of A includes a cycle $q_0 \xrightarrow{e_1} \dots \xrightarrow{e_k} q_k$ (i.e., such that $q_0 = q_k$) where $L(q_0) = \{\{f\}\}$.*

Theorem 9.3 directly results from the fact that diagnosability holds iff there is no critical pair. If the twin plant contains a cycle in its ambiguous region, this cycle (together with the prefix that led to it) can be turned into a critical pair. Conversely, if a critical pair exists then the twin plant exhibits this pair; because the state space is finite, a (infinite) critical pair must eventually loop back to a state previously visited, which implies the existence of a cycle in the ambiguous region of the twin plant.

Example We illustrate the use of Theorem 9.3 with the examples provided before. Figure 9.9 presents the twin plants of the DESs depicted in Fig. 9.8.

First concentrate on the second twin plant: a transition labeled with o_a loops on state $\langle 0\emptyset, 2f\rangle$. This transition corresponds to the two transitions $0\emptyset \xrightarrow{o\_a} 0\emptyset$ and $2f \xrightarrow{o\_a} 2f$ from the original model. The infinite path $\langle 0\emptyset, 0\emptyset\rangle \xrightarrow{f\_1,r} \langle 0\emptyset, 1f\rangle \xrightarrow{o\_a} \langle 0\emptyset, 2f\rangle(\xrightarrow{o\_a} \langle 0\emptyset, 2f\rangle)^\star$ demonstrates that the faulty and infinite path $0\emptyset \xrightarrow{f\_1} 1f \xrightarrow{o\_a} 2f(\xrightarrow{o\_a} 2f)^\star$ produces the same observation as the nominal path $0\emptyset \xrightarrow{o\_a} 0\emptyset(\xrightarrow{o\_a} 0\emptyset)^\star$, which proves that the system is not diagnosable.

The first twin plant does not include cycles in its ambiguous part. We conclude that the corresponding DES model is diagnosable.

Finally, we provide in Fig. 9.10 the twin plant for (the state-fault reformulation of) the system of Fig. 9.1. This twin plant does include cycles in its ambiguous part. Again, we conclude that the system of Fig. 9.1 is not diagnosable. ◇

a. Twin plant of model of Figure 9.8.a.

b. Twin plant of model of Figure 9.8.b.

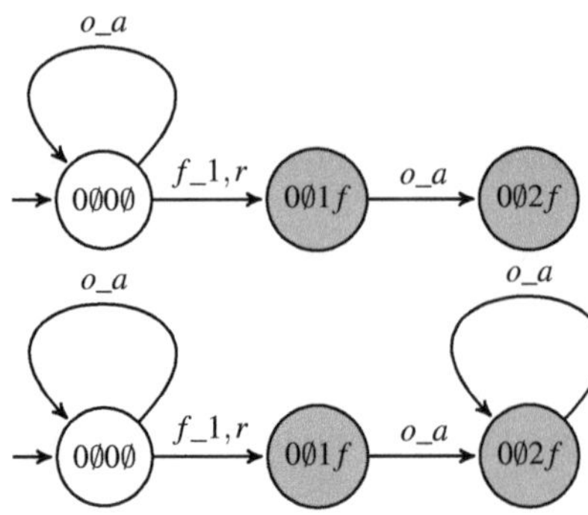

Fig. 9.9 Twin plant for the examples of Fig. 9.8. Ambiguous states are represented with a gray filling

We finish this section on the twin plant with some complexity results. The twin plant is quadratic in the size of the original model. Furthermore, determining whether there is a cycle in an automaton is quadratic in the size of the twin plant. We conclude that testing diagnosability is only polynomial for a model given as a single explicit automaton.

9.4.4 Other Benefits of Diagnosability

Diagnosability checking has been presented as a "validation" process that provides a guarantee to the system designer. In other words, it acts as a final procedure. However, diagnosability can be used *during* the design process in order to help the designer decide which sensors are required.

Ye et al. [41] discussed the problem of finding an *optimal sensor placement* on a system modeled as a DES. In the context of this problem, every sensor corresponds to an event: this event is observable if and only if the sensor is installed on the system. Optimal sensor placement is then the problem of finding a minimal subset of sensors that guarantees system diagnosability. Diagnosability is monotonic with respect to observability, i.e., if a system with observable events $\mathcal{E}_o$ is diagnosable, then the same system with a superset of observable events $\mathcal{E}_o{}' \supseteq \mathcal{E}_o$ is also diagnosable. Therefore, given a set of observable events that makes the system diagnosable, it is possible to iterate over these events and verify whether the system remains diagnosable when an event becomes unobservable.

Furthermore, as already remarked, the diagnosis problem is hard in the computational sense. Diagnosability checking can be used to identify what parts of the system need to be monitored in order to guarantee diagnosability: the complexity of diagnosis being a function of the size of the model, its computational requirements can be reduced if diagnosability is satisfied when only a small part of the system is con-

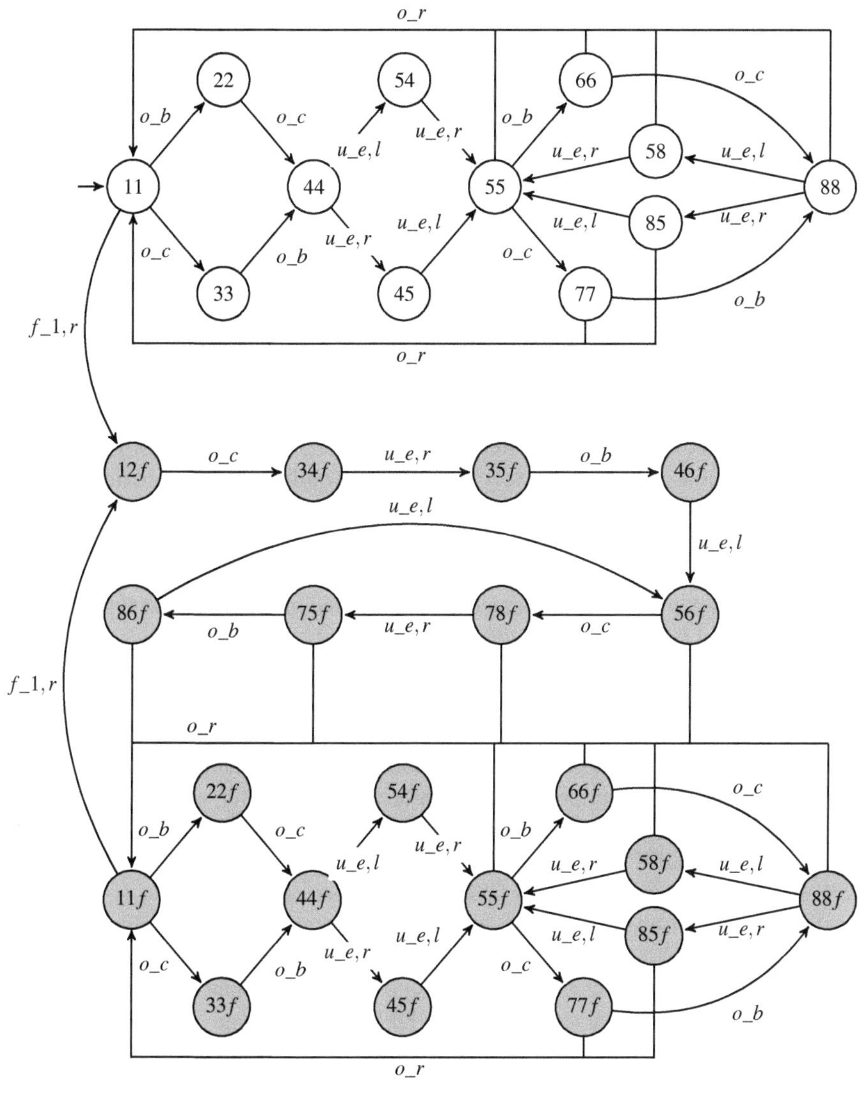

Fig. 9.10 Twin plant relevant to the automatic reformulation (Fig. 9.5) of the DES in Fig. 9.1

sidered. For instance, in a very large system such as a power transmission network or a telecommunication network, which may comprise thousands of components, diagnosing accurately a fault on a specific piece of equipment may only require to look at the observable messages produced by a handful of components surrounding this piece of equipment. A diagnosability procedure was proposed by Pencolé that identifies such a part [28].

9.5 Advanced Topics

In the previous sections of this chapter, we presented only the basics of the theory of diagnosis of DESs. We now briefly survey some extensions, providing pointers to the relevant literature.

9.5.1 Distributed Modeling

One of the objectives of model-based diagnosis and fault detection and identification is to make it easy to adapt the diagnostic engine to changes in the system. In particular, we expect a plug-and-play approach where the modeling task consists in combining building blocks that match the system component.

The definition of a DES in Sect. 9.2.1 is monolithic, i.e., it contains a single automaton to represent the whole system. This is impractical, both for computational and modeling reasons: nontrivial systems have at least millions of states and it is unreasonable to assume that such models can be written by hand without errors. Instead, component types are modeled separately, and instantiated and synchronized when the model for the whole system is proposed.

This approach was advocated as early as the seminal work from Sampath et al. [31]. Compositional modeling was proposed also for *active systems* [2, 21, 23, 24], a framework for DESs where it is possible to model components individually along with the way they communicate (synchronously or asynchronously, with a first-in-first-out or last-in-first-out procedure, with or without a buffer, etc.).

In essence, a distributed DES is a collection of models $\{A^1, \ldots, A^n\}$, where each model is a *local* DES $A^i = \langle \mathcal{Q}^i, \mathcal{T}^i, \mathcal{E}^i, \mathcal{E}_o{}^i, \mathcal{E}_f{}^i, q^i_{\text{init}} \rangle$. The DES A^i is supposed to represent either a physical component or some abstract aspect of the system (such as the fact that an event occurred on a physical component and is about to affect another component). The state space of the system is the Cartesian product of the state space of each DES: $\mathcal{Q} = \mathcal{Q}^1 \times \cdots \times \mathcal{Q}^n$, which means that a state of the system is a tuple $\langle q^1, \ldots, q^n \rangle$ that associates each DES A^i with a state q^i. The set of events of the system is the union of the sets of each DES. A transition $\langle q^1, \ldots, q^n \rangle \xrightarrow{e} \langle q'^1, \ldots, q'^n \rangle$ exists iff, for each DES A^i that includes e (i.e., such that $e \in \mathcal{E}^i$), there exists a transition $q^i \xrightarrow{e} q'^i$.

Example The models of the three (interacting) components (a pump, a valve, and a controller) of a distributed DES, as proposed by Sampath et al. [31], are shown in Fig. 9.11. This example models a simplified HVAC system. The initial state of the system is $\langle POFF, VC, C1 \rangle$, which indicates that the pump is initially off, the valve is closed, and the controller is in its first state. Event pump_failed_off is enabled from this state, as it is only mentioned by the model of the pump, and therefore affects just this component: the state reached is $\langle PFOFF, VC, C1 \rangle$. Notice that, while the event stop_pump triggers a transition exiting the states $PFOFF$ and $POFF$ of

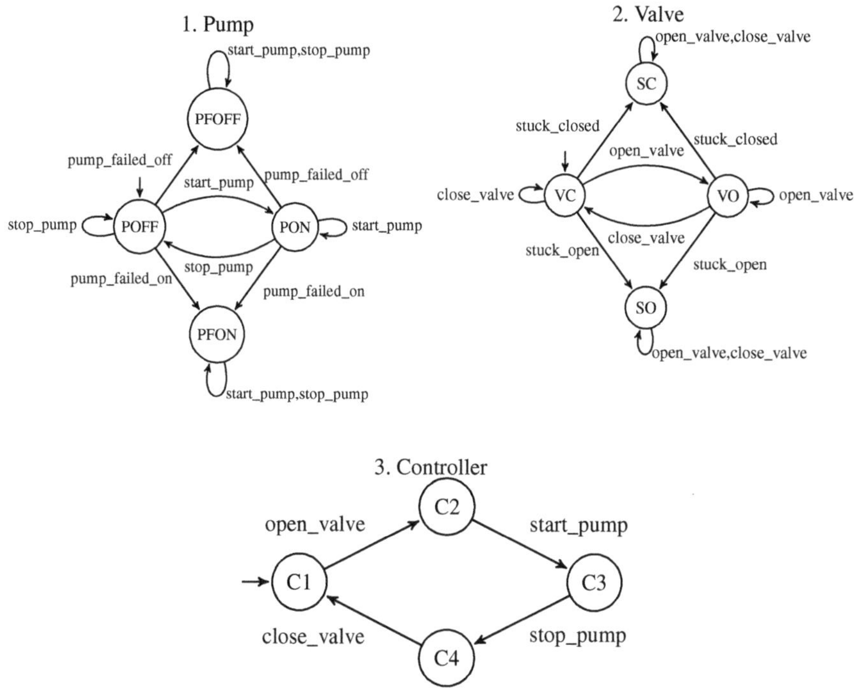

Fig. 9.11 A distributed DES with three components

component pump, such an event is enabled neither in the initial state of the whole system nor in state $\langle POFF, VC, C1\rangle$ because this event affects the controller but is not allowed in the initial state of the controller. ◊

9.5.2 Decentralized and Distributed Methods for Diagnosis

Being equipped with a distributed model is tremendously helpful to make diagnosis techniques scale.

The definition of good Boolean variables for the symbolic representation of a distributed DES is much easier than for a monolithic model of the same DES. Indeed, a Boolean state variable should only model an aspect from one of the local DES. This way, if an event only affects a small number of components, then the formula that represents the effects of this event will mention only the variables that are used to model the component DESs that include this event.

Lamperti and Zanella [21, 24] proposed a diagnosis procedure that does not require to build explicitly the complete behavioral model of the distributed system. Instead, given a diagnosis problem, they exploit the given observation to build a model on-the-fly, by merging only the parts of the behavioral model that is relevant for diagnosis.

Pencolé and Cordier [29] proposed *i.* to perform a local diagnosis, i.e., compute the local explanations for each local DES and their observed events, and *ii.* to merge these "local diagnoses" incrementally until the explanations are computed for the whole system. This procedure allows them to prune away large parts of the search space early and scale diagnosis of DESs to large telecommunication networks.

Su and Wonham [34], as well as Kan John and Grastien [19], proposed a different approach, where the local diagnoses, instead of being merged, are "projected" on the events shared among components, and this information is then synchronized with the other local diagnoses.

Finally, notice that diagnosability can be used to determine which subsystem is sufficient to diagnose precisely a fault. Pencolé proposed [28] to verify diagnosability locally, i.e., by assuming that the distributed model is $\{A_1, \ldots, A_m\}$ instead of $\{A_1, \ldots, A_n\}$ where $m \ll n$. If a fault is diagnosable in this subsystem, then it is unnecessary to consider the whole system for this fault; if every fault is diagnosable in a small subsystem, then the complexity of diagnosis remains small. Grastien proposed [12] to study the system designs at an early stage in order to guarantee that such properties will be satisfied regardless of the final implementation.

In some contexts, there is no central diagnoser for the system, either because the system is naturally distributed, or because there is no central authority, or even because some of the parties involved in the system do not want to share private information.

In such decentralized contexts, each component has its own diagnoser, which is in charge of diagnosing the component itself [40]. These local diagnosers exchange messages in which they share their conclusions about how the components they monitor communicated. There is generally no central clock, which means that there is no total order between the local observations and their messages. Interestingly, this implies that diagnosability is non-decidable [33]: there exists no procedure that can verify whether a system is diagnosable in a decentralized setting.

9.5.3 Petri Nets

Petri nets [26, 30] are another very popular modeling tool for DESs. They are primarily used to model flows of resources. A petri net consists in a collection of *places* and *transitions*. A state of the system is defined as the number of *tokens* in each place: tokens can be interpreted as resources. Each transition links a number of places (the resources consumed by the transition) to a number of places (the resources generated by the transition).

Under some assumptions (for instance, that the number of tokens on a given place is bounded, i.e., the petri nets is *safe*), it is possible to build a finite state machine (i.e., an automaton) equivalent to the petri net. Petri nets, however, can also very compactly represent concurrent behaviors, where events affect different parts of the system. This property has been exploited for diagnosis [3].

9.5.4 Uncertain Observations

Considering that an observation is a sequence of observed events is often too strong an assumption. The flow of observed events generally includes uncertainty that needs to be handled.

Lamperti and Zanella identified [20, 22], among others, two main types of uncertainty. *Logical uncertainty* represents the uncertainty about the content of an observed event, for instance, because its reading is hard to interpret owing to noise. *Temporal uncertainty* represents the uncertainty about the emission order between events, in general, because the transmission delay of observed events is nontrivial compared to the dynamics of the system. An observation is then defined as a collection of partially ordered "fragments"; each fragment is labeled with a number of events, among which there is the one that has actually occurred; the partial order indicates the known temporal relations between the occurrences of these events. From this observation, an *index space* is drawn, this being a deterministic automaton whose language consists in all possible sequences of observed events that match the observation. The diagnostic algorithm then guarantees that all these sequences are considered, since we do not know which is the one that actually occurred.

Example On top of Fig. 9.12 an uncertain observation is shown. In this example, an event is first observed, although, due to logical uncertainty, it is impossible to decide whether this event is *o_r* or *o_c*. Then two fragments are observed, for which the order is uncertain. The first of these fragments is the certain occurrence of *o_c*; the

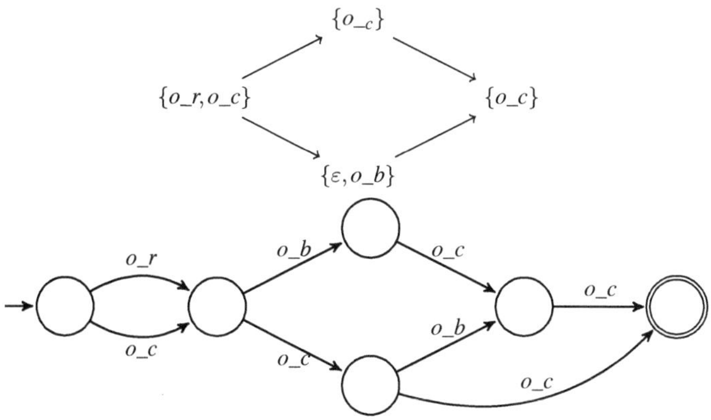

Fig. 9.12 Uncertain observation (above) and its index space (below)

second is the possible occurrence of o_b (ε indicates that maybe no observable event occurred). Finally, event o_c is observed.

Irrespective of the model, there are six possible sequences of observable events that match this uncertain observation:

$$o\_r, o\_b, o\_c, o\_c$$
$$o\_r, o\_c, o\_b, o\_c$$
$$o\_r, o\_c, o\_c$$
$$o\_c, o\_b, o\_c, o\_c$$
$$o\_c, o\_c, o\_b, o\_c$$
$$o\_c, o\_c, o\_c.$$

Lamperti and Zanella presented a systematic method to generate the *index space*: in the current example, such an automaton, which represents precisely these sequences, is displayed on the bottom of Fig. 9.12. The index space is then used to compute all the explanations of the model that are compatible with the uncertain observation. ◊

Diagnosability can also be checked in uncertain environments. It is possible to verify how much uncertainty can be tolerated before diagnosability becomes compromised [35].

9.5.5 Sensor Placement

Sensor placement is the problem of finding where sensors should be placed so that diagnosability is satisfied, while minimizing the cost of operating or purchasing/installing these sensors. Sensor placement can be easily defined as an optimization problem on top of the diagnosability checking procedure [4].

In a similar context, another optimization consists in turning the sensors off when they are not useful, so as to reduce the power consumption, increase their lifetime, and decrease the cost associated with the processing of the events. This leads to the definition of "dynamic diagnosers" that not only provide the diagnosis but also the set of sensors that can be turned on/off. The problem was studied by Thorsley and Teneketzis [36] and, independently, by Cassez and Tripakis [6].

Example Consider the DES model in Fig. 9.13 and assume that all events but the faulty one f_1 are observable. The DES is clearly diagnosable: a fault is always followed by the occurrence of o_c. Assume, however, that monitoring the event o_c is very expensive, either because it requires a lot of energy or because it hampers somehow the operations of the system. In this context, it is reasonable to only monitor events o_a and o_b, and turn on the sensor for o_c whenever o_b occurred once (or any specified number of times). ◊

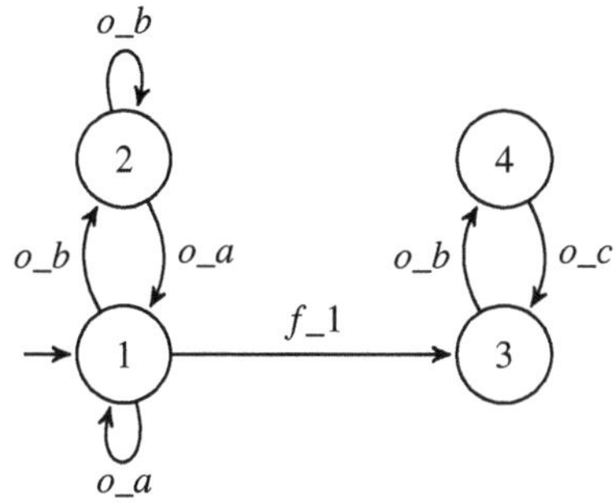

Fig. 9.13 A DES where dynamic sensors can be used

9.5.6 *Probability and Time*

Probabilities can be introduced in the system model in order to provide a more accurate view of the diagnosis. This approach has not been very popular because it involves a number of difficulties. First, it is very hard to obtain precise estimates of the probabilities of the occurrence of any event in any state. Second, distributed models are not appropriate to represent the probability of concurrent events: if the system contains two components A and B, for which the next event is guaranteed to be a and b respectively, what is the probability that the next event will be a? Finally, probabilistic diagnosis is computationally much more expensive. Probabilistic diagnosability has been studied by Nouioua and Dague [27].

Timed automata are an extension of DESs by Alur and Dill [1] to reason about systems for which time is a relevant feature. Under some restrictions on the use of time, timed automata can be treated as DESs, and diagnosis of timed automata has been proposed [37]. However, the approach based on the diagnoser cannot be used because it is impossible to determine a timed automaton [38].

9.5.7 *Explainable Diagnosis*

In some contexts, it can be useful to provide the system manager not only with a diagnosis but also with some justification or explanation of the diagnosis. This has the dual benefit of improving the awareness of the manager and increasing his confidence in the diagnosis.

Diagnosis of DESs has been defined as the projection on the fault events of all behaviors that are consistent with the observation. Nontrivial systems can have a large flow of observed events and it can be difficult for a computer to "explain" how it reached its conclusion. Christopher and Grastien [7] proposed to highlight a "critical" part of the observations, sufficient to diagnose accurately the system: ideally, this critical part should be small enough that a human manager is able to handle it.

Example Consider again the DES of Fig. 9.1. In a large flow of observed events, the reception of the sequence of the following three ones, o_r, o_c, o_c, with no event in between, is sufficient to diagnose the occurrence of a fault. ◊

9.6 Conclusion

Diagnosis of DESs is an important task in the real world as several systems, such as digital circuits, can most naturally be modeled as DESs, and also many others can be modeled as DESs at some abstraction level. In the literature, model-based diagnosis of DESs is typically applied to intelligent alarm processing in a remote control room, where the received alarms are the observable events taken as input by the diagnosis process. For instance, in [23] the protection devices of power transmission lines are considered as a case study for diagnosis and monitoring of DESs. Analogously, experimental diagnosis results inherent to the real alarm log from the operations center of a company that owns and operates an electricity transmission network in Australia are presented in [14, 15]. In addition, the contribution in [14] addresses the set of event logs recorded on the ground during test flights of an autonomous unmanned helicopter (UAV), while in [3, 29] the alarms received by the supervision center of a telecommunication network are taken into account.

In this chapter, we introduced the problem of model-based diagnosis when the model is a DES. A DES is essentially an automaton whose states represent system states and transitions represent how the system state evolves over time. Diagnosis is performed by first computing the system behaviors (trajectories on the model) that agree with the observations (transition labels), and then checking whether these behaviors are nominal or faulty. We discussed how this is implemented practically, in particular, to address the issue of complexity.

We then discussed diagnosability of DES, i.e., the property that the occurrence of a fault will always be detected and identified by the diagnostic engine. We showed that diagnosability can be verified in polytime, and we discussed the benefits of diagnosability in relation to complexity.

Finally, we introduced some more advanced topics that make the approach applicable to realistic scenarios.

References

1. Alur, R., Dill, D.: A theory of timed automata. Theor. Comput. Sci. (TCS) **126**(2), 183–235 (1994)
2. Baroni, P., Lamperti, G., Pogliano, P., Zanella, M.: Diagnosis of large active systems. Artif. Intell. **110**(1), 135–183 (1999)
3. Benveniste, A., Fabre, E., Haar, S., Jard, C.: Diagnosis of asynchronous discrete-event systems: a net unfolding approach. IEEE Trans. Autom. Control **48**(5), 714–727 (2003)

4. Brandán Briones, L., Lazovik, A., Dague, P.: Optimal observability for diagnosability. In: Nineteenth International Workshop on Principles of Diagnosis (DX-08), pp. 31–38, Blue Mountains, New South Wales, Australia (2008)
5. Bryant, R.: Graph-based algorithms for Boolean function manipulation. IEEE Trans. Comput. (TC) **35**(8), 677–691 (1986)
6. Cassez, F., Tripakis, S.: Fault diagnosis with dynamic observers. In: Ninth International Workshop on Discrete Event Systems (WODES-08), pp. 212–217, Goteborg, Sweden (2008)
7. Christopher, C., Grastien, Al.: Formulating event-based critical observations in diagnostic problems. In: 26th International Workshop on Principles of Diagnosis (DX-15), pp. 119–126, Paris, France (2015)
8. Cimatti, A., Pecheur, C., Cavada, R.: Formal verification of diagnosability via symbolic model checking. In: Proceedings of the 18th International Joint Conference on Artificial Intelligence, pp. 363–369, Acapulco, Mexico (2003)
9. Clarke, Ed., Grumberg, O., Peled, D.: Model Checking. MIT Press (2000)
10. Cordier, M., Largouët, C.: Using model-checking techniques for diagnosing discrete-event systems. In: Twelfth International Workshop on Principles of Diagnosis DX'01, pp. 39–46, San Sicario, Italy (2001)
11. Darwiche, A., Marquis, P.: A knowledge compilation map. J. Artif. Intell. Res. **17**, 229–264 (2002)
12. Grastien, A.: Diagnosis properties by design. In: 22nd International Workshop on Principles of Diagnosis (DX-11), pp. 155–158, Murnau, Germany (2011) (Open problems track)
13. Grastien, A., Anbulagan, Rintanen, J., Kelareva, E.: Diagnosis of discrete-event systems using satisfiability algorithms. In: 22nd National Conference on Artificial Intelligence—AAAI'07, pp. 305–310, Vancouver, BC, Canada (2007)
14. Grastien, A., Haslum, P.: Diagnosis as planning: two case studies. In: Scheduling and Planning Applications Workshop—SPARK'11, pp. 37–44, Freiburg, Germany (2011)
15. Grastien, A., Haslum, P., Thiébaux, S.: Conflict-based diagnosis of discrete event systems: theory and practice. In: 13th International Conference on Principles of Knowledge Representation and Reasoning—KR 2012, pp. 4989–4996, Rome, Italy (2012)
16. Grastien, A., Torta, G.: Reformulation for the diagnosis of discrete-event systems. In: 21st International Workshop on Principles of Diagnosis (DX-10), pp. 63–70, Portland, Oregon, USA (2010)
17. Huang, J., Darwiche, A.: The language of search. J. Artif. Intell. Res. **29**, 191–219 (2007)
18. Jiang, S., Huang, Z., Chandra, V., Kumar, R.: A polynomial algorithm for testing diagnosability of discrete event systems. IEEE Trans. Autom. Control **46**(8), 1318–1321 (2001)
19. Kan John, P., Grastien, A.: Local consistency and junction tree for diagnosis of discrete-event systems. In: Eighteenth European Conference on Artificial Intelligence (ECAI-08), pp. 209–213, Patras, Greece (2008)
20. Lamperti, G., Zanella, M.: Diagnosis of discrete-event systems from uncertain temporal observations. Artif. Intell. **137**(1–2), 91–163 (2002)
21. Lamperti, G., Zanella, M.: Diagnosis of Active Systems—Principles and Techniques. The Kluwer International Series in Engineering and Computer Science, vol. 741. Kluwer Academic Publishers, Dordrecht, The Netherlands (2003)
22. Lamperti, G., Zanella, M.: On processing temporal observations in monitoring of discrete-event systems. In: Manolopoulos, Y., Filipe, J., Constantopoulos, P., Cordeiro, J. (eds.) Enterprise Information Systems VIII. Lecture Notes in Business Information Processing, vol. 3, pp. 135–146. Springer, Berlin, Heidelberg (2008)
23. Lamperti, G., Zanella, M.: Monitoring of active systems with stratified uncertain observations. IEEE Trans. Syst. Man Cybern. Part A Syst. Hum. **41**(2), 356–369 (2011)
24. Lamperti, G., Zanella, M., Zhao, X.: Introduction to Diagnosis of Active Systems. Springer International Publishing (2018)
25. McDonald, J., Burt, G., Young, D.: Alarm processing and fault diagnosis using knowledge based systems for transmission and distribution network control. IEEE Trans. Power Syst. (TPS) **7**(3), 1292–1298 (1992)

26. Murata, T.: Petri nets: properties, analysis and applications. Proc. IEEE **77**(4), 541–580 (1989)
27. Nouioua, F., Dague, P.: A probabilistic analysis of diagnosability in discrete event systems. In: Eighteenth European Conference on Artificial Intelligence (ECAI-08) (2008)
28. Pencolé, Y.: Diagnosability analysis of distributed discrete event systems. In: Sixteenth European Conference on Artificial Intelligence (ECAI-04), pp. 173–178, Valencia, Spain (2004)
29. Pencolé, Y., Cordier, M.O.: A formal framework for the decentralised diagnosis of large scale discrete event systems and its application to telecommunication networks. Artif. Intell. (AIJ) **164**(1–2), 121–170 (2005)
30. Petri, C.A.: Kommunikation mit Automaten. Ph.D. thesis, Universität Hamburg (1962)
31. Sampath, M., Sengupta, R., Lafortune, S., Sinnamohideen, K., Teneketzis, D.: Diagnosability of discrete-event systems. IEEE Trans. Autom. Control **40**(9), 1555–1575 (1995)
32. Schumann, A., Pencolé, Y., Thiébaux, S.: A spectrum of symbolic on-line diagnosis approaches. In: Proceedings of the 22nd Conference on Artificial Intelligence—AAAI-07, pp. 335–340, Vancouver, BC, Canada (2007)
33. Sengupta, R., Tripakis, S.: Decentralized diagnosability of regular languages is undecidable. In: 43rd IEEE Conference on Decision and Control (CDC-04), pp. 423–428 (2004)
34. Su, R., Wonham, W.: Global and local consistencies in distributed fault diagnosis for discrete-event systems. IEEE Trans. Autom. Control **50**(12), 1923–1935 (2005)
35. Su, X., Zanella, M., Grastien, A.: Diagnosability of discrete-event systems with uncertain observations. In: 25th International Joint Conference on Artificial Intelligence, pp. 1265–1271 (2016)
36. Thorsley, D., Teneketzis, D.: Active acquisition of information for diagnosis and supervisory control of discrete event systems. J. Discrete Event Dyn. Syst. (JDEDS) **17**, 531–583 (2007)
37. Tripakis, S.: Fault diagnosis for timed automata. In: Seventh International Symposium Formal Techniques in Real-Time and Fault-Tolerant Systems (FTRTFT-02), pp. 205–221, Oldenburg, Germany (2002)
38. Tripakis, S.: Folk theorems on the determinization and minimization of timed automata. In: First International Workshop on Formal Modeling and Analysis of Timed Systems (FORMATS-03), pp. 182–188, Marseille, France (2003)
39. Uztok, U., Darwiche, A.: A top-down compiler for sentential decision diagrams. In: Proceedings of the 24th International Joint Conference on Artificial Intelligence, pp. 3141–3148, Buenos Aires, Argentina (2015)
40. Wang, Y., Yoo, T.S., Lafortune, S.: New results on decentralized diagnosis of discrete-event systems. In: 42nd Annual Allerton Conference on Communication, Control, and Computing (AACCCC-04), Monticello, United States (2004)
41. Ye, L., Dague, P., Yan, Y.: A distributed approach for pattern diagnosability. In: 20th International Workshop on Principles of Diagnosis (DX-09), pp. 179–186, Stockholm, Sweden (2009)
42. Yoo, T., Lafortune, S.: Polynomial-time verification of diagnosability of partially observed discrete-event systems. IEEE Trans. Autom. Control **47**(9), 1491–1495 (2002)

Part II
Advanced Approaches

Chapter 10
Fault Diagnosis Using Set-Membership Approaches

Vicenç Puig and Masoud Pourasghar

10.1 Introduction

As discussed in Chap. 4, model-based fault detection of dynamic processes is based on the use of models (i.e., *analytical redundancy*) to check the consistency of observed behaviours. However, when building a model of a dynamic process to monitor its behaviour, there is always some mismatch between the modelled and real behaviour due to the fact that some effects are neglected, some non-linearities are linearised in order to simplify the model, some parameters have tolerance when are compared between several units of the same component, some errors in parameters (or in the structure) of the model are introduced in the model identification process, etc. These modelling errors introduce some uncertainty in the model. Usually, this uncertainty can be bounded and included in the fault detection model. There are several ways of considering the uncertainty associated with the model depending if is located in the parameters (*structured*) or in the model structure (*non-structured*). In FD literature, a fault diagnosis algorithm able to handle uncertainty is called *robust*. The *robustness* of a fault diagnosis algorithm is the degree of sensitivity to faults compared to the degree of sensitivity to uncertainty [11]. Research on robust fault diagnosis methods has been very active in the FDI community these last years. One of the most developed families of approaches, called *active*, is based on generating residuals which are insensitive to uncertainty, while at the same time sensitive to faults. This approach has been extensively developed by several researchers using different techniques: unknown input observers, robust parity equations, H_∞, etc. In [11], there is an excellent survey of this active approach. On the other hand, there is a second family of approaches, called *passive*, which enhances the robustness of the fault detection system at the decision-making stage by propagating the uncertainty to the residuals and generating an *adaptive threshold*. Seminal papers suggesting

V. Puig (✉) · M. Pourasghar
Advanced Control Systems (SAC) Group, Universitat Politècnica de Catalunya (UPC), Barcelona, Spain
e-mail: vicenc.puig@upc.edu

T. Escobet et al. (eds.), *Fault Diagnosis of Dynamic Systems*,
https://doi.org/10.1007/978-3-030-17728-7_10

this approach are [22] in the time domain and [14] in the frequency domain. This passive approach has been developed these last years by several researches but still is under research, see for example, to see some recent results: [2, 5, 16, 20, 33, 35, 37, 38, 41], among other. This approach has also been integrated with *Qualitative Reasoning* tools (coming from AI community) see for example: the tools *CA∼EN* [15, 44], *SQualTrack* [5] or *MOSES* [39], are some examples. For a more detailed review the reader is referred to [4].

This chapter will review the passive approach when considering the nominal model plus the uncertainty on every parameter bounded by intervals. This type of uncertainty modelling provides a type of models known as *interval models*. Noise will be also be considered to be unknown-but-bounded and modelled in a deterministic framework. The use of interval models has received several names depending on the field of application: in circuit analysis it is known as worst-case (or tolerance analysis), in automatic control as *set-membership* (also known, in this field as *error-bounded* approach) and in qualitative reasoning as semi-quantitative. In the automatic control literature, the *set-membership* approach applied to parameter and state estimation has been treated extensively in [29] while their application to control can be found in [1, 6]. The worst-case analysis of circuits has been treated in [26] and in several research papers appearing in circuits journals and conferences. Finally, the semi-quantitative approach is treated in [27] and in several papers appearing in artificial intelligence journals and conferences.

This chapter also reviews the different approaches that can be used to identify interval models for fault detection. This research has started with the seminal work of [34]. Then, the application of set-membership methods to fault detection. Finally, the chapter presents the application of several set-membership approaches to the tank system introduced in Chap. 2.

The remainder of the chapter is organized as follows: Sect. 10.2 introduces the use of interval models of dynamic systems for fault detection. In Sect. 10.3, fault detection using the interval approach is recalled, while Sect. 10.4 presents the fault detection using the error-bounding approach. Section 10.5 reviews the methods for interval and error-bounding identification using real data. Section 10.6 presents the application of set-membership methods for fault detection four-tank case study. Finally, the major conclusions are summarised in Sect. 10.7.

10.2 Interval Models for Fault Detection

10.2.1 Interval Models of Dynamic Systems

Considering that the system to be monitored can be described by a MIMO linear uncertain dynamic model in discrete-time and state-space form as follows

$$\begin{aligned} \mathbf{x}(k+1) &= \mathbf{A}(\boldsymbol{\theta})\mathbf{x}(k) + \mathbf{B}(\boldsymbol{\theta})\mathbf{u}(k) + \mathbf{w}(k) \\ \mathbf{y}(k) &= \mathbf{C}(\boldsymbol{\theta})\mathbf{x}(k) + \mathbf{D}(\boldsymbol{\theta})\mathbf{u}(k) + \mathbf{v}(k) \end{aligned} \tag{10.1}$$

where $\mathbf{y}(k) \in \Re^{n_y}$, $\mathbf{u}(k) \in \Re^{n_u}$, $\mathbf{x}(k) \in \Re^{n_x}$ are the system output, input and the state-space vectors, respectively; $\mathbf{w}(k) \in \Re^{n_x}$ and $\mathbf{v}(k) \in \Re^{n_y}$ are the disturbances and noises, respectively, both assumed unknown but bounded, i.e., $w_i \in \left[\underline{\delta}_i, \overline{\delta}_i\right]$ and $v_i \in \left[\underline{\sigma}_i, \overline{\sigma}_i\right]$; the state, input, output and direct transmission matrices are $\mathbf{A}(\boldsymbol{\theta}) \in \Re^{n_x \times n_x}$, $\mathbf{B}(\boldsymbol{\theta}) \in \Re^{n_x \times n_u}$, $\mathbf{C}(\boldsymbol{\theta}) \in \Re^{n_y \times n_x}$ and $\mathbf{D}(\boldsymbol{\theta}) \in \Re^{n_y \times n_u}$, respectively; $\boldsymbol{\theta} \in \Re^{n_\theta}$ is the vector of uncertain parameters where Θ is a bounded set (of interval box type) such that and in particular for each component $\theta_i \in \left[\underline{\theta}_i, \overline{\theta}_i\right]$, $i = 1, \ldots, n_\theta$. This is why the resulting model is known as an interval model. The set Θ contains all possible values of $\boldsymbol{\theta}$ when the system operates normally. Notice that when the parameters $\boldsymbol{\theta}$ are scheduled with the operating point using some known scheduling function and variable then system (10.1) is known as a *linear parameter varying* (LPV) system [40]. Intervals for uncertain parameters can also be inferred from real data as will be discussed in Sect. 10.5.

The system in Eq. (10.1) can, alternatively, be expressed in input-output form using the shift operator q^{-1} and assuming zero initial conditions as follows:

$$\mathbf{y}(k) = \mathbf{M}(q^{-1}, \boldsymbol{\theta})\mathbf{u}(k) \tag{10.2}$$

where $\mathbf{M}(q^{-1}, \boldsymbol{\theta})$ is given by

$$\mathbf{M}(q^{-1}, \boldsymbol{\theta}) = \mathbf{C}(\boldsymbol{\theta})(q\mathbf{I} - \mathbf{A}(\boldsymbol{\theta}))^{-1}\mathbf{B}(\boldsymbol{\theta}) + \mathbf{D}(\boldsymbol{\theta})$$

10.2.2 Interval Models for Fault Detection

The principle of model-based fault detection is to test whether the system measurements are consistent with the behaviour described by a model of the faultless system. Consistent means that measured system behaviour agrees with the behaviour estimated using the model. If the measurements are inconsistent with this model, the existence of a fault is proved. The residual vector, known also as *analytical redundant relation* (ARR), defined as the difference between measured $\mathbf{y}(k)$ and predicted system outputs $\hat{\mathbf{y}}(k)$

$$\mathbf{r}(k) = \mathbf{y}(k) - \hat{\mathbf{y}}(k) \tag{10.3}$$

is usually used to check the consistency. Ideally, the residuals should only be affected by the faults. However, the presence of disturbances, noise and modelling errors causes the residuals to become nonzero and thus interferes with the detection of faults. Therefore, the fault detection procedure must be robust against these undesired effects [11]. In case of modelling a dynamic system using an interval model, the predicted output is described by a set that can be bounded at any iteration by an interval

$$\hat{y}_i(k) \in [\underline{\hat{y}_i}(k), \overline{\hat{y}_i}(k)] \tag{10.4}$$

in a non-faulty case. Such interval is computed independently for each output (neglecting couplings between outputs) as follows

$$\underline{\hat{y}_i}(k) = \min_{\theta \in \Theta}(\hat{y}_i(k,\boldsymbol{\theta})) \text{ and } \overline{\hat{y}_i}(k) = \max_{\theta \in \Theta}(\hat{y}_i(k,\boldsymbol{\theta})) \tag{10.5}$$

Such interval can be computed using the algorithm based on numerical optimization presented in [36]. Then, the fault detection test is based on propagating the parameter uncertainty to the residual, and checking if

$$\mathbf{y}(k) \in \left[\underline{\hat{\mathbf{y}}}(k) - \sigma, \overline{\hat{\mathbf{y}}}(k) + \sigma\right] \tag{10.6}$$

where σ is the noise bound. Equivalently, previous test can be formulated in terms of the residual checking if

$$0 \in \left[\underline{\mathbf{r}}(k), \overline{\mathbf{r}}(k)\right] = \mathbf{y}(k) - \left[\underline{\hat{\mathbf{y}}}(k) - \sigma, \overline{\hat{\mathbf{y}}}(k) + \sigma\right] \tag{10.7}$$

holds or not. In case it does not hold a fault can be indicated. This test is named as *direct test*. Alternatively, the *inverse test* consists in checking if there exists some parameter value in the parameter uncertainty set Θ such that model (10.2) is consistent with the system measurements. More formally, to check if

$$\exists \boldsymbol{\theta} \in \Theta \mid \hat{\mathbf{y}}(k,\boldsymbol{\theta}) \in \left[\mathbf{y}(k) - \sigma, \mathbf{y}(k) + \sigma\right] \tag{10.8}$$

holds or not. In the case that this condition is not satisfied, a discrepancy between measurements and the model is detected and a fault should be indicated. This test can be implemented with the parameter estimation algorithms used in the error-bounding approach [29] as will be discussed later in this chapter. The direct test is related to the use of parity equation or observer methods while the inverse test is related to parameter estimation methods. According to [24], parity equations and observer approaches are more suitable for additive faults while parameter estimation approach is better suited for multiplicative (parametric) faults.

10.3 Fault Detection Using the Interval Approach

10.3.1 Fault Detection Using Interval Observers

The system described by (10.1) can be monitored using a linear observer with *Luenberger structure*. The resulting *interval observer* can be written as:

$$\begin{aligned} \hat{\mathbf{x}}(k+1,\boldsymbol{\theta}) &= \mathbf{A}(\boldsymbol{\theta})\hat{\mathbf{x}}(k) + \mathbf{B}(\boldsymbol{\theta})\mathbf{u}(k) + \mathbf{w}(k) \\ &\quad + \mathbf{L}(\mathbf{y}(k) - \hat{\mathbf{y}}(k)) \\ \hat{\mathbf{y}}(k,\boldsymbol{\theta}) &= \mathbf{C}(\boldsymbol{\theta})\hat{\mathbf{x}}(k) + \mathbf{v}(k) \end{aligned} \tag{10.9}$$

where $\hat{\mathbf{x}}(k, \boldsymbol{\theta})$ is the estimated state-space vector and $\hat{\mathbf{y}}(k, \boldsymbol{\theta})$ is the estimated output vector for a given value of $\boldsymbol{\theta} \in \Theta$ tanking into account process and sensor noise bounds. The observer gain matrix $\mathbf{L} \in \Re^{nx \times ny}$ is designed to stabilize the matrix $\mathbf{A}_o(\boldsymbol{\theta})$ and to guarantee a desired performance regarding fault detection for all $\boldsymbol{\theta} \in \Theta$ [12]. Alternatively, the observer given by (10.9) can be expressed in input-output form using the q-transform and considering zero initial conditions as follows:

$$\hat{\mathbf{y}}(k) = \mathbf{G}(q^{-1}, \boldsymbol{\theta})\mathbf{u}(k) + \mathbf{H}(q^{-1}, \boldsymbol{\theta})\mathbf{y}(k) \tag{10.10}$$

where

$$\begin{aligned}
\mathbf{G}(q^{-1}, \boldsymbol{\theta}) &= \mathbf{C}(\boldsymbol{\theta}) \left(q\mathbf{I} - \mathbf{A}_o(\boldsymbol{\theta})\right)^{-1} \mathbf{B}(\boldsymbol{\theta}) \\
\mathbf{H}(q^{-1}, \boldsymbol{\theta}) &= \mathbf{C}(\boldsymbol{\theta}) \left(q\mathbf{I} - \mathbf{A}_o(\boldsymbol{\theta})\right)^{-1} \mathbf{L} \\
\mathbf{A}_o(\boldsymbol{\theta}) &= \mathbf{A}_o(\boldsymbol{\theta}) - \mathbf{L}\mathbf{C}(\boldsymbol{\theta})
\end{aligned}$$

Interval observation requires solving the optimization problems introduced in Eq. (10.5) using Eq. (10.10). In order to preserve *uncertain parameter time-invariance* and to avoid the *wrapping effect*[1] [36], the observer output prediction in Eq. (10.5) is substituted by

$$\hat{\mathbf{y}}\left(k\right) = \mathbf{C}\left(\boldsymbol{\theta}\right) \mathbf{A}_0\left(\boldsymbol{\theta}\right)^k \mathbf{x}_0 + \mathbf{C}\left(\boldsymbol{\theta}\right) \sum_{j=0}^{k-1} \mathbf{A}_0\left(\boldsymbol{\theta}\right)^{(k-1-j)} \mathbf{B}\left(\boldsymbol{\theta}\right) \mathbf{u}\left(j\right) \tag{10.11}$$

When proceeding in this way, the optimization problems in Eq. (10.11) will not be convex because of the non-linearity with respect to parameters. Therefore, the existence of a unique optimum is not guaranteed. In order to guarantee that the global optimum is reached, a global optimization algorithm must be used. In particular, a branch and bound interval arithmetic global optimization based on *Hansen's algorithm* [21] can be used. An additional computational problem appears when using Eq. (10.11) since the degree of the polynomial in the objective function increases with time. This implies that the amount of computation needed is also increasing with time, being impossible to operate over a large time period. This problem can be solved if the interval system (10.1) is asymptotically stable [36]. In this case, the predicted system output at time k depends, approximately, only on the inputs that occurred in a *time sliding window* with a length ℓ (whose value is of the order of the settling time) and the state at the beginning of such window. Then, Eq. (10.11) can be approximated by limiting the computation to a finite time horizon as it has been proposed in [36].

---

[1]The problem of wrapping is related to the use of a crude approximation of set of states associated with the interval simulation. If at each iteration, the true solution set is wrapped into its interval hull, since the overestimation of the wrapped set is proportional to its radius, a spurious growth of the enclosures can result if the composition of wrapping and mapping is iterated.

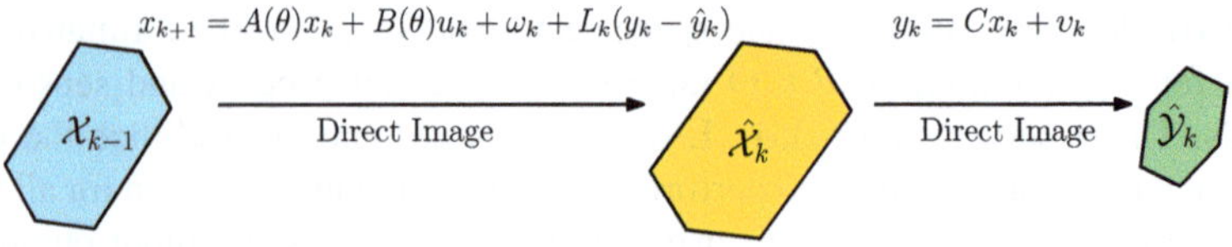

Fig. 10.1 Interval observer

Fig. 10.2 Zonotope

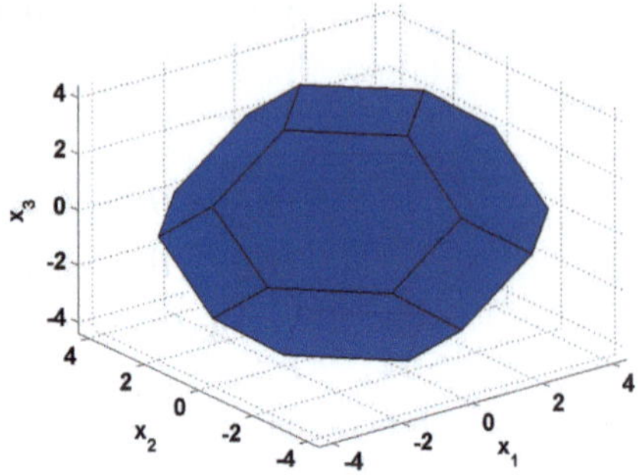

In case that uncertain parameters are considered *time-varying*, an iterative algorithm can be used that obtains the set of uncertain states at time k, $\mathbb{X}_k$ from the set $\mathbb{X}_{k+1}$ using the algorithm presented in Fig. 10.1 [19].

To implement such algorithm, the set of uncertain should be approximated since the exact set of estimated states would be difficult to compute. Several geometrical shapes has been proposed in the literature ranging from parallelotopes [13] or ellipsoids [28] to zonotopes as proposed by [3]. A zonotope $\mathbb{X}$ of order m can be viewed as the Minkowski sum of m segments:

$$\mathbb{X} = \mathbf{p} \oplus \mathbf{H}B^m = \left\{\mathbf{p} + \mathbf{Hz} : \mathbf{z} \in B^m\right\} \tag{10.12}$$

where the segments are defined by the columns of matrix $\mathbf{H}$ and B^m is a unitary box composed of m unitary intervals. The order m is a measure for the geometrical complexity of the zonotopes (see Fig. 10.2 for a tridimensional zonotope).

Zonotope arithmetic possesses a set of operations (as sum, affine transformation, intersection) that can be very efficiently implemented since they only involve operations with matrices [3].

10.3.2 Interval ARMA Parity Equations

In case the observer gain in (10.9) is taken equal to zero ($\mathbf{L} = 0$), the observer becomes an *interval simulator*, since the output prediction is based only in the inputs and previous output predictions, and (10.10) becomes: $\hat{\mathbf{y}}(k) = \mathbf{M}(q^{-1}, \boldsymbol{\theta})\mathbf{u}(k)$ while the residual is given by

$$\mathbf{r}(k) = \mathbf{y}(k) - \hat{\mathbf{y}}(k) = \mathbf{y}(k) - \mathbf{M}(q^{-1}, \boldsymbol{\theta})\mathbf{u}(k) \tag{10.13}$$

According to [17], Eq. (10.13) corresponds with an *ARMA primary parity equation*or *residual*. This is an *open-loop approach*. Interval simulation requires solving the optimization problems following the same strategy as in the case of the interval observer but using the system matrices (10.1). In order to reduce computing complexity, as in the observer case, a time window could also be used. In this case, this approach is known as ℓ-order ARMA parity equation [43].

10.3.3 Interval MA Parity Equations

On the other hand, in case observer gain in (10.8) is designed such that on the poles are at the origin (*deadbeat observer*), the observer becomes an *interval predictor*, since the output prediction is based only in measured inputs and outputs and follows the real system output after the minimum number of samples. The prediction Eq. (10.10) is moving average (MA) and follows a *closed-loop approach*. Thus, the corresponding residuals are called *MA primary parity equations* or *residuals* [17]. The optimization problems (10.5) that must be solved now are linear with respect to the parameters, and, therefore, convex. This means that there exist very efficient algorithms to solve them (as the *simplex algorithm*). Because of the linearity, the existence of a unique optimum is guaranteed being located in one of the vertices of the parameter uncertainty intervals. Interval prediction is not affected by the problem of wrapping because the predicted output is based on the previous output measurements instead of the interval of the previous predicted outputs [35]. Thus, interval prediction considers uncertain parameters as time varying. But, time invariance in uncertain parameters is wanted to be preserved, a ℓ-order MA parity equation should be used [43]. Finally, [33] has, recently, proposed a method to obtain the interval parity equations directly from state-space using the *Chow-Wilsky scheme*.

10.3.4 Comparison

In [35], the behaviour of the different interval fault detection approaches considered so far are studied and compared using the FD benchmark proposed in DAMADICS project. Table 10.1 summarises the results of this comparison. This table can be used as a guideline to decide in which applications an approach is more suitable than the others. Prediction and simulation approaches have antagonistic properties: prediction, because of its deadbeat observer behaviour, does not suffer from the wrapping effect, has low computational complexity, has low sensitivity to unmodeled dynamics but can suffer from the sensor following fault effect and has high sensitivity

Table 10.1 Interval-based fault detection approaches features

Issue		Simulator	Observer	Predictor
Wrapping effect		Yes	Yes	No
Computational complexity		High	High	Low
Unmodeled dynamics sensitivity		High	Medium	Low
Initial conditions sensitivity		High	Medium	Low
Fault sensitivity	Actuator	Dynamic	Dynamic	Constant
	Sensor	Constant	Pulse	Deadbeat
Noise sensitivity	Process	LP filter	LP filter	Gain
	Sensor	Gain	HP Filter	HP filter

to sensor noise. On the other side, the simulation approach has the opposite properties, presenting good performance to detect sensor faults in noisy systems. Finally, the observer approach is in the middle, with the advantage that since it has one more degree of freedom (the observer gain), can be designed trying to minimize the bad effects and maximize the good effects of the other two approaches.

10.4 Fault Detection Using Error-Bounding Approach

Alternatively to the interval approach presented in previous section, the error-bounding approach relies on checking whether the measured sequence of system inputs and outputs available at every time instant k could have been generated by the model (2) and parameter values in the parameter uncertainty set Θ [31]. This approach is related with the inverse test described in Sect. 10.2.

10.4.1 Fault Detection Test in the Parameter Space

The inverse test involves checking if there exists a parameter in the parameter uncertainty set Θ_k such that model (10.2) is consistent with the systems measurements. This test can be easily implemented using the error-bounding parameter estimation procedure described in Sect. 10.5 since it can operate in the recursive form as follows:

$$\Theta_{k+1} = \Theta_k \cap \mathbb{F}_k \tag{10.14}$$

where:

$$\mathbb{F}_k = \left\{ \boldsymbol{\theta} \in \Re^{n_\theta} | \mathbf{y}(k) - \sigma \leq \mathbf{M}(q^{-1}, \boldsymbol{\theta})\mathbf{u}(k) \leq \mathbf{y}(k) + \sigma \right\}$$

is the strip of consistent parameters with the current measurements. In fault detection using the inverse test, the model is assumed invalidated and fault is indicated if

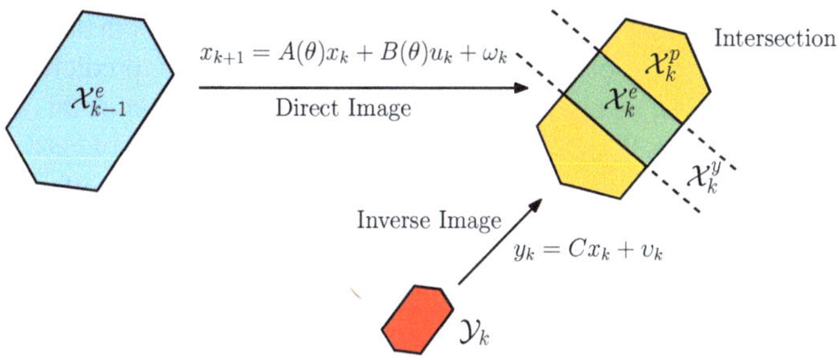

Fig. 10.3 Error-bounding state estimation

$\Theta_{k+1} = \emptyset$ [23]. Once the fault is indicated, the feasible parameter set Θ_k should be reset to a set that contains all possible values even in faulty situation. Then, the faulty feasible parameter set can be identified (fault isolation) and the fault size can be estimated by comparing the feasible parameter set before and after fault detection using, for example, the distance between centres of these sets (fault estimation).

Despite of outer approximation is the most used in fault detection due to it contains all the consistent models, the inner approximation, that contains only consistent parameters, can complement the use of outer approximation in order to improve fault detection behavior.

10.4.2 Fault Detection Test in the State Space

An error-bounded state estimator assumes a priori bounds on noise and uncertain parameters and constructs sets of estimated states that are consistent with the a priori bounds and current measurements. Several researchers as [8, 13, 28, 42] and [25], among others, have addressed this issue. Consider the system given by (10.1), an initial compact set $\mathbb{X}_o$ and a sequence of measured inputs and outputs, the uncertain state set at time k using the error-bounding approach can computed using the algorithm presented in Fig. 10.3. A fault is detected when $\mathbb{X}_k^e = \mathbb{X}_k^p \cap \mathbb{X}_k^y = \emptyset$ [18, 32].

10.5 Identification for Robust Fault Detection

10.5.1 Model Parametrisation

One of the key points in model based fault detection is how models are calibrated to fit real data taken from the monitored system in non-faulty situations. Identification should deliver a calibrated nominal model plus its modelling error in the form of interval parameters, that it will provide an interval of confidence for predicted behaviour, i.e., the interval model, as already discussed in the introduction of this chapter. To this aim, several authors [9, 10, 34] have suggested an adaptation of classical system identifications methods to provide the nominal model plus the uncertainty intervals for parameters that guarantee that all recorded data from the system in non-faulty scenarios will be included in the interval model. These algorithms are based on using classical identification methods (for example, least-squares) to provide the nominal estimate for system parameters. Then, the intervals of uncertainty for parameters are adjusted until all the measured data is covered by the model prediction interval. These algorithms proceed considering that the interval model (10.1) to be identified can be expressed in regressor form as follows

$$y(k) = \varphi^T(k)\boldsymbol{\theta} + v(k) = \hat{y}(k) + v(k) \tag{10.15}$$

where: $\varphi(k)$ is the regressor vector of dimension n_θ which can contain any function of inputs $u(k)$ and outputs $y(k)$; $v(k)$ is additive noise bounded by a constant $|v(k)| \leqslant \sigma$; $\boldsymbol{\theta} \in \Theta_k$ is the parameter vector of dimension n_θ and Θ_k is the set that bounds parameter values. This set can again be approximated by ellipsoids, parallelotopes or zonotopes [29]. In case that this set is described by a zonotope centered in the nominal model, it can be parameterised as follows [7]

$$\Theta_k = \boldsymbol{\theta}^0 \oplus \mathbf{H}B^n = \left\{\boldsymbol{\theta}^0 + \mathbf{Hz} : \mathbf{z} \in B^n\right\} \tag{10.16}$$

Notice that a particular case corresponds to the case the parameter set Θ_k is an interval box:

$$[\theta_i] = \left[\theta_i^{\min}, \theta_i^{\max}\right] = \left[\theta_i^0 - \lambda_i, \theta_i^0 + \lambda_i\right] \tag{10.17}$$

with $i = 1, \ldots, n_\theta$. This set can be viewed as a zonotope with $\mathbf{H}$ equal to a $n_\theta \times n_\theta$ diagonal matrix:

$$\mathbf{H} = diag(\lambda_1, \lambda_2, \ldots, \lambda_{n_\theta}) \tag{10.18}$$

Given a sequence of M regressor vector values '(k) in a fault free scenario and a model parameterised as in (10.15), the aim is to estimate model parameters and their uncertainty (model set) following either an *interval* or *error-bounding parameter estimation* approach.

10.5.2 Interval Parameter Estimation

In this case, the set of uncertain parameters Θ_k should be obtained in such a way that all measured data in a fault free scenario will be covered by the predicted output interval produced by using model (10.15) and the uncertainty parameter set, that is:

$$\overline{\hat{y}}(k) \geqslant y(k) - \sigma \text{ and } \underline{\hat{y}}(k) \leqslant y(k) + \sigma \;\; \forall k = 1, \ldots, M \tag{10.19}$$

where:

$$\overline{\hat{y}}(k) = \max \left(\varphi^T(k)\boldsymbol{\theta} \right) \text{ with } \boldsymbol{\theta} \in \Theta_k \tag{10.20a}$$

$$\underline{\hat{y}}(k) = \min \left(\varphi^T(k)\boldsymbol{\theta} \right) \text{ with } \boldsymbol{\theta} \in \Theta_k \tag{10.20b}$$

This type of model identification was first suggested in [34] in the context of fault detection using a direct test and an interval LTI model in prediction.

Considering that the parameter set Θ_k can be described as the zonotope (10.16) and proceeding as in [34], the maximum and minimum interval prediction bounds provided by model (10.15) are given by

$$\overline{\hat{y}}(k) = \hat{y}^0(k) + \left\| \varphi^T(k)\mathbf{H} \right\|_1 \tag{10.21a}$$

$$\underline{\hat{y}}(k) = \hat{y}^0(k) - \left\| \varphi^T(k)\mathbf{H} \right\|_1 \tag{10.21b}$$

where $\hat{y}^0(k)$ is the model output prediction with nominal parameters, i.e., $\hat{y}^0(k) = \varphi^T(k)\boldsymbol{\theta}^0$ where $\boldsymbol{\theta}^0 = (\theta_1^0, \ldots, \theta_{n_\theta}^0)$.

Notice that in the particular case of interval parameters:

$$\left\| \varphi^T(k)\mathbf{H} \right\|_1 = \sum_{i=1}^{n} \lambda_i \left| \varphi_i(k) \right| \tag{10.22}$$

replacing Eqs. (10.21a) and (10.21b) in inclusion conditions (10.19), the optimal zonotope that fulfills the *interval prediction condition* can be computed solving the following optimization problem

$$\begin{aligned} &\min_{\mathbf{H}} \; f(\Theta_{\mathbf{k}}(\mathbf{H})) \\ &\text{subject to:} \\ &\left\| \varphi^T(k)\mathbf{H} \right\|_1 \geqslant \left| y(k) - \hat{y}^0(k) \right| - \sigma \quad \forall k = 1, \ldots, M \end{aligned} \tag{10.23}$$

In this optimization problem, the cost function f is usually the interval prediction thickness that can be calculated as

$$\sum_{k=1}^{N} (\overline{\hat{y}}(k) - \underline{\hat{y}}(k)) = 2 \sum_{k=1}^{N} \left\| \varphi^T(k)\mathbf{H} \right\|_1 \tag{10.24}$$

In order to reduce the complexity of optimization problem (10.23), the zonotope that bounds Θ_k can be parameterised such that $\mathbf{H} = \lambda \mathbf{H}_0$, corresponding with a zonotope with predefined shape (determined by $\mathbf{H}_0$) and a scalar λ. Then, in this case interval prediction thickness (10.24) is given by

$$\sum_{k=1}^{N} (\overline{\hat{y}}(k) - \underline{\hat{y}}(k)) = 2|\lambda| \sum_{k=1}^{N} \left\| \varphi^T(k)\mathbf{H}_0 \right\|_1 = f(|\lambda|) \tag{10.25}$$

and restrictions of *Algorithm* 10.23 can be expressed as follows:

$$\lambda \left\| \varphi^T(k)\mathbf{H}_0 \right\|_1 \geqslant \left| y(k) - \hat{y}^0(k) \right| - \sigma \tag{10.26}$$

leading to

$$\lambda \geqslant \frac{\left| y(k) - \hat{y}^0(k) \right| - \sigma}{\left\| \varphi^T(k)\mathbf{H}_0 \right\|_1} \tag{10.27}$$

such that optimization problem (10.23) can be rewritten

$$\begin{aligned} &\min_{\lambda} \; 2|\lambda| \sum_{k=1}^{N} \left\| \varphi^{\mathrm{T}}(k)\mathbf{H}_0 \right\|_1 \\ &\text{subject to:} \\ &\lambda \geq \frac{\left| y(k) - \hat{y}^0(k) \right| - \sigma}{\left\| \varphi^T(k)\mathbf{H}_0 \right\|_1} \quad \forall k = 1, \ldots, M \end{aligned} \tag{10.28}$$

The optimal solution provided by such algorithm is:

$$\lambda = \sup_{k \in \{1,\ldots,M\}} \left(\frac{\left| y(k) - \hat{y}^0(k) \right| - \sigma}{\left\| \varphi^T(k)\mathbf{H}_0 \right\|_1} \right) \tag{10.29}$$

10.5.3 Error-Bounding Parameter Estimation

On the other hand, the set of uncertain parameters Θ_k using a error-bounded parameter estimation approach is obtained in such a way that the predicted behaviour is consistent with all the measured data in a fault-free scenario. In this case the obtained model satisfies that the predicted behaviour is always inside the interval of possible measurements. That is:

$$\hat{y}(k) - \sigma \leqslant y(k) \leqslant \hat{y}(k) + \sigma \quad \forall k = 1, \ldots, M \tag{10.30}$$

where: $\hat{y}(k) = \varphi^T(k)\boldsymbol{\theta}$ and $\boldsymbol{\theta} \in \Theta_k$.

Algorithms for identifying such kind of model are also known as *bounded-error parameter estimation* algorithms. In [29] there is a survey of such methods.

Using this approach, the parameter set Θ_k that contains all models consistent with data, known as *Feasible Parameter Set* (FPS), is defined as follows:

$$\mathbb{FPS} = \{\boldsymbol{\theta} \in \Theta_k | y(k) - \sigma \le \boldsymbol{\varphi}^T(k)\boldsymbol{\theta} \le y(k) + \sigma, \\ k = 1, \ldots, M\} \tag{10.31}$$

The exact description of $\mathbb{FPS}$ is in general not simple. For this reason existing algorithms usually approximate the $\mathbb{FPS}$ using an inner/outer simpler shapes as boxes, ellipsoids or zonotopes [29]. The approximation set is called approximated feasible parameter set (AFPS). In this chapter, algorithms that provide inner/outer AFPS using zonotopes in case of using the model parameterised as in (10.15) are presented.

10.5.3.1 Outer Approximations

Outer approximation algorithms find the parameter set Θ_k of minimum volume such that $\mathbb{FPS} \subseteq \Theta_k$. This kind of algorithms usually implies an excessive computational cost and recursive forms have been proposed as the one described in [7]. This recursive approach is based in computing iteratively the AFPS using zonotopes and related operations as follows:

$$\mathbb{AFPS}_{k+1} = \mathbb{AFPS}_k \cap \mathbb{F}_k \tag{10.32}$$

where $\mathbb{F}_k = \left\{\boldsymbol{\theta} \in \mathbb{R}^{n_\theta} | y(k) - \sigma \leqslant \boldsymbol{\varphi}^T(k)\boldsymbol{\theta} \leqslant y(k) + \sigma\right\}$

10.5.3.2 Inner Approximations

Inner approximation algorithms find the parameter set Θ_k of maximum volume such that $\Theta_k \subseteq \mathbb{FPS}$.

A bounded-error inner approximation using zonotopes parameterised as in Eq. (10.16) for models expressed as in (10.15) can be obtained in a similar way as proposed in optimization (10.28). The inner approximation algorithm comes from fact the FPS conditions (10.31) can be bounded by:

$$y(k) - \sigma \le \underline{\hat{y}}(k) \le \boldsymbol{\varphi}^T(k)\boldsymbol{\theta} \le \overline{\hat{y}}(k) \le y(k) + \sigma$$

where $\underline{\hat{y}}(k)$ and $\overline{\hat{y}}(k)$ are defined as in (10.20a)–(10.20b), respectively and, in the case of is Θ_k a zonotope, calculated as in (10.21a)–(10.21b). Then, the maximum inner zonotope, centered in $\boldsymbol{\theta}^0$, can be computed solving the following optimization problem

$$\begin{aligned} &\max_{\mathbf{H}} \; f(\Theta_{\mathbf{k}}(\mathbf{H})) \\ &\text{subject to:} \\ &\left\|\boldsymbol{\varphi}^T(k)\mathbf{H}\right\|_1 \le \sigma - \left|y(k) - \hat{y}^0(k)\right| \quad \forall k = 1, \ldots, M \end{aligned} \tag{10.33}$$

where the cost function f in the error-bounded approach is usually the volume of the zonotope defined by (10.16). This volume only depends of matrix $\mathbf{H}$ and of B^n with a volume equal to 2^n. In the particular case, $\mathbf{H}$ is a square matrix ($n_\theta = n$), the volume is given by $vol(\Theta_k) = 2^n\ |\det(\mathbf{H})|$. See [30] for more details.

Similarly to optimization problem (10.23), to reduce the computational complexity, the particular case $\mathbf{H} = \lambda\mathbf{H}_0$ will be considered. Then, if H_0 is a square matrix $vol(\Theta_k) = |2\lambda|^n\ |\det(\mathbf{H}_0)|$ and restrictions of optimization (10.33) can be expressed as:

$$\left\|\varphi^T(k)\mathbf{H}_0\right\|_1 \leq \sigma - \left|y(k) - \hat{y}^0(k)\right| \tag{10.34}$$

leading to

$$\lambda \leq \frac{\sigma - \left|y(k) - \hat{y}^0(k)\right|}{\left\|\varphi^T(k)\mathbf{H}_0\right\|_1} \tag{10.35}$$

such that it can be rewritten as

$$\begin{aligned} &\max_{\lambda}\ vol(\Theta_k) = f(|\lambda|) \\ &\text{subject to:} \\ &\lambda \leq \frac{\sigma - \left|y(k) - \hat{y}^0(k)\right|}{\left\|\varphi^T(k)\mathbf{H}_0\right\|_1} \quad \forall k = 1, \ldots, M. \end{aligned}$$

The optimal solution provided by such algorithm is:

$$\lambda = \inf_{k\in\{1,\ldots,M\}} \left(\frac{\sigma - \left|y(k) - \hat{y}^0(k)\right|}{\left\|\varphi^T(k)\mathbf{H}_0\right\|_1}\right) \tag{10.36}$$

10.6 Case Study

10.6.1 Description

In this section, the three-tank system that is described in Chap. 2 is considered to illustrate the approaches presented in previous sections.

As it can be seen from the schematic diagram of the system setup in Fig. 2.3, the input of the three-tank system is the pump flow rate that is determined by the voltage of the pump (actuator) and the output of the process is the water level of Tank 3 that is obtained as voltages from the measurement devices. Moreover, the state of the system is the water level of the tanks.

The methods presented in this chapter uses a linearised model of this system obtained around the following operating point:

- $h_1^* = 0.51\,\text{cm}$,

- $h_2^* = 0.50\,\text{cm}$,
- $h_3^* = 0.35\,\text{cm}$,

that is expressed in discrete-time as in (10.1) using the Euler discretization with a sampling time of 1 s. The linearized model of this system can be written in the discrete-time state-space form as

$$\mathbf{h}_{k+1} = \mathbf{A}\mathbf{h}_k + \mathbf{B}\mathbf{u}_k + \mathbf{E}_\omega \omega_{(k)}, \tag{10.37}$$

$$\mathbf{y}_k = \mathbf{C}\mathbf{h}_k + \mathbf{E}_\upsilon \upsilon_{(k)}, \tag{10.38}$$

where

$$\mathbf{A} = \begin{bmatrix} 0.9712 & 0 & 0 \\ 0.0288 & 0.9638 & 0.0074 \\ 0 & 0.0074 & 0.9877 \end{bmatrix}, \qquad \mathbf{B} = \begin{bmatrix} 64.9351 \\ 0 \\ 0 \end{bmatrix}, \qquad \mathbf{C} = \begin{bmatrix} 0 & 0 & 1 \end{bmatrix}. \tag{10.39}$$

Uncertainty in the parameters is considered by enlarging the nominal values with a interval of $\pm 1\%$ of the nominal value.

Moreover, state disturbances ω and measurement noise υ bounded using unitary intervals influencing all the state-space directions and the output are modeled respectively with $\mathbf{E}_\omega$ and $\mathbf{E}_\upsilon$ as

$$\mathbf{E}_\omega = \begin{bmatrix} 0.05 & 0 & 0 \\ 0 & 0.05 & 0 \\ 0 & 0 & 0.05 \end{bmatrix}, \qquad \mathbf{E}_\upsilon = \begin{bmatrix} 0.08 \end{bmatrix}. \tag{10.40}$$

10.6.2 State Estimation

First, state estimation using interval observer and error-bounding approaches will be compared. In this regard, the state estimation zonotopes obtained from interval observer and error-bounding approaches in steady-state operation of the system are presented in Figs. 10.4 and 10.5, respectively.

Moreover, the time evolution of the state estimation bounds can be computed at each time instant. The interval hull of the state estimation is projected into the output space by using the system matrix **C**. Figure 10.6 is obtained from the projection of the state estimation into the output space in time. Note from Fig. 10.6 that the maximum and the minimum bounds of the obtained zonotopic state estimations using both interval observer and error-bounding approaches are almost the same. However, the results obtained from the simulation of error-bounding approach is a bit more conservative in comparison with the obtained results from the interval observer approach. A possible explanation for this might be related to the information of the output measurement since there is one-time-instant delay between the results obtained using the interval observer and error-bounding approaches.

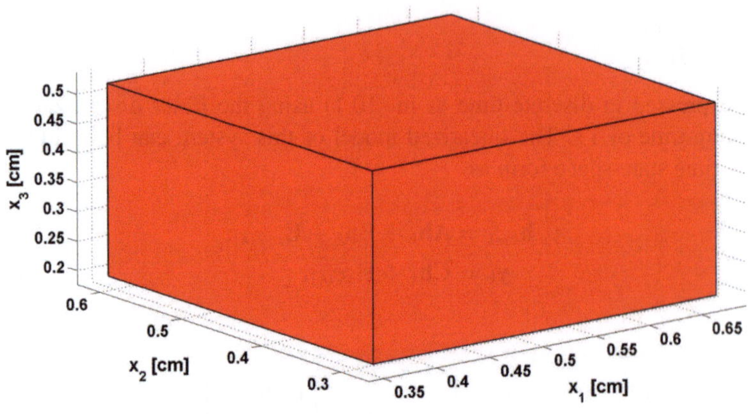

Fig. 10.4 State estimation using interval observer approach in steady state

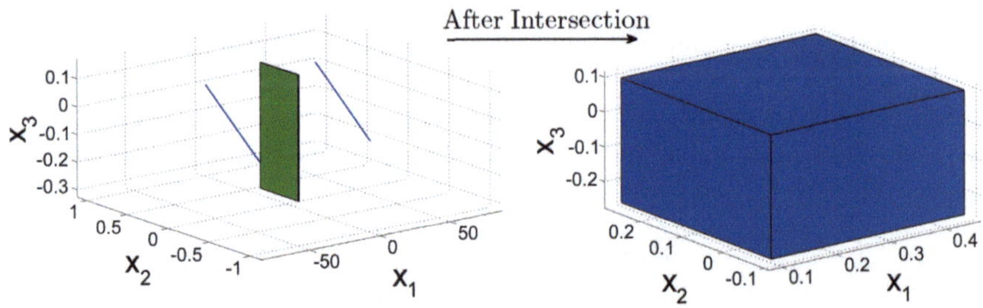

Fig. 10.5 State estimation using error-bounding approach in steady state

10.6.3 Application to FD

In this section, the extension of the comparison of both the interval observer and error-bounding approaches is performed when applied to fault detection. For the purpose of further analysis, two fault scenarios are considered as

- sensor fault that affects the measurement of the output of the system,
- actuator fault, which affects the system input.

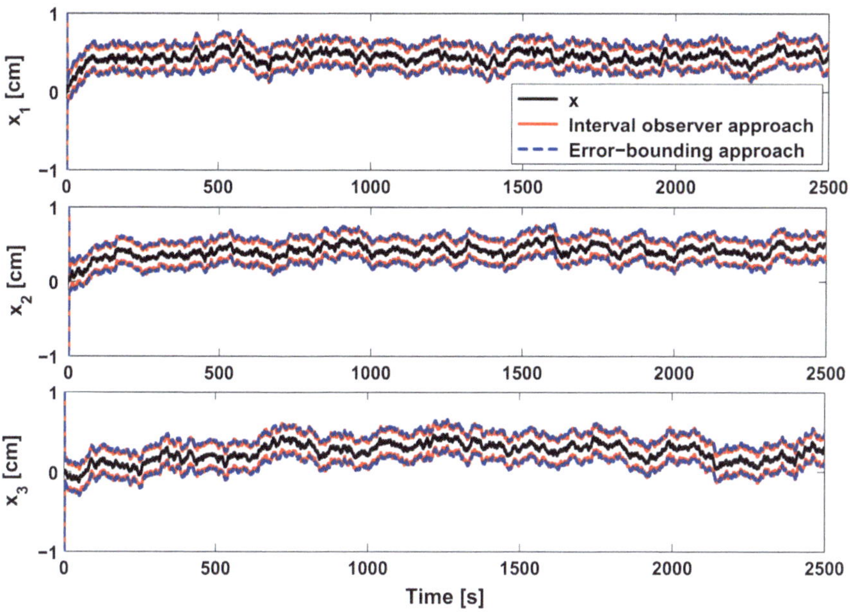

Fig. 10.6 Projection of the state estimations into the output space

Thus, the dynamical model (10.37) affected by different actuator and sensor faults can be rewritten as

$$\mathbf{h}(k+1) = \mathbf{A}\mathbf{h}(k) + \mathbf{B}\mathbf{u}(k) + \mathbf{E}_\omega\omega(k) + \mathbf{F}_a\mathbf{f}_a(k), \tag{10.41}$$

$$\mathbf{y}(k) = \mathbf{C}\mathbf{h}(k) + \mathbf{E}_\upsilon\upsilon(k) + \mathbf{F}_y\mathbf{f}_y(k), \tag{10.42}$$

where vectors $\mathbf{f}_a \in \Re^{n_u}$ and $\mathbf{f}_y \in \Re^{n_y}$ denote the actuator and output sensor faults with their associated matrices $\mathbf{F}_a \in \Re^{n_x \times n_u}$ and $\mathbf{F}_y \in \Re^{n_y \times n_y}$, respectively.

Regarding the disturbances, (10.40) is used to simulate all the possible disturbances acting in all the directions.

10.6.3.1 Sensor Fault

The dynamical model (10.41) can be rewritten to deal with the output sensor fault as

$$\mathbf{h}(k+1) = \mathbf{A}\mathbf{h}(k) + \mathbf{B}\mathbf{u}(k) + \mathbf{E}_\omega\omega(k), \tag{10.43}$$

$$\mathbf{y}(k) = \mathbf{C}\mathbf{h}(k)k + \mathbf{E}_\upsilon\upsilon(k) + \mathbf{F}_y\mathbf{f}_y(k), \tag{10.44}$$

In this fault scenario, the occurrence of the sensor fault $\mathbf{f}_y$ is simulated at $k = 1000$ and it remains until the end of the simulation. Figure 10.7 shows the projection of

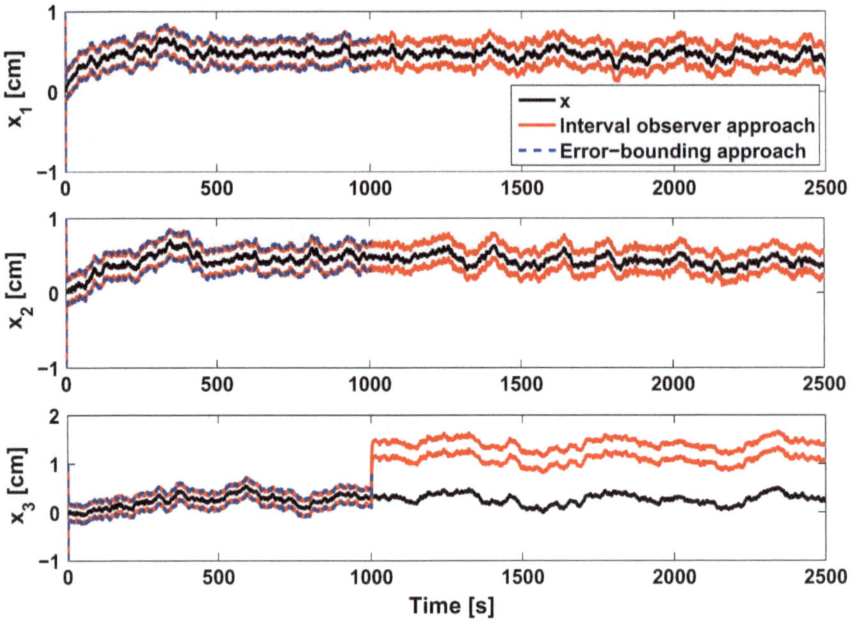

Fig. 10.7 State estimation in the case of a sensor fault

the computed state-bounding zonotope into the state space when the actuator fault occurs at time instant $k = 1000$. The state estimation in this figure is obtained by considering both interval observer and error-bounding approaches.

As it is mentioned before, the fault is simulated after time instant $k = 1000$. Thus, before this time instant, the system is only affected by the effect of model uncertainty, disturbances and noises. Consequently, the bounds that are obtained at the first 1000 time instants show the effect of both disturbances and noise. As it can be seen from the time instant $k = 0$ until $k = 1000$ where the system is only affected by the disturbance, both approaches can properly follow the system.

After time instant $k = 1000$, the effect of the sensor fault can be seen on the state estimation in Fig. 10.7. The inconsistency of the observation by the model using both interval observer and error-bounding approaches and the nominal behaviour of the system allows to detect the fault. Furthermore, comparing the approaches after the fault occurrence reveals the improvement feature of interval observer approach in comparison with error-bounding approach. As it can be seen in Fig. 10.7, after detecting the fault with both approaches, monitoring the system with error-bounding approach should be stopped since the intersection between the strip computed using the measurements and the prediction step is empty. This fact can be observed from Figs. 10.8 and 10.9 where the strip and prediction zonotope are shown before and after fault occurrence, respectively. Thus, the system could could only be monitored using interval observer approach after the fault detection.

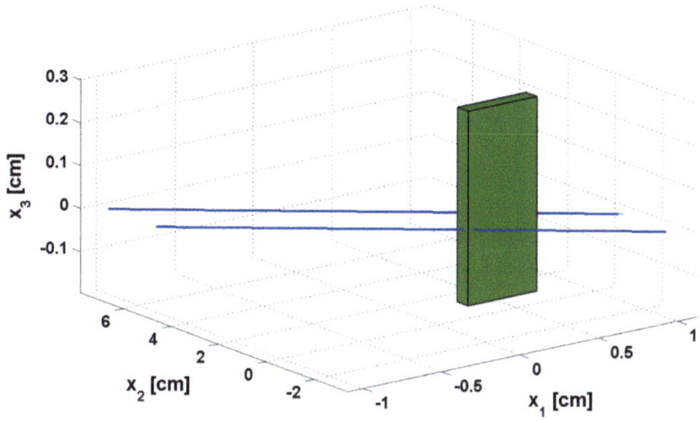

Fig. 10.8 Before occurrence of the fault

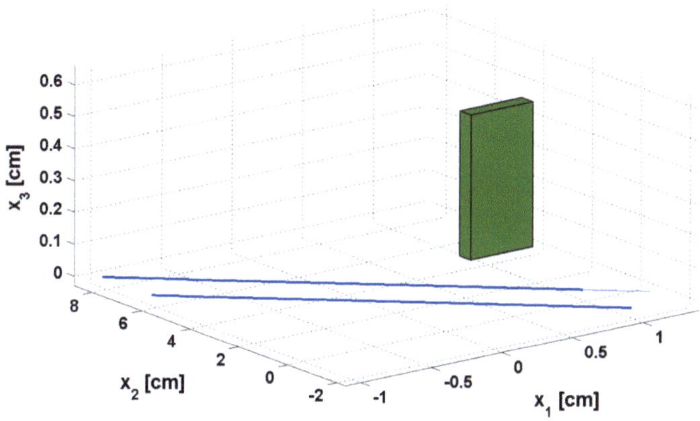

Fig. 10.9 After occurrence of the fault

Finally, the fault detection test is also applied using both approaches considering the residual. As it can be seen from Fig. 10.10, zero is included in both residual zonotopes generated by interval observer and error-bounding approaches before the occurrence of the fault. But, after the fault occurrence at $k = 1000$, both zonotopes move and the fault can be detected since zero is outside of the zonotopes bounding the residual. It is worth mentioning that in the case of sensor fault, because of the fault reinjection through the observer structure, leads to the transient behavior presented in Fig. 10.10 that it does not allow to detect the presence of the fault after its detection (fault following effect). On the other hand, as it is mention before after detecting the fault the error-bounding approach should be stopped for monitoring the system. Furthermore, as it can be seen in Fig. 10.10, the interval observer approach can detect the fault a bit sooner than error-bounding zonotope since its residual generation is done using different measurement information coming from the previous time

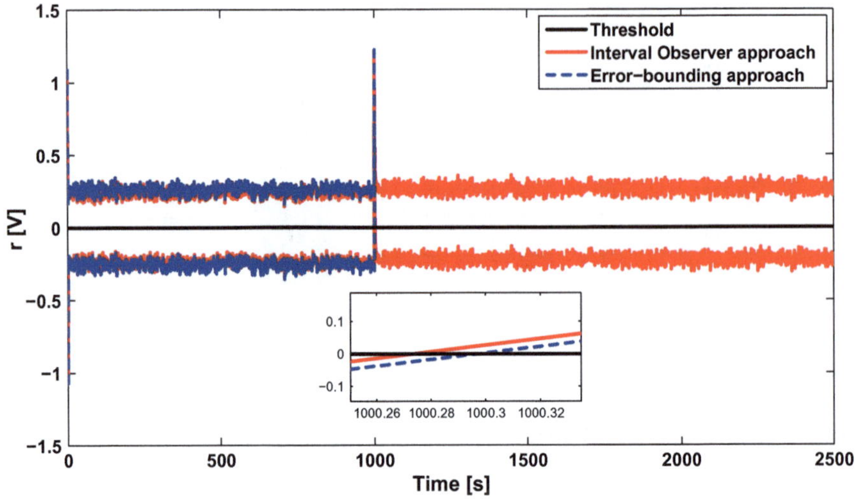

Fig. 10.10 Residual bounds in the case of actuator fault

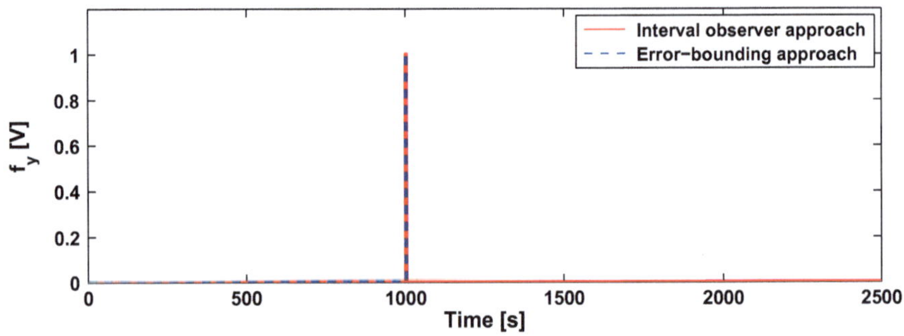

Fig. 10.11 Fault detection test result in the case of sensor fault

instant. Moreover, Fig. 10.11 shows the fault detection test result obtained from both approaches: 0 means no-fault in the system is detected and 1 means the fault is detected.

10.6.3.2 Actuator Fault

The dynamical model (10.41) can be rewritten in the case of actuator fault as

$$\mathbf{h}(k+1) = \mathbf{A}\mathbf{h}(k) + \mathbf{B}\mathbf{u}(k) + \mathbf{E}_{\omega}\omega(k) + \mathbf{F}_a\mathbf{f}_a(k), \tag{10.45}$$

$$\mathbf{y}(k) = \mathbf{C}\mathbf{h}(k) + \mathbf{E}_{\upsilon}\upsilon(k), \tag{10.46}$$

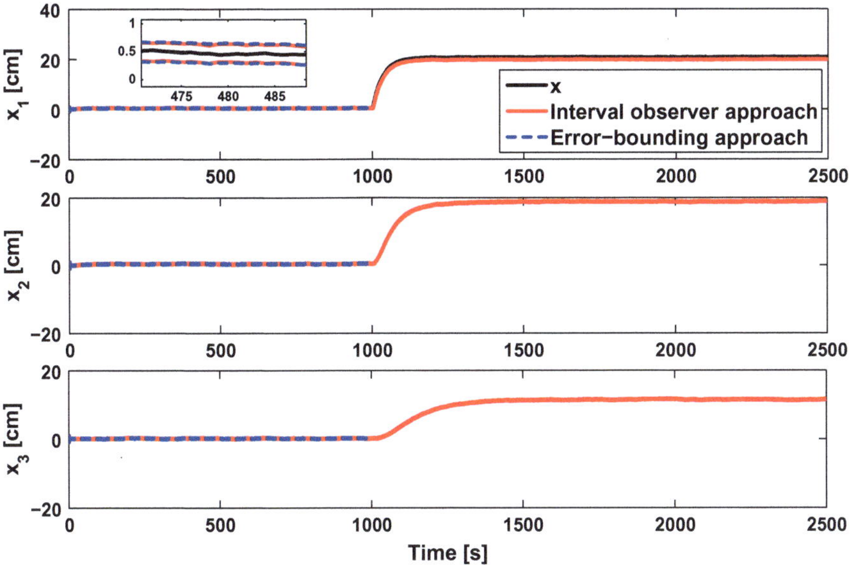

Fig. 10.12 State estimation in the case of an actuator fault

Similarly to the sensor case, from time step $k = 0$ until $k = 1000$, the actuator is healthy and working properly. As it can be seen in Fig. 10.12, the system is only influenced by the effect of uncertainties and both interval observer and error-bounding approaches can properly follow the system. But, after $k = 1000$, the actuator fault is injected as indicated to the system and it remains until the end of the simulation.

As it is mentioned before, since the intersection between the strip and the predicted zonotope is empty, the state estimation should be stopped using error-bounding approach. Therefore, as can be seen in Figs. 10.13 and 10.14, which obtained before and after fault occurrence, respectively, the empty intersection is obtained when fault occurs in the case of error-bounding approach. Thus, the fault is detected and following the system behaviour is stopped using this approach. On the other hand, the interval observer is still able to monitor the system behaviour after the fault differently from the output sensor case.

Furthermore, due to the occurrence of the actuator fault in the system, the residual zonotope is affected. Therefore, this fault is detected whenever zero is not included by the residual set. Therefore, as shown in Fig. 10.15, the fault is detected using both approaches. Similar to the case of sensor fault, the error-bounding approach can detect the fault with a small delay since it used the different information that the interval observer approach (the information with one step delay).

Moreover, Fig. 10.16 shows the fault detection test result obtained from both approaches: 0 means no-fault in the system and 1 means the fault is detected.

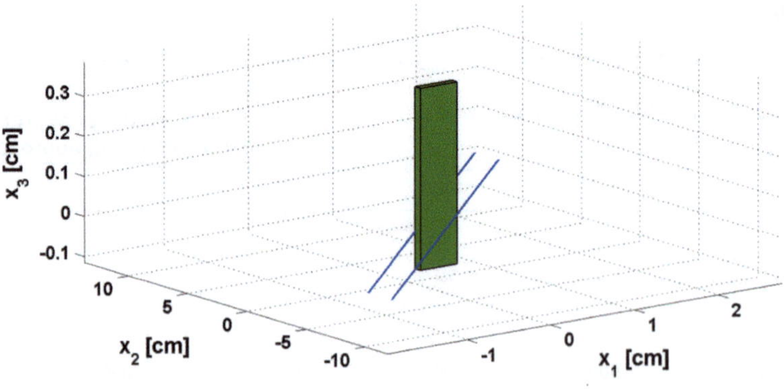

Fig. 10.13 Before occurrence of the actuator fault

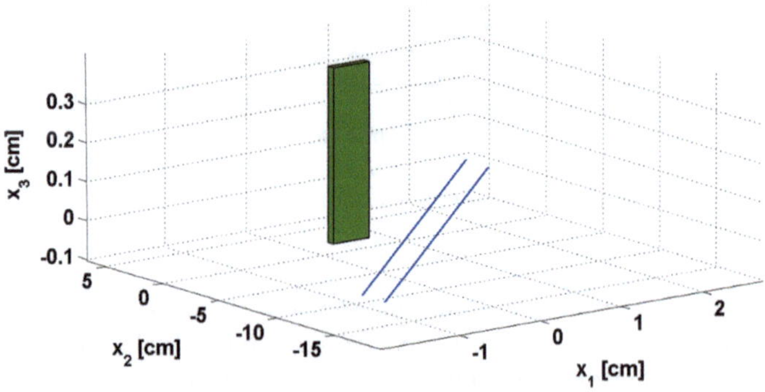

Fig. 10.14 After occurrence of the actuator fault

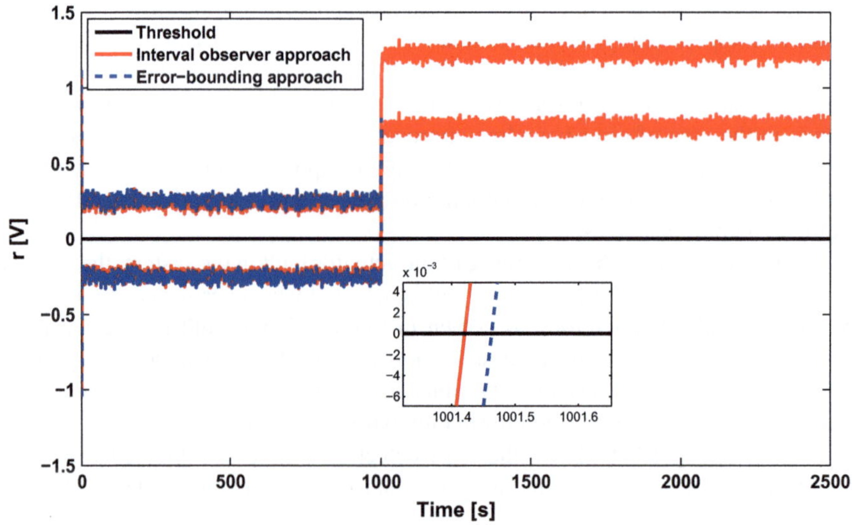

Fig. 10.15 Residual bounds in the case of an actuator fault

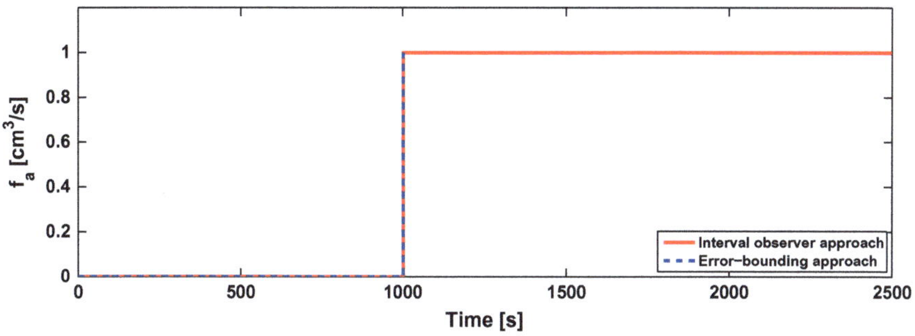

Fig. 10.16 Fault detection test result in the case of an actuator fault

10.7 Conclusions

This chapter has reviewed the use of set-membership methods in robust fault diagnosis. Alternatively to the statistical methods, set-membership methods use a deterministic unknown-but-bounded description of noise and parametric uncertainty (interval models). Using approximating sets to enclose the exact set of possible behaviours (in parameter or state space), these methods allows to check the consistency between observed and predicted behaviour. When an inconsistency is detected a fault can be indicated, otherwise nothing can be stated. The same principle has been used to estimate interval models for fault detection and to develop methods for fault tolerance evaluation. Finally, the application of these methods to the tank system introduced in Chap. 2 has been used to exemplify the successful use of the proposed set-membership methods in fault diagnosis.

Acknowledgements This work has been partially funded by the Spanish State Research Agency (AEI) and the European Regional Development Fund (ERFD) through the projects DEOCS (ref. MINECO DPI2016-76493) and SCAV (ref. MINECO DPI2017-88403-R). This work has also been partially funded by AGAUR of Generalitat de Catalunya through the Advanced Control Systems (SAC) group grant (2017 SGR 482) and by Agència de Gestió d'Ajuts Universitaris i de Recerca.

References

1. Ackermann, J.: Robust Control: The Parameter Space Approach. Springer, London, UK (2002)
2. Adrot, O., Flaus, J.M.: Fault detection based on uncertain models with bounded parameters and bounded parameter variations. In: Proceedings of the 17th IFAC World Congress, Seoul, Korea (2008)
3. Alamo, T., Bravo, J., Camacho, E.: Guaranteed state estimation by zonotopes. Automatica **41**(6), 1035–1043 (2005)
4. Armengol, J., Travé-Massuyès, L., Vehí, J., de la Rosa, J.L.: A survey on interval model simulators and their properties related to fault detection. Annu. Rev. Control **24**, 31–39 (2000)
5. Armengol, J., Vehí, J., Sainz, M., Herrero, P., Gelso, E.: Squaltrack: a tool for robust fault detection. IEEE Trans. Syst. Man Cybern. Part B **39**(2), 475–488 (2008)

6. Bhattacharyya, S., Chapellat, H., Keel, L.: Robust control: the parametric approach. Prentice Hall PTR, New Jersey (1995)
7. Bravo, J., Alamo, T., Camacho, E.: Bounded error identification of systems with time-varying parameters. IEEE Trans. Autom. Control **51**, 1144–1150 (2006)
8. Calafiore, G.: A set-valued non-linear filter for robust localization. In: Proceedings of European Control Conference (ECC'01), Porto, Portugal (2001)
9. Calafiore, G., Campi, M.C., Ghaoui, L.E.: Identification of reliable predictor models for unknown systems: a data-consistency approach based on learning theory. In: Proceedings of the 15th IFAC World Congress, Barcelona, Spain (2002)
10. Campi, M., Calafiore, G., Garatti, S.: Interval predictor models: identification and reliability. Automatica **45**(8), 382–391 (2009)
11. Chen, J., Patton, R.: Robust Model-Based Fault Diagnosis for Dynamic Systems. Kluwer Academic Publishers, Dordrecht (1999)
12. Chilali, M., Gahinet, P.: H_∞ design with pole placement constraints: an LMI approach. IEEE Trans. Autom. Control **41**(3), 358–367 (1996)
13. Chisci, L., Garulli, A., Zappa, G.: Recursive state bounding by parallelotopes. Automatica **32**, 1049–1055 (1996)
14. Emami-Naeini, A., Akhter, M., Rock, S.: Effect of model uncertainty on failure detection: the threshold selector. IEEE Trans. Autom. Control **AC-33**, 1106–1115 (1988)
15. Escobet, T., Travé-Massuyès, L., Tornil, S., Quevedo, J.: Fault detection of a gas turbine fuel actuator based on qualitative causal models. In: Proceedings of European Control Conference (ECC'01), Porto, Portugal (2001)
16. Fagarasan, I., Ploix, S., Gentil, S.: Causal fault detection and isolation based on a set-membership approach. Automatica **40**, 2099–2110 (2004)
17. Gertler, J.: Fault Detection and Diagnosis in Engineering Systems. Marcel Dekker, New York (1998)
18. Guerra, P., Puig, V., Ingimundarson, A.: Robust fault detection using a consistency-based state estimation. In: Proceedings of IEEE European Control Conference (ECC'07), Kos, Greece (2007)
19. Guerra, P., Puig, V., Witczak, M.: Robust fault detection with unknown-input interval observers using zonotopes. In: Proceedings of 17th IFAC World Congress, Seoul, Korea (2008)
20. Hamelin, F., Sauter, D.: Robust fault detection in uncertain dynamic systems. Automatica **36**(11), 1747–1754 (2000)
21. Hansen, E.: Global Optimization Using Interval Analysis. Marcel Dekker, New York (1992)
22. Horak, D.: Failure detection in dynamic systems with modelling errors. AIAA J. Guid. Control Dyn. **11**(6), 508–516 (1988)
23. Ingimundarson, A., Bravo, J., Puig, V., Alamo, T., Guerra, P.: Robust fault detection using zonotope-based set-membership consistency test. J. Adapt. Control Signal Process. **23**(4), 311–330 (2008)
24. Isermann, R.: Fault Diagnosis Systems: An Introduction from Fault Detection to Fault Tolerance. Springer, New York (2006)
25. Kieffer, M., Jaulin, L., Walter, E.: Guaranteed recursive non-linear state bounding using interval analysis. Int. J. Adapt. Control Signal Process. **16**(3), 193–218 (2002)
26. Kolev, L.: Interval Methods for Circuit Analysis. World Scientific, Singapore (1993)
27. Kuipers, B.: Qualitative Reasoning - Modelling and Simulation with Incomplete Knowledge. MIT Press, Cambridge, MA (1994)
28. Maksarov, D., Norton, J.: State bounding with ellipsoidal set description of the uncertainty. Int. J. Control **65**(5), 847–866 (1996)
29. Milanese, M., Norton, J., Piet-Lahanier, H., Walter, E.: Bounding Approaches to System Identification. Plenum Press, New York (1996)
30. Montgomery, H.: Computing the volume of a zonotope. Am. Math. Mon. **96**, 431 (1989)
31. Ocampo, C., Tornil, S., Puig, V.: Robust fault detection using interval constraints satisfaction and set computations. In: Proceedings of IFAC SAFEPROCESS'06, Beijing, China (2006)

32. Planchon, P., Lunze, J.: Robust diagnosis using state estimation. In: Proceedings of IFAC SAFEPROCESS'06, Beijing, China (2006)
33. Ploix, S., Adrot, O.: Parity relations for linear uncertain dynamic systems. Automatica **42**(6) (2006)
34. Ploix, S., Adrot, O., Ragot, J.: Parameter uncertainty computation in static linear models. In: Proceedings of the 38th IEEE Conference on Decision and Control, Phoenix, Arizona, USA (1999)
35. Puig, V., Quevedo, J., Escobet, T., Nejjari, F., de las Heras, S.: Passive robust fault detection of dynamic processes using interval models. IEEE Trans. Control Syst. Technol. **16**(5), 1083–1089 (2008)
36. Puig, V., Saludes, J., Quevedo, J.: Worst-case simulation of discrete linear time-invariant interval dynamic systems. Reliab. Comput. **9**(4), 251–290 (2003)
37. Puig, V., Stancu, A., Escobet, T., Nejjari, F., Quevedo, J., Patton, R.: Passive robust fault detection using interval observers: application to the DAMADICS benchmark problem. Control Eng. Pract. **14**(6), 621–633 (2006)
38. Rambeaux, F., Hamelin, F., Sauter, D.: Optimal thresholding for robust fault detection of uncertain systems. Int. J. Robust Nonlinear Control **10**(14), 1155–1173 (2000)
39. Rinner, B., Weiss, U.: Online monitoring by dynamically refining imprecise models. IEEE Trans. Syst. Man Cybern. Part B **34**(4), 1811–1822 (2004)
40. Rugh, W., Shamma, J.: A survey of research on gain-scheduling. Automatica **36**(10), 1401–1425 (2000)
41. Sainz, M., Armengol, J., Vehí, J.: Fault detection and isolation of the three-tank system using the modal interval analysis. J. Process Control **12**(2), 325–338 (2002)
42. Shamma, J.: Approximate set-value observer for nonlinear systems. IEEE Trans. Autom. Control **42**(5), 648–658 (1997)
43. Tornil, S., Escobet, T., Travé-Massuyès, L.: Robust fault detection using interval models. In: Proceedings of European Control Conference (ECC'03), Cambridge, UK (2003)
44. Travé-Massuyes, L., Escobet, T., Pons, R., Tornil, S.: The CA EN diagnosis system and its automatic modelling method. Computación y Sistemas **5**(2), 648–658 (2001)

Chapter 11
Selected Estimation Strategies for Fault Diagnosis of Nonlinear Systems

Marcin Witczak and Marcin Pazera

11.1 Introduction

Fault diagnosis boils down to deciding if a fault occurs (fault detection), locating a faulty component (fault isolation), and assessing the size of a fault (fault identification and estimation) [10, 79]. This causes that fault diagnosis can be perceived as a three-step procedure, which covers fault detection, isolation, and identification. As can be observed in the literature, the problem of fault detection and isolation (FDI) has been studied widely (see [15, 21, 22, 32, 39, 79, 80] and the reference therein).

Due to the above-described reasons, the fault identification/estimation task has attained less attention. However, last-year dynamical development of the active Fault-Tolerant Control (FTC) necessitates the need for fault identification [10]. Indeed, when a fault accommodation strategy is needed, an active FTC requires online fault diagnosis including not only FDI but also fault identification. In other words, the lack of such knowledge causes that an effective compensation of the fault effect is simply impossible.

In light of the above fact, several books have been published in the last decade on the emerging problem of FTC, which deal with fault identification and estimation problems. In particular, [33], which is mainly devoted to fault diagnosis and its applications, provides some general rules for hardware-redundancy-based FTC. On the other hand, the concepts of achieving passive FTC are introduced in [45], where the authors also investigate the problem of performance and stability of FTC under imperfect fault diagnosis. In particular, they consider (under some assumptions) the effect of delayed fault detection and imperfect fault identification. However, the fault

M. Witczak (✉) · M. Pazera
Institute of Control and Computation Engineering, University of Zielona Góra,
Zielona Góra, Poland
e-mail: M.Witczak@issi.uz.zgora.pl

M. Pazera
e-mail: M.Pazera@issi.uz.zgora.pl

T. Escobet et al. (eds.), *Fault Diagnosis of Dynamic Systems*,
https://doi.org/10.1007/978-3-030-17728-7_11

diagnosis scheme is treated separately during the design and no real integration of fault diagnosis and FTC is proposed. FTC is also treated in [51], where a number of practical FTC case studies are presented, i.e., a winding machine, a three-tank system, and an active suspension system. The problem of fault estimation, which can also be perceived as the estimation of an unknown input, has been addressed using different strategies. Some of these strategies deserve particular attention, namely: augmenting the state vector by an unknown input, two-stage Kalman filter [34], minimum variance input and state estimator [23, 35], adaptive estimation [87], sliding mode high-gain observers [72], and finally, an $\mathcal{H}_\infty$ approach [50].

The fault estimation can also be formulated as a parameter estimation problem [62] leading to the application of any parameter estimation algorithms such as least squares, generalized/extended least squares or instrumental variables. Recently, a lot of effort has been devoted to extend the previous approaches to nonlinear systems. For example, [18] proposed a unified framework based on a model reference approach for nonlinear systems that can be represented by means of Takagi-Sugeno models. In [65], a fault estimation scheme for nonlinear systems that can be modeled in Linear Parameter Varying (LPV) form is presented. In [69], an observer scheme that estimates simultaneously the state and the fault is presented. On the other hand, the robustness of the fault estimation algorithm can be stated as the degree to which the sensitivity of the fault estimates is invariant in the presence of model parameter variations, disturbances, uncertainty, and noise.

For nonlinear systems, the observer-based FDI approaches have gained a lot of interest [7, 58] and FDI formulations for some classes of nonlinear systems have been derived. In [26], state-affine nonlinear systems have been handled, and in [27, 57], the class of input-affine systems has been considered, among others. The work [56] presents a detailed geometric description of how to tackle the residual generation problem for nonlinear systems. On the other hand, [44] presents a procedure to design a bank of extended $\mathcal{H}_\infty$ observers for sensor FDI for a certain class of nonlinear systems. There are also approaches that employ soft computing techniques, e.g., neural networks [49].

The main objective of this chapter is to present the main principles regarding classical fault detection and isolation with analytical techniques. It should be pointed out that this historical review is based on the material originally presented in [79]. These techniques are mainly based on the concept of the so-called residual, which is defined as a difference between the behavior of the system and the model. The remaining part of the chapter deals with techniques, which can be used for simultaneous state and fault estimation. In particular, the presentation starts with actuator fault estimation scheme. Subsequently, simultaneous sensor and actuator fault estimation are discussed. It should be pointed out that the material used for describing fault estimation part was originally presented in [54, 55, 81]. Finally, a process fault estimation strategy is portrayed. The last part of the chapter presents an illustrative example concerning various fault estimations of the three-tank system.

11.2 Methods

11.2.1 Parameter Estimation

Similarly as in the case of linear-in-parameter systems, the FDI problem boils down to estimating the parameters of the model of the system. The system can be generally described by

$$y_k = \mathbf{g}\left(\boldsymbol{\phi}_k, \mathbf{p}_k\right) + v_k, \tag{11.1}$$

while $\boldsymbol{\phi}_k$ may contain the previous or current system input $\mathbf{u}_k$, the previous system or model output and the previous prediction error. The approach presented here inherits all the drawbacks and advantages of its linear counterparts presented in the preceding chapters. The FDI scheme is also the same. Another problem arises because $\mathbf{g}(\cdot)$ is nonlinear in its parameters. In this case, nonlinear parameter estimation techniques should be applied [74]. For complex models, this may cause serious difficulties with a fast reaction on faults; consequently, fault detection cannot be performed effectively and reliably. Irrespective of the above difficulties, there are, of course, some studies in which such an approach is utilized, e.g., [73].

11.2.2 Parity Relation

An extension of the parity relation to nonlinear polynomial dynamic systems was proposed in [24]. In order to describe this approach, let us consider a system described by the state-space equations

$$\mathbf{x}_{k+1} = \mathbf{g}\left(\mathbf{x}_k, \mathbf{u}_k, \mathbf{f}_k\right), \tag{11.2}$$

$$\mathbf{y}_k = \mathbf{h}(\mathbf{x}_k, \mathbf{u}_k, \mathbf{f}_k), \tag{11.3}$$

where $\mathbf{g}(\cdot)$ and $\mathbf{h}(\cdot)$ are assumed to be polynomials. Equations. (11.2)–(11.3) can always be expressed on a time window $[k - S, k]$. As a result, the following structure can be obtained:

$$\mathbf{y}_{k-S,k} = \mathbf{H}\left(\mathbf{x}_{k-S}, \mathbf{u}_{k-S,k}, \mathbf{f}_{k-S,k}\right), \tag{11.4}$$

with $\mathbf{u}_{k-S,k} = \mathbf{u}_{k-S}, \ldots, \mathbf{u}_k$ and $\mathbf{f}_{k-S,k} = \mathbf{f}_{k-S}, \ldots, \mathbf{f}_k$. In order to check the consistency of the model equations, the state variables have to be eliminated. This results in the following equation:

$$\boldsymbol{\phi}(\mathbf{y}_{k-S,k}, \mathbf{u}_{k-S,k}, \mathbf{f}_{k-S,k}) = \mathbf{0}. \tag{11.5}$$

Table 11.1 Principle of fault isolation with the nonlinear parity relation

Fault location	Nonzero element of $\mathbf{z}_{f,k}$	Nonzero element of $\mathbf{z}_{b,k}$
ith sensor	z_f^i	All elements dependent on y_i
ith actuator	All elements dependent on u_i	z_b^i

Since $\mathbf{g}(\cdot)$ and $\mathbf{h}(\cdot)$ are assumed to be polynomials, elimination theory can be applied to transform (11.4) into (11.5). Knowing that the $\boldsymbol{\phi}_i(\cdot)$s are polynomials and therefore they are expressed as sums of monomials, it seems natural to split the expression (11.5) into two parts, i.e.,

$$\mathbf{z}_k = \boldsymbol{\phi}_1(\mathbf{y}_{k-S,k}, \mathbf{u}_{k-S,k}), \tag{11.6}$$

$$\mathbf{z}_k = \boldsymbol{\phi}_2(\mathbf{y}_{k-S,k}, \mathbf{u}_{k-S,k}, \mathbf{f}_{k-S,k}). \tag{11.7}$$

The right-hand side of (11.6) contains all the monomials in $\mathbf{y}_{k-S,k}$ and $\mathbf{u}_{k-S,k}$ only, while (11.7) contains all the monomials involving at least one of the components of $\mathbf{f}_{k-S,k}$. The above condition ensures that $\mathbf{z}_k = \mathbf{0}$ in the fault-free case. Since the fault signal $\mathbf{f}_{k-S,k}$ is not measurable, only Eq. (11.6) can be applied to generate the residual $\mathbf{z}_k$ and, consequently, to detect faults.

One drawback of this approach is that it is limited to polynomial models or, more precisely, to models for which the state vector $\mathbf{x}_k$ can be eliminated. Another drawback is that it is assumed that a perfect model is available, i.e., there is no model uncertainty. This may cause serious problems while applying the approach to real systems.

Parity relation for a more general class of nonlinear systems was proposed by Krishnaswami and Rizzoni [41]. There are two residual vectors, namely, the forward $\mathbf{z}_{f,k}$ residual vector and the backward $\mathbf{z}_{b,k}$ residual vector. These residuals are generated using the forward and inverse (backward) models, respectively. Based on these residual vectors, fault detection can (theoretically) be easily performed, while fault isolation should be realized according to Table 11.1. The authors suggest an extension of the proposed approach to cases where model uncertainty is considered.

Undoubtedly, strict existence conditions for an inverted model as well as possible difficulties with the application of the known identification techniques make the usefulness of this approach for a wide class of nonlinear systems questionable.

Another parity relation approach for nonlinear systems was proposed by Shumsky [67]. The concepts of the parity relation and parameter estimation fault detection techniques are combined. In particular, the parity relation is used to detect offsets in the model parameters. The necessary condition is that there exists a transformation $\mathbf{x}_k = \boldsymbol{\xi}(\mathbf{u}_k, \ldots, \mathbf{u}_{k+S}, \mathbf{y}_k, \ldots, \mathbf{y}_{k+S})$, which may cause serious problems in many practical applications. Another inconvenience is that the approach inherits most drawbacks concerning parameter estimation based fault detection techniques.

11.2.3 Observers

Model linearization is a straightforward way of extending the applicability of linear techniques to nonlinear systems. On the other hand, it is well known that such approaches work well when there is no large mismatch between the linearized model and the nonlinear system. Two types of linearization can be distinguished, i.e., linearization around the constant state and linearization around the current state estimate. It is obvious that the second type of linearization usually yields better results. Unfortunately, during such a linearization, the influence of terms higher than linear is usually neglected (as in the case of the extended Luenberger observer and the extended Kalman filter). One way out from this problem is to improve the performance of linearization-based observers. Another way is to use linearization-free approaches. Unfortunately, the application of such observers is limited to certain classes of nonlinear systems.

Generally, FDI principles of nonlinear observer-based schemes are not different than those of linear ones. Apart from this similarity, the design complexity and feasibility of such FDI schemes are usually far more sophisticated.

11.2.3.1 Extended Luenberger Observers and Kalman Filters

Let us consider a nonlinear discrete-time system described by the following state-space equations:

$$\mathbf{x}_{k+1} = \mathbf{g}\left(\mathbf{x}_k, \mathbf{u}_k\right) + \mathbf{L}_{1,k}\mathbf{f}_k, \tag{11.8}$$

$$\mathbf{y}_{k+1} = \mathbf{h}(\mathbf{x}_{k+1}) + \mathbf{L}_{2,k+1}\mathbf{f}_{k+1}. \tag{11.9}$$

In order to apply the Luenberger observer, it is necessary to linearize the Eqs. (11.8) and (11.9) around either a constant value (e.g., $\mathbf{x}_=\mathbf{0}$) or the current state estimate $\hat{\mathbf{x}}_k$. The latter approach seems to be more appropriate as it improves its approximation accuracy as $\hat{\mathbf{x}}_k$ tends to $\mathbf{x}_k$. In this case, the approximation can be realized as follows:

$$\mathbf{A}_k = \left.\frac{\partial \mathbf{g}\left(\mathbf{x}_k, \mathbf{u}_k\right)}{\partial \mathbf{x}_k}\right|_{\mathbf{x}_k=\hat{\mathbf{x}}_k}, \quad \mathbf{C}_k = \left.\frac{\partial \mathbf{h}(\mathbf{x}_k)}{\partial \mathbf{x}_k}\right|_{\mathbf{x}_k=\hat{\mathbf{x}}_k}. \tag{11.10}$$

As a result of using the Luenberger observer, the state estimation error takes the form

$$\begin{aligned}\mathbf{e}_{k+1} = {} & [\mathbf{A}_{k+1} - \mathbf{K}_{k+1}\mathbf{C}_k]\mathbf{e}_k + \mathbf{L}_{1,k}\mathbf{f}_k - \mathbf{K}_{k+1}\mathbf{L}_{2,k}\mathbf{f}_k + \\ & + \mathbf{o}(\mathbf{x}_k, \hat{\mathbf{x}}_k),\end{aligned} \tag{11.11}$$

where $\mathbf{o}(\mathbf{x}_k, \hat{\mathbf{x}}_k)$ stands for the linearization error caused by the introduction of (11.10).

Because of a highly time-varying nature of $\mathbf{A}_k$ and $\mathbf{C}_k$ as well as the linearization error $\mathbf{o}(\mathbf{x}_k, \hat{\mathbf{x}}_k)$, it is usually very difficult to obtain an appropriate form of the gain matrix $\mathbf{K}_{k+1}$. This is the main reason why this approach is rarely used in practice.

As the Kalman filter constitutes a stochastic counterpart of the Luenberger observer, the extended Kalman filter can also be designed for the following class of nonlinear systems:

$$\mathbf{x}_{k+1} = \mathbf{g}\left(\mathbf{x}_k, \mathbf{u}_k\right) + \mathbf{L}_{1,k}\mathbf{f}_k + \mathbf{w}_k, \tag{11.12}$$

$$\mathbf{y}_{k+1} = \mathbf{h}(\mathbf{x}_{k+1}) + \mathbf{L}_{2,k+1}\mathbf{f}_{k+1} + \mathbf{v}_{k+1}, \tag{11.13}$$

while, similarly to the linear case, $\mathbf{w}_k$ and $\mathbf{v}_k$ are zero-mean white noise sequences. Using the linearization (11.10) and neglecting the influence of the linearization error, it is straightforward to use the Kalman filter algorithm. The main drawback of such an approach is that it works well only when there is no large mismatch between the model linearized around the current state estimate and the nonlinear behavior of the system.

The EKF can also be used for deterministic systems, i.e., as an observer for the system (11.8)–(11.9) (see [11] and the references therein). In this case, the noise covariance matrices can be set almost arbitrarily. As was proposed in [11], this possibility can be used to increase the convergence of an observer.

Apart from difficulties regarding linearization errors, similarly to the case of linear systems, the presented approaches do not take model uncertainty into account. This drawback disqualifies those techniques for most practical applications, although there are cases for which such techniques work with acceptable efficiency, e.g., [40].

11.2.3.2 Observers for Lipschitz Systems

Let us consider a class of nonlinear systems which can be described by the following state-space equations:

$$\mathbf{x}_{k+1} = \mathbf{A}\mathbf{x}_k + \mathbf{B}\mathbf{u}_k + \mathbf{h}(\mathbf{y}_k, \mathbf{u}_k) + \mathbf{g}\left(\mathbf{x}_k, \mathbf{u}_k\right), \tag{11.14}$$

$$\mathbf{y}_{k+1} = \mathbf{C}\mathbf{x}_{k+1}, \tag{11.15}$$

and $\mathbf{g}\left(\mathbf{x}_k, \mathbf{u}_k\right)$ satisfies

$$\|\mathbf{g}\left(\mathbf{x}_1, \mathbf{u}\right) - \mathbf{g}\left(\mathbf{x}_2, \mathbf{u}\right)\|_2 \leq \gamma \|\mathbf{x}_1 - \mathbf{x}_2\|_2, \ \forall \mathbf{x}_1, \mathbf{x}_2, \mathbf{u}, \tag{11.16}$$

where $\gamma > 0$ stands for the Lipschitz constant. Many nonlinear systems can be described by (11.14), e.g., sinusoidal nonlinearities satisfy (11.16), even polynomial nonlinearities satisfy (11.16) assuming that $\mathbf{x}_k$ is bounded. This means that (11.14)–(11.15) can be used for describing a wide class of nonlinear systems, which is very important from the point of view of potential industrial applications.

The first solution for state estimation of the continuous-time counterpart of (11.14)–(11.15) was developed by Thau [71]. Assuming that the pair $(\mathbf{A}, \mathbf{C})$ is observable, Thau proposed a convergence condition, but he did not provide an effective design procedure for the observer. In other words, in light of this approach, the observer has to be designed with a trial-and-error procedure that amounts to solving a large number of Lyapunov equations and then checking the convergence conditions. Many different authors followed a similar procedure but they proposed less restrictive convergence conditions (see, e.g., [63]). Finally, in [2, 60, 61], the authors proposed a more effective observer design. In particular, in [61], the authors employed the concept of the distance to the unobservability of the pair $(\mathbf{A}, \mathbf{C})$, and proposed an iterative coordinate transformation technique reducing the Lipschitz constant. In [2], the authors employed and improved the results of [61], but the proposed design procedure does not seem straightforward. In [60], the author reduced the observer design problem to a global optimization one. The main disadvantage of this approach is that the proposed algorithm does not guarantee obtaining a global optimum. Thus, many trial-and-error steps have to be carried out to obtain a satisfactory solution. Recently, in [59], the authors proposed the so-called dynamic observer with a mixed binary search and the $\mathcal{H}_\infty$ optimization procedure.

Unfortunately, the theory and practice concerning observers for discrete-time systems (11.14)–(11.15) are significantly less mature than those for their continuous-time counterparts. Indeed, there are few papers only [12, 76] dealing with discrete-time observers. The authors of the above works proposed different parameterizations of the observer, but the common disadvantage of these approaches is that a trial-and-error procedure has to be employed that boils down to solving a large number of Lyapunov equations. Moreover, the authors do not provide convergence conditions similar to those for continuous-time observers [63, 71].

11.2.3.3 Coordinate-Change-Based Observers

Another possible approach can be implemented by a suitable nonlinear change of coordinates to bring the original system into a linear (or pseudo-linear) one. Let us consider the following nonlinear system:

$$\mathbf{x}_{k+1} = \mathbf{g}\left(\mathbf{x}_k, \mathbf{u}_k\right) + \mathbf{L}_{1,k}\mathbf{f}_k, \tag{11.17}$$

$$\mathbf{y}_{k+1} = \mathbf{h}(\mathbf{x}_{k+1}) + \mathbf{L}_{2,k+1}\mathbf{f}_{k+1}. \tag{11.18}$$

For design purposes, let us assume that $\mathbf{f}_k = \mathbf{0}$. The basic idea underlying coordinate-change-based observers is to determine the coordinate change (at least locally defined) of the form

$$\begin{aligned} \mathbf{s} &= \boldsymbol{\phi}(\mathbf{x}), \\ \bar{\mathbf{y}} &= \boldsymbol{\varphi}(\mathbf{y}), \end{aligned} \tag{11.19}$$

such that in the new coordinates (11.19), the system (11.17)–(11.18) is described by

$$\mathbf{s}_{k+1} = \mathbf{A}(\mathbf{u}_k)\mathbf{s}_k + \boldsymbol{\psi}(\mathbf{y}_k, \mathbf{u}_k), \tag{11.20}$$

$$\bar{\mathbf{y}}_{k+1} = \mathbf{C}\mathbf{s}_{k+1}, \tag{11.21}$$

where $\boldsymbol{\psi}(\cdot)$ is a nonlinear (in general) function. There is no doubt that the observer design problem is significantly simplified when (11.20)–(11.21) are used instead of (11.17)–(11.18). The main drawback of such an approach is related to strong design conditions that limit its application to particular classes of nonlinear systems (for an example regarding single-input single-output systems, the reader is referred to [13]).

11.2.3.4 Observers for Bilinear and Low-Order Polynomial Systems

A polynomial (and, as a special case, bilinear) system description is a natural extension of linear models. Designs of observers for bilinear and low-order polynomial systems have been considered in a number of papers [8, 25, 30, 36, 66, 83]. Let us consider a bilinear system modeled by the following state space equations:

$$\mathbf{x}_{k+1} = \mathbf{A}_k\mathbf{x}_k + \sum_{i=1}^{r} \mathbf{A}^i u_{i,k}\mathbf{x}_k + \mathbf{B}\mathbf{u}_k + \mathbf{L}_1\mathbf{f}_k, \tag{11.22}$$

$$\mathbf{y}_{k+1} = \mathbf{C}\mathbf{x}_{k+1} + \mathbf{D}\mathbf{u}_k + \mathbf{L}_2\mathbf{f}_{k+1}. \tag{11.23}$$

Hou and Pugh [30] established the necessary conditions for the existence of the observer for the continuous-time counterpart of (11.22)–(11.23). Moreover, they proposed a design procedure involving the transformation of the original system (11.22)–(11.23) into an equivalent, quasi-linear one. An observer for systems which can be described by state-space equations consisting of both linear and polynomial terms was proposed in [8, 66].

11.3 Robustness Issues

Irrespective of the linear or nonlinear FDI technique being employed, FDI performance will be usually impaired by the lack of robustness to model uncertainty. Indeed, the model reality mismatch may cause very undesirable situations such as undetected faults or false alarms. This may lead to serious economical losses or even catastrophes.

Taking into account the above conditions, a large amount of knowledge on designing robust fault diagnosis systems has been accumulated through the literature since the beginning of the 1980s. For a comprehensive survey regarding such techniques, the reader is referred to the excellent monographs [16, 22, 39, 53]. Thus, the subject of this section is to outline the main issues of robust fault diagnosis.

Let us start with above-described parameter estimation techniques. The main assumption underlying the FDI approaches presented in these sections was the fact that perfect parameter estimates can be obtained. This is, of course, very hard to attain in practice. Indeed, the fact that the measurements used for parameter estimation can be corrupted by noise and disturbances contributes directly to the so-called parameter uncertainty. This means that there is no unique $\hat{\mathbf{p}}$ that is consistent with a given set of measurements, but there is a parameter set $\mathbb{P}$ that satisfies this requirement. This parameter set is usually called the confidence region or the feasible parameter set [48, 74]. Such a set can be determined online for linear-in-parameter systems using either statistical [74] or bounded-error approaches [48, 74, 77]. In both the cases, fault diagnosis tasks can be realized in two different ways. The first one boils down to

$$\mathbf{p}_{\text{nom}} \notin \mathbb{P}_k \quad \Rightarrow \quad \mathbf{f}_k \neq \mathbf{0}, \tag{11.24}$$

where $\mathbb{P}_k$ is the parameter confidence region associated with k input–output measurements. The second approach is implemented in such a way as the knowledge regarding $\mathbb{P}_k$ is used to calculate the so-called output confidence interval:

$$y_k^N \leq y_k \leq y_k^M. \tag{11.25}$$

This confidence interval is then used for fault detection, i.e.,

$$y_k < y_k^N \quad \text{or} \quad y_k > y_k^M \quad \Rightarrow \quad \mathbf{f}_k \neq \mathbf{0}. \tag{11.26}$$

The advantage of (11.26) over (11.24) is that no knowledge regarding the nominal (fault-free) parameter vector $\mathbf{p}_{\text{nom}}$ is required.

In the case of fault isolation, the feasible parameter set $\mathbb{P}$ can be used for calculating the parameter confidence intervals:

$$p_{i,k}^N \leq p_i \leq p_{i,k}^M, \quad i = 1, \dots, n_p. \tag{11.27}$$

These intervals can then be employed for fault isolation:

$$p_{i,\text{nom}} < p_{i,k}^N \quad \text{or} \quad p_{i,\text{nom}} > p_{i,k}^M \quad \Rightarrow \quad f_{i,k} \neq \mathbf{0}, \quad i = 1, \dots, s = n_p. \tag{11.28}$$

In order to extend the above strategies for nonlinear-in-parameter systems, it is necessary to use various linearization strategies, which usually impair the reliability and effectiveness of FDI.

An alternative approach to robust parameter estimation based fault diagnosis was proposed in [52, 86]. The authors considered the continuous-time counterpart of the following system:

$$\mathbf{x}_{k+1} = \mathbf{g}\left(\mathbf{x}_k, \mathbf{u}_k\right) + \mathbf{h}(\mathbf{x}_k, \mathbf{u}_k, k) + \mathbf{L}(k - k_0)\boldsymbol{\phi}(\mathbf{x}_k, \mathbf{u}_k), \tag{11.29}$$

while $\mathbf{L}(k - k_0)$ is a matrix function representing the time profiles of the faults (with k_0 being an unknown time), and $\mathbf{h}(\mathbf{x}_k, \mathbf{u}_k, k)$ is the modeling error satisfying

$$|h_i(\mathbf{x}_k, \mathbf{u}_k, k)| \le \bar{h}, \quad i = 1, \dots, n, \tag{11.30}$$

where $\bar{h}$ is a known bound. A change in the system behavior caused by the faults is modeled by $\boldsymbol{\phi}(\mathbf{x}_k, \mathbf{u}_k)$, which belongs to a finite set of functions:

$$\mathbb{F} = \left\{\boldsymbol{\phi}^1(\mathbf{x}_k, \mathbf{u}_k), \dots, \boldsymbol{\phi}^s(\mathbf{x}_k, \mathbf{u}_k)\right\}. \tag{11.31}$$

Each fault function $\boldsymbol{\phi}^i(\mathbf{x}_k, \mathbf{u}_k)$, $i = 1, \dots, s$ is assumed to be a parametric fault, i.e., a fault with a known nonlinear structure but with unknown parameter vectors. To tackle parametric fault detection, the authors proposed the so-called detection and approximation observer while fault isolation was realized with a bank of nonlinear adaptive observers.

As can be observed in the literature [16, 22, 39, 53], the most common approach to robust fault diagnosis is to use robust observers. This is mainly because of the fact that the theory of robust observers is relatively well developed in the control engineering literature. Indeed, the most common approaches to representing model uncertainty in robust observers for linear systems can be divided into five categories:

Norm-bounded model uncertainty: it corresponds to a system description whose matrices are modeled in the form of a known matrix $\mathbf{M}_0$ and an additive uncertainty term $\boldsymbol{\delta}_M$ satisfying $\|\boldsymbol{\delta}_M\| \le 1$. Thus, the matrices describing the system are of the following form:

$$\mathbf{M} = \mathbf{M}_0 + \boldsymbol{\delta}_M = \mathbf{M}_0 + \mathbf{HFE}, \tag{11.32}$$

where $\mathbf{F}$ and $\mathbf{E}$ are known constant matrices, and

$$\mathbf{F}^T\mathbf{F} \preceq \mathbf{I}. \tag{11.33}$$

Polytopic model uncertainty: it corresponds to a system description whose matrices are contained in a polytope of matrices, i.e.,

$$\mathbf{M} \in \mathrm{Co}\,(\mathbf{M_1}, \dots, \mathbf{M_N}), \tag{11.34}$$

or, equivalently, $\mathbf{M} \in \mathbb{M}$ with

$$\mathbb{M} = \left\{\mathbf{X} : \mathbf{X} = \sum_{i=1}^{N} \alpha_i \mathbf{M}_i,\ \alpha_i \ge 0,\ \sum_{i=1}^{N} \alpha_i = 1\right\}. \tag{11.35}$$

Affine model uncertainty: it corresponds to a system description whose matrices are modeled as a collection of fixed affine functions of varying parameters

$p_1, \ldots, p_{n_p}$, i.e.,

$$\mathbf{M}(\mathbf{p}) = \mathbf{M}_0 + p_1\mathbf{M}_1 + \cdots + p_{n_p}\mathbf{M}_{n_p}, \tag{11.36}$$

whereas $\mathbf{M}_i$, $i = 0, \ldots, n_p$ are known matrices, and

$$\underline{p}_i \leq p_i \leq \bar{p}_i, \quad i = 1, \ldots, n_p. \tag{11.37}$$

Interval model uncertainty: it corresponds to a system description whose matrices $\mathbf{M}$ are not known precisely but their elements are described by known intervals, i.e.,

$$\underline{m}_{i,j} \leq m_{i,j} \leq \bar{m}_{i,j}. \tag{11.38}$$

Unknown input model uncertainty: it corresponds to the system description in which model uncertainty is modeled by an unknown additive term, i.e.,

$$\mathbf{x}_{k+1} = \mathbf{A}\mathbf{x}_k + \mathbf{B}\mathbf{u}_k + \mathbf{E}\mathbf{d}_k, \tag{11.39}$$

where $\mathbf{d}_k$ is an unknown input and $\mathbf{E}$ denotes its distribution matrix which is known, i.e., it can be efficiently estimated with one of the approaches described in [16, 39].

As can be found in the literature [16, 22, 32, 39, 53], the most popular approach is to use unknown input model uncertainty. The observer resulting from such an approach is called the Unknown Input Observer (UIO).

Although the origins of UIOs can be traced back to the early 1970s (cf. the seminal work of Wang et al. [75]), the problem of designing such observers is still of paramount importance both from the theoretical and practical viewpoints. A large amount of knowledge on using these techniques for model-based fault diagnosis has been accumulated through the literature for the last three decades (see [16, 39] and the references therein). Generally, design problems regarding UIOs can be divided into three distinct categories:

Design of UIOs for linear deterministic systems: Apart from the seminal paper of Wang et al. [75], it is worth noting a few pioneering and important works in this area, namely, the geometric approach by Bhattacharyya [9], the inversion algorithm by Kobayashi and Nakamizo [37], the algebraic approach by Hou and Müller [28], and, finally, the approach by Chen, Patton and Zhang [17]. The reader is also referred to the recently published developments, e.g., [31].

Design of UIOs for linear stochastic systems: Most design techniques concerning such a class of linear systems make use of ideas for linear deterministic systems along with the Kalman filtering strategy. Thus, the resulting approaches can be perceived as Kalman filters for linear systems with unknown inputs. The representative approaches of this group were developed by Chen, Patton and Zhang [16, 17], Darouach and Zasadzinski [19], Hou and Patton [29], and, finally, Keller and

Darouach [34].

A significantly different approach was proposed in [78]. Instead of using the Kalman filter like approach, the author employed the bounded-error state estimation technique [46], but the way of decoupling the unknown input remained the same as that in [34].

Design of UIOs for nonlinear systems: The design approaches developed for nonlinear systems can generally be divided into three categories:

Nonlinear state transformation based techniques: apart from a relatively large class of systems to which they can be applied, even if the nonlinear transformation is possible, it sometimes leads to another nonlinear system and hence the observer design problem remains open (see [4, 64] and the references therein).

Linearization-based techniques: such approaches are based on a similar strategy as that for the extended Kalman filter [39]. In [78, 82], the author proposed an Extended Unknown Input Observer (EUIO) for nonlinear systems. He also proved that the proposed observer is convergent under certain conditions.

Observers for particular classes of nonlinear systems: for example, UIOs for polynomial and bilinear systems [8, 25] or UIOs for Lipschitz systems [38, 59].

To illustrate the general principle of the UIO, let us consider a linear system described by the following state-space equations:

$$\mathbf{x}_{k+1} = \mathbf{A}_k\mathbf{x}_k + \mathbf{B}_k\mathbf{u}_k + \mathbf{E}_k\mathbf{d}_k + \mathbf{L}_{1,k}\mathbf{f}_k, \tag{11.40}$$

$$\mathbf{y}_{k+1} = \mathbf{C}_{k+1}\mathbf{x}_{k+1} + \mathbf{L}_{2,k+1}\mathbf{f}_{k+1}, \tag{11.41}$$

where the term $\mathbf{E}_k\mathbf{d}_k$ stands for model uncertainty as well as real disturbances acting on the system, and $\text{rank}(\mathbf{E}_k) = q$. The general structure of an unknown input observer can be given as follows [16]:

$$\mathbf{s}_{k+1} = \mathcal{F}k + 1\mathbf{s}_k + \mathbf{T}_{k+1}\mathbf{B}_k\mathbf{u}_k + \mathbf{K}_{k+1}\mathbf{y}_k, \tag{11.42}$$

$$\hat{\mathbf{x}}_{k+1} = \mathbf{s}_{k+1} + \mathbf{H}_{k+1}. \tag{11.43}$$

If the following relations hold true:

$$\mathbf{K}_{k+1} = \mathbf{K}_{1,k+1} + \mathbf{K}_{2,k+2}, \tag{11.44}$$

$$\mathbf{T}_{k+1} = \mathbf{I} - \mathbf{H}_{k+1}\mathbf{C}_{k+1} \tag{11.45}$$

$$\mathcal{F}k + 1 = \mathbf{A}_k - \mathbf{H}_{k+1}\mathbf{C}_{k+1}\mathbf{A}_k - \mathbf{K}_{1,k+1}\mathbf{C}_k, \tag{11.46}$$

$$\mathbf{K}_{2,k+1} = \mathcal{F}k + 1\mathbf{H}_k, \tag{11.47}$$

then (assuming the fault-free mode, i.e., $\mathbf{f}_k = \mathbf{0}$) the state estimation error is

$$\mathbf{e}_{k+1} = \mathcal{F}k + 1\mathbf{e}_k + [\mathbf{I} - \mathbf{H}_{k+1}\mathbf{C}_{k+1}]\mathbf{E}_k\mathbf{d}_k. \tag{11.48}$$

From the above equation, it is clear that to decouple the effect of an unknown input from the state estimation error (and, consequently, from the residual), the following relation should be satisfied:

$$[\mathbf{I} - \mathbf{H}_{k+1}\mathbf{C}_{k+1}]\mathbf{E}_k = \mathbf{0}. \tag{11.49}$$

The necessary condition for the existence of a solution to (11.49) is $\text{rank}(\mathbf{E}_k) = \text{rank}(\mathbf{C}_{k+1}\mathbf{E}_k)$ [16, p. 72, Lemma 3.1], and a special solution is

$$\mathbf{H}^*_{k+1} = \mathbf{E}_k \left[(\mathbf{C}_{k+1}\mathbf{E}_k)^T \mathbf{C}_{k+1}\mathbf{E}_k\right]^{-1} (\mathbf{C}_{k+1}\mathbf{E}_k)^T. \tag{11.50}$$

The remaining task is to design the matrix $\mathbf{K}_{1,k+1}$ so as to ensure the convergence of the observer. This can be realized in a similar way as it is done in the case of the Luenberger observer. Finally, the state estimation error and the residual are given by

$$\begin{aligned} \mathbf{e}_{k+1} &= \mathcal{F}k + 1\mathbf{e}_k + \mathbf{T}_{k+1}\mathbf{L}_{1,k}\mathbf{f}_k \\ &\quad - \mathbf{H}_{k+1}\mathbf{L}_{2,k+1}\mathbf{f}_{k+1} - \mathbf{K}_{1,k+1}\mathbf{L}_{2,k}\mathbf{f}_k, \end{aligned} \tag{11.51}$$

$$\mathbf{z}_{k+1} = \mathbf{C}_{k+1}\mathbf{e}_{k+1} + \mathbf{L}_{2,k+1}\mathbf{f}_{k+1}. \tag{11.52}$$

Since the Kalman filter constitutes a stochastic counterpart of the Luenberger observer, there can also be developed a stochastic counterpart of the UIO [16], i.e., an observer which can be applied to the following class of systems:

$$\mathbf{x}_{k+1} = \mathbf{A}_k\mathbf{x}_k + \mathbf{B}_k\mathbf{u}_k + \mathbf{E}_k\mathbf{d}_k + \mathbf{L}_{1,k}\mathbf{f}_k + \mathbf{w}_k, \tag{11.53}$$

$$\mathbf{y}_{k+1} = \mathbf{C}_k\mathbf{x}_{k+1} + \mathbf{L}_{2,k+1}\mathbf{f}_{k+1} + \mathbf{v}_{k+1}. \tag{11.54}$$

Apart from the robustness properties, another reason why UIOs are very popular in fault diagnosis schemes is the fact that they can be effectively applied to sensor and actuator fault isolation. First, the sensor fault isolation scheme is described. In this case, the actuators are assumed to be fault free, and hence, for each of the observers, the system can be characterized as follows:

$$\mathbf{x}_{k+1} = \mathbf{g}(\mathbf{x}_k) + \mathbf{h}(\mathbf{u}_k) + \mathbf{E}_k\mathbf{d}_k, \tag{11.55}$$

$$\mathbf{y}^j_{k+1} = \mathbf{C}^j_{k+1}\mathbf{x}_{k+1} + \mathbf{f}^j_{k+1}, \qquad j = 1, \ldots, m, \tag{11.56}$$

$$y_{j,k+1} = \mathbf{c}_{j,k+1}\mathbf{x}_{k+1} + f_{j,k+1}, \tag{11.57}$$

where, similarly as it was done in [16], $\mathbf{c}_{j,k} \in \Re^n$ is the jth row of the matrix $\mathbf{C}_k$, $\mathbf{C}^j_k \in \Re^{m-1\times n}$ is obtained from the matrix $\mathbf{C}_k$ by deleting the jth row, $\mathbf{c}_{j,k}$, $y_{j,k+1}$ is the jth element of $\mathbf{y}_{k+1}$, and $\mathbf{y}^j_{k+1} \in \Re^{m-1}$ is obtained from the vector $\mathbf{y}_{k+1}$ by deleting the jth component $y_{j,k+1}$. Thus, the problem reduces to designing m UIOs. Therefore, each residual generator (observer) is driven by all inputs and all outputs but one. When all sensors but the jth one are fault free and all actuators are fault free, then the residual $\mathbf{z}_k = \mathbf{y}_k - \hat{\mathbf{y}}_k$ will satisfy the following isolation logic:

$$\begin{cases} \|\mathbf{z}_k^j\| < T_j^H \\ \|\mathbf{z}_k^l\| \geq T_l^H, \end{cases} \qquad l = 1, \ldots, j-1, j+1, \ldots, m, \tag{11.58}$$

while T_i^H denotes a prespecified threshold. Similarly to the case of the sensor fault isolation scheme, in order to design the actuator fault isolation scheme, it is necessary to assume that all sensors are fault free. Moreover, the term $\mathbf{h}(\mathbf{u}_k)$ in

$$\mathbf{x}_{k+1} = \mathbf{g}\left(\mathbf{x}_k\right) + \mathbf{h}(\mathbf{u}_k) + \mathbf{E}_k\mathbf{d}_k, \tag{11.59}$$

$$\mathbf{y}_{k+1} = \mathbf{C}_{k+1}\mathbf{x}_{k+1} \tag{11.60}$$

should have the following structure:

$$\mathbf{h}(\mathbf{u}_k) = \mathbf{B}(\mathbf{u}_k)\mathbf{u}_k, \tag{11.61}$$

where the ith column of $\mathbf{B}(\mathbf{u}_k)$ is a nonlinear function of the form $\mathbf{b}_i(\mathbf{u}_k^i)$ and $\mathbf{u}_k^i \in \Re^{r-1}$ is obtained from $\mathbf{u}_k$ by deleting its ith component, $u_{i,k}$.

In this case, for each of the observers, the system can be characterized as follows:

$$\begin{aligned} \mathbf{x}_{k+1} &= \mathbf{g}\left(\mathbf{x}_k\right) + \mathbf{h}^i(\mathbf{u}_k^i + \mathbf{f}_k^i) + \mathbf{h}_i(u_{i,k} + f_{i,k}) + \mathbf{E}_k\mathbf{d}_k \\ &= \mathbf{g}\left(\mathbf{x}_k\right) + \mathbf{h}^i(\mathbf{u}_k^i + \mathbf{f}_k^i) + \mathbf{E}_k^i\mathbf{d}_k^i, \end{aligned} \tag{11.62}$$

$$\mathbf{y}_{k+1} = \mathbf{C}_{k+1}\mathbf{x}_{k+1}, \qquad i = 1, \ldots, r, \tag{11.63}$$

with

$$\mathbf{h}^i(\mathbf{u}_k^i + \mathbf{f}_k^i) = \mathbf{B}^i(\mathbf{u}_k)(\mathbf{u}_k^i + \mathbf{f}_k^i), \tag{11.64}$$

$$\mathbf{h}_i(u_{i,k} + f_{i,k}) = \mathbf{b}_i(\mathbf{u}_k^i)(u_{i,k} + f_{i,k}), \tag{11.65}$$

while $\mathbf{B}^i(\mathbf{u}_k)$ is obtained from $\mathbf{B}(\mathbf{u}_k)$ by deleting its ith column, and

$$\mathbf{E}_k^i = [\mathbf{E}_k \ \mathbf{b}_i(\mathbf{u}_k^i)], \quad \mathbf{d}_k^i = \begin{bmatrix} \mathbf{d}_k \\ u_{i,k} + f_{i,k} \end{bmatrix}. \tag{11.66}$$

11.4 Fault Estimation Strategies

The objective of the preceding sections was to discuss FDI methods from the historical perspective. However, as it was already mentioned, recent developments of FTC cased and increased research attention concerning fault estimation. It is also important to underline that the resulting fault estimates can be efficiently used for fault detection and isolation [80]. In particular, it is split into actuator, actuator, and sensor as well as process fault estimation approaches, respectively. Finally, the last part of the chapter presents an illustrative example concerning selected fault estimation of a three-tank system. It should be pointed out that the material used for describing fault estimation part was originally presented in [54, 55, 81].

11.4.1 Actuator Fault Estimation Under Unknown Inputs

Let us consider a nonlinear discrete-time system

$$\mathbf{x}_{k+1} = \mathbf{A}\mathbf{x}_k + \mathbf{B}\mathbf{u}_k + \mathbf{D}\mathbf{d}_k + \mathbf{g}\left(\mathbf{x}_k, \mathbf{u}_k\right) + \mathbf{B}\mathbf{f}_k + \mathbf{W}_1\mathbf{w}_k, \tag{11.67}$$

$$\mathbf{y}_k = \mathbf{C}\mathbf{x}_k + \mathbf{W}_2\mathbf{w}_k, \tag{11.68}$$

where $\mathbf{x}_k \in \mathbb{X} \subset \Re^n$ is the state, $\mathbf{u}_k \in \mathbb{U} \in \Re^r$ stands for the input, $\mathbf{y}_k \in \Re^m$ denotes the output, $\mathbf{f}_k \in \Re^r$ stands for the fault, $\mathbf{d}_k \in \Re^q$ is the unknown input disturbance, and $\mathbf{w}_k \in l_2$ is a an exogenous disturbance vector satisfying

$$l_2 = \left\{\mathbf{w} \in \Re^n |\ \|\mathbf{w}\|_{l_2} < +\infty\right\},\ \|\mathbf{w}\|_{l_2} = \left(\sum_{k=0}^{\infty} \|\mathbf{w}_k\|^2\right)^{\frac{1}{2}}. \tag{11.69}$$

Finally, $\mathbf{W}_1 \in \Re^{n\times n}$ and $\mathbf{W}_2 \in \Re^{m\times n}$ stand for the exogenous disturbance distribution matrices.

Following the original work [81], the following assumptions are considered:

Assumption 1: There exists a matrix $\mathbf{M}$ such that

$$\left(\mathbf{g}\left(\mathbf{a},\mathbf{u}\right) - \mathbf{g}\left(\mathbf{b},\mathbf{u}\right)\right)^T \left(\mathbf{a}-\mathbf{b}\right) \le (\mathbf{a}-\mathbf{b})^T\mathbf{M}(\mathbf{a}-\mathbf{b}),$$
$$\forall \mathbf{a},\mathbf{b} \in \mathbb{X},\ \mathbf{u} \in \mathbb{U}. \tag{11.70}$$

Assumption 2: There exists a matrix $\mathbf{M}$ such that

$$\left(\mathbf{g}\left(\mathbf{a},\mathbf{u}\right) - \mathbf{g}\left(\mathbf{b},\mathbf{u}\right)\right)^T \left(\mathbf{g}\left(\mathbf{a},\mathbf{u}\right) - \mathbf{g}\left(\mathbf{b},\mathbf{u}\right)\right) \le (\mathbf{a}-\mathbf{b})^T\mathbf{M}^T\mathbf{M}(\mathbf{a}-\mathbf{b}),$$
$$\forall \mathbf{a},\mathbf{b} \in \mathbb{X},\ \mathbf{u} \in \mathbb{U}. \tag{11.71}$$

Assumption 3: The fault satisfies

$$\boldsymbol{\varepsilon}_k = \mathbf{f}_{k+1} - \mathbf{f}_k, \quad \boldsymbol{\varepsilon}_k \in l_2. \tag{11.72}$$

Assumption 4: The following rank condition is satisfied

$$\operatorname{rank}(\mathbf{D}) = \operatorname{rank}(\mathbf{C}\mathbf{D}) = q, \quad q \le m. \tag{11.73}$$

It is worth noting that if $\mathbf{M}^T\mathbf{M} = \gamma^2\mathbf{I}$, then the relation underlying *Assumption 2* (cf. [68, 85]) becomes a usual Lipschitz condition [1, 47, 59, 60] with γ being a Lipschitz constant (cf. previous sections for a brief outline of such observers). This appealing property makes the employed strategy more general than those presented in the literature [1, 47, 59, 60]. Moreover, a significant progress was recently obtained in the observer design for nonlinear systems by introducing the so-called *one-sided* Lipschitz condition [88], which means that a wider spectrum of systems can be tackled with the new approach. Indeed, if $\mathbf{M} = \zeta \mathbf{I}$, then the relation underlying *Assumption 1* becomes a usual one-sided Lipschitz condition, which is imposed along with the usual Lipschitz condition (see [88] for further details and explanations). Thus, it is evident that this appealing property makes again the employed strategy more general than those presented in the literature (see, [88] and the references therein). Finally, *Assumption 3* is required to attain a suitable fault estimation quality while *Assumption 4* is used to decouple the effect of an unknown input (see, e.g., [14, 23] for further details). Subsequently, using the Differential Mean Value Theorem (DMVT) [84], it can be shown that

$$\mathbf{g}\,(\mathbf{a}, \mathbf{u}) - \mathbf{g}\,(\mathbf{b}, \mathbf{u}) = \mathbf{M}_{x,u}(\mathbf{a} - \mathbf{b}), \tag{11.74}$$

with

$$\mathbf{M}_{x,u} = \begin{bmatrix} \frac{\partial g_1}{\partial x}(\mathbf{c_1}, \mathbf{u}) \\ \vdots \\ \frac{\partial g_n}{\partial x}(\mathbf{c_n}, \mathbf{u}) \end{bmatrix}, \tag{11.75}$$

where $\mathbf{c}_1, \ldots, \mathbf{c}_n \in \mathrm{Co}(\mathbf{a}, \mathbf{b})$, $\mathbf{c}_i \neq \mathbf{a}$, $\mathbf{c}_i \neq \mathbf{b}$, $i = 1, \ldots, n$. Assuming that

$$\bar{a}_{i,j} \geq \frac{\partial g_i(\mathbf{x})}{\partial x_j} \geq \underline{a}_{i,j}, \quad i = 1, \ldots, n, \quad j = 1, \ldots, n, \tag{11.76}$$

it is clear that there exists a matrix $\mathbf{M} \in \mathbb{M}$ such that

$$\mathbb{M} = \left\{ \mathbf{M} \in \Re^{n \times n} | \bar{a}_{i,j} \geq m_{i,j} \geq \underline{a}_{i,j},\ i, j = 1, \ldots, n, \right\}, \tag{11.77}$$

which means that conditions of *Assumptions 1* and 2 can be satisfied.

Given the above preliminaries and assumptions, the objective of the subsequent part is to provide a state and fault estimation strategy for the class of nonlinear discrete-time systems (11.67)–(11.68). The main advantage of the proposed approach is the fact that apart from estimating simultaneously the fault and the state, it is able to decouple the effect of unknown inputs and minimize the influence of external disturbances within the $\mathcal{H}_\infty$ framework.

Considering the system model (11.67)–(11.68), the problem is to design an observer that will be able to estimate simultaneously the state $\mathbf{x}_k$ and the fault $\mathbf{f}_k$ and decouple the effect of the unknown input $\mathbf{d}_k$. For that purpose, the following structure is proposed [81]:

$$\mathbf{z}_{k+1} = \mathbf{N}\mathbf{z}_k + \mathbf{G}\mathbf{u}_k + \mathbf{L}\mathbf{y}_k + \mathbf{T}\mathbf{B}\hat{\mathbf{f}}_k + \mathbf{T}\mathbf{g}\left(\hat{\mathbf{x}}_k, \mathbf{u}_k\right), \tag{11.78}$$

$$\hat{\mathbf{x}}_k = \mathbf{z}_k - \mathbf{E}\mathbf{y}_k, \tag{11.79}$$

$$\hat{\mathbf{f}}_{k+1} = \hat{\mathbf{f}}_k + \mathbf{F}(\mathbf{y}_k - \mathbf{C}\hat{\mathbf{x}}_k). \tag{11.80}$$

with the convergence and disturbance attenuation rules given by the following theorem.

Theorem 11.1 *For a prescribed disturbance attenuation level μ, the observer design problem for the system (11.67)–(11.68) is solvable if there exist $\mathbf{N}$, $\mathbf{U}$, $\mathbf{P} \succ \mathbf{0}$, $\alpha > 0$, $\beta > 0$ such that for all $\mathbf{M} \in \mathbb{M}$, the following condition is satisfied:*

$$\begin{bmatrix} \mathbf{I}-\mathbf{P}+\alpha\mathbf{V}^T(\mathbf{M}+\mathbf{M}^T)\mathbf{V}^T & -\alpha\mathbf{V}^T & \mathbf{0} & \bar{\mathbf{A}}^T\mathbf{P}-\bar{\mathbf{C}}^T\mathbf{N}^T & \mathbf{V}^T\mathbf{M}^T\mathbf{U}^T \\ -\alpha\mathbf{V} & -\beta\mathbf{I} & \mathbf{0} & \mathbf{Y}^T\mathbf{P} & \mathbf{0} \\ \mathbf{0} & \mathbf{0} & -\mu^2\mathbf{I} & \bar{\mathbf{W}}^T\mathbf{P}-\bar{\mathbf{V}}^T\mathbf{N}^T & \mathbf{0} \\ \mathbf{P}\bar{\mathbf{A}}-\mathbf{N}\bar{\mathbf{C}} & \mathbf{P}\mathbf{Y} & \mathbf{P}\bar{\mathbf{W}}-\mathbf{N}\bar{\mathbf{V}} & -\mathbf{P} & \mathbf{0} \\ \mathbf{U}\mathbf{M}\mathbf{V} & \mathbf{0} & \mathbf{0} & \mathbf{0} & \beta\mathbf{I}-\mathbf{U}-\mathbf{U}^T \end{bmatrix} \prec \mathbf{0}, \tag{11.81}$$

with $\mathbf{N} = \mathbf{P}\bar{\mathbf{K}}$.

Finally, the design procedure boils down to solving (11.81) for $\mathbf{M} \in \mathbb{M}$ with respect to $\mathbf{N}$, $\mathbf{U}$, $\mathbf{P}$, α, β and then calculating

$$\bar{\mathbf{K}} = \mathbf{P}^{-1}\mathbf{N}. \tag{11.82}$$

Note that $\mathbb{M}$, defined by (11.77), can be equivalently described by

$$\mathbb{M} = \left\{ \mathbf{M}(\alpha) : \quad \mathbf{M}(\alpha) = \sum_{i=1}^{N} \alpha_i \mathbf{M}_i, \ \sum_{i=1}^{N} \alpha_i = 1, \ \alpha_i \geq 0 \right\}, \tag{11.83}$$

where $N = 2^{n^2}$. Note that this is a general description, which does not take into account that some elements of $\mathbf{M}$ may be constant. In such cases, N is given by $N = 2^{(n-c)^2}$, where c stands for the number of constant elements of $\mathbf{M}$.

Thus, solving (11.81) with respect to $\mathbf{N}$, $\mathbf{U}$, $\mathbf{P}$, α, β is equivalent to solving (for $i = 1, \ldots, N$)

$$\begin{bmatrix} \mathbf{I}-\mathbf{P}+\alpha\mathbf{V}^T(\mathbf{M}_i+\mathbf{M}_i^T)\mathbf{V}^T & -\alpha\mathbf{V}^T & \mathbf{0} & \bar{\mathbf{A}}^T\mathbf{P}-\bar{\mathbf{C}}^T\mathbf{N}^T & \mathbf{V}^T\mathbf{M}_i^T\mathbf{U}^T \\ -\alpha\mathbf{V} & -\beta\mathbf{I} & \mathbf{0} & \mathbf{Y}^T\mathbf{P} & \mathbf{0} \\ \mathbf{0} & \mathbf{0} & -\mu^2\mathbf{I} & \bar{\mathbf{W}}^T\mathbf{P}-\bar{\mathbf{V}}^T\mathbf{N}^T & \mathbf{0} \\ \mathbf{P}\bar{\mathbf{A}}-\mathbf{N}\bar{\mathbf{C}} & \mathbf{P}\mathbf{Y} & \mathbf{P}\bar{\mathbf{W}}-\mathbf{N}\bar{\mathbf{V}} & -\mathbf{P} & \mathbf{0} \\ \mathbf{U}\mathbf{M}_i\mathbf{V} & \mathbf{0} & \mathbf{0} & \mathbf{0} & \beta\mathbf{I}-\mathbf{U}-\mathbf{U}^T \end{bmatrix} \prec \mathbf{0}, \tag{11.84}$$

and then determining

$$\bar{\mathbf{K}} = \mathbf{P}^{-1}\mathbf{N}. \tag{11.85}$$

Having the above-described fault estimation strategy, actuator faults and the system state can be estimated simultaneously. However, such a task can be realized efficiently iff sensors are assumed to be fault free. Thus, the objective of the subsequent section is to present an approach capable of dealing with such an issue [55].

11.4.2 Simultaneous Sensor and Actuator Fault Estimation

As it was shown in numerous books and papers [3, 6, 20, 42, 43], the above nonlinear system can be efficiently modeled using Takagi-Sugeno models. Indeed, even the approach presented in the preceding section can be used for settling such a task. By introducing possible sensor and actuator faults, external exogenous disturbances as well as output equation, the Takagi-Sugeno model can be rewritten as

$$\mathbf{x}_{k+1} = \mathbf{A}(\alpha)\mathbf{x}_k + \mathbf{B}(\alpha)\mathbf{u}_k + \mathbf{B}(\alpha)\mathbf{f}_{a,k} + \mathbf{W}_1\mathbf{w}_k \tag{11.86}$$

$$\mathbf{y}_k = \mathbf{C}\mathbf{x}_k + \mathbf{f}_{s,k} + \mathbf{W}_2\mathbf{w}_k, \tag{11.87}$$

where $\alpha \in \Re^p$ is a vector of premise variables depending on the system inputs inputs and outputs [70], $\mathbf{f}_{a,k} \in \mathbb{F}_a \subset \Re^r$ represents the actuator fault while $\mathbf{f}_{s,k} \in \mathbb{F}_s \subset \Re^m$ denotes the sensor fault vector, and $\mathbf{w}_k$ represents an exogenous disturbance vector with $\mathbf{W}_1$ and $\mathbf{W}_2$ distribution matrices. It is assumed that $\mathbf{w}_k$ satisfies

$$l_2 = \left\{\mathbf{w} \in \Re^n \mid \|\mathbf{w}\|_{l_2} < +\infty\right\}, \tag{11.88}$$

$$\|\mathbf{w}\|_{l_2} = \left(\sum_{k=0}^{\infty} \|\mathbf{w}_k\|^2\right)^{\frac{1}{2}}. \tag{11.89}$$

It can be easily shown that $\mathbf{w}_k$ can be split in such a way as $\mathbf{w}_k = \left[\mathbf{w}_{1,k}^T, \mathbf{w}_{2,k}^T\right]^T$ where $\mathbf{w}_{1,k}$ and $\mathbf{w}_{2,k}$ are process and measurement uncertainties, respectively.

Thus, the problem is to estimate the state along with actuator and sensor faults.

To solve the above defined problem of concomitant estimation of the state $\mathbf{x}_k$ as well as actuator $\mathbf{f}_{a,k}$ and sensor $\mathbf{f}_{s,k}$ faults, the following estimator is proposed [55]:

$$\begin{aligned}\hat{\mathbf{x}}_{k+1} &= \mathbf{A}(\alpha)\hat{\mathbf{x}}_k + \mathbf{B}(\alpha)\mathbf{u}_k + \mathbf{B}(\alpha)\hat{\mathbf{f}}_{a,k} \\ &+ \mathbf{K}_x\left(\mathbf{y}_k - \mathbf{C}\hat{\mathbf{x}}_k - \hat{\mathbf{f}}_{s,k}\right),\end{aligned} \tag{11.90}$$

$$\hat{\mathbf{f}}_{a,k+1} = \hat{\mathbf{f}}_{a,k} + \mathbf{K}_a\left(\mathbf{y}_k - \mathbf{C}\hat{\mathbf{x}}_k - \hat{\mathbf{f}}_{s,k}\right), \tag{11.91}$$

$$\hat{\mathbf{f}}_{s,k+1} = \hat{\mathbf{f}}_{s,k} + \mathbf{K}_s\left(\mathbf{y}_k - \mathbf{C}\hat{\mathbf{x}}_k - \hat{\mathbf{f}}_{s,k}\right), \tag{11.92}$$

where $\mathbf{K}_x$, $\mathbf{K}_a$, and $\mathbf{K}_s$ are the gain matrices for the state, actuator, and sensor fault, respectively. The convergence and disturbance attenuation of the above estimator obey the rules provided by the following theorem [55].

Theorem 11.2 *For a prescribed disturbance attenuation level* $\mu > 0$, *the* $\mathcal{H}_\infty$ *observer design problem for the system* (11.86)–(11.87) *and the observer* (11.90)–(11.92) *is solvable if there exist matrices* $\mathbf{P}$, $\mathbf{U}$, *and* $\mathbf{N}$ *such that the following constraints are satisfied:*

$$\begin{bmatrix} \mathbf{I}-\mathbf{P} & \mathbf{0} & \bar{\mathbf{A}}(\alpha)^T\mathbf{U}-\bar{\mathbf{C}}^T\mathbf{N}^T \\ \mathbf{0} & -\mu^2\mathbf{I} & \bar{\mathbf{W}}\mathbf{U}-\bar{\mathbf{V}}^T\mathbf{N}^T \\ \mathbf{U}\bar{\mathbf{A}}(\alpha)-\mathbf{N}\bar{\mathbf{C}} & \mathbf{U}\bar{\mathbf{W}}-\mathbf{N}\bar{\mathbf{V}} & \mathbf{P}-\mathbf{U}-\mathbf{U}^T \end{bmatrix} \prec 0. \tag{11.93}$$

The final design procedure of the robust TS-based observer can be summarized as follows:

1. Select μ and solve LMI (11.93) to get matrices $\mathbf{N}$, $\mathbf{U}$ and $\mathbf{P}$,
2. Calculate $\bar{\mathbf{K}} = \mathbf{U}^{-1}\mathbf{N}$,
3. Determine gain matrices $\bar{\mathbf{K}} = \left[\mathbf{K}_x^T, \mathbf{K}_a^T, \mathbf{K}_s^T\right]^T$.

It should be noted that any system is traditionally divided into three parts, namely, sensors, actuators, and the process itself. In spite of the fact that mostly the goal is to estimate sensors and actuator faults, there can be applications for which process fault estimation is also of interest. Thus, the subsequent section tackles with a process fault estimation directly.

11.4.3 Process Fault Estimation

Let us consider a nonlinear discrete-time system:

$$\mathbf{x}_{k+1} = \mathbf{A}\mathbf{x}_k + \sum_{i=1}^{n_f} \mathbf{A}_{f,i}\mathbf{f}_{i,k}\mathbf{x}_k + \mathbf{B}\mathbf{u}_k + \mathbf{g}(\mathbf{x}_k) + \mathbf{W}_1\mathbf{w}_k, \tag{11.94}$$

$$\mathbf{y}_k = \mathbf{C}\mathbf{x}_k + \mathbf{W}_2\mathbf{w}_k, \tag{11.95}$$

where $\mathbf{x}_k \in \mathbb{X} \subset \Re^n$, $\mathbf{y}_k \in \Re^m$, $\mathbf{u}_k \in \Re^r$, $\mathbf{f}_k \in \Re^{n_f}$ are the state, output, control input, and process fault vectors, respectively, and $\mathbf{g}(\mathbf{x}_k)$ is the nonlinear function which describes the behavior of the system with respect to state. Note that $\mathbf{A}_{f,i}$ denote the distribution matrix of ith process fault $\mathbf{f}_i$, i.e., it describes the way in which the fault influences the system matrix. Moreover, $\mathbf{W}_1$ and $\mathbf{W}_2$ denote the noise distribution matrices and $\mathbf{w}_k$ expresses the exogenous disturbance vector. Furthermore, $\mathbf{w}_k$ can be split as follows: $\mathbf{w}_k = \left[\mathbf{w}_{1,k}^T, \mathbf{w}_{2,k}^T\right]^T$ where $\mathbf{w}_{1,k}$ and $\mathbf{w}_{2,k}$ represent the process and measurement uncertainties, respectively.

The problem is to design an observer which makes it possible to simultaneously estimate the state and process fault. Following [54], the following substitution is applied:

$$\mathbf{z}_{i,k} = \mathbf{f}_i \mathbf{x}_k, \tag{11.96}$$

which leads to the new representation of the system:

$$\mathbf{x}_{k+1} = \mathbf{A}\mathbf{x}_k + \sum_{i=1}^{n_f} \mathbf{A}_{f,i}\mathbf{z}_{i,k} + \mathbf{B}\mathbf{u}_k + \mathbf{g}(\mathbf{x}_k) + \mathbf{W}_1\mathbf{w}_k, \tag{11.97}$$

$$\mathbf{y}_k = \mathbf{C}\mathbf{x}_k + \mathbf{W}_2\mathbf{w}_k. \tag{11.98}$$

Thus, the problem is to design an estimator that will be able to estimate $\mathbf{z}_{i,k}$ and $\mathbf{x}_k$ simultaneously. While the idea to estimate the fault is based on recovering it from $\mathbf{z}_{i,k}$ and $\mathbf{x}_k$. For the aim of further deliberations, let us consider the following assumptions:

Assumption 1:

$$\begin{aligned} \boldsymbol{\varepsilon}_{i,k} &= \mathbf{z}_{i,k+1} - \mathbf{z}_{i,k}, \\ \boldsymbol{\varepsilon}_{i,k} \in \mathcal{E}_{\boldsymbol{\varepsilon},i}, \quad \mathcal{E}_{\boldsymbol{\varepsilon},i} &= \{\boldsymbol{\varepsilon} : \boldsymbol{\varepsilon}^T \mathbf{Q}_{\boldsymbol{\varepsilon},i} \boldsymbol{\varepsilon} \leq 1\}, \mathbf{Q}_{\boldsymbol{\varepsilon},i} \succ 0. \end{aligned} \tag{11.99}$$

Assumption 2:

$$\mathbf{w}_k \in \mathcal{E}_{\mathbf{w}}, \quad \mathcal{E}_{\mathbf{w}} = \{\mathbf{w} : \mathbf{w}^T \mathbf{Q}_{\mathbf{w}} \mathbf{w} \leq 1\}, \mathbf{Q}_{\mathbf{w}} \succ 0. \tag{11.100}$$

To handle the aforementioned problem of simultaneous estimation of the state $\mathbf{x}_k$ and $\mathbf{z}_{i,k}$, the process fault estimator of the following structure is employed:

$$\hat{\mathbf{x}}_{k+1} = \mathbf{A}\hat{\mathbf{x}}_k + \sum_{i=1}^{n_f} \mathbf{A}_{f,i}\hat{\mathbf{z}}_{i,k} + \mathbf{B}\mathbf{u}_k + \mathbf{g}\left(\hat{\mathbf{x}}_k\right) + \mathbf{K}\left(\mathbf{y}_k - \mathbf{C}\hat{\mathbf{x}}_k\right), \tag{11.101}$$

$$\hat{\mathbf{z}}_{i,k+1} = \hat{\mathbf{z}}_{i,k} + \mathbf{L}_i\left(\mathbf{y}_k - \mathbf{C}\hat{\mathbf{x}}_k\right), \qquad i = 1 \ldots s. \tag{11.102}$$

Note that ith fault estimate obeys

$$\hat{\mathbf{z}}_{j,k} = \hat{\mathbf{f}}^j_{i,k}\hat{\mathbf{x}}_{j,k}, \quad i = 1, \ldots, s, \quad j = 1, \ldots, n. \tag{11.103}$$

Indeed, there is no single estimate $\hat{\mathbf{f}}_{i,k}$ satisfying n equations of the above form (11.103). Thus, rather than having one single estimate, a set of n estimates will be obtained $\hat{\mathbf{f}}^j_{i,k}$, $j = 1, \ldots, n$. Thus, the final step is to calculate the Mean Value (MV) of these estimates

$$\hat{\mathbf{f}}_{i,k} = \frac{1}{n}\sum_{j=1}^{n} \hat{\mathbf{f}}^j_{i,k}. \tag{11.104}$$

For the purpose of subsequent analysis, the following definitions are recalled [5]:

Definition 11.1: System $\mathbf{x}_{k+1} = \mathbf{A}\mathbf{x}_k + \mathbf{B}\mathbf{v}_k$ is strictly quadratically bounded for all allowable $\mathbf{v}_k \in \mathbb{E}_v$, if $\mathbf{x}_k^T\mathbf{P}\mathbf{x}_k > 1$ implies $\mathbf{x}_{k+1}^T\mathbf{P}\mathbf{x}_{k+1} < \mathbf{x}_k^T\mathbf{P}\mathbf{x}_k$ for any $\mathbf{x}_k \in \mathbb{E}_v$.

It should be highlighted that the strict quadratic boundedness ensures that $\mathbf{x}_{k+1}^T \mathbf{P} \mathbf{x}_{k+1} < \mathbf{x}_k^T \mathbf{P} \mathbf{x}_k$ for any $\mathbf{v}_k \in \mathbb{E}_v$ when $\mathbf{x}_k^T \mathbf{P} \mathbf{x}_k > 1$.

Based on the above definition the following theorem was formulated [54], which shapes the convergence and disturbance attenuation of the process fault estimator.

Theorem 11.3 *The process fault estimator is strictly quadratically bounded for all* $\mathbf{v}_k \in \mathbb{E}_v$ *and* $\mathbf{M}_x \in \mathbb{M}$ *if there exist matrices* $\mathbf{P} \succ 0$, $\mathbf{U}$ *and a scalar* $\alpha \in (0, 1)$, *such that the following inequality is satisfied:*

$$\begin{bmatrix} -\mathbf{P} + \alpha \mathbf{P} & \mathbf{0} & \bar{\mathbf{A}}_x^T \mathbf{P} - \bar{\mathbf{C}}^T \mathbf{U}^T \\ \mathbf{0} & -\alpha \mathbf{Q}_v & \bar{\mathbf{W}}^T \mathbf{P} - \bar{\mathbf{V}}^T \mathbf{U}^T \\ \mathbf{P} \bar{\mathbf{A}}_x - \mathbf{U} \bar{\mathbf{C}} & \mathbf{P} \bar{\mathbf{W}} - \mathbf{U} \bar{\mathbf{V}} & -\mathbf{P} \end{bmatrix} \preceq 0. \tag{11.105}$$

Finally, the design procedure brings down to solve (11.105) and then calculate

$$\bar{\mathbf{K}} = \begin{bmatrix} \mathbf{K} \\ \mathbf{L}_1 \\ \mathbf{L}_2 \\ \vdots \\ \mathbf{L}_{n_f} \end{bmatrix} = \mathbf{P}^{-1} \mathbf{U}. \tag{11.106}$$

Solving (11.105) is equivalent to solve (for $i = 1, \ldots, N$)

$$\begin{bmatrix} -\mathbf{P} + \alpha \mathbf{P} & \mathbf{0} & \bar{\mathbf{A}}_{x,i}^T \mathbf{P} - \bar{\mathbf{C}}^T \mathbf{U}^T \\ \mathbf{0} & -\alpha \mathbf{Q}_v & \bar{\mathbf{W}}^T \mathbf{P} - \bar{\mathbf{V}}^T \mathbf{U}^T \\ \mathbf{P} \bar{\mathbf{A}}_{x,i} - \mathbf{U} \bar{\mathbf{C}} & \mathbf{P} \bar{\mathbf{W}} - \mathbf{U} \bar{\mathbf{V}} & -\mathbf{P} \end{bmatrix} \preceq 0, \tag{11.107}$$

and then determine

$$\bar{\mathbf{K}} = \begin{bmatrix} \mathbf{K} \\ \mathbf{L}_1 \\ \mathbf{L}_2 \\ \vdots \\ \mathbf{L}_{n_f} \end{bmatrix} = \mathbf{P}^{-1} \mathbf{U}. \tag{11.108}$$

11.5 Illustrative Examples

The objective of this section is to show the performance of the simultaneous sensor and actuator fault estimation scheme using the three-tank system, which is depicted in Fig. 11.1.

Such a system was designed to allow to verify in the laboratory conditions the methodologies developed for the control and fault diagnosis of the linear and non-

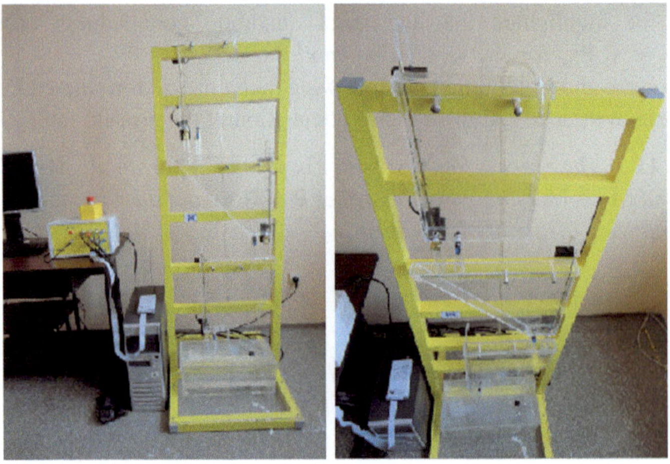

Fig. 11.1 Three-tank system

linear systems. The considered system consists of three cascaded connected separate tanks. The tanks contain the drain valves as well as electro-valves. Moreover, they contain the level sensors which are based on a hydraulic pressure measurement. Those tanks are differently shaped what implies the nonlinearities of the system. The liquid level in each tank varies in the range 0–0.35 [m]. The three-tank system is fed with a DC water pump which is used to fill the upper tank by the liquid. The communication with the multi-tank system is assured by the application of a PC-based digital controller through the dedicated I/O board and the power interface and MATLAB/Simulink environment. The sampling time of the system is set to 0.01 [s] Moreover, the disturbances influencing the system are distributed through

$$\mathbf{W}1 = 0.01\mathbf{I}_{3\times 3}, \qquad \mathbf{W}2 = 0.01\mathbf{I}_{3\times 3}, \tag{11.109}$$

The nonlinear model of the multi-tank system is described by the set of following relations [80] (see Fig. 11.2):

$$\begin{cases} h_1(k+1) = h_1(k) - c_1 h_1(k)^{\beta} + bu(k) \\ h_2(k+1) = h_2(k) + (c_3 + c_4 h_2(k))^{-1}\left(c_2 h_1(k)^{\alpha} - c_5 h_2(k)^{\beta}\right) \\ h_3(k+1) = h_3(k) + \left(c_7 - (c_8 - h_3(k))^2\right)^{-0.5}\left(c_6 h_2(k)^{\beta} - c_9 h_3(k)^{\beta}\right) \end{cases} \tag{11.110}$$

where the real-data-based parameters of the multi-tank system were identified as: $b = 1.14$, $\beta = 0.5$, $c_1 = 1.15 \cdot 10^{-4}$, $c_2 = 1.01 \cdot 10^{-6}$, $c_3 = 3.5 \cdot 10^{-3}$, $c_4 = 3.48 \cdot 10^{-2}$, $c_5 = 1.20 \cdot 10^{-6}$, $c_6 = 3.42 \cdot 10^{-5}$, $c_7 = 1.33 \cdot 10^{-1}$, $c_8 = 3.5 \cdot 10^{-1}$ and $c_9 = 2.80 \cdot 10^{-5}$. The multi-tank model (11.110) can be expressed in the form (11.86), with $\mathbf{x}_k = [h_1(k), h_2(k), h_3(k)]^T$, $u_k = u(k)$, and

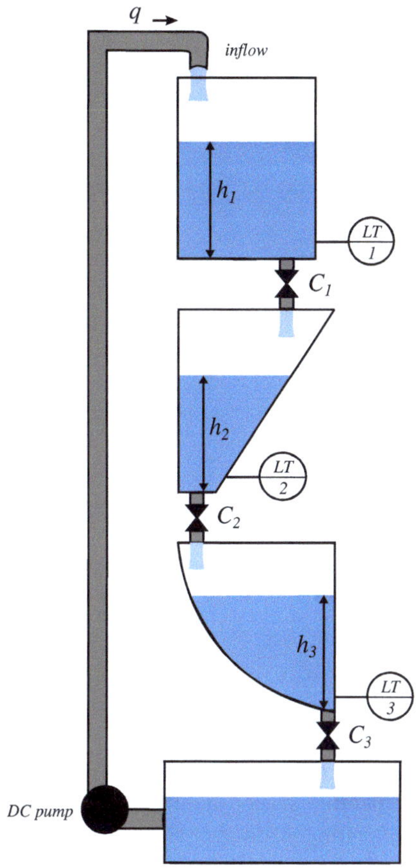

Fig. 11.2 Three-tank system conceptual representation

$$A(\alpha) = \begin{pmatrix} s^{(1)} & 0 & 0 \\ s^{(2)} & s^{(3)} & 0 \\ 0 & s^{(4)} & s^{(5)} \end{pmatrix} \qquad B = \begin{pmatrix} b \\ 0 \\ 0 \end{pmatrix}$$

where

$$\begin{aligned} s^{(1)} &= 1 - c_1 h_1(k)^{\beta-1} \\ s^{(2)} &= c_2 \left(c_3 + c_4 h_2(k)\right)^{-1} h_1(k)^{\beta-1} \\ s^{(3)} &= 1 - c_5 \left(c_3 + c_4 h_2(k)\right)^{-1} h_2(k)^{\beta-1} \\ s^{(4)} &= c_6 \left(c_7 - (c_8 - h_3(k))^2\right)^{-0.5} h_2(k)^{\beta-1} \\ s^{(5)} &= 1 - c_9 \left(c_7 - (c_8 - h_3(k))^2\right)^{-0.5} h_3(k)^{\beta-1} \end{aligned}$$

Moreover, let us consider the following fault scenarios:

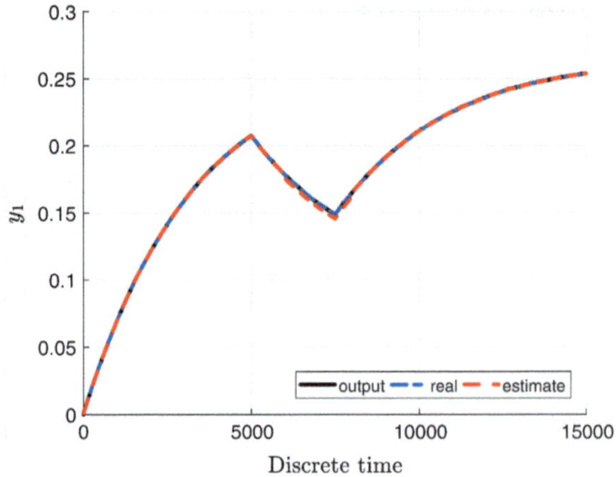

Fig. 11.3 The response of the system—first tank

$$\mathbf{f}_{a,k} = \begin{cases} -0.45 \cdot \mathbf{u}_k, & 5000 \leq k \leq 7500, \\ 0, & \text{otherwise}, \end{cases} \tag{11.111}$$

$$\begin{aligned} \mathbf{f}_{s,1,k} &= 0, \\ \mathbf{f}_{s,2,k} &= \begin{cases} \mathbf{y}_{2,k} + 0.04, & 6000 \leq k \leq 8000, \\ 0, & \text{otherwise}. \end{cases} \\ \mathbf{f}_{s,3,k} &= \begin{cases} \mathbf{y}_{3,k} - 0.1, & 7000 \leq k \leq 9000, \\ 0, & \text{otherwise}. \end{cases} \end{aligned} \tag{11.112}$$

which means that the actuator fault can be regarded as an intermittent one. Note that temporary sensor faults occurred in two of three sensors, and they were partly at the same time. Moreover, the sensor faults appear during the actuator malfunction. This is a realistic situation that might happen in the industrial conditions. Figures 11.3, 11.4, and 11.5 present the response of the states for the first, second, and third tank, respectively, as well as their estimates (red dashed line) and measured by a sensor (black solid line). In these plots, it can be seen that the estimates are following the real states irrespective to the fact of fault occurrence. It can be said that the state estimate is fault resistant.

Figures 11.6, 11.7, and 11.8 present the sensor faults and their estimates. The observer estimates the sensor faults with very high accuracy. Finally, Fig. 11.9 presents the estimation results of actuator fault case.

The results obtained with the described approach clearly show its quality and recommend its straightforward implementation in fault-tolerant control systems.

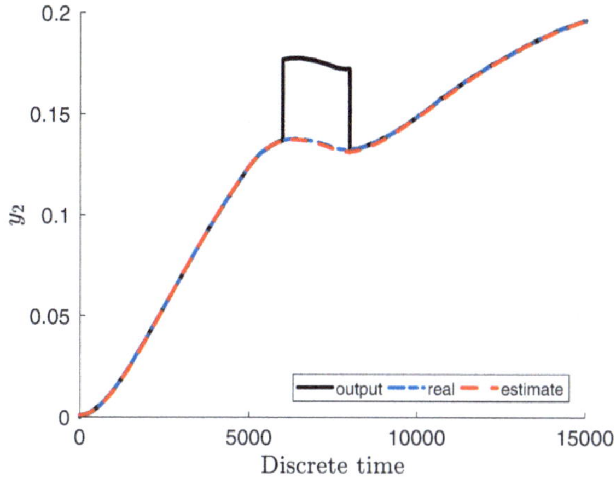

Fig. 11.4 The response of the system—second tank

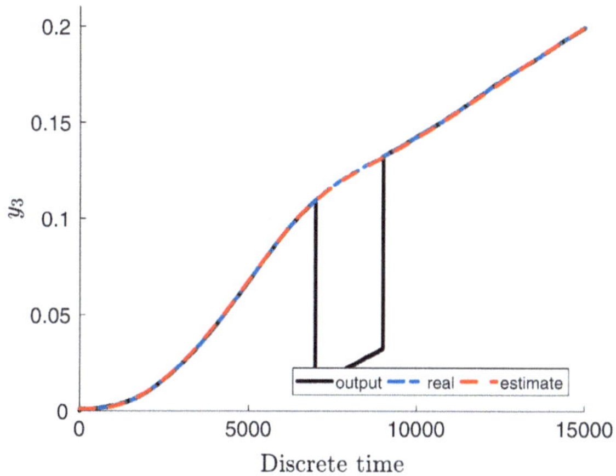

Fig. 11.5 The response of the system—third tank

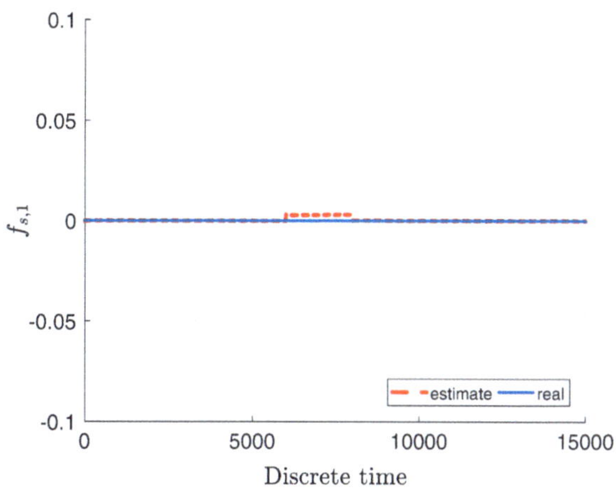

Fig. 11.6 The sensor fault for the first tank and its estimate

Fig. 11.7 The sensor fault for the second tank and its estimate

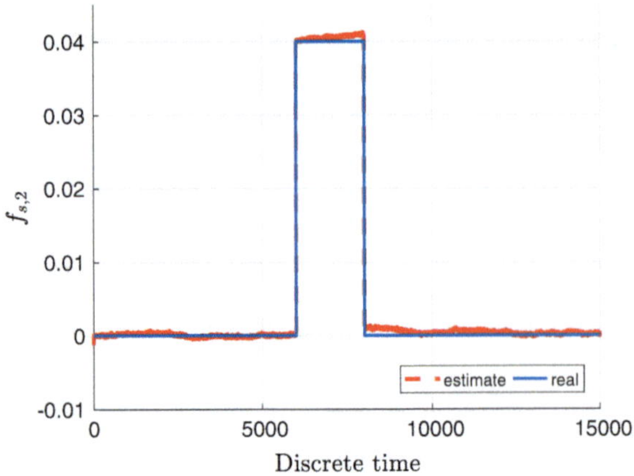

Fig. 11.8 The sensor fault for the third tank and its estimate

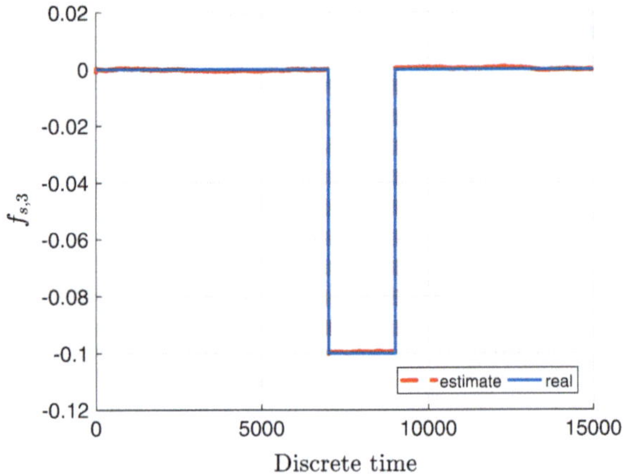

Fig. 11.9 The actuator fault and its estimate

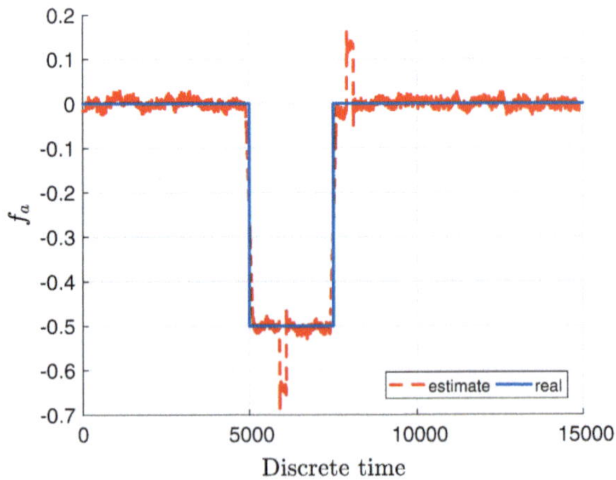

11.6 Conclusions

The objective of this chapter was to review standard estimation strategies for fault detection and isolation of a class of nonlinear systems are briefly discuss their advantages and drawback. In particular, the classical residual-based framework techniques like parameter estimation, parity relation, and observers were briefly reviewed. A particular attention was given to the last group of approaches and a brief nonlinear observer review was provided. The rest of the chapter was devoted to the direct fault estimation strategies. It was split into actuator, actuator and sensor as well as process fault estimation approaches, respectively. Finally, the last part of the chapter presents an illustrative example concerning the selected fault estimation of a three-tank system.

Acknowledgements The work was supported by the National Science Centre of Poland under grant: UMO-2017/27/B/ST7/00620.

References

1. Abbaszadeh, M., Marquez, H.: LMI optimization approach to robust $\mathcal{H}_\infty$ observer design and static output feedback stabilization for non-linear uncertain systems. Int. J. Robust Nonlinear Control **19**(3), 313–340 (2008)
2. Aboky, C., Sallet, G., Vivalda, J.C.: Observers for Lipschitz nonlinear systems. Int. J. Control **75**(3), 204–212 (2002)
3. Abonyi, J., Babuska, R.: Local and global identification and interpretation of parameters in takagi-sugeno fuzzy models. In: The Ninth IEEE International Conference on Fuzzy Systems, 2000. FUZZ IEEE 2000, vol. 2, pp. 835–840 (2000). https://doi.org/10.1109/FUZZY.2000.839140
4. Alcorta Garcia, E., Frank, P.M.: Deterministic nonlinear observer-based approaches to fault diagnosis. Control Eng. Pract. **5**(5), 663–670 (1997)
5. Alessandri, A., Baglietto, M., Battistelli, G.: Design of state estimators for uncertain linear systems using quadratic boundedness. Automatica **42**(3), 497–502 (2006)
6. Alexiev, K., Georgieva, O.: Improved fuzzy clustering for identification of takagi-sugeno model. In: Intelligent Systems, 2004. Proceedings. 2004 2nd International IEEE Conference, vol. 1, pp. 213–218 (2004). https://doi.org/10.1109/IS.2004.1344669
7. Amato, F., Cosentino, C., Mattei, M., Paviglianiti, G.: A mixed direct/functional redundancy scheme for the fdi on a small commercial aircraft. In: IFAC Symposium Fault Detection Supervision and Safety of Technical Processes, SAFEPROCESS, pp. 331–339 (2003)
8. Ashton, S.A., Shields, D.N., Daley, S.: Design of a robust fault detection observer for polynomial non-linearities. In: Proceedings of 14th IFAC World Congress. Beijing, P.R. China (1999). CD-ROM
9. Bhattacharyya, S.P.: Observer design for linear systems with unknown inputs. IEEE Trans. Autom. Control **23**(3), 483–484 (1978)
10. Blanke, M., Kinnaert, M., Lunze, J., Staroswiecki, M.: Diagnosis and Fault-Tolerant Control, 2nd edn. Springer, Berlin (2006)
11. Boutayeb, M., Aubry, D.: A strong tracking extended Kalman observer for nonlinear discrete-time systems. IEEE Trans. Autom. Control **44**(8), 1550–1556 (1999)
12. Busawon, K., Saif, M., Farza, M.: A discrete-time observer for a class of non-linear systems. In: Proceedings of 36th IEEE Conference on Decision and Control, CDC, pp. 4796–4801 (1997)

13. Califano, C., Monaco, S., Normand-Cyrot, D.: On the observer design in discrete-time. Syst. Control Lett. **49**(4), 255–266 (2003)
14. Chadli, M., Karimi, H.: Robust observer design for unknown inputs Takagi-Sugeno models. IEEE Trans. Fuzzy Syst. **21**(1), 158–164 (2013)
15. Chen, J., Patton, R.J.: Robust Model-Based Fault Diagnosis for Dynamic Systems. Kluwer Academic Publishers, London (1999)
16. Chen, J., Patton, R.J.: Robust Model Based Fault Diagnosis for Dynamic Systems. Kluwer Academic Publishers, London (1999)
17. Chen, J., Patton, R.J., Zhang, H.: Design of unknown input observers and fault detection filters. Int. J. Control **63**(1), 85–105 (1996)
18. Ichalal, D., Marx, B., Ragot, J., Maquin, D.: Fault detection, isolation and estimation for Takagi-Sugeno nonlinear systems. J. Franklin Inst. **351**(7), 3651–3676 (2014)
19. Darouach, M., Zasadzinski, M.: Unbiased minimum variance estimation for systems with unknown exogenous inputs. Automatica **33**(4), 717–719 (1997)
20. Deng, S., Yang, L.: Reliable control design of discrete-time takagi-sugeno fuzzy systems with actuator faults. Neurocomputing **173, Part 3**, 1784–1788 (2016). https://doi.org/10.1016/j.neucom.2015.09.053
21. Ding, S.: Model-Based Fault Diagnosis Techniques: Design Schemes. Algorithms and Tools. Springer, Berlin (2008)
22. Gertler, J.: Fault Detection and Diagnosis in Engineering Systems. Marcel Dekker, New York (1998)
23. Gillijns, S., De Moor, B.: Unbiased minimum-variance input and state estimation for linear discrete-time systems. Automatica **43**, 111–116 (2007)
24. Guernez, C., Cassar, J.P., Staroswiecki, M.: Extension of parity space to non-linear polynomial dynamic systems. In: Proceedings of 3rd IFAC Symposium Fault Detection, Supervision and Safety of Technical Processes, SAFEPROCESS'97, vol. 2, pp. 861–866. Hull, UK (1997)
25. Hac, A.: Design of disturbance decoupled observers for bilinear systems. ASME J. Dyn. Syst. Meas. Control **114**, 556–562 (1992)
26. Hammouri, H., Kabore, P., Othman, S., Biston, J.: Observer-based approach to fault detection and isolation for nonlinear systems. IEEE Trans. Autom. Control **44**(10), 1879–1884 (1999)
27. Hammouri, H., Kinnaert, M., Yaagoubi, E.H.E.: Observer-based approach to fault detection and isolation for nonlinear systems. IEEE Trans. Autom. Control **44**, 1879–1884 (1999)
28. Hou, M., Mueller, P.C.: Design of observers for linear systems with unknown inputs. IEEE Trans. Autom. Control **37**(6), 871–875 (1992)
29. Hou, M., Patton, R.: Optimal filtering for systems with unknown inputs. IEEE Trans. Autom. Control **43**(3), 445–449 (1998)
30. Hou, M., Pugh, A.C.: Observing state in bilinear systems: an UIO approach. In: Proceedings of IFAC Symposium Fault Detection, Supervision and Safety of Technical Processes: SAFEPROCESS'97, vol. 2, pp. 783–788. Hull, UK (1997)
31. Hui, S., Zak, S.: Observer design for systems with unknown input. Int. J. Appl. Math. Comput. Sci. **15**(4), 431–446 (2005)
32. Iserman, R.: Fault Diagnosis Systems. An Introduction from Fault Detection to Fault Tolerance. Springer, New York (2006)
33. Isermann, R.: Fault Diagnosis Applications: Model Based Condition Monitoring, Actuators, Drives, Machinery, Plants, Sensors, and Fault-tolerant Systems. Springer, Berlin (2011)
34. Keller, J.Y., Darouach, M.: Two-stage Kalman estimator with unknown exogenous inputs. Automatica **35**(2), 339–342 (1999)
35. Kemir, K., Ben Hmida, F., Ragot, J., Gossa, M.: Novel optimal recursive filter for state and fault estimation of linear systems with unknown disturbances. Int. J. Appl. Math. Comput. Sci. **21**(4), 629–638 (2011)
36. Kinneart, M.: Robust fault detection based on observers for bilinear systems. Automatica **35**(11), 1829–1842 (1999)
37. Kobayashi, N., Nakamizo, R.: An observer design for linear systems with unknown inputs. Int. J. Control **17**(3), 471–479 (1982)

38. Koenig, D., Mammar, S.: Desing of a class of reduced unknown inputs non-linear observer for fault diagnosis. In: Proceedings of American Control Conference, ACC. Arlington, USA (2002)
39. Korbicz, J., Kościelny, J., Kowalczuk, Z., Cholewa, W. (eds.): Fault Diagnosis. Models, Artificial Intelligence, Applications. Springer, Berlin (2004)
40. Kowalczuk, Z., Gunawickrama, K.: Leak detection and isolation for transission pipelines via non-linear state estimation. In: Proceedings of 4th IFAC Symposium Fault Detection, Supervision and Safety of Technical Processes, SAFEPROCESS 2000, vol. 2, pp. 943–948. Budapest, Hungary (2000)
41. Krishnaswami, V., Rizzoni, G.: Non-linear parity equation residual generation for fault detection and isolation. In: Proceedings of 2nd IFAC Symposium Fault Detection, Supervision and Safety of Technical Processes, SAFEPROCESS'94, vol. 1, pp. 317–322. Espoo, Finland (1994)
42. Kroll, A., DÃijrrbaum, A.: On control-specific derivation of affine takagi-sugeno models from physical models: assessment criteria and modeling procedure. In: Computational Intelligence in Control and Automation (CICA), pp. 23–30 (2011). https://doi.org/10.1109/CICA.2011.5945746
43. Li, L., Ding, S.X., Yang, Y., Zhang, Y.: Robust fuzzy observer-based fault detection for nonlinear systems with disturbances. Neurocomputing **174, Part B**, 767–772 (2016). https://doi.org/10.1016/j.neucom.2015.09.102
44. Mattei, M., Gaetano, P., Valerio, S.: Nonlinear observers with $\mathcal{H}_\infty$ performance for sensor fault detection and isolation: a linear matrix inequality design procedure. Control Eng. Pract. **13**, 1271–1281 (2006)
45. Mahmoud, M., Jiang, J., Zhang, Y.: Active Fault Tolerant Control Systems: Stochastic Analysis and Synthesis. Springer, Berlin (2003)
46. Maksarow, D., Norton, J.: State bounding with ellipsoidal set description of the uncertainty. Int. J. Control **65**(5), 847–866 (1996)
47. Marquez, H.: Nonlinear Control Systems. Analysis and Design. Wiley, New Jersey (2003)
48. Milanese, M., Norton, J., Piet-Lahanier, H., Walter, E.: Bounding Approaches to System Identification. Plenum Press, New York (1996)
49. Mrugalski, M.: An unscented Kalman filter in designing dynamic GMDH neural networks for robust fault detection. Int. J. Appl. Math. Comput. Sci. **23**(1), 157–169 (2013)
50. Nobrega, E., Abdalla, M., Grigoriadis, K.: Robust fault estimation of unceratain systems using an LMI-based approach. Int. J. Robust Nonlinear Control **18**(7), 1657–1680 (2008)
51. Noura, H., Theilliol, D., Ponsart, J., Chamseddine, A.: Fault-tolerant Control Systems: Design and Practical Applications. Springer, Berlin (2003)
52. Parisini, T., Polycarpou, M., Sanguineti, M., Vemuri, A.: Robust parametric and non-parametric fault diagnosis in nonlinear input-output systems. In: Proceedings of 36th Conference Decision and Control, pp. 4481–4482. San Diego, USA (1997)
53. Patton, R.J., Frank, P.M., Clark, R.N.: Issues of Fault Diagnosis for Dynamic Systems. Springer, Berlin (2000)
54. Pazera, M., Korbicz, J.: A process fault estimation strategy for non-linear dynamic systems. In: Journal of Physics: Conference Series, vol. 783, p. 012003. IOP Publishing (2017)
55. Pazera, M., Witczak, M., Buciakowski, M., Mrugalski, M.: Simultaneous estimation of multiple actuator anc sensor faults for takagi-sugeno fuzzy systems. In: 2017 22nd International Conference on Methods and Models in Automation and Robotics (MMAR), pp. 939–944. IEEE (2017)
56. Persis, C.D., Isidori, A.: A geometric approach to nonlinear fault detection and isolation. IEEE Trans. Autom. Control **46**(6), 853–865 (2001)
57. Persis, C.D., Santis, R.D., Isidori, A.: An $\mathcal{H}_\infty$-suboptimal fault detection filter for bilinear systems. Proc. Nonlinear Control Year **2000**, 331–339 (2000)
58. Persis, C.D., Santis, R.D., Isidori, A.: Nonlinear actuator fault detection and isolation for a vtol aircraft. In: Proceedings of the 2001 American Control Conference. Accessed from 25–27 June 2001

59. Pertew, A.M., Marquez, H.J., Zhao, Q.: $\mathcal{H}_\infty$ synthesis of unknown input observers for non-linear lipschitz systems. Int. J. Control **78**(15), 1155–1165 (2005)
60. Rajamani, R.: Observers for Lipschitz non-linear systems. IEEE Trans. Autom. Control **43**(3), 397–401 (1998)
61. Rajamani, R., Cho, Y.M.: Existence and design of observers for nonlinear systems: relation to distance to unobservability. Int. J. Control **69**(5), 717–731 (1998)
62. Rotondo, D., Nejjari, F., Puig, V.: A virtual actuator and sensor approach for fault tolerant control of lpv systems. J. Process Control **24**(3), 203–222 (2014)
63. Schreier, G., Ragot, J., Patton, R.J., Frank, P.M.: Observer design for a class of non-linear systems. In: Proceedings of 3rd IFAC Symposium Fault Detection, Supervision and Safety of Technical Processes, SAFEPROCESS'97, vol. 1, pp. 483–488. Hull, UK (1997)
64. Seliger, R., Frank, P.: Robust observer-based fault diagnosis in non-linear uncertain systems. In: Patton, F., Clark (eds.) Issues of Fault Diagnosis for Dynamic Systems. Springer, Berlin (2000)
65. Seron, M.M., Doná, J.A.D.: Robust fault estimation and compensation for LPV systems under actuator and sensor faults. Automatica **52**, 294–301 (2015)
66. Shields, D.N., Ashton, S.: A fault detection observer method for non-linear systems. In: Proceedings of 4th IFAC Symp. Fault Detection, Supervision and Safety of Technical Processes, SAFEPROCESS 2000, vol. 1, pp. 226–231. Budapest, Hungary (2000)
67. Shumsky, A.: Robust residual generation for diagnosis of non-linear systems: parity relation approach. In: Proceedings of 3rd IFAC Symposium Fault Detection, Supervision and Safety of Technical Processes, SAFEPROCESS'97, vol. 2, pp. 867–872. Hull, UK (1997)
68. Stipanovic, D., Siljak, D.: Robust stability and stabilization of discrete-time non-linear: the LMI approach. Int. J. Control **74**(5), 873–879 (2001)
69. Tabatabaeipour, S.M., Bak, T.: Robust observer-based fault estimation and accommodation of discrete-time piecewise linear systems. J. Frankl. Inst. **351**(1), 277–295 (2014)
70. Takagi, T., Sugeno, M.: Fuzzy identification of systems and its applications to modeling and control. IEEE Trans. Syst. Man Cybern. **SMC-15**(1), 116–132 (1985)
71. Thau, F.E.: Observing the state of nonlinear dynamic systems. Int. J. Control **17**(3), 471–479 (1973)
72. Veluvolu, K., Kim, M., Lee, D.: Nonlinear sliding mode high-gain observers for fault estimation. Int. J. Syst. Sci. **42**(7), 1065–1074 (2011)
73. Walker, B., Kuang-Yang, H.: FDI by extended Kalman filter parameter estimation for an industrial actuator benchmark. Control Eng. Pract. **3**(12), 1769–1774 (1995)
74. Walter, E., Pronzato, L.: Identification of Parametric Models from Experimental Data. Springer, London (1996)
75. Wang, S.H., Davison, E.J., Dorato, P.: Observing the states of systems with unmeasurable disturbances. IEEE Trans. Autom. Control **20**(5), 716–717 (1975)
76. Wang, Z., Unbehauen, H.: A class of non-linear observers for discrete-time systems with parametric uncertainty. Int. J. Syst. Sci. **31**(1), 19–26 (2000)
77. Watkins, J., Yurkovich, S.: Set-membership strategies for fault detection and isolation. In: Proceedings of IEEE–SMC Symposium on Modelling, Analysis and Simulation, CESA, pp. 824–830. Lille, France (1996)
78. Witczak, M.: Identification and Fault Detection of Non-linear Dynamic Systems. Lecture Notes in Control and Computer Science, vol. 1. University of Zielona Góra Press, Zielona Góra (2003)
79. Witczak, M.: Modelling and Estimation Strategies for Fault Diagnosis of Non-linear Systems. Springer, Berlin (2007)
80. Witczak, M.: Fault Diagnosis and Fault-Tolerant Control Strategies for Non-Linear Systems: Analytical and Soft Computing approaches. Springer International Publishing, Heidelberg (2014)
81. Witczak, M., Buciakowski, M., Puig, V., Rotondo, D., Nejjari, F.: An lmi approach to robust fault estimation for a class of nonlinear systems. Int. J. Robust Nonlinear Control **26**(7), 1530–1548 (2016)

82. Witczak, M., Obuchowicz, A., Korbicz, J.: Genetic programming based approaches to identification and fault diagnosis of non-linear dynamic systems. Int. J. Control **75**(13), 1012–1031 (2002)
83. Yu, D., Shileds, D.N.: Bilinear fault detection observer and its application to a hydraulic system. Int. J. Control **64**, 1023–1047 (1996)
84. Zemouche, A., Boutayeb, M.: Observer design for Lipschitz non-linear systems: the discrete time case. IEEE Trans. Circuits Syst. - II: Express Briefs **53**(8), 777–781 (2006)
85. Zemouche, A., Boutayeb, M., Iulia Bara, G.: Observer for a class of Lipschitz systems with extension to $\mathcal{H}_\infty$ performance analysis. Syst. Control Lett. **57**(1), 18–27 (2008)
86. Zhang, H.Y., Chan, C.W., Cheung, K.C., J., H.: Nonlinear adaptive observer design based on b-spline neural network. In: 14-th IFAC World Congress, pp. 213–217. Beijing, P.R. China (1999)
87. Zhang, X., Polycarpou, M., Prisini, T.: Fault diagnosis of a class of nonlinear uncertain systems with Lipschitz nonlinearities using adaptive estimation. Automatica **46**(2), 290–299 (2010)
88. Zhao, Y., Tao, J., Shi, N.: A note on observer design for one-sided lipschitz nonlinear systems. Syst. Control Lett. **59**(1), 66–71 (2010)

Chapter 12
Model-Based Diagnosis with Probabilistic Models

Gregory Provan

12.1 Introduction

The literature on stochastic methods for diagnosis is extensive, covering domains such as medicine [59], process control [10], and electronics [12]. The use of uncertainty in modeling and inference arises due to several factors, including model uncertainty, measurement noise, external disturbance, and stochastic outcomes.[1]

Given this uncertainty, several methods have been used to address stochastic behaviors. Two areas have created several stochastic frameworks:

- Control theory has developed different methods for diagnosing dynamical systems. These methods are known as Fault Detection and Isolation (FDI), e.g., banks of Kalman filters [10];
- Artificial Intelligence, which has developed methods based on expert systems and Probabilistic Graphical Models (PGMs) [44], e.g., Bayesian Networks(BNs).

Researchers have used two main techniques for representing uncertainty: (1) adding noise to the process or measurements, e.g., as in Kalman filters; and/or (2) representing transitions in the system dynamics as stochastic, e.g., as in Hidden Markov Models (HMMs) and Dynamic BNs.

This chapter focuses on defining the underlying science for Stochastic Model-Based Diagnosis (SMBD), rather than categorizing the many approaches. We describe the basic issues of model representation and inference. We present the most important Bayesian representations used for designing and diagnosing models with

[1]Given two seemingly identical artifacts, applying the same inputs can result in different outcomes; this is especially true in medicine. Faults may be due to unseen defects that are very difficult to uncover.

G. Provan (✉)
Computer Science Department, University College Cork, Cork, Ireland
e-mail: g.provan@cs.ucc.ie

T. Escobet et al. (eds.), *Fault Diagnosis of Dynamic Systems*,
https://doi.org/10.1007/978-3-030-17728-7_12

inherent uncertainty, together with their associated inference methods. We discuss the strengths and limitations of each approach we introduce.

Since this is a tutorial introduction to probabilistic representations for MBD, we do not cover the details of methods for optimizing inference, such as robust fault detection algorithms [10]. We provide pointers to the literature in the section on related work (Sect. 12.2).

This chapter focuses on probabilistic representations, since they all have a common underlying mathematical basis and computational framework. We exclude non-probabilistic methods for uncertainly, such as rule-based and fuzzy methods. We refer the reader to [21, 46] for more information on fuzzy techniques and [36, 43] for more information on rule-based techniques.

Our contributions are as follows:

- We formulate stochastic MBD (SMBD) as a stochastic filtering task [53].
- We summarize the general task of stochastic filtering, and the use of Factor Graphs [39] as a computational tool for inference.
- We show how well-known FDI techniques, such as HMMs, Kalman Filters, Bayesian Networks, and Particle Filters, are instances of this general approach.
- We illustrate the use of some approaches on the multi-tank benchmark.

12.2 State of the Art

This section provides a brief summary of the state of the art in diagnosis with uncertainty. We refer the reader to [22, 30, 61] for more detailed surveys of FDI, and [56] for a comparison of FDI and AI approaches to diagnosis.

SMBD has been worked on by multiple communities using several approaches. We review some of the most significant approaches here. We divide our summary into two sections: (1) encoding uncertainty directly into models (which is the focus of this work); and (2) using uncertainty to increase inference efficiency in fault isolation.

12.2.1 Stochastic Modeling Approaches

Stochastic FDI approaches: Reference [9] provides an overview of FDI in the presence of uncertainty, and of robust methods. Reference [28] defines an approach for computing parity equations in systems in which modeling uncertainty is represented as multiplicative disturbances, and faults are represented as discrepancies in a set of underlying parameters.

Papers addressing particular modeling assumptions include HMM models [54], hidden semi-Markov models [8], Bayesian multi-models [6], and particle filters for diagnosing nonlinear systems [63].

Applications of SMBD include industrial domain diagnosis using Kalman filters [5], diagnosis of multiphase batch operation [8], and particle filters for real-time diagnostics [58].

Very little research has focused on models in which stochastic transitions are represented as stochastic differential equations, due to the challenges in modeling and inference. One approach [64] develops stochastic differential equation models for power systems diagnosis. Our proposed framework does not cover stochastic differential equation systems that do not make Markov assumptions.

Robust FDI approaches: Given the noise and uncertainty inherent in diagnostic modeling and inference, robust methods have been developed to decouple nonzero residuals arising due to modeling errors/noise and due to faults. These disturbance decoupling methods focus on being able to compute residuals that are sensitive to faults but insensitive to model inaccuracy and noise [10].

Typical models for process and measurement noise (zero mean white noise) cannot capture real systems, which have non-Gaussian, time-varying noise distributions. Instead of making assumptions on stochastic distributions, it is possible to bound the noise and using inference methods such as the set-membership approach [57, 62]. The limitation of this approach is a possibility of higher false negatives, since residuals can be smaller than the residual uncertainty due to model uncertainty.

Reference [9] proposes observer-based methods for solving optimal filtering and robust fault diagnosis problems, given stochastic systems with unknown disturbances. The optimal observer generates disturbance-decoupled state estimates with minimum variance, assuming time-varying systems with both noise and unknown disturbances. Reference [51] proposes robust fault diagnosis algorithms for nonlinear systems based on constraint satisfaction techniques.

Stochastic AI methods: AI researchers have addressed the SMBD task using graphical models [44] in the majority of cases. Many researchers have developed BN structures tailored to static SMBD, such as [37, 50]. Researchers have subsequently developed a tool for building diagnostic BNs [35], which has been commercialized (https://www.bayesfusion.com/). Dynamic models have also been developed, using Dynamic BNs (DBNs), e.g., [38]. DBNs have been applied to many domains, such as CNC machine tools [55] process control systems [41], and applications in dependability, risk analysis, and maintenance [60]. When models are nonlinear, particle filters [24] have been applied. Reference [7] describes the application of exact Bayesian and particle filtering techniques to stochastic hybrid systems.

Medical diagnostics has employed a wide range of AI approaches for modeling [59]. This community has developed graphical models for SMBD, both for static [2, 45] and dynamic approaches [3, 49].

Non-probabilistic approaches: Researchers have developed techniques for models outside of the Bayesian framework we address. For example, diagnosis methods have been analyzed of stochastic Petri nets [1] and stochastic Automata [29, 40].

Fuzzy methods have been developed for MBD, as in [20, 42], and differ from our framework primarily in the use of fuzzy algebras for uncertainty quantification and analysis.

12.2.2 Stochastic Search Approaches

The AI community has adopted a second main approach to using uncertainty for diagnostic inference, employing the uncertainty measure as a search heuristic. This contrasts with the previously mentioned approach that incorporates the uncertainty measure into the underlying representation, graphical models, and then uses message-passing algorithms for stochastic fault isolation. This search-based approach uses probabilities not as an inherent part of the model, but for increasing the efficiency of fault isolation using symbolic models. This work, developed within the MBD framework, typically assumes a symbolic model together with some measure of uncertainty over the fault space. One common approach is to assume that all components fail independently with equal probability, and that components fail with very small probability [33]. This approach has been used successfully in the General Diagnostic Engine (GDE) [19], which uses the ATMS [17], a symbolic inference tool, as a core part of its inference engine. In GDE, probabilistic methods defined a means of focusing search diagnostic inference on the most likely diagnoses [18, 33], thereby reducing inference complexity significantly.

The use of probabilistic methods for focusing has been formalized in [34], which showed that probabilistic methods for focusing search in the ATMS are an instance of a general algebraic approach to inference based on semirings. Reference [26] describes methods to add uncertainty beyond the standard failure independence assumptions.

An alternative framework to the ATMS is Disjunction Negation Normal Form (DNNF) [16], which has been used with qualitative probabilities for diagnostics inference [15]. The DNNF approach is similar to the ATMS approach in that both methods "compile" the symbolic model into a representation that speeds up diagnostic inference (at the cost of space); both methods use probabilistic methods to reduce the space of diagnostic hypotheses considered during fault isolation.

More recently, stochastic randomization has been applied to symbolic diagnostic inference. Reference [25] describes an algorithm for randomized search over a symbolic set of diagnoses that sacrifices completeness for computational efficiency.

From the surface, it appears as if these methods for using probabilities for search rather than as a modeling formalism lead to technically different algorithms. However, from a theoretical standpoint, both approaches are instances of using semiring algebras for inference [34]. Reference [47] has characterized how using probabilities both for as a model specification and as a means of inference are instances of semiring algebras, and both approaches can employ message-passing inference (based on semirings) for fault isolation.

Table 12.1 Notation used in the chapter

$\mathbf{x}_{1:k}$	$\{\mathbf{x}_1, \mathbf{x}_2, \ldots, \mathbf{x}_k\}$ vector of states over time
$\mathbf{y}_{1:k}$	$\{\mathbf{y}_1, \mathbf{y}_2, \ldots, \mathbf{y}_k\}$ vector of observations over time
$\phi_{1:k}$	$\{\phi_1, \phi_2, \phi_3, \ldots, \phi_k\}$ vector of modes over time
$\mathbf{u}_{1:k}$	$\{\mathbf{u}_1, \mathbf{u}_2, \ldots \mathbf{u}_k\}$ vector of control inputs over time
v_n	Process noise
w_n	Measurement noise

12.3 Problem Statement

This section places the SMBD problem within the context of stochastic (Bayesian) filtering [53]. We introduce our notation in Sect. 12.3.1, and then define the task of stochastic filtering in Sect. 12.3.2. We show how SMBD is an instance of stochastic filtering in Sect. 12.3.4.

12.3.1 Notation

We use the following notation throughout the chapter. We first introduce a nonlinear system with process and observation noise (Table 12.1).

Definition 12.1 (*Nonlinear System* (*continuous-time*)) We model a system Ψ using a set of N interacting variables that define a set of stochastic relations of the form

$$\dot{\mathbf{x}}_t = f(\mathbf{x}_t, \mathbf{u}_t, \phi_t, v_t), \tag{12.1}$$

$$\mathbf{y}_t = g(\mathbf{x}_t, \mathbf{u}_t, \phi_t, w_t), \tag{12.2}$$

where $\mathbf{x}_t$ is the state variable at time t, $\dot{\mathbf{x}}_t$ is the derivative of $\mathbf{x}_t$ with respect to t, $\mathbf{u}_t$ is the vector of inputs (controls), ϕ is the set of modes, θ is the vector of model parameters, $\mathbf{y}_t$ is the output vector, and v_t and w_t are process and observation noise, respectively.

For computational reasons, many approaches focus on filtering models with a discrete-time temporal representation, given by

Definition 12.2 (*Nonlinear System* (*discrete-time*))

$$\mathbf{x}_{k+1} = f(\mathbf{x}_k, \mathbf{u}_k, \phi_k, v_k), \tag{12.3}$$

$$\mathbf{y}_k = g(\mathbf{x}_k, \mathbf{u}_k, \phi_k, w_k), \tag{12.4}$$

where v_k and w_k can be viewed as white noise random sequences with unknown statistics in the discrete-time domain.

We can characterize Eqs. 12.3 and 12.4 using a stochastic representation in which we explicitly define the uncertainty in state transitions and in measurements. To do this, we rewrite Eqs. 12.3 and 12.4 into conditional probability equations as follows:

Definition 12.3 (*Stochastic System (discrete-time)*)

$$\mathbf{x}_{k+1} : P(\mathbf{x}_{k+1}|\mathbf{x}_k, \mathbf{u}_k, \phi_k), \tag{12.5}$$

$$\mathbf{y}_k : P(\mathbf{y}_k|\mathbf{x}_k, \mathbf{u}_k, \phi_k). \tag{12.6}$$

Equation 12.5 characterizes the state transition probability, and Eq. 12.6 describes the stochastic measurement model. We explicitly characterize the noise models v_k and w_k in terms of the state and measurement distributions.

12.3.2 Stochastic Filtering

Given the model described by Eqs. 12.5 and 12.6, our stochastic diagnosis problem consists of computing the health state $P(\phi_k|\mathbf{y}_{1:k})$ given that we have observed $\mathbf{y}_{1:k}$. To perform this inference, we use Bayes' rule, which yields

$$P(\phi_k|\mathbf{y}_{1:k}) = \frac{P(\phi_k, \mathbf{y}_{1:k})}{P(\mathbf{y}_{1:k})} = \frac{P(\mathbf{y}_{1:k}|\phi_k)P(\phi_k)}{P(\mathbf{y}_{1:k})}. \tag{12.7}$$

We now describe this in terms of *stochastic filtering*, which solves a statistical inversion problem. We estimate an unknown vector-valued time series $\phi_{1:k} \equiv \{\phi_1, \phi_2, \phi_3, \ldots\}$ which is observed through a set of noisy measurements $\mathbf{y}_{1:k} \equiv \{\mathbf{y}_1, \mathbf{y}_2, \ldots\}$. In the following, we simplify our notation for the purposes of exposition, suppressing $\mathbf{u}$, v_n and w_n.

Definition 12.4 (*Statistical inversion problem*) Statistical inversion estimates the hidden states $\phi_{1:k}$ from the observed measurements $\mathbf{y}_{1:k}$. To do this, we use Bayes' rule to compute the joint posterior distribution of the states given the measurements as

$$P(\phi_{1:k}|\mathbf{y}_{1:k}) \propto \frac{P(\phi_{1:k})P(\mathbf{y}_{1:k}|\phi_{1:k})}{P(\mathbf{y}_{1:k})}., \tag{12.8}$$

where

- $P(\phi_{1:k})$ is the prior distribution defined by the dynamic model,
- $P(\mathbf{y}_{1:k}|\phi_{1:k})$ is the likelihood model for the measurements,
- $P(\mathbf{y}_{1:k})$ is the normalization constant defined as

$$P(\mathbf{y}_{1:k}) = \int P(\mathbf{y}_{1:k}|\phi_{1:k})P(\phi_{1:k})d\phi_{1:k}. \tag{12.9}$$

Although the posterior density provides a complete solution of the stochastic filtering problem, the problem still remains intractable since the density is a function rather than a finite-dimensional point estimate.

Unfortunately, computing the full posterior formulation is intractable, and it must be recomputed each time we obtain a new measurement. This is further complicated by the fact that with each new time step, the computational complexity of that step increases due to the increased dimensionality of the full posterior distribution. Hence, additional information or restrictive approximations on the full posterior computation (or other modeling assumptions) are necessary to achieve computational tractability.

12.3.3 Recursive Filtering

For problems, e.g., control of dynamical systems, where we make observations at every time step, recursive filtering enables us to monitor the system for faults recursively over time using the incoming measurements and the process model. Recursive methods consist of two parts: prediction and update [53].

- *Prediction* consists of estimating the current state given the previous measurements, i.e., $p(\mathbf{x}_k|\mathbf{y}_{1:k-1})$. We compute this using

$$p(\mathbf{x}_k|\mathbf{y}_{1:k-1}) = \int p(\mathbf{x}_k|\mathbf{x}_{k-1})p(\mathbf{x}_{k-1}|\mathbf{y}_{1:k-1})\, d\mathbf{x}_{k-1}. \tag{12.10}$$

- *Update* consists of computing

$$p(\mathbf{x}_k|\mathbf{y}_{1:k}) = \frac{p(\mathbf{y}_k|\mathbf{x}_k)p(\mathbf{x}_k|\mathbf{y}_{1:k-1})}{p(\mathbf{y}_k|\mathbf{y}_{1:k-1})} \propto p(\mathbf{y}_k|\mathbf{x}_k)p(\mathbf{x}_k|\mathbf{y}_{1:k-1}). \tag{12.11}$$

  The denominator is given by

$$p(\mathbf{y}_k|\mathbf{y}_{1:k-1}) = \int p(\mathbf{y}_k|\mathbf{x}_k)p(\mathbf{x}_k|\mathbf{y}_{1:k-1})d\mathbf{x}_k. \tag{12.12}$$

  Equation 12.12 is constant relative to x, so we can substitute it for a coefficient α; in practice, we compute the numerator and normalize it since its integral must be unity.

12.3.4 Stochastic Filtering for Diagnostics

Diagnosis is an instance of general stochastic filtering where we aim to compute the health state (mode) ϕ_t given a sequence of observations $\mathbf{y}_{1:k}$, rather than a generic state $\mathbf{x}_t$. A mode ϕ consists of a collection of states, i.e., $\phi \subseteq \mathcal{X}$. We assume that the set of modes $\Phi = \{\phi_1, \ldots, \phi_q\}$ partition the state space, as shown in Fig. 12.1.

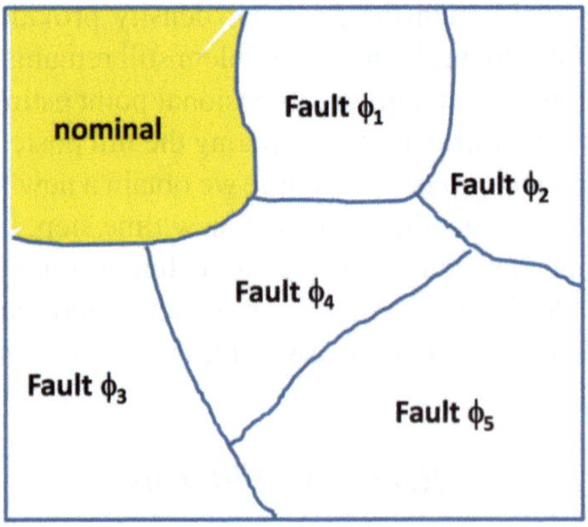

Fig. 12.1 Depiction of partitioning of 2D state space $\mathcal{X}$ by modes

In the following, we assume that we will estimate fault-mode ϕ. Our SMBD task is to estimate diagnosis $\phi_{1:k}$ based on measurements $\mathbf{y}_{1:k}$ and controls $\mathbf{u}_{1:k}$ (if applicable).

We can specify two computational diagnostic tasks within this framework:

- *Diagnosis*: a diagnosis is given by $P(\phi_k|\mathbf{y}_{1:k})$ (Eq. 12.8), and can be computed by a filtering algorithm in the Bayes filtering framework.
- *Most Likely Explanation (MLE)*: the MLE ϕ^* corresponds to the most likely sequence of fault modes, denoted as

$$\phi^* = \arg\max_{\phi_{1:k}} P(\phi_{1:k}|\mathbf{y}_{1:k}),$$

  and can be computed by the Viterbi algorithm in the Bayes filtering framework [53].

Additionally, we can define a *prognostic task* as that of forecasting the future health of the system. We can write this as $P(\phi_{k+q}|\mathbf{y}_{1:k})$, for $q \in \mathbb{Z}^+$, and can be computed by a prediction algorithm in the Bayes filtering framework.

12.4 Proposed Approach

This section describes how we can use stochastic methods for diagnosis. We describe the factor graph representation in Sect. 12.4.1, and then focus on the inference task (based on factor graphs) in Sect. 12.4.2.

12.4.1 Factor Graph Representation

We can define all the SMBD models considered in this chapter as Probabilistic Graphical Models (PGMs) [44]. We will use the factor graph approach [39], since

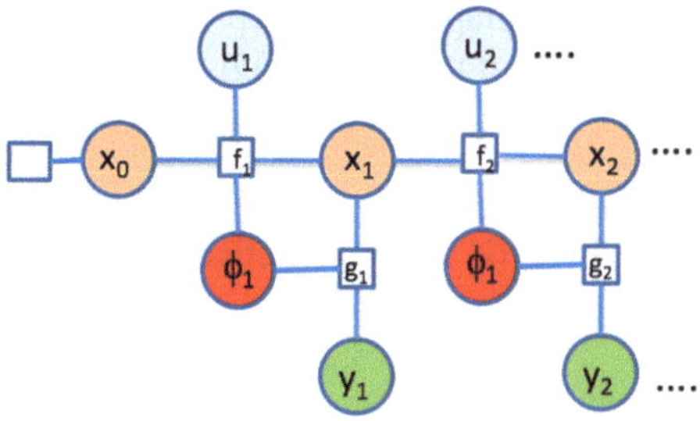

Fig. 12.2 Factor graph for discrete-time state space model for $k = 1,\ 2$

it provides both a framework for representing models that specify probability distributions and a computational framework for performing inference on these models.

A *Factor Graph* (FG) is a graphical model that represents factorizations of arbitrary multivariate functions Ξ and dependencies among variables $\mathbf{Z}$ of a model Ψ. We assume that we have a function Ξ that we can represent as a collection of subfunctions. We illustrate a factor graph using instantiations for discrete-time state equations for $k = 1,\ 2$:

$$\begin{aligned}\mathbf{x}_k &= f(\mathbf{x}_{k-1}, \mathbf{u}_k, \phi_k),\\ \mathbf{y}_k &= g(\mathbf{x}_k, \phi_k).\end{aligned}$$

For $k = 1$, we can rewrite our state equations in an alternative form:

$$\begin{aligned}\mathbf{x}_1 &= f(\mathbf{x}_0, \mathbf{u}_1, \phi_1) \simeq f_1(\mathbf{x}_1, \mathbf{x}_0, \mathbf{u}_0, \phi_0)\\ \mathbf{y}_1 &= g(\mathbf{x}_1, \phi_1) \simeq g_1(\mathbf{y}_1, \mathbf{x}_1, \phi_1).\end{aligned}$$

Analogously, for $k = 2$, we can write the equations as $f_2(\mathbf{x}_2, \mathbf{x}_1, \mathbf{u}_2, \phi_2)$ and $g_2(\mathbf{y}_2, \mathbf{x}_2, \phi_2)$. As a consequence, we can write the system equations Ξ for $k = 0,\ 1$ as the composition of f_1, f_2, g_1, g_2, i.e., $\Xi = \bigotimes_{k=1}^{2} f_i \otimes g_i$, where $\otimes$ denotes function composition. Figure 12.2 depicts the factor graph for Ξ. We represent variables by circular nodes and functions by square nodes. An edge joins variables that participate in particular functions, e.g., $\mathbf{y}_2, \mathbf{x}_2, \phi_2$ participate in function g_2.

We now formally define a factor graph. We first define the neighbors $\mathbf{Nb}_{\mathcal{G}}(\xi)$ of a factor node ξ in a factor graph $\mathcal{G}$ as the variables connected to ξ in $\mathcal{G}$. We can view an FG as specifying an abstract functional decomposition, as well as a factorization of the joint probability $P(\mathbf{Z})$. Both definitions are provided below.

Definition 12.5 (*Factor Graph* (*Function Decomposition*)) Given a decomposable function Ξ over variables $\mathbf{Z}$, a FG is a bipartite graph $\mathcal{G} = (\mathbf{N}, \mathbf{E})$ with a node for each variable and factor in the model. Every factor $\xi \in \Xi$ corresponds to a set of edges in $\mathbf{E}$, where an edge connects ξ to its neighboring variables $\mathbf{Nb}_{\mathcal{G}}(\xi)$:

$$\Xi = \bigotimes_{\xi \in \Xi} \xi(\mathbf{Nb}_{\mathcal{G}}(\xi)).$$

A factor graph can also represent the joint probability $P(\mathbf{Z})$, as defined in terms of the product of the factors [39]. In this case, the factor ξ is a probability distribution, and we need to use a normalization constant.

Definition 12.6 (*Factor Graph* (*Probability Decomposition*)) A factor graph is a tuple $\mathcal{F} = (\mathcal{G}, \Xi)$, and it defines the joint probability distribution

$$P_{\mathcal{F}}(\mathbf{Z}) = \frac{1}{\Upsilon} \prod_{\xi \in \Xi} P(\mathbf{Nb}_{\mathcal{G}}(\xi)), \tag{12.13}$$

where Υ is the normalization constant given as

$$\Upsilon = \sum_{z \in \mathbf{val}(\mathbf{Z})} \prod_{\xi \in \Xi} P(\mathbf{Nb}_{\mathcal{G}}(\xi)), \tag{12.14}$$

where random variable Z takes a set of values denoted as $val(Z)$.

Inference in FGs takes place through message-passing, with forward and backward messages transmitted through the FG in a precise sequence [39]. This inference framework enables us to perform all necessary classes of stochastic filtering for diagnostics. Reference [48] describes FG templates for diagnostics and the computations necessary for diagnostics inference.

12.4.2 Examples of Models and Inference

Table 12.2 compares several SMBD models, ranging from simple linear (HMM) to nonlinear (Particle Filter). The first two approaches, HMM and KF, employ a first-order Markov assumption and use a tree-structured graphical model. The main difference between the two models is that an HMM typically assumes discrete-valued variables with a model specified by a Markov system, whereas the KF assumes a Linear-Gaussian Model, i.e., continuous-valued variables constrained by a Linear Time-Invariant (LTI) model with Gaussian noise on the model and the observations. Both approaches have poly-time complexity due to the restriction to tree-structured graphical models.

The next two models, BN and DBN, generalize the tree-structured graphical model to a DAG, which causes inference complexity to become NP-complete. The BN assumes atemporal behavior, whereas the DBN allows system dynamics. The Particle Filter is the only model that allows nonlinear systems with arbitrary topologies, and it adopts a sample-based inference approach to approximate nonlinear systems.

This SMDB task is computationally infeasible for arbitrary system topologies: it is NP-hard in the *atemporal* case for arbitrary network topologies [13], and the task remains NP-hard even for approximate inference [14]. When we include time, the task becomes even more challenging: with additional time steps, the dimensionality

Table 12.2 Key properties of different model representations, given n state variables

Method	Uncertainty	Distribution	Structure	Equations	Inference complexity
HMM	Process	Discrete/continuous	Tree	–	$O(n)$
Kalman filter	Noise	Gaussian	Tree	Linear-Gaussian Model	$O(n^3)$
BN	Process	Discrete/continuous	DAG	–	NP-hard
DBN	Process	Discrete/continuous	DAG	–	NP-hard
Particle filter	Process	Continuous	Arbitrary	Linear or nonlinear	NP-hard

of the full posterior distribution also increases, leading to increased computational complexity of each time step.

To overcome these computational limitations, a variety of restrictions are typically enforced. We will examine the most common restricted problem formulations that have poly-time complexity, which include restricted topologies, i.e., trees, and linear Gaussian models, resulting in the Hidden Markov Model (HMM) and Kalman Filter (KF) models. We also examine more complex models, namely the Bayesian network and its temporal extension, the Dynamic Bayesian network,as well as a method for nonlinear models, namely the Particle Filter.

12.4.2.1 Hidden Markov Model (HMM)

A Hidden Markov Model (HMM) is a finite-state Markov model that has been used widely for diagnosis. Figure 12.3 depicts the PGM of an HMM, showing an example of the structure underlying an HMM. In this figure, the factor nodes explicitly show the conditional probabilities. At each time step, an HMM has a hidden (fault) node (which is discrete-valued) and an observation node (which is discrete- or continuous-valued). The HMM has a tree-structured graphical model, leading to efficient diagnostic inference. We can represent the joint distribution for an HMM using

$$P(\mathbf{x}_{0:k}, \mathbf{y}_{1:k}) = P(\mathbf{x}_0) \prod_{i=1}^{k} P(\mathbf{x}_i | \mathbf{x}_{i-1}) P(\mathbf{y}_i | \mathbf{x}_i). \tag{12.15}$$

To use an HMM for diagnostic purposes, we need to define the hidden state variable $\mathbf{x}_k$ as the set of failure modes ϕ_k, and then we can compute the most likely failure state using

$$\phi^* = \arg\max_{\phi} P(\phi_n, \mathbf{y}_n | \Theta). \tag{12.16}$$

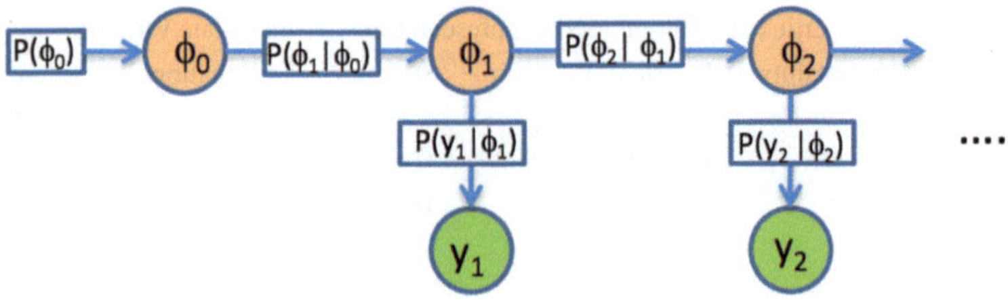

Fig. 12.3 HMM structure for stochastic diagnosis. We explicitly represent the conditional probabilities that define the factors

To perform this computation, we must estimate distributions for $P(\mathbf{x}_t|\mathbf{x}_{t-1})$ and $P(\mathbf{y}_t|\mathbf{x}_t)$, for $t = 1, \ldots, n$, using an algorithm like the Viterbi algorithm [53].

12.4.2.2 Linear-Gaussian Model/Kalman Filter

This section describes a restriction to a linear dynamic system with Gaussian noise on process and observations, and its well-known filter, the Kalman Filter (KF) [31]. A Linear-Gaussian Model (LGM) is a model that has topology identical to that of the HMM, except that each node has a linear-Gaussian distribution. The KF defines a method of performing inference in this model.

Definition 12.7 (*Discrete-Time Linear-Gaussian System*)

$$\mathbf{x}_{k+1} = \mathbf{A}\mathbf{x}_k + \mathbf{B}\mathbf{u}_k + \mathbf{v}_k + \Lambda\phi_t, \tag{12.17}$$

$$\mathbf{y}_k = \mathbf{C}\mathbf{x}_k + \mathbf{w}_k, \tag{12.18}$$

where $\mathbf{x}_k$ is an n-dimensional vector, $\mathbf{A}$, $\mathbf{B}$ and $\mathbf{C}$ are matrices with suitable dimensions, $\mathbf{x}_0$ has mean $\bar{\mathbf{x}}_0$ and covariance Σ_0, $\mathbf{d}(t)$ and $\mathbf{v}(t)$ are Gaussian white noise sequences, and we have additive faults with matrix Λ.

The Kalman Filter is a recursive algorithm for state estimation that has been used throughout a range of applications, particularly in process control and other control domains [11]. For a linear stochastic system in the input/output model, the Kalman filter is optimal [32], i.e., it is equivalent to an optimal predictor.

The Kalman Filter has an elegant Bayesian interpretation [11]: it reduces to a maximum a posteriori (MAP) solution [52] that maximizes $P(x_k, \phi_k|y_k, u_k)$ (or equivalently $log P(x_k|y_k)$). For diagnostics, it solves our direct-estimation SMBD task exactly for Linear-Gaussian Models.

Several multistep approaches have applied Kalman Filters to diagnosing linear-Gaussian dynamic systems; see, e.g., [27]. Diagnostic inference entails designing a state estimator with minimum estimation error. We then monitor the innovation (measurement residual) process or the prediction errors. If we design the filter based on an observer with Kalman gain matrix $\mathbf{K}$, we obtain the equations shown in Eq. 12.19. In this case, we compare the measured value $\mathbf{y}_k$ and the computed value $\hat{\mathbf{y}}_k$.

Definition 12.8 (*Kalman Filter with Observer*)

$$\begin{aligned}\mathbf{x}_{k+1} &= \mathbf{A}\mathbf{x}_k + \mathbf{K}(\mathbf{y}_k - \mathbf{C}\mathbf{x}_k) + \mathbf{B}\mathbf{u}_k + \mathbf{v}_k + \Lambda\phi_t,\\ \mathbf{y}_k &= \mathbf{C}\mathbf{x}_k + \mathbf{w}_k,\end{aligned} \tag{12.19}$$

where $\mathbf{K}$ is the Kalman gain matrix.

12.4.2.3 Bayesian Network

A Bayesian Network $\mathcal{B}$ is a static PGM in which the structure is given by of a directed acyclic graph (DAG), rather than the tree structure of the HMM or LGM. Whereas an HMM or LGM effectively denotes two variable types (fault or observation), a BN allows arbitrary variable types. Hence, we can model many different latent state variables in order to specify a model in more detail.

We define the distribution for the Bayesian Network (BN) using the set $\mathcal{Z}$ of variables to denote the BN variables, which comprise state and observation variables. The BN structure enables us to specify the required distribution for the model, i.e.,

$$P(\mathcal{Z}) = \prod_{z \in \mathcal{Z}} P(z|\pi(z)), \tag{12.20}$$

where $\pi(z)$ are the parents of z in the structure graph G.

12.4.2.4 Dynamic Bayesian Network

A discrete-time Dynamic Bayesian Network (DBN) [44] is a temporally generalized version of a BN in which a BN fragment is replicated over several discrete-valued time steps. The DBN models probability distributions over semi-infinite collections of random variables and can encode arbitrary distributions and probabilistic factorizations, given a DAG structure.

Definition 12.9 (*Dynamic Bayesian Network*) A Dynamic Bayesian Network (DBN) is a temporal structure (over discrete time steps $0, 1, \ldots, k$) in which we have

- a BN $\mathcal{B}_i$ for time step $i = 0, 1, \ldots, k$;
- an initialization (prior) at $k = 0$, i.e., over $\mathcal{B}_0$;
- a transition model between time slices, i.e., $P(\mathcal{B}_k|\mathcal{B}_{k-1}, \ldots, \mathcal{B}_0)$.

If we adopt a Markov assumption, then the transition model is given by $P(\mathcal{B}_k|\mathcal{B}_{k-1})$.

12.4.2.5 Nonlinear Model/Particle Filter

The Particle Filter (PF) is a Bayesian filter for nonlinear systems. Compared to the Kalman filter (KF) and its extensions (e.g., extended or unscented KF), a PF can provide more accurate state estimation results in many cases, such as for systems with strong nonlinearities and/or non-Gaussian distributions.

For a nonlinear system, computing the analytic solution for the posterior distribution $P(\mathbf{x}_{0:k}|\mathbf{y}_{1:k})$ is often difficult. Instead of analytically calculating $P(\mathbf{x}_{0:k}|\mathbf{y}_{1:k})$, a PF approximates it with a set of N particles $\mathbf{x}^i_{0:k}$, $i = 1, \ldots, N$, where the initial particles, i.e., $\mathbf{x}^i_0$, are drawn from $P(\mathbf{x}_0)$.

A well-known sampling method, importance sampling, draws particles using an importance distribution given by

$$q(\mathbf{x}_{0:k}|\mathbf{y}_{1:k}) = q(\mathbf{x}_{0:k-1}|\mathbf{y}_{1:k-1})q(\mathbf{x}_k|\mathbf{x}_{0:k-1}, \mathbf{y}_{1:k}). \tag{12.21}$$

Using Bayes' rule, we can compute the importance weight for each particle in a recursive form, i.e.,

$$w^i_k = \frac{P(\mathbf{x}_{0:k}|\mathbf{y}_{1:k})}{q(\mathbf{x}_{0:k}|\mathbf{y}_{1:k})} \tag{12.22}$$

$$\propto w^i_{k-1} \frac{p\left(\mathbf{y}_k|\mathbf{x}^i_k\right) p\left(\mathbf{x}^i_k|\mathbf{x}^i_{k-1}\right)}{q\left(\mathbf{x}^i_k|\mathbf{x}^i_{0:k-1}, \mathbf{y}_{1:k}\right)}, \tag{12.23}$$

where w^i_k is the importance weight. We denote the normalized w^i_k as $\tilde{w}^i_k$. Thus, we can approximate the posterior distribution $P(\mathbf{x}_{0:k}|\mathbf{y}_{1:k})$ by particles such that we obtain

$$P(\mathbf{x}_{0:k}|\mathbf{y}_{1:k}) \approx \sum_{i=1}^{N} \tilde{w}^i_k \delta\left(\mathbf{x}_{0:k} - \mathbf{x}^i_{0:k}\right),$$

where $\delta(\cdot)$ is the Dirac delta function.

If the importance distribution satisfies

$$q(\mathbf{x}_k|\mathbf{x}_{0:k-1}, \mathbf{y}_{1:k}) = q(\mathbf{x}_k|\mathbf{x}_{k-1}, \mathbf{y}_k),$$

then

$$\tilde{w}^i_k \propto \tilde{w}^i_{k-1} \frac{p\left(\mathbf{y}_k|\mathbf{x}^i_k\right) p\left(\mathbf{x}^i_k|\mathbf{x}^i_{k-1}\right)}{q\left(\mathbf{x}^i_k|\mathbf{x}^i_{k-1}, \mathbf{y}_k\right)}$$

and thus $P(\mathbf{x}_k|\mathbf{y}_{1:k})$ can be approximated such that

$$P(\mathbf{x}_k|\mathbf{y}_{1:k}) \approx \sum_{i=1}^{N} \tilde{w}^i_k \delta\left(\mathbf{x}_k - \mathbf{x}^i_k\right).$$

Using this approximation of $P(\mathbf{x}_k|\mathbf{y}_{1:k})$, we can compute the states $\mathbf{x}_k$ using methods such as Maximum Mean-Squared Error (MMSE) or maximum *a posteriori* (MAP).

The proposed Sequential Importance Sampling (SIS) approach suffers from an unavoidable problem, particle degeneracy, where the weights of all but one particle become very small after several iterations [23]. We can monitor the effective sample size from the renormalized weights using N_{eff} [4], which is given by

$$N_{\text{eff}} = \frac{1}{\sum_{i=1}^{N} \left(\tilde{w}_k^i\right)^2}.$$

The smaller the effective sample size, the greater the degree of degeneracy. We combat the particle degeneracy problem by using a resampling procedure, in which we eliminate the particles with small weights after resampling, and create many offspring for the particles with large weights. We thus approximate the posterior distribution $P(\mathbf{x}_k|\mathbf{y}_{1:k})$ using these offspring particles as

$$P(\mathbf{x}_k|\mathbf{y}_{1:k}) \approx \frac{1}{N} \sum_{i=1}^{N} N_k^i \delta\left(\mathbf{x}_k - \mathbf{x}_k^i\right),$$

where N_k^i is the number of offspring for the parent particle $\mathbf{x}_k^i$.

12.5 Simulations and Results

This section compares and contrasts several SMBD representations using the multi-tank model as the application system. We assume that our task is to compute $P(\phi_t|\mathbf{y}_{1:k})$ for dynamical systems or $P(\phi|\mathbf{y})$ for static systems. In the following examples, we assume that faults are permanent: once a fault occurs the system will not revert to a nominal state. Table 12.3 defines the notation for our models.

12.5.1 Comparison of Approaches

In the following, we focus on examples of pipe blockage as faults. We create all models (structure and distributions) by hand; it is possible to manually create model

Table 12.3 Notation used for stochastic models of three-tank system; we suppress temporal indices for ease of notation

Variable	Notation	Description
$\mathbf{x}$	$\{h_1, h_2, h_3\}$	Tank heights
$\mathbf{y}$	$\{y_1, y_2, y_3\}$	$(h_{T_{1,m}}, q_{23,m}, h_{T_{3,m}})$
$\mathbf{u}$	$\{q_i\}$	T_1 inflow
ϕ	ϕ^N,	Nominal mode
	$\phi^{F_1}, \phi^{F_2}, \phi^{F_3}$,	Flow blockage from tank $i = 1, 2, 3$
	$\phi^{T_1}, \phi^{T_2}, \phi^{T_3}$	Tank leakage in tank $i = 1, 2, 3$

structure and then estimate distributions from data, or to learn the structure and distributions from data. The worked examples described here can be implemented in any of the popular tools for Bayesian Networks, e.g., Genie/SMILE (https://www.bayesfusion.com), Hugin (https://www.hugin.com) or Netica (https://www.norsys.com). For applications that use Kalman filters or particle filters, a Factor Graph tool is more suited than a Bayesian Network; examples include OpenGM2 (https://hci.iwr.uni-heidelberg.de/opengm2) or the tool FACTORIE (http://factorie.cs.umass.edu).

12.5.1.1 Hidden Markov Model

Figure 12.3 depicts the structure of the HMM that we build. To create an HMM for the multi-tank model, we need to represent two distributions for each time step k: (1) fault state, $P(\phi_k|\phi_{k-1})$; and (2) observation state, $P(\mathbf{y}_k|\phi_k)$, for $k = 1, \ldots, N$. We first define our states, and then the distributions over the state transitions.

Fault state: An HMM defines the state ϕ_k using discrete-valued variables, so we define for ϕ_k a multivariate distribution over $\phi^N, \ \phi^{F_1}, \ \phi^{F_2}, \ \phi^{F_3}$, corresponding to nominal and leak faults in $T_1, \ T_2, \ T_3$, respectively, where each fault can take values of $\{true, false\}$.

Observation state: Here, we must define a distribution over the observation, conditioned on the fault state. We assume each observation is discrete-valued, i.e., each observation takes values $\{low, nominal, high\}$.

To use an HMM, we assume that we will have the initial state $\mathbf{x}_0$, and will then estimate the distributions $P(\phi_k|\phi_{k-1})$ and $P(\mathbf{y}_k|\phi_k)$, for $t = 1, \ldots, N$; if we assume that the distributions do not change over time, then we just need to estimate a single distribution for each of $P(\phi_k|\phi_{k-1})$ and $P(\mathbf{y}_k|\phi_k)$.

Given observations $\mathbf{y}_t$, for $t = 1, 2, \ldots$, we can then solve this system using the forward algorithm [44]. This model can compute a distribution over ϕ_t as its output.

We specify an HMM where both ϕ and $\mathbf{y}$ are discrete-valued. We assume that $y_1, \ y_3$ take values from the set $\{L, N\}$ denoting low and nominal, respectively, and y_2 takes values from the set $\{l, h\}$ denoting low and high, respectively. The measurement $\mathbf{y}$ can then take on one of 2^3 possible discrete values.

If we restrict our analysis to leaks in the tanks, our fault vector at step k is $\phi_k = \{\phi_k^N, \phi_k^{T_1}, \phi_k^{T_2}, \phi_k^{T_3}\}$. The resulting transition matrix for $P(\phi_k|\phi_{k-1})$ is

$$
\begin{array}{c c}
 & \begin{array}{cccc} \phi_{k-1}^N & \phi_{k-1}^{T_1} & \phi_{k-1}^{T_2} & \phi_{k-1}^{T_3} \end{array} \\
\begin{array}{c} \phi_k^N \\ \phi_k^{T_1} \\ \phi_k^{T_2} \\ \phi_k^{T_3} \end{array} &
\begin{bmatrix} 0.97 & 0 & 0 & 0 \\ 0.01 & 1 & 0 & 0 \\ 0.01 & 0 & 1 & 0 \\ 0.01 & 0 & 0 & 1 \end{bmatrix}
\end{array}.
$$

This Conditional Probability Table (CPT) states that there is a probability of 0.01 of transitioning from a nominal to a fault state, and that once a fault state occurs it is an

Table 12.4 HMM diagnostic outputs for three-tank system with 3 time steps. Abnormal reading at $k = 3$ shows $\phi_k^{T_3}$ identified to be the most likely fault

Scenario	Fault distribution		
$(y_1\ y_2\ y_3)$	$P(\phi_k^{T_1})$	$P(\phi_k^{T_2})$	$P(\phi_k^{T_3})$
$k = 1$: (nom, nom, nom)	0.002	0.002	0.001
$k = 2$: (nom, nom, high)	0.002	0.002	0.001
$k = 3$: (nom, high, high)	0.11	0.11	0.23

absorbing state. If we also consider pipe faults, then the fault vector will increase in size from 4×4 to 7×7.

$$P(\mathbf{y}_k|\phi_k): \begin{array}{r|cccc|} & \phi_k^N & \phi_k^{T_1} & \phi_k^{T_2} & \phi_k^{T_3} \\ \hline \mathbf{y}_k = NlN & 0 & 0 & 0 & 0 \\ \mathbf{y}_k = NlL & 0 & 0 & 0 & 0 \\ \mathbf{y}_k = LlN & 0 & 0 & 0 & 0 \\ \mathbf{y}_k = LlL & 0.95 & 0.2 & 0.2 & 0.05 \\ \mathbf{y}_k = NhN & 0.02 & 0.2 & 0.2 & 0.05 \\ \mathbf{y}_k = NhL & 0.01 & 0.2 & 0.2 & 0.25 \\ \mathbf{y}_k = LhN & 0.01 & 0.2 & 0.2 & 0.25 \\ \mathbf{y}_k = LhL & 0.01 & 0.2 & 0.2 & 0.4 \\ \hline \end{array}$$

Given this simple model, we can compute distributions for the scenario shown in Table 12.4. We set $P(\phi_0) = (1\,0\,0\,0)$ as our initialization. Here, we define a scenario by the readings for $\mathbf{y}$ over 3 time steps, $k = 1, 2, 3$. We denote a fault in tank at step k using $P(\phi_k = f)$. An abnormal reading at $k = 3$ results in $\phi_k^{T_3}$ indicated as the most likely fault.

The HMM has the advantage of being simple to define and computationally tractable. However, as models become more complex, the CPTs grow exponentially in the number of health states and observation states. In our example, we only show single-fault scenarios. The CPTs for multiple-fault scenarios need to be explicitly defined; for example, if we included all 6 faults (3 tank leakage faults and 3 pipe blockage faults), and considered double-fault scenarios, this would require a 21×8 matrix, given 6 single-fault and 15 double-fault possibilities. This exponential growth of CPTs can be alleviated by using hierarchical HMMs, or by using DBNs, both of which allow us to factorize the large CPTs in terms of more, smaller CPTs.

12.5.1.2 Bayesian Network

A BN is an atemporal graphical model that uses a DAG for its structure. As such, the BN is a strict generalization of the tree structure for any time point of an HMM. Figure 12.4 shows the structure that we develop for illustrative purposes: each tank has an inflow/outflow: q_i/q_{1o} for tank 1, q_{12}/q_{2o} for tank 2, and q_{23}/q_{3o} for tank 3.

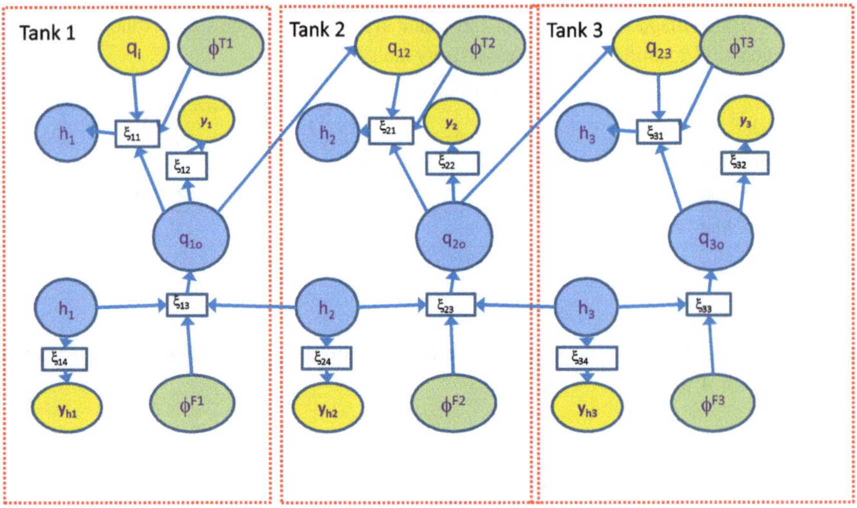

Fig. 12.4 BN structure for stochastic diagnosis of tank system

The BN model that we define represents a more complex model than the HMM model. For example, rather than being constrained to have a single variable to represent all health states, we can represent each fault using a distinguished variable; hence each tank has a variable for tank leak faults and a variable for valve faults. In this BN model, we use the following discrete values for tank k: $q_{i,in}$ takes values in the domain {nominal, high}, and both $q_{i,out}$ and $\dot{h}_i$ take values in the domain {0, low, nominal}. We assume that all fault variables take values of failed (f) or nominal (OK).

We denote the generic inflow/outflow as $q_{i,in}/q_{i,out}$ for tank i. For tank i, we specify the factor as follows:

- factor ξ_{i1} defines the CPT $P(\dot{h}_i|q_{i,in}, q_{i,out}, \phi^{T_i})$;
- factor ξ_{i2} defines the CPT $P(y_i|q_{io})$;
- factor ξ_{i3} defines the CPT $P(q_{i,out}|h_i, h_{i+1}, \phi^{F_i})$ for tanks $i = 1, 2$ and $P(q_{3o}|h_3, \phi^{F_3})$ for tank 3;
- factor ξ_{i4} defines the CPT $P(y_{h_i}|h_i)$.

We can define the factors ξ_{i1}, ξ_{i2} and ξ_{i4} to be the same for each of the tanks; only factor ξ_{i3} will differ for each tank in the system (Table 12.5).

The distributions for $P(y_i|q_{i,out})$ and $P(y_{h_i}|h_i)$ are as given below, for tanks $i = 1, 2, 3$. These distributions encode noise on the observation, stating that the correct observation has probability of 0.95.

	low	*nominal*	0
low	0.95	0.025	0.01
nominal	0.04	0.95	0.04
0	0.01	0.025	0.95

Table 12.5 Distribution for $P(\dot{h}_i | q_{i,in}, q_{i,out}, \phi^{T_i})$, for tank $i = 1, 2, 3$

$q_{i,in}$	*nominal*						*high*					
$q_{i,out}$	*nom*		*low*		0		*nom*		*low*		0	
ϕ^{Ti}	*OK*	*f*	*OK*	*f*	*OK*	*f*	*OK*	*f*	*OK*	*f*	*OK*	*f*
$\dot{h}_i = nom$	0.95	0.09	0.10	0.05	0.01	0.80	0.80	0.30	0.70	0.10	0.02	0.70
$\dot{h}_i = low$	0.05	0.01	0	0	0.99	0	0.20	0	0.30	0	0.98	0.20
$\dot{h}_i = 0$	0	0.90	0.90	0.95	0	0.20	0	0.70	0	0.90	0	0.10

Table 12.6 BN diagnostic outputs for three-tank system with 1 time step

Scenario	Fault distribution		
q_i, (y_1 y_2 y_3)	$P(\phi^{F1} = f)$	$P(\phi^{F2} = f)$	$P(\phi^{F3} = f)$
1: nom, (nom, low, low)	0.466	0.045	0.10
2: nom, (nom, nom, low)	0.031	0.39	0.04

Figure 12.4 shows the BN for the 3-tank system, with health-variable nodes in green, observation/control-variable nodes in yellow, and latent-variable nodes in blue. The failure nodes each represent a single-fault variable, so the BN can compute all fault combinations automatically; in contrast, in an HMM, we need to explicitly define the fault combinations of interest in a CPT. In this model, we assume that all faults are mutually conditionally independent, so probabilities of multiple-fault scenarios are determined by taking the product of the priors on the individual faults: e.g., $P(\phi^1, \phi^2) = \prod P(\phi^1)P(\phi^2)$. In addition, this model can automatically compensate for missing sensor readings, e.g., y_1 is missing. In an HMM, we would need to add in specific rows in a CPT to denote a missing reading for y_1 while y_2, y_3 are both present.

We perform inference, given instantiation $\mathbf{y} = \{y_1, y_2, y_3\}$, using either exact or approximate inference [44], to generate distributions over each of ϕ^N, ϕ^{T_1}, ϕ^{T_2}, ϕ^{T_3}.

Table 12.6 reports diagnostic results for two scenarios. We assume we know the control input q_i and the measurement $\mathbf{y}$. This is a static model, so each scenario consists of a single $\mathbf{y}$ measurement. Scenario 1 indicates anomalous pressure readings in tanks 2 and 3, leading to assigning $P(\phi^{F_1} = f)$ of 0.466. Scenario 1 indicates anomalous pressure readings in tank 3, leading to assigning $P(\phi^{F_2} = f)$ a value of 0.39. This static model is relatively under-sensed, since we have a single observation for each of the pressure sensors.

12.5.1.3 Dynamic Bayesian Network

Figure 12.5 depicts a DBN version of the tank example, where we show two time slices of the BN from the previous section. The arcs between two time steps are denoted by dotted green arcs. Extending this model to a DBN can increase the diagnostics accuracy, since we can now measure $\mathbf{y}$ across multiple time steps.

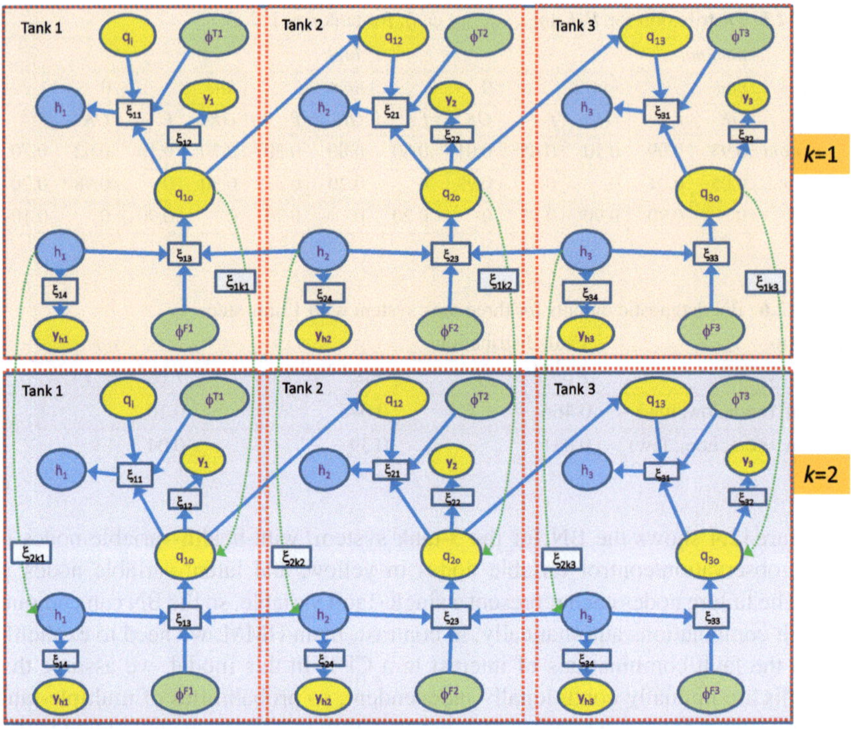

Fig. 12.5 Dynamic BN across two time steps

This network extends the BN of Sect. 12.5.1.2 with inter-slice distributions, as shown in Fig. 12.5 as the factors ξ_{1ki}, ξ_{2ki}, for tanks $i = 1, 2, 3$ at time step k. ξ_{1k1} corresponds to $P(q_{i,k}|q_{i,k-1})$, i.e., the dependence of flow q_i at time step k on q_i at time step $k-1$; ξ_{1k2} corresponds to $P(h_{i,k}|h_{i,k-1})$

We can apply the DBN to Scenario 2 of Sect. 12.5.1.2, as shown in Table 12.7. The certainty of the diagnosis can improve over time, since there are multiple observations for a permanent fault.

The DBN is a direct temporal extension of a BN in which we assume that the system is homogeneous (CPTs do not change over time). A DBN can model dynamical systems within these assumptions; however, the computational demands of a DBN increase over time, since we replicate the underlying BN for every new time step.

A DBN can also be seen as a generalization of an HMM. An HMM encodes fault state transitions, $P(\phi_k|\phi_{k-1})$, and observation state conditional dependencies, $P(\mathbf{y}_k|\phi_k)$, for $k = 1, \ldots, K$. A DBN can encode fault state transitions by defining inter-temporal arcs between slices for every fault variable. In this tank model, we would thus define factors for $P(\phi_k^{T_i}|\phi_{k-1}^{T_i})$ and $P(\phi_k^{F_i}|\phi_{k-1}^{F_i})$, for tank $i = 1, 2, 3$ and time steps $k = 1, \ldots, K$. The DBN enables the representation of significantly more fine-grained models than does an HMM. Also, the inter-temporal distributions in a

Table 12.7 DBN diagnostic outputs for three-tank system with 4 time steps

Time step	Observation	Fault distribution		
	q_i, (y_1 y_2 y_3)	$P(\phi^{F1} = f)$	$P(\phi^{F2} = f)$	$P(\phi^{F3} = f)$
1	nom, (nom, nom, nom)	0.003	0.002	0.003
2	nom, (nom, nom, low)	0.031	0.39	0.04
3	nom, (nom, nom, low)	0.031	0.44	0.08
4	nom, (nom, nom, 0)	0.031	0.78	0.14

DBN will in many cases have smaller CPTs than those of an HMM; for example, an HMM represents $P(\phi_k|\phi_{k-1})$ as a single large CPT whereas a DBN represents this as a set of smaller tables for every $\phi \in \boldsymbol{\phi}$.

12.6 Conclusions

We have presented a tutorial introduction to stochastic diagnosis using probabilistic representations. We have shown how the most widely used methods for representing uncertainty and for performing inference in the presence of uncertainty are instances of stochastic filtering, which provides a unifying framework for these approaches. We have illustrated our approach using the tank benchmark, developing models for HMM and BN representations.

This chapter has demonstrated that stochastic (probabilistic) approaches developed in FDI and Dx communities are instances of the same underlying framework, complementing the relationships demonstrated for analytical redundancy relations [56]. This framework can form the basis for cross-pollination of techniques developed in the FDI and Dx communities, e.g., enabling robust methods to be applied to DBN models.

Acknowledgements Supported by SFI grants 12/RC/2289 and 13/RC/2094.

References

1. Aghasaryan, A., Fabre, E., Benveniste, A., Boubour, R., Jard, C.: A petri net approach to fault detection and diagnosis in distributed systems. ii. extending viterbi algorithm and hmm techniques to petri nets. In: Proceedings of the 36th IEEE Conference on Decision and Control 1997, vol. 1, pp. 726–731. IEEE, San Diego, CA, USA (1997)
2. Andreassen, S., Jensen, F.V., Olesen, K.G.: Medical expert systems based on causal probabilistic networks. Int. J. Bio Med. Comput. **28**(1–2), 1–30 (1991)
3. Arroyo-Figueroa, G., Sucar, L.E.: A temporal bayesian network for diagnosis and prediction. In: Proceedings of the Fifteenth Conference on Uncertainty in Artificial Intelligence, pp. 13–20. Morgan Kaufmann Publishers Inc. (1999)

4. Arulampalam, M.S., Maskell, S., Gordon, N., Clapp, T.: A tutorial on particle filters for online nonlinear/non-gaussian bayesian tracking. IEEE Trans. Signal Process. **50**(2), 174–188 (2002)
5. Auger, F., Hilairet, M., Guerrero, J.M., Monmasson, E., Orlowska-Kowalska, T., Katsura, S.: Industrial applications of the kalman filter: a review. IEEE Trans. Ind. Electron. **60**(12), 5458–5471 (2013)
6. Berec, L.: A multi-model method to fault detection and diagnosis: Bayesian solution. An introductory treatise. Int. J. Adapt. Control Signal Process. **12**(1), 81–92 (1998)
7. Blom, H.A., Bloem, E.A.: Exact bayesian and particle filtering of stochastic hybrid systems. IEEE Trans. Aerosp. Electron. Syst. **43**(1) (2007)
8. Chen, J., Jiang, Y.C.: Development of hidden semi-markov models for diagnosis of multiphase batch operation. Chem. Eng. Sci. **66**(6), 1087–1099 (2011)
9. Chen, J., Patton, R.J.: Optimal filtering and robust fault diagnosis of stochastic systems with unknown disturbances. IEE Proc. Control Theory Appl. **143**(1), 31–36 (1996)
10. Chen, J., Patton, R.J.: Robust Model-Based Fault Diagnosis for Dynamic Systems, vol. 3. Springer Science & Business Media, Berlin (2012)
11. Chen, Z., et al.: Bayesian filtering: from Kalman filters to particle filters, and beyond. Statistics **182**(1), 1–69 (2003)
12. Chiang, L.H., Russell, E.L., Braatz, R.D.: Fault Detection and Diagnosis in Industrial Systems. Springer Science & Business Media, Berlin (2001)
13. Cooper, G.F.: The computational complexity of probabilistic inference using bayesian belief networks. Artif. Intell. **42**(2), 393–405 (1990)
14. Dagum, P., Luby, M.: Approximating probabilistic inference in bayesian belief networks is NP-hard. Artif. Intell. **60**(1), 141–153 (1993)
15. Darwiche, A.: Model-based diagnosis using structured system descriptions. J. Artif. Intell. Res. **8**, 165–222 (1998)
16. Darwiche, A.: Decomposable negation normal form. J. ACM (JACM) **48**(4), 608–647 (2001)
17. De Kleer, J.: An assumption-based TMS. Artif. Intell. **28**(2), 127–162 (1986)
18. De Kleer, J.: Focusing on probable diagnoses. AAAI **91**, 842–848 (1991)
19. De Kleer, J., Williams, B.C.: Diagnosing multiple faults. Artif. Intell. **32**(1), 97–130 (1987)
20. Dexter, A.: Fuzzy model based fault diagnosis. IEE Proc. Control Theory Appl. **142**(6), 545–550 (1995)
21. Dexter, A.L., Benouarets, M.: Model-based fault diagnosis using fuzzy matching. IEEE Trans. Syst. Man Cybern. Part A Syst. Hum. **27**(5), 673–682 (1997)
22. Ding, S.: Model-based fault diagnosis techniques: design schemes, algorithms, and tools. Springer Science & Business Media, Berlin (2008)
23. Doucet, A., Godsill, S., Andrieu, C.: On sequential monte carlo sampling methods for bayesian filtering. Stat. Comput. **10**(3), 197–208 (2000)
24. Doucet, A., Johansen, A.M.: A tutorial on particle filtering and smoothing: fifteen years later. Handbook of Nonlinear Filtering, vol. 12, pp. 656–704 (2009)
25. Feldman, A., Provan, G., Van Gemund, A.: Approximate model-based diagnosis using greedy stochastic search. J. Artif. Intell. Res. **38**, 371–413 (2010)
26. Flesch, I., Lucas, P.J., van der Weide, T.P.: Conflict-based diagnosis: adding uncertainty to model-based diagnosis. In: Proceedings of the 20th International Joint Conference on Artificial Intelligence (IJCAI-07), vol. 2007, pp. 380–385, Hyderabad, India (2007)
27. Garcia, E.A., Frank, P.: Deterministic nonlinear observer-based approaches to fault diagnosis: a survey. Control Eng. Pract. **5**(5), 663–670 (1997)
28. Hagenblad, A., Gustafsson, F., Klein, I.: A comparison of two methods for stochastic fault detection: the parity space approach and principal components analysis. IFAC Proc. Vol. **36**(16), 1053–1058 (2003)
29. Hofbaur, M.W., Williams, B.C.: Mode estimation of probabilistic hybrid systems. In: International Workshop on Hybrid Systems: Computation and Control, pp. 253–266. Springer, Berlin (2002)
30. Hwang, I., Kim, S., Kim, Y., Seah, C.E.: A survey of fault detection, isolation, and reconfiguration methods. IEEE Trans. Control Syst. Technol. **18**(3), 636–653 (2010)

31. Kalman, R.E.: A new approach to linear filtering and prediction problems. J. Basic Eng. **82**(1), 35–45 (1960)
32. Kalman, R.E., Bucy, R.S.: New results in linear filtering and prediction theory. J. Fluids Eng. **83**(1), 95–108 (1961)
33. de Kleer, J.: Using crude probability estimates to guide diagnosis. Artif. Intell. **45**(3), 381–391 (1990)
34. Kohlas, J., Anrig, B., Haenni, R., Monney, P.A.: Model-based diagnostics and probabilistic assumption-based reasoning. Artif. Intell. **104**(1–2), 71–106 (1998)
35. Kraaijeveld, P., Druzdzel, M.J., Onisko, A., Wasyluk, H.: Genierate: an interactive generator of diagnostic bayesian network models. In: Proceeding 16th International Workshop Principles Diagnosis, pp. 175–180. Citeseer, Pacific Grove, CA, USA (2005)
36. Kramer, M.A., Palowitch, B.: A rule-based approach to fault diagnosis using the signed directed graph. AIChE J **33**(7), 1067–1078 (1987)
37. Lampis, M., Andrews, J.: Bayesian belief networks for system fault diagnostics. Qual. Reliab. Eng. Int. **25**(4), 409–426 (2009)
38. Lerner, U., Parr, R., Koller, D., Biswas, G., et al.: Bayesian fault detection and diagnosis in dynamic systems. AAAI/IAAI, 531–537 (2000)
39. Loeliger, H.A.: An introduction to factor graphs. IEEE Signal Process. Mag. **21**(1), 28–41 (2004)
40. Lunze, J., Schröder, J.: State observation and diagnosis of discrete-event systems described by stochastic automata. Discret. Event Dyn. Syst. **11**(4), 319–369 (2001)
41. Moya, N., Biswas, G., Alonso-Gonzalez, C.J., Koutsoukos, X.: Structural observability. application to decompose a system with possible conflicts. In: Proceedings of the 21st International Workshop on Principles of Diagnosis, pp. 241–248 (2010)
42. Palma, J., Juarez, J.M., Campos, M., Marin, R.: Fuzzy theory approach for temporal model-based diagnosis: an application to medical domains. Artif. Intell. Med. **38**(2), 197–218 (2006)
43. Pau, L.: Survey of expert systems for fault detection, test generation and maintenance. Expert Syst. **3**(2), 100–110 (1986)
44. Pernkopf, F., Peharz, R., Tschiatschek, S.: Introduction to probabilistic graphical models. In: Academic Press Library in Signal Processing, vol. 1, pp. 989–1064. Elsevier, Amsterdam (2014)
45. Pradhan, M., Provan, G., Middleton, B., Henrion, M.: Knowledge engineering for large belief networks. In: Proceedings of the Tenth International Conference on Uncertainty in Artificial Intelligence, pp. 484–490. Morgan Kaufmann Publishers Inc. (1994)
46. Prasath, V., Lakshmi, N., Nathiya, M., Bharathan, N., Neetha, P.: A survey on the applications of fuzzy logic in medical diagnosis. Int. J. Sci. Eng. Res. **4**(4), 1199–1203 (2013)
47. Provan, G.: A general characterization of model-based diagnosis. In: ECAI 2016: 22nd European Conference on Artificial Intelligence. IOS Press, The Hague, The Netherlands (2016)
48. Provan, G.: A graphical framework for stochastic model-based diagnosis. In: 3rd Conference on Control and Fault-Tolerant Systems (SysTol) 2016, pp. 566–571. IEEE, Barcelona, Spain (2016)
49. Provan, G.M., Clarke, J.R.: Dynamic network construction and updating techniques for the diagnosis of acute abdominal pain. IEEE Trans. Pattern Anal. Mach. Intell. **15**(3), 299–307 (1993)
50. Przytula, K.W., Thompson, D.: Construction of bayesian networks for diagnostics. In: Proceedings of the 2000 IEEE Aerospace Conference, vol. 5, pp. 193–200. IEEE, Big Sky, MT, USA (2000)
51. Puig, V., Escobet, T., Ocampo-Martinez, C., Tornil-Sin, S.: Robust fault diagnosis of non-linear systems using constraints satisfaction. IFAC Proc. Vol. **42**(8), 1138–1143 (2009)
52. Rauch, H.E., Striebel, C., Tung, F.: Maximum likelihood estimates of linear dynamic systems. AIAA J. **3**(8), 1445–1450 (1965)
53. Särkkä, S.: Bayesian Filtering and Smoothing, vol. 3. Cambridge University Press, Cambridge (2013)

54. Smyth, P.: Hidden markov models for fault detection in dynamic systems. Pattern Recognit. **27**(1), 149–164 (1994)
55. Tobon-Mejia, D., Medjaher, K., Zerhouni, N.: CNC machine tool's wear diagnostic and prognostic by using dynamic Bayesian networks. Mech. Syst. Signal Process. **28**, 167–182 (2012)
56. Travé-Massuyès, L.: Bridging control and artificial intelligence theories for diagnosis: a survey. Eng. Appl. Artif. Intell. **27**, 1–16 (2014)
57. Veres, S., Norton, J.: Parameter-bounding algorithms for linear errors-in-variables models. In: Bounding Approaches to System Identification, pp. 275–288. Springer, Berlin (1996)
58. Verma, V., Gordon, G., Simmons, R., Thrun, S.: Real-time fault diagnosis. IEEE Robot. Autom. Mag. **11**(2), 56–66 (2004)
59. Wagholikar, K.B., Sundararajan, V., Deshpande, A.W.: Modeling paradigms for medical diagnostic decision support: a survey and future directions. J. Med. Syst. **36**(5), 3029–3049 (2012)
60. Weber, P., Medina-Oliva, G., Simon, C., Iung, B.: Overview on bayesian networks applications for dependability, risk analysis and maintenance areas. Eng. Appl. Artif. Intell. **25**(4), 671–682 (2012)
61. Willsky, A.S.: A survey of design methods for failure detection in dynamic systems. Automatica **12**(6), 601–611 (1976)
62. Witczak, M., Korbicz, J., Patton, R.J.: A bounded-error approach to designing unknown input observers. IFAC Proc. Vol. **35**(1), 437–442 (2002)
63. Yin, S., Zhu, X.: Intelligent particle filter and its application to fault detection of nonlinear system. IEEE Trans. Ind. Electron. **62**(6), 3852–3861 (2015)
64. Zhao, J., Xu, Y., Luo, F., Dong, Z., Peng, Y.: Power system fault diagnosis based on history driven differential evolution and stochastic time domain simulation. Inf. Sci. **275**, 13–29 (2014)

Chapter 13
Mode Detection and Fault Diagnosis in Hybrid Systems

Hamed Khorasgani and Gautam Biswas

13.1 Introduction

This chapter primarily reviews methods for online monitoring, fault detection, and fault isolation of complex hybrid systems, such as automobiles [39], aircraft [13], and spacecraft [6, 12]. Hybrid systems mix continuous behaviors with discrete mode transitions that may be attributed to configuration changes in the system, or to simplifying assumptions, where complex nonlinearities are substituted by a sequence of simpler piecewise linear behaviors [33]. As a result, mathematical models that describe hybrid system behaviors are complex, and this makes the design of hybrid diagnosers a challenging task.

The hybrid diagnosis task combines tracking of the combined discrete and continuous evolution of system behavior, detection of changes from nominal behavior, and then isolating the cause for the deviations [34, 35]. Tracking mode changes itself can be a challenging task [30], and it becomes much more difficult when faults occur in the system. This is because: (i) faults cause changes in the system behavior trajectory, therefore, they are hard to track with the fault-free model; and (ii) differentiating between changes in behavior caused by faults from those caused by mode transitions is hard. Furthermore, approaches that attempt to extend continuous system diagnosis by pre-enumerating all of the operational modes of a system are computationally intractable. To overcome these difficulties, researchers have developed a number of methodologies for hybrid system diagnosis. These include state estimation [19], parity equations [10], fault signatures based on temporal causal graphs [34], and coupling of methods for continuous and discrete-event diagnosis [3].

H. Khorasgani
Big Data Laboratory, R&D, Hitachi, Santa Clara, CA, USA
e-mail: hamed.khorasgani@hal.hitachi.com

G. Biswas (✉)
Institute for Software Integrated Systems, Vanderbilt University, Nashville, TN, USA
e-mail: gautam.biswas@Vanderbilt.Edu

T. Escobet et al. (eds.), *Fault Diagnosis of Dynamic Systems*,
https://doi.org/10.1007/978-3-030-17728-7_13

Mode detection is an essential step when tracking hybrid systems behavior. Domlan et al. [14] have developed sufficient conditions for mode detectability of linear hybrid systems. However, this result does not extend to hybrid systems that exhibit nonlinear behaviors in one or more of their operating modes. In this chapter, we extend our previous work [21] to develop an algorithm that selects a minimal set of equations to address the mode identification problem in structurally feasible nonlinear hybrid systems [15, 40]. This reduces our hybrid diagnosis task to a mode identification + continuous diagnosis problem.

In this chapter, we adopt a structural approach to developing a mode detection algorithm. We extend the well-known Minimal Structurally Overdetermined (MSO) approach for continuous systems [25] to systems of equations that contain continuous and discrete variables for describing hybrid systems behavior [32, 33]. Using the mixed continuous–discrete equations, we derive hybrid minimal structurally overdetermined (HMSO) sets for fault detection and isolation in hybrid systems.

The mode detection module can track the operating mode of the system in the presence of system faults. When the operating mode is detected, the corresponding diagnosis methodology efficiently picks a minimal set of HMSOs that guarantee complete fault diagnosability in the current mode. Our previous work developed an algorithm for finding a set of residuals with minimum cardinality that satisfies a pre-specified diagnosability performance [21]. We formulated the HMSO selection as a Binary Integer Linear Programming (BILP) problem. This is equivalent to finding an optimal solution for the set covering problem, which is known to be NP-hard [24], and, therefore, any algorithm for finding a set of residuals with minimal cardinality and required diagnosability performance has exponential computational complexity. We overcome this problem by developing a greedy search algorithm to find a minimal set of HMSOs that guarantee complete fault detectability and isolability in the current mode. Therefore, the selected HMSOs set may not have the minimum cardinality number. A larger number of HMSOs increases the computational cost for residual generation, but this increase in the computational complexity is negligible compared to the computational complexity of finding an optimal solution for the set covering problem.

The rest of this chapter is organized as follows. Section 13.2 briefly reviews previous work in hybrid systems monitoring and diagnosis. A formal definition of hybrid systems and the running example, a four-tank system, is presented in Sect. 13.3. The problems of mode detection and FDI for hybrid systems and our general approach to address these problems are formally described in Sect. 13.4. Section 13.6 presents our algorithm for mode detection. The fault diagnosis approach in hybrid systems is presented in Sect. 13.7. Section 13.8 presents the case study and Sect. 13.9 presents the conclusions of the chapter.

13.2 Background and Literature Review

As discussed earlier, a number of approaches have been developed for diagnosis of hybrid systems. State estimation approaches employ multiple model-estimation schemes to track the likely trajectories of a hybrid system [1]. However, the tracking complexity grows exponentially as the number of system modes increase, and the complexity is further compounded, when one considers trajectories that include possible faults in the system.

Adaptive multiple model-estimation schemes that track a subset of the most likely modes at a given time step have been proposed to solve this problem [27]. An efficient approach estimates the discrete modes and continuous state variables simultaneously [16, 20]. Williams and Hofbaur's estimation approach tracks only a set of highly likely paths for diagnostic analysis [19]. However, these estimation methods may fail to detect faults because their initial probability of occurrence is low.

Narasimhan and Biswas [34] have used a Hybrid Bond Graph (HBG) representation [33] that captures discrete switches in model configuration at the component level to avoid pre-enumeration of all the system modes to maintain detection performance. They define the system mode by the state of all of the discrete switches associated with the system. They assume accurate mode tracking under nominal conditions, and incremental online generation of potential fault signatures for a particular mode after a mode transition. This reduces the complexity of tracking operating modes after fault detection. However, they had to implement a roll back function to account for the fact that a fault may be detected only after a certain number of unknown mode transitions take place after the actual fault occurrence, and then a quick roll forward to catch up with the current measurements. In addition, their approach recomputes the fault signatures after every predicted mode transition, which adds to the computational complexity of the algorithm. Bregon et al. [9] extended the Possible Conflicts (PCs) approach for continuous systems [37] to decompose a hybrid system into smaller subsystems. This increases the efficiency of the algorithm by avoiding causality reassignment for the complete bond graph model.

Cocquempot et al. [11] used parity equations and Analytical Redundancy Relations (ARRs) to detect the current operating mode, and then applied a proper set of residuals to isolate faults in each operating mode. They also derive the necessary and sufficient conditions for discernable modes under no fault conditions, but do not extend this work to fault conditions [10]. These methods also suffer from the pre-enumeration of all possible modes problem, which is computationally expensive, and in some cases infeasible.

Bayoudh et al. [4] developed parameterized ARRs to account for different modes of a class of hybrid systems. They consider faults as new modes in the system, and use the ARRs to track mode transitions. Low et al. [28, 29] developed the concept of Global ARRs (GARRs) for fault detection in hybrid systems. GARRs are analytical redundancy relations between continuous and known discrete parameters that are valid over all of the hybrid system operating modes. Levy et al. [26] proposed an integrated approach, which combines GARRs and a discrete monitoring approach [38] for mode detection.

13.3 Hybrid Systems Modeling

This section provides a brief overview of hybrid systems modeling and behaviors. The running example, a hybrid four-tank system is also introduced in this section. We adopt the Mosterman and Biswas [33] approach to represent the mode transition functions in hybrid modeling. The equation set $\mathcal{E}$ is extended to contain both continuous and discrete variables, much like the mixed logic-dynamic approach proposed by Bemporad and Morari [5].

Formally, we define a hybrid system $\mathcal{H}$ as follows:

Definition 13.1 (*Hybrid system model*) A hybrid system model $\mathcal{H}$ is a tuple: $(\mathcal{X}, \Sigma, \mathcal{T}, \mathcal{E}, \mathcal{M})$, where $\mathcal{X}$ represents the set of continuous variables in the system, $\Sigma = \Sigma_1 \cup \Sigma_2$ represents the set of discrete variables; Σ_1 are variables whose values are defined by controlled mode transitions, i.e., signals that are generated external to the system, e.g., by a controller; Σ_2 are a set of variables whose values are defined by autonomous mode transitions, i.e., they are based on values associated with continuous variables in the system; $\mathcal{T} : \mathcal{X} \to \Sigma_2$ represents the set of conditions on continuous variables that define autonomous mode transitions; and $\mathcal{E} : \mathcal{X} \times \Sigma \to \mathcal{X}$ represents the set of equations that define the hybrid system behavior. The total number of modes in the hybrid system, $\mathcal{M}$ is exponential in the number of discrete variables, Σ, i.e., $\mathcal{M} = 2^{|\Sigma|}$.

We extend the hybrid model $\mathcal{H}$ to support diagnosis by considering measurements and faults in the system. The hybrid system model for diagnosis, $\mathcal{H}_\Gamma$, is defined as

Definition 13.2 (*Hybrid system model for diagnosis*) A hybrid system model for diagnosis is $\mathcal{H}_\Gamma = \mathcal{H} \cup \mathcal{Y} \cup \mathcal{Z} \cup \mathcal{F}$, where $\mathcal{Y}$ is a set of continuous measurements that are made on the system, i.e., $\mathcal{Y} = \Phi(\mathcal{X})$; $\mathcal{Z}$ represents discrete measurements, where $\mathcal{Z} \subseteq \Sigma$; and $\mathcal{F}$ is the set of fault parameters that are of diagnostic interest.

We use a configured four-tank system, shown in Fig. 13.1, as a running example throughout this chapter. Tanks 1 and 3 have inflows, $q_{in_{1,m}}$ and $q_{in_{3,m}}$, respectively, that we assume are measured, therefore, know. For Tanks 1 and 4, connecting pipes to the adjacent tanks are located at the bottom of the tanks. For Tanks 2 and 3, the connecting pipes are located at known heights, h_1 and h_2, respectively. The flow through the connecting pipe between Tanks 1 and 2 is controlled by an on/off valve, whose state is controlled by an external signal, generated by a controller. The controller function depends on the liquid level in Tanks 1 and 2, which are measured values, $h_{T_{1,m}}$ and $h_{T_{2,m}}$, respectively. The flow through this valve is a measured variable, $q_{12,m}$.

For the valve on the outlet pipe of Tank 4, we assume the discrete on/off state of the valve is determined by the known external control signal $\sigma_{6,m}$. The direction and amount of flow between Tanks 2 and 3, and between Tanks 3 and 4, depend on the liquid levels in the tanks. The liquid levels in Tanks 2 and 4 are measured and represented by the variables $h_{T_{2,m}}$ and $h_{T_{4,m}}$, respectively. In addition, the flow through the two connecting pipes are also measured, and represented by $q_{23,m}$ and $q_{34,m}$, respectively. All of the continuous and discrete measurements are shown as encircled variables in the figure.

The equations, transition conditions, and the output variables corresponding to the hybrid system diagnosis model for our running example are listed below. $\mathcal{E} = \{e_i | 1 \leq i \leq 21\}$ defines the set of equations, $\mathcal{T} = \{t_{1:5}\}$ is the set of transitions, $\mathcal{X} = \{\dot{h}_{T_{1:4}}, h_{T_{1:4}}, q_{1:4}, q_{in_{1,3}}\}$ defines the set of continuous variables, $\Sigma = \{\sigma_{1:6}\}$ is the set of discrete variables, and $\mathcal{F} = \{Leakage_{T1}, Stuck_{q_{12}}, Leakage_{T2}, Stuck_{q_{23}}, Leakage_{T4}\}$ represents the set of faults associated with the hybrid system model. The system has six binary variables and, therefore, 2^6 possible modes of operation.

$$
\begin{aligned}
e_1 &: \dot{h}_{T_1}(t) = \frac{1}{A_{T1}}(q_{in_1}(t) - q_{12}(t) - Leakage_{T1}\sqrt{2gh_{T1}(t)}) \\
e_2 &: h_{T_1}(t) = \int \dot{h}_{T_1}(t)\,dt \\
e_3 &: q_{12}(t) = (1 - Stuck_{q_{12}})\sigma_1(t) S_{P1} sign(h_{T_1}(t) - h_{T_2}(t))\sqrt{2g|h_{T_1}(t) - h_{T_2}(t)|} \\
t_1 &: \sigma_1(t) = \begin{cases} 1 & h_{T_1}(t) \geq h_{T_2} \\ 0 & h_{T_1}(t) < h_{T_2} \end{cases} \\
e_4 &: \dot{h}_{T_2}(t) = \frac{1}{A_{T2}}(q_{12}(t) - q_{23}(t) - Leakage_{T2}\sqrt{2gh_{T2}(t)}) \\
e_5 &: h_{T_2}(t) = \int \dot{h}_{T_2}(t)\,dt \\
e_6 &: q_{23}(t) = (1 - Stuck_{q_{23}}) S_{P2} sign(\sigma_2(t) h_{T_2}(t) - \sigma_3(t) h_{T_3}(t))\sqrt{2g|\sigma_2(t) h_{T_2}(t) - \sigma_3(t) h_{T_3}(t)|} \\
t_2 &: \sigma_2(t) = \begin{cases} 1 & h_{T_2}(t) \geq h_1 \\ 0 & h_{T_2}(t) < h_1 \end{cases} \\
t_3 &: \sigma_3(t) = \begin{cases} 1 & h_{T_3}(t) \geq h_1 \\ 0 & h_{T_3}(t) < h_1 \end{cases} \\
e_7 &: \dot{h}_{T_3}(t) = \frac{1}{A_{T3}}(q_{in_3}(t) + q_{23}(t) - q_{34}(t)) \\
e_8 &: h_{T_3}(t) = \int \dot{h}_{T_3}(t)\,dt \\
e_9 &: q_{34}(t) = S_{P3} sign(\sigma_4(t) h_{T_3}(t) - \sigma_5(t) h_{T_4}(t))\sqrt{2g|\sigma_4(t) h_{T_3}(t) - \sigma_5(t) h_{T_4}(t)|} \\
t_4 &: \sigma_4(t) = \begin{cases} 1 & h_{T_3}(t) \geq h_2 \\ 0 & h_{T_3}(t) < h_2 \end{cases} \\
t_5 &: \sigma_5(t) = \begin{cases} 1 & h_{T_4}(t) \geq h_2 \\ 0 & h_{T_4}(t) < h_2 \end{cases} \\
e_{10} &: \dot{h}_{T_4}(t) = \frac{1}{A_{T4}}(q_{34}(t) - q_{40}(t) - Leakage_{T4}\sqrt{2gh_{T4}(t)}) \\
e_{11} &: q_{40}(t) = S_{P4}\sigma_6(t)\sqrt{2gh_{T_4}(t)} \\
e_{12} &: h_{T_4}(t) = \int \dot{h}_{T_4}(t)\,dt.
\end{aligned}
\tag{13.1}
$$

In the equations, h_{T_i} represents the liquid level in Tank i, and q_{ij} represents the flow from Tank i to Tank j. q_{in_i} represents the inflow into Tank i, and the on/off state of valve i is represented by $\sigma_i = 1$ (on), and $\sigma_i = 0$ (off). The area of Tank i is represented as A_{Ti}. The height of the pipes, h_1 and h_2, are assumed to be constant and known. The measurements are presented in (13.2).

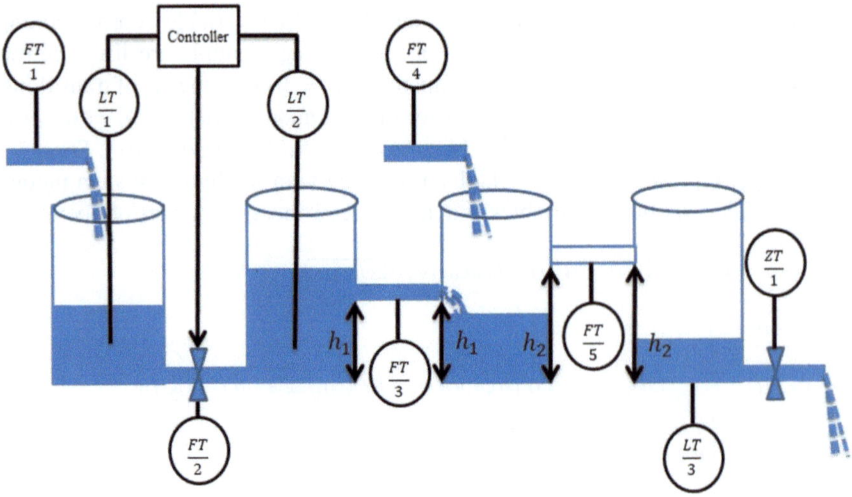

Fig. 13.1 Running example: hybrid four-tank system

$$\begin{aligned} e_{13} &: q_{in_1}(t) = q_{in_{1,m}}(t) \\ e_{14} &: h_{T_1}(t) = h_{T_{1,m}}(t) \\ e_{15} &: q_{12}(t) = q_{12,m}(t) \\ e_{16} &: h_{T_2}(t) = h_{T_{2,m}}(t) \\ e_{17} &: q_{23}(t) = q_{23,m}(t) \end{aligned} \qquad \begin{aligned} e_{18} &: q_{in_3}(t) = q_{in_{3,m}}(t) \\ e_{19} &: q_{34}(t) = q_{34,m}(t) \\ e_{20} &: h_{T_4}(t) = h_{T_{4,m}}(t) \\ e_{21} &: \sigma_6(t) = \sigma_{6,m}(t). \end{aligned} \tag{13.2}$$

$\mathcal{Y} = \{q_{in_{1,m}}, h_{T_{1,m}}, q_{12,m}, h_{T_{2,m}}, q_{23,m}, q_{in_{3,m}}, q_{34,m}, h_{T_{4,m}}\}$ defines the set of continuous measurements, $\mathcal{Z} = \{\sigma_{6,m}\}$ is the set of discrete measurements.

13.4 Hybrid Systems Diagnosis

In this section, we formally introduce the mode estimation, and the fault detection and isolation problems for hybrid systems.

13.4.1 Mode Detection and Mode Observability in Hybrid Systems

This subsection introduces the basic concepts and definitions associated with structural mode detection and fault detection and isolation in hybrid systems. The

Dulmage–Mendelsohn (DM) decomposition divides the set of equations of a system model into three parts: (1) under determined, (2) exactly determined, and (3) over determined [15]. Over determined sets form the basis for fault detection and isolation (e.g., [25]). We extend the DM decomposition approach [15] to include mode detection. This requires knowing the values of all of the discrete variables (e.g., $\Sigma = \{\sigma_{1:6}\}$ for the four-tank system example). To calculate the value of the discrete variable σ_i, we need a subset of determined transition and behavior equations (sets $\mathcal{T}$ and $\mathcal{E}$) that include σ_i and a sequential ordering for computing σ_i. For example, to compute σ_1 in our running example (13.1), we can use e_{14} and e_{16} to compute h_{T_1}, and h_{T_2}, respectively, and substitute for h_{T_1}, and h_{T_2}, in t_1 to compute σ_1.

The next step is the make the modes detectable in the presence of faults. To accomplish this, we introduce the concepts of Structurally Determined (SD) sets and Minimal Structurally Determined (MSD) sets.

Definition 13.3 (*Structurally Determined Set in Hybrid Systems*) Consider a set of equations and transitions and their associated continuous variables, discrete variables, and faults: $(\mathcal{E}, \mathcal{T}, \mathcal{X}, \Sigma, \mathcal{F})$. The set of equations and transitions is Structurally Determined (SD) if the cardinality of the set $\mathcal{E}$ plus cardinality of $\mathcal{T}$ is greater than or equal to the sum of the cardinalities of the sets $\mathcal{X}$, Σ, and $\mathcal{F}$, i.e., $|\mathcal{E}| + |\mathcal{T}| \geq |\mathcal{X}| + |\Sigma| + |\mathcal{F}|$.

Definition 13.4 (*Minimal Structurally Determined Set in Hybrid Systems*) A set of structurally determined equations is Minimal Structurally Determined (MSD) if it has no subset of structurally determined equations.

Consider the four-tank system represented by Eq. (13.1). A minimal structurally determined set in this system is $MSD_1 = (\mathcal{E}_1, \mathcal{T}_1, \mathcal{X}_1, \Sigma_1, \mathcal{F}_1)$, where $\mathcal{E}_1 = \{e_{14}, e_{16}\}$, $\mathcal{T}_1 = \{t_1\}$, $\mathcal{X}_1 = \{h_{T_1}, h_{T_2}\}$, $\Sigma_1 = \{\sigma_1\}$, and $\mathcal{F}_1 = \{\}$. For the sake of brevity, in the rest of the chapter we simply say a specific equation, transition, variable, or fault is a member of a MSD (e.g., $\sigma_1 \in MSD_1$). MSDs represent solvable sets of variables in the system and they can be used for mode detection. We define a detectable discrete variable in a hybrid system as follows.

Definition 13.5 (*Detectable discrete variable in hybrid systems*) A discrete variable $\sigma \in \Sigma$ is detectable for a diagnostic hybrid system, $\mathcal{H}_d$, if there is a minimal structurally determined set MSD_i in the system, such that $\sigma \in MSD_i$.

For example, discrete variable σ_1 in Eq. (13.1) is detectable because $\sigma_1 \in MSD_1$. Babaali and Egerstedt [2] define mode observability based on continuous variables trajectories in different modes. In this work, we define a mode observable diagnostic hybrid system model as follows.

Definition 13.6 (*Mode Observable Diagnostic Hybrid system*) A hybrid system $\mathcal{H}_d = (\mathcal{X}, \Sigma, \mathcal{T}, \mathcal{E}, \mathcal{M}, \mathcal{Y}, \mathcal{Z}, \mathcal{F})$ is mode observable if all the discrete variables $\sigma_i \in \Sigma$ are detectable (i.e., they are directly observed, or their values are computable given $\mathcal{H}_d$).

13.4.2 Fault Detection and Isolation in Hybrid Systems

For hybrid systems, mode detection is a first step in fault detection and isolation. Our mode detection scheme is based on the MSD approach, making the assumption that the hybrid system is mode observable. In addition, we assume faults and mode changes do not occur at the same time, and a fault is detected and isolated in the same mode in which it initially occurs. Therefore, our approach to fault detection and isolation requires the synchronous solution of the mode identification, fault detection, and fault isolation tasks. We describe our diagnosis algorithm in greater detail in the next section.

We formally define the concepts of Hybrid Structurally Overdetermined (HSO) and Hybrid Minimal Structurally Overdetermined (HMSO) sets for hybrid system diagnosis next.

Definition 13.7 (*Hybrid Structural Overdetermined Set*) Consider a system represented by a set of equations with associated continuous variables, discrete variables, and faults: $(\mathcal{E}, \mathcal{X}, \Sigma, \mathcal{F})$. This set of equations is Hybrid Structurally Overdetermined (HSO) if the cardinality of the set $\mathcal{E}$ is greater than the cardinality of set $\mathcal{X}$, i.e., $|\mathcal{E}| > |\mathcal{X}|$ and all the $\sigma \in \Sigma$ are detectable.

Definition 13.8 (*Hybrid Minimal Structurally Overdetermined Set*) A HSO is Hybrid Minimal Structurally Overdetermined (HMSO) if it has no subset of hybrid structurally overdetermined equations.

Consider the four-tank system in Eq. (13.1). A hybrid minimal structurally overdetermined set in this system is $\mathbf{HMSO_1} = (\mathcal{E}_2, \mathcal{X}_2, \Sigma_2, \mathcal{F}_2)$, where $\mathcal{E}_2 = \{e_1, e_2, e_{13}, e_{14}, e_{15}\}$, $\mathcal{X}_2 = \{\dot{h}_{T_1}, h_{T_1}, q_{12}, q_{in_1}\}$, $\Sigma_2 = \{\}$, and $\mathcal{F}_2 = \{Leakage_{T1}\}$. Note that $\mathbf{HMSO_1}$ does not include a discrete variable. HMSOs capture the redundancies in the hybrid system model, and can be used for fault detection and isolation.

We define fault detectability in hybrid systems as follows.

Definition 13.9 (*Detectable fault in hybrid systems*) A fault $f \in \mathcal{F}$ is detectable in hybrid system $\mathcal{H}_d$ if there is a hybrid minimal structurally overdetermined set $\mathbf{HMSO_i}$ in the system, such that $f \in \mathbf{HMSO_i}$.

For example, consider fault $Leakage_{T1}$ that appears in Eq. (13.1) of the four-tank system. $Leakage_{T1}$ is detectable because $Leakage_{T1} \in HMSO_1$.

Definition 13.10 (*Isolable faults in hybrid systems*) A fault $f_i \in \mathcal{F}$ is isolable from fault $f_j \in \mathcal{F}$ if there exists a hybrid minimal structurally overdetermined set $\mathbf{HMSO_i}$ in $\mathcal{H}_d$, such that $f_i \in \mathbf{HMSO_i}$ and $f_j \notin \mathbf{HMSO_i}$.

As an example $Leakage_{T1}$ is isolable from $Stuck_{q_{12}}$, because $Leakage_{T1} \in \mathbf{HMSO_1}$ and $Stuck_{q_{12}} \notin \mathbf{HMSO_1}$.

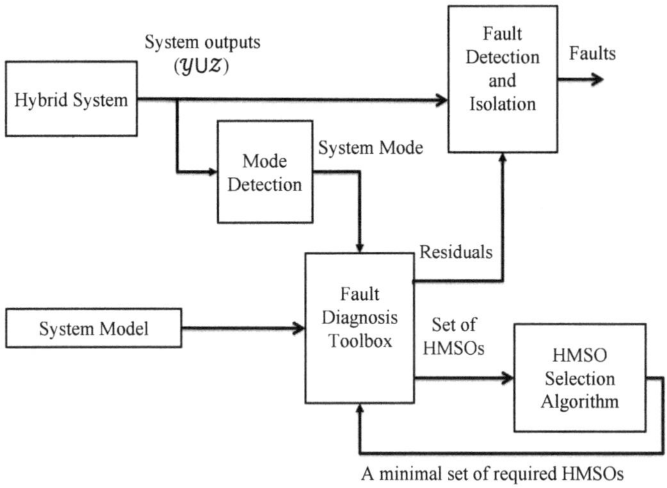

Fig. 13.2 Fault detection and isolation in hybrid systems

13.5 Problem Formulation and Solution

Figure 13.2 describes our approach for hybrid systems diagnosis. The first step involves detecting the operating mode of the hybrid system, i.e., computing the value of its discrete variables. We assume the hybrid system is mode observable and develop an algorithm to find a Minimal Structural Determined (MSD) set. Using the MSD, we solve for the value of each discrete variable to determine the current operating mode of the system. As discussed, this result is correct even if faults have occurred in the system.

Definition 13.11 (*Mode detection problem*) Let Σ denote the set of discrete variables in a diagnostic hybrid system, $\mathcal{H}_d$. Our goal is to develop an algorithm that selects a minimal structurally determined set MSD_i for each $\sigma_i \in \Sigma$ such that $\sigma_i \in MSD_i$, i.e.,

$$\begin{aligned} \forall \sigma_i \in \Sigma \quad & \text{find}\, MSD_i, \\ \text{such that} \quad & \sigma_i \in MSD_i. \end{aligned} \tag{13.3}$$

The mode detection unit in Fig. 13.2 is designed to track the operating mode of the system in a continuous manner. When the hybrid system transitions to a new mode, the diagnosis algorithm generates a new set of residuals to support fault detection and isolation for that mode. When the hybrid system revisits a mode that was active earlier, the algorithm can reduce computation time by reviving a cashed set of residuals that was generated for this mode. When our mode detection algorithm reports a mode transition, we use the fault diagnosis toolbox[1] developed by Frisk et al. [18] to generate the entire set of HMSOs for the new operating mode. We then select a minimal set of HMSOs for fault detection and isolation in this mode.

The next step involves selecting a minimal set of HMSOs that guarantee complete fault detectability and isolability in each operating mode of the hybrid system. The fault detection and isolation problem is defined as follows. Given the hybrid system model for diagnosis, $\mathcal{H}_\Gamma$, with a set of faults $\mathcal{F}$, we assume that the set of HMSOs in each mode, m, $\mathbf{HMSO_m} = \{HMSO_{m_1}, HMSO_{m_2}, \ldots, HMSO_{m_r}\}$, is sufficient to detect and uniquely isolate all of the faults in that mode, $F_m \subset \mathcal{F}$. The set of fault candidates may not be the same in all the operating modes. Our goal is to develop an algorithm that selects a minimal subset of HMSOs, $\mathbf{HMSO^*_m}$ for each mode, m that guarantee fault detectability for each fault $f_i \in F_m$ and fault isolability for each pair of faults $f_j \in F_m$ and $f_k \in F_m$.

More formally, for each operation mode we define *minimal HMSO set* as a minimal set of HMSOs that can be used to achieve the complete structural diagnosability.

Definition 13.12 (*Minimal HMSO set for operation mode m*) $\mathbf{HMSO^*_m} \subset \mathbf{HMSO_m}$ is a minimal set of HMSOs for diagnosing the faults F_m in mode m, if $\mathbf{HMSO^*_m}$ fulfills the following diagnosability performance requirements:

$$\begin{aligned} &\forall f_i \& f_j \in F_m : \\ &\exists HMSO_k \in \mathbf{HMSO^*_m} : \\ &f_i \in HMSO_k, \\ &f_j \notin HMSO_k, \end{aligned} \tag{13.4}$$

and all proper subsets of $\mathbf{HMSO^*_m}$ do not.

Like before, we use the fault diagnosis toolbox (see Fig. 13.2) to derive the residual set from the selected HMSOs, $\mathbf{HMSO^*_m}$ for fault diagnosis in each mode. As discussed, HMSO-selection and residual generation only occurs when a mode is visited for the first time. Otherwise, the saved mode information is retrieved for diagnostic analysis. In the next section, we present each step of this algorithm in greater detail.

13.6 Mode Detection Algorithm

We discuss our proposed approach to finding a minimal set of constraints for detecting each discrete mode change during the hybrid system operation. We illustrate the procedure by solving this problem for the running example, and then generalize this approach and present an algorithm that solves this problem.

[1] https://faultdiagnosistoolbox.github.io/.

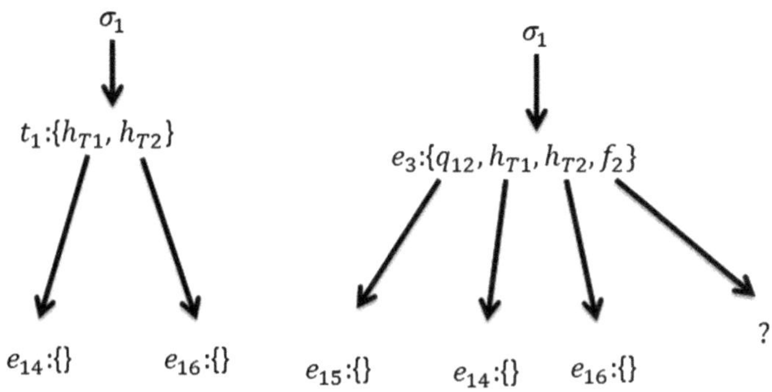

Fig. 13.3 Detecting σ_1

The tank system model includes six discrete variables. To detect each $\sigma_i \in \Sigma$, we have to find a determined set of equations and transitions (MSD) that include σ_i. We illustrate this for σ_1 by starting with all of the system equations and transitions in $\mathcal{H}_d$ that include σ_1. The relevant equations and transitions are e_3, and t_1 as shown in Fig. 13.3. Then for the additional variables and faults in each equation, we need to add other equations so that they may be substituted out to generate an equation with only one unknown variable. For t_1, we need to add two new constraints: one with h_{T_1} and the other with h_{T_2}. e_{14} and e_{16} represent measurement equations and h_{T_1} and h_{T_2} are the only unknown variables in these equations, respectively. Therefore, these equations do not add new variables to t_1, and $\mathcal{E}_1 = \{e_{14}, e_{16}\}$ plus t_1 is a minimal structurally determined set that makes σ_1 detectable.

However, Fig. 13.3 shows that there are no additional equations that we can add to e_3 to generate a structurally determined set. More formally, Fig. 13.3 depicts the matching algorithm presented as Algorithm 13.1. If we initialize the algorithm with the set of unknown variables and faults, $\mathcal{U}$, (in this example h_{T_1} and h_{T_2} are the unknown variables) it returns a set of complete matching of variables and equations in the subsystem that includes the unknown variables.

Algorithm 13.1: Count-Matchings.

input: current matching M
input: sets of determined variables $\mathcal{D}$ and undetermined variables $\mathcal{U}$
if $\mathcal{U} = \emptyset$ **then**
  return M as a feasible (minimal) matching.
for each $x \in \mathcal{U}$ **do**
  for each y which can determine x **do**
    Let M' be $M \cup \{x \rightarrow y\}$
    Let $\mathcal{D}'$ be $\mathcal{D} \cup \{x\}$.
    Let $\mathcal{U}'$ be $\mathcal{U} \setminus \{x\}$.
    Add all the undetermined variables of y to $\mathcal{U}'$.
    COUNT- MATCHINGS(M', $\mathcal{D}'$, $\mathcal{U}'$)

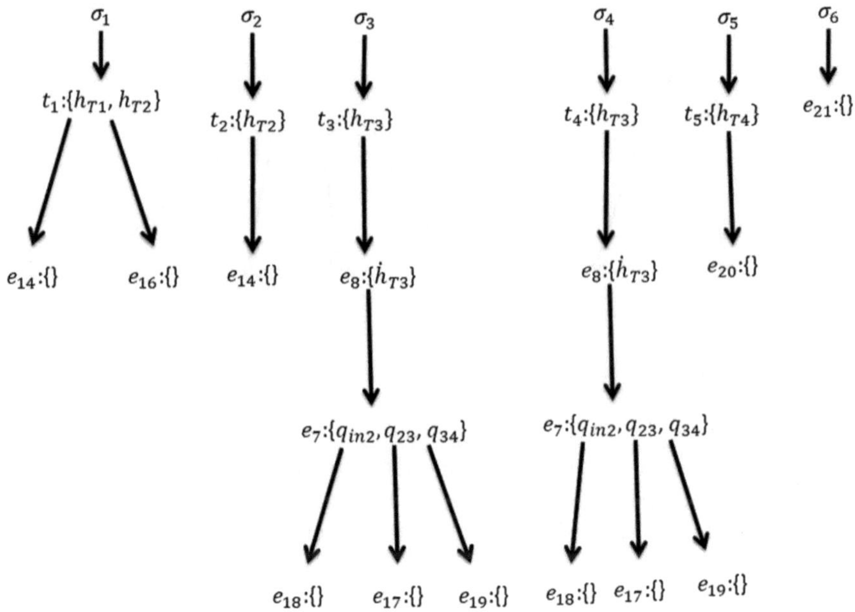

Fig. 13.4 Detecting σ_1–σ_6

We use Algorithm 13.1 to find a MSD set to detect each discrete variable in the running example (see Fig. 13.4). Algorithm 13.2 summarizes the procedure.

Algorithm 13.2: Mode Detection.

1: **input:** $(\mathcal{X}, \Sigma, \mathcal{E}, \mathcal{T}, \mathcal{Y}, \mathcal{Z}, \mathcal{M}, \mathcal{F})$
2: **for each** $\sigma \in \Sigma$ **do**
3: $\quad \mathcal{C}_\sigma$ = the set of equations and transitions that include σ
4: $\quad$ **for each** $c \in \mathcal{C}_\sigma$ **do**
5: $\quad\quad$ Let M be (c, σ)
6: $\quad\quad$ Let $\mathcal{D}$ be σ and $\mathcal{U}$ be the rest of variables in c
7: $\quad\quad$ MSD_σ = Count-Matchings$(M, \mathcal{D}, \mathcal{U})$
8: $\quad\quad$ **if** $MSD_\sigma \neq \emptyset$ **then**
9: $\quad\quad\quad$ **select** MSD_σ as a feasible MSD for σ.

Table 13.1 shows the set of constraints in each MSD.

To compute the value of each discrete variable, we assign a reverse causality to the graphs shown in Fig. 13.4 and solve the MSDs for the discrete variables using the known variables. Equation (13.5) shows the solution for each discrete variable.

Table 13.1 Set of selected MSDs

HMSOs	Set of equations	Discrete variable
MSD_1	t_1, e_{14}, e_{16}	σ_1
MSD_2	t_2, e_{16}	σ_2
MSD_3	$t_3, e_8, e_7, e_{18}, e_{17}, e_{19}$	σ_3
MSD_4	$t_4, e_8, e_7, e_{18}, e_{17}, e_{19}$	σ_4
MSD_5	t_5, e_{20}	σ_5
MSD_6	e_{21}	σ_6

$$
\begin{aligned}
\sigma_1(t) &= \begin{cases} 1 & h_{T_{1,m}}(t) \geq h_{T_{2,m}}(t) \\ 0 & h_{T_{1,m}}(t) < h_{T_{2,m}}(t) \end{cases} \\
\sigma_2(t) &= \begin{cases} 1 & h_{T_{2,m}}(t) \geq h_1 \\ 0 & h_{T_{2,m}}(t) < h_1 \end{cases} \\
\sigma_3(t) &= \begin{cases} 1 & \int \frac{1}{A_{T3}}(q_{in_{3,m}}(t) + q_{23,m}(t) - q_{34,m}(t)) \geq h_1 \\ 0 & \int \frac{1}{A_{T3}}(q_{in_{3,m}}(t) + q_{23,m}(t) - q_{34,m}(t)) < h_1 \end{cases} \\
\sigma_4(t) &= \begin{cases} 1 & \int \frac{1}{A_{T3}}(q_{in_{3,m}}(t) + q_{23,m}(t) - q_{34,m}(t)) \geq h_2 \\ 0 & \int \frac{1}{A_{T3}}(q_{in_{3,m}}(t) + q_{23,m}(t) - q_{34,m}(t)) < h_2 \end{cases} \\
\sigma_5(t) &= \begin{cases} 1 & h_{T_{4,m}}(t) \geq h_2 \\ 0 & h_{T_{4,m}}(t) < h_2 \end{cases} \\
\sigma_6(t) &= \sigma_{6,m}(t).
\end{aligned}
\tag{13.5}
$$

13.7 Fault Detection and Isolation Algorithm

In this section, we present our algorithm to select a minimal HMSOs set that guarantees complete diagnosability in each operating mode. We then discuss the residual generation and the fault detection and isolation algorithm for hybrid systems.

13.7.1 Selecting a Minimal HMSO Set for FDI

The FDI algorithm assumes that the discrete variables values have been computed and the hybrid system mode is known. Therefore, we can remove the discrete variables from the list of unknown variables. Let us assume that we generate r HMSOs for the hybrid system diagnosis model, $\mathcal{H}_d$, in mode m, $\mathbf{HMSO_m} = \{HMSO_1^m, HMSO_2^m, \ldots, HMSO_r^m\}$, given its set of measurements $\mathcal{Y} \cup \mathcal{Z}$. Our goal is to design a greedy search algorithm for selecting a minimal set of residuals

$\mathbf{HMSO^*_m} \subseteq \mathbf{HMSO_m}$ in a way that makes all the system faults in this mode, F_m, structurally diagnosable. Algorithm 13.3 sorts the HMSOs by number of equations in the first step (line 5). By applying the sorting step first, the greedy search selects HMSOs with the smallest number of equations. The smaller the number of equations that make up an HMSO, the smaller are the number of system model parameters in this HMSO (this is a heuristic). As a result, the HSMO is likely to be more robust against model uncertainties. Moreover, the computational complexity of solving for unknown variables and deriving residuals from HMSOs with fewer equations is likely to be lower than HMSOs that include more equations (note, this is again a heuristic).

Algorithm 13.3: HMSO-Selection.

1: **input:** set of selected HMSOs for the current mode $\mathbf{HMSO^*_m}$
2: **input:** set of HMSOs in the current mode $\mathbf{HMSO_m}$
3: **input:** set of undetectable faults UD
4: **input:** set of pair of faults that are not isolable UI
5: sort $\mathbf{HMSO_m}$ by number of equations
6: **if** $UD = \emptyset$ and $UI = \emptyset$ **then**
7:    **return** $\mathbf{HMSO^*_m}$ as a minimal HMSO set.
8: $HMSO_{new}$= Find-HMSO($\mathbf{HMSO_m}$,$\mathbf{HMSO^*_m}$,UD,UI)
9: **for each** $f \in UD$ if $f \in HMSO_{new}$ **do**
10:    Let UD be $UD \setminus \{f\}$
11: **for each** $(f_i, f_j) \in UD$ if $f_i \in HMSO_{new}$ and $f_j \notin HMSO_{new}$ **do**
12:    Let UI be $UI \setminus \{(f_i, f_j)\}$
13: Let $\mathbf{HMSO^*_m}$ be $\mathbf{HMSO^*_m} \cup \{HMSO_{new}\}$.
14: Let $\mathbf{HMSO_m}$ be $\mathbf{HMSO_m} \setminus \{HMSO_{new}\}$.
15: HMSO- SELECTION($\mathbf{HMSO_m}$, $\mathbf{HMSO^*_m}$, UD, UI)

When the system transitions to a mode m for the first time, the set of selected HMSOs for this mode, $\mathbf{HMSO^*_m}$, is empty, and therefore, no fault is detectable or isolable in this mode. At each step, the function Find-HMSO adds a HMSO candidate, $HMSO_{new}$, to $\mathbf{HMSO^*_m}$ that makes a fault detectable, if not isolable from other faults that can occur in this mode. The algorithm keeps adding HMSOs to $\mathbf{HMSO^*_m}$ till all the faults are detectable and isolable. The selected $\mathbf{HMSO^*_m}$ is used to generate the set of residuals for this mode. The residuals for each visited mode is cached to be used when the system returns to the mode.

Algorithm 13.4: Find-HMSO.

input: set of selected HMSOs for the current mode $HMSO^*_m$
input: set of HMSOs in the current mode $HMSO_m$
input: set of undetectable faults UD
input: set of pair of faults that are not isolable UI
for each $HMSO \in HMSO_m$ if ($\exists f \in HMSO$ and $f \in UD$) or ($\exists f_i \in HMSO$ and $\exists f_j \notin HMSO$ and $(f_i, f_j) \in UI$) **do**
   return $HMSO$ as the selected HMSO

As an example, consider the case where the discrete variables in the running example take on values: $\Sigma = \{1, 0, 0, 1, 1, 0\}$. If we number the system modes from 0 to 63, the mode number for the running example is 38. In this operating mode, the system equations are

$$\begin{aligned}
e_1 &: \dot{h}_{T_1}(t) = \frac{1}{A_{T1}}(q_{in_1}(t) - q_{12}(t) - Leak_{T1}\sqrt{2gh_{T1}(t)})\\
e_2 &: h_{T_1}(t) = \int \dot{h}_{T_1}(t)\,dt\\
e_3 &: q_{12}(t) = (1 - Stuck_{q_{12}})S_{P1}sign(h_{T_1}(t) - h_{T_2}(t))\sqrt{2g|h_{T_1}(t) - h_{T_2}(t)|}\\
e_4 &: \dot{h}_{T_2}(t) = \frac{1}{A_{T2}}(q_{12}(t) - q_{23}(t) - Leak_{T2}\sqrt{2gh_{T2}(t)})\\
e_5 &: h_{T_2}(t) = \int \dot{h}_{T_2}(t)\,dt\\
e_6 &: q_{23}(t) = 0\\
e_7 &: \dot{h}_{T_3}(t) = \frac{1}{A_{T3}}(q_{in_2}(t) + q_{23}(t) - q_{34}(t))\\
e_8 &: h_{T_3}(t) = \int \dot{h}_{T_3}(t)\,dt\\
e_9 &: q_{34}(t) = S_{P3}sign(h_{T_3}(t) - h_{T_4}(t))\sqrt{2g|h_{T_3}(t) - h_{T_4}(t)|}\\
e_{10} &: \dot{h}_{T_4}(t) = \frac{1}{A_{T4}}(q_{34}(t) - q_{40}(t) - Leak_{T4}\sqrt{2gh_{T4}(t)})\\
e_{11} &: q_{40}(t) = 0\\
e_{12} &: h_{T_4}(t) = \int \dot{h}_{T_4}(t)\,dt.
\end{aligned} \tag{13.6}$$

The system measurements are

$$\begin{aligned}
e_{13} &: q_{in_1}(t) = q_{in_{1,m}}(t) & e_{17} &: q_{23}(t) = q_{23,m}(t)\\
e_{14} &: h_{T_1}(t) = h_{T_{1,m}}(t) & e_{18} &: q_{in_2}(t) = q_{in_{3,m}}(t)\\
e_{15} &: q_{12}(t) = q_{12,m}(t) & e_{19} &: q_{34}(t) = q_{34,m}(t)\\
e_{16} &: h_{T_2}(t) = h_{T_{2,m}}(t) & e_{20} &: h_{T_4}(t) = h_{T_{4,m}}(t).
\end{aligned} \tag{13.7}$$

The set of system faults in this operating mode is $F_{38} = \{Leak_{T1}, Stuck_{q_{12}}, Leak_{T2}, Leak_{T4}\}$. Note that in this operation mode $Stuck_{q_{23}}$ is not among the system faults. In fact, when $\sigma_2 = 0$ the valve flow $q_{23} = 0$ independent of the pipe condition, R_{R_2}. In this situation, a fault in R_{R_2} does not have any effect on the system dynamics. To detect $Stuck_{q_{23}}$ the diagnoser needs to wait for a mode change in the hybrid system or adopt an active fault diagnosis approach [31]. The fault diagnosis toolbox, produces 47 HMSOs for the running example in this operating mode, i.e., $\mathbf{HMSO_{38}}$

Table 13.2 Set of selected HMSOs for FDI in mode 1010

HMSOs	Set of equations	Set of faults
$HMSO_1^{38}$	$e_3, e_{14}, e_{15}, e_{16}$	$Stuck_{q12}$
$HMSO_2^{38}$	$e_{10}, e_{11}, e_{12}, e_{19}, e_{20}$	$Leak_{T4}$
$HMSO_3^{38}$	$e_4, e_5, e_{15}, e_{16}, e_{17}$	$Leak_{T2}$
$HMSO_4^{38}$	$e_1, e_2, e_{13}, e_{14}, e_{15}$	$Leak_{T1}$

Table 13.3 Selected residuals for FDI in mode 38

	$Detection$	$Leak_{T1}$	$Stuck_{q12}$	$Leak_{T2}$	$Stuck_{q23}$	$Leak_{T4}$
$Leak_{T1}$	r_4	X	r_4	r_4	r_4	r_4
$Stuck_{q12}$	r_1	r_1	X	r_1	r_1	r_1
$Leak_{T2}$	r_3	r_3	r_3	X	r_3	r_3
$Leak_{T4}$	r_2	r_2	r_2	r_2	r_2	X

$= \{HMSO_1^{38}, HMSO_2^{38}, ..., HMSO_{47}^{38}\}$. Table 13.2 shows the selected HMSOs and their sets of equations and faults in mode 38. To implement the fault detection and isolation approach, we generate residuals in the manner shown next.

13.7.2 Generating Residuals for Hybrid Systems

Algorithm 13.3 selects a minimal set of HMSOs for fault detection and isolation. When the system operating mode is detected, implying that the discrete variables in the hybrid system are known, we can use the fault diagnosis toolbox to generate a residual from each HMSO. Consider the set of HMSOs in Table 13.2. The residual generated by the fault diagnosis toolbox from each HMSO is presented as follows.

$$\begin{aligned}
HMSO_1^{38} &: r_1(t) = q_{12,m}(t) - S_{P1}sign(h_{T_{1,m}}(t) - h_{T_{2,m}}(t))\sqrt{2g|h_{T_{1,m}}(t) - h_{T_{2,m}}(t)|}\\
HMSO_2^{38} &: r_2(t) = h_{T_{4,m}}(t) - \frac{1}{A_{T_4}}\int q_{34,m}(t)\\
HMSO_3^{38} &: r_3(t) = h_{T_{2,m}}(t) - \frac{1}{A_{T_2}}\int q_{12,m}(t) - q_{23,m}(t)\\
HMSO_4^{38} &: r_4(t) = h_{T_{1,m}}(t) - \frac{1}{A_{T_1}}\int q_{in_{1,m}}(t) - q_{12,m}(t).
\end{aligned} \tag{13.8}$$

To demonstrate how this set of four residuals is sufficient to detect and isolate all of the system faults in the current mode, we present the residuals for detecting individual faults and isolating pairs of faults in Table 13.3. The detection column lists the residuals that can be used to detect the system faults, and the residuals in row f_i and column f_j can be applied to differentiate between faults f_i and f_j.

13.7.3 Designing the Fault Diagnoser

Figure 13.2 shows the application of the generated residuals to detect and isolate system faults in each operation mode. To apply the residuals for FDI, we need to go beyond structural analysis, take into account parameter values, and also consider sensor noise and uncertainties in the system model. Residuals represent redundancies in the system equations, and they can form the basis for fault detection and isolation. Ideally, each residual should compute to a value of zero in the fault-free case, and residuals sensitive to a fault become nonzero when the fault occurs. In practice, due to model uncertainties and measurement noise, a residual may deviate from zero when no faults have occurred.

To address this problem, we apply a Z-test [8] to determine where the change in the residual value, r, is statistically significant. We consider the last N_2 residual values to compute the mean value of residual distribution (assumed to be a normal distribution):

$$\mu_r(k) = \frac{1}{N_2} \sum_{i=k-N_2+1}^{k} r(i). \tag{13.9}$$

The last N_1 samples (typically, $N_1 \gg N_2$) to compute the variance:

$$\sigma_r^2(k) = \frac{1}{k - N_1 - 1} \sum_{i=k-N_1+1}^{k} (r(i) - \mu_r(k))^2. \tag{13.10}$$

The confidence level for the Z-test, α, determines the bounds, z_-, and z_+, and, therefore, the sensitivity of the residuals.

$$P(z_- < (r(k) - \mu_r(k)) < z_+) = 1 \mp \alpha. \tag{13.11}$$

The Z-test is implemented as follows:

$$\begin{aligned} z_- < r(k) - \mu_r(k) < z_+ &\rightarrow \text{NF} \\ \text{Otherwise} &\rightarrow \text{Fault}. \end{aligned} \tag{13.12}$$

13.8 Case Study

We apply our structural mode detection and fault detection and isolation approaches to the Reverse Osmosis (RO) system [7], which is a hybrid nonlinear system.

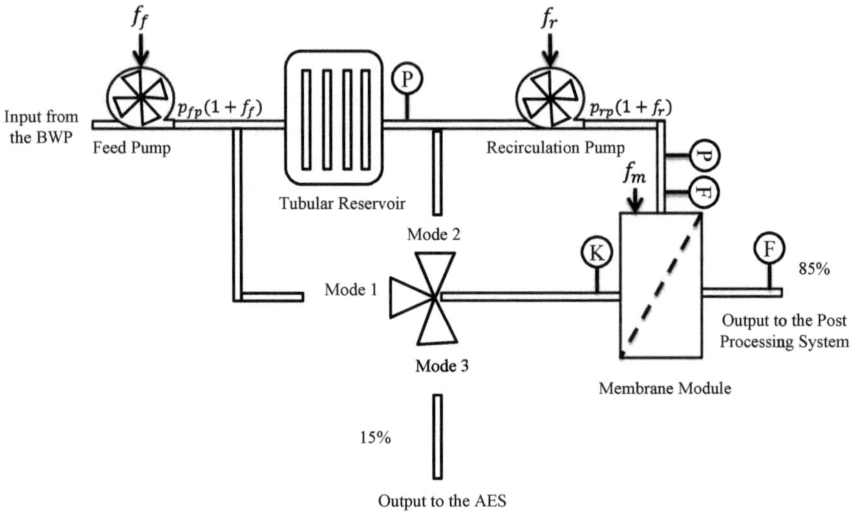

Fig. 13.5 Reverse Osmosis System (RO)

13.8.1 The RO System Model

The RO system, shown in Fig. 13.5, is a part of the Advanced Water Recovery System (AWRS), which was designed and built at the NASA Johnson Space Center for long duration manned space missions. The AWRS converts wastewater to potable water for the astronauts in microgravity conditions. The RO system is designed to remove inorganic impurities from the wastewater.

This system operates in three modes controlled by a three-way valve. In the first mode of operation (valve setting 1), the water circulates in the longer loop. The accumulation of impurities in the membrane increases membrane resistance, R_{memb}, which decreases the output flow rate to the post-processing system. After a specific period of time, the system switches to the secondary mode (valve setting 2). In this mode, the recirculation pump circulate the water in a smaller secondary loop, which increases the output flow rate.

As clean water leaves the RO system, the concentration of brine, e_{Cbrine}, in the residual water increases. High concentration of brine leads to increases in R_{memb}, which decreases the system performance significantly.Again, after a predetermined time interval, the system switches to the purge mode (valve setting 3). In this mode, the recirculation pump is turned off, and concentrated brine is pushed out to the Air Evaporation System (AES).

Using first principles, we have created the hybrid model of the RO system [21] as shown below:

$$e_{RO1} : \dot{q}_{fp}(t) = \frac{1}{I_{fp}}(-R_{fp}q_{fp}(t) - p_{tr}(t) + p_{fp}(t)(1 - f_f))$$

$$e_{RO2}: \dot{p}_{tr}(t) = \frac{1}{C_{tr}}(q_{fp}(t) + \sigma_1(t)\frac{p_{memb}(t) - p_{tr}(t)}{R_{return_l}} - q_{rp}(t) +$$

$$\sigma_2(t)\frac{p_{memb}(t) - p_{tr}(t)}{R_{return_s}})$$

$$e_{RO3}: q_{rp}(t) = \frac{\sigma_1(t) + \sigma_2(t)}{I_{rp}} \int (-R_{rp}q_{rp}(t) - R_{forward}q_{rp}(t) - p_{memb}(t)$$

$$+ p_{rp}(t)(1 - f_r)) + (1 - \sigma_1(t) - \sigma_2(t))\frac{p_{tr}(t) - p_{memb}(t)}{R_{forward}}$$

$$e_{RO4}: \dot{p}_{memb}(t) = \frac{1}{C_{memb}}(q_{rp} - \frac{p_{memb}}{R_{memb}(t)(1 + f_m)} - \sigma_1\frac{p_{memb} - p_{tr}}{R_{return_l}}$$

$$- \sigma_2\frac{p_{memb} - p_{tr}}{R_{return_s}} - (1 - \sigma_1 - \sigma_2)\frac{p_{pmemb}}{R_{return_{AES}}})$$

$$e_{RO5}: \dot{e}_{Cbrine}(t) = \frac{1}{1.667 \times 10^{-8} C_{brine}}(\sigma_1(t)\frac{p_{memb}(t) - p_{tr}(t)}{R_{return_l}}$$

$$+ \sigma_2(t)\frac{p_{memb}(t) - p_{tr}(t)}{R_{return_s}} + (1 - \sigma_1(t) - \sigma_2(t))\frac{p_{memb}(t)}{R_{return_{AES}}})$$

$$e_{RO6}: \dot{e}_{Ck}(t) = \frac{q_{rp}(t)}{C_k}(6e_{Cbrine}(t) + 0.1)/(1.667 \times 10^{-8}),$$

where q_{fp} is the volume flow rate generated by the feed pump, p_{tr} is the pressure of the fluid in the tubular reservoir, q_{rp} is the volume flow rate due to the recirculation pump, and p_{memb} represents the pressure of fluid at the membrane through which the clean water passes (but leaves the impurities behind). Two abstract variables, e_{Cbrine} and e_{Ck}, capture the dynamics of the impurities in the fluid. σ_1 and σ_2 are the discrete variables. $\sigma_1 = 1, \sigma_2 = 0$ indicate that the system is in the first mode of operation, $\sigma_1 = 0, \sigma_2 = 1$ indicates that the system is in the second mode of operation, and $\sigma_1 = \sigma_2 = 0$ indicates that the system is in the third mode of operation. The efficiency decreases in the feed pump, f_f and the recirculation pump, f_r, and the membrane clogging factor f_m represent the faults in the system. The system inputs are the feed pump pressure, p_{fp} and the recirculation pump pressure, p_{rp}. The RO system parameter values and the input signals used in this case study are presented in Table 13.4.

There are five sensors in the system. The system inputs are also assumed to be known. To complete the set of constraints, we present the measured variables, the differential constraints, and the dynamic value for the membrane resistance below.

$$e_{RO7}: p_{tr}(t) = y_1(t)$$

$$e_{RO8}: p_{mem}(t) = y_2(t)$$

$$e_{RO9}: q_{fp}(t) = y_3(t)$$

$$e_{RO10}: e_{Cbrine}(t) = y_4(t)$$

$$e_{RO11}: e_{Ck}(t) = y_5(t)$$

$$e_{RO12}: p_{fp}(t) = u_1(t)$$

$$e_{RO13}: p_{rp}(t) = u_2(t)$$

Table 13.4 RO system parameters and inputs

Parameter	Name	Value and unit (SI)
Feed pump inertance	I_{fp}	0.1 N s^2/m^5
Recirculation pump inertance	I_{rp}	2 N s^2/m^5
Feed pump energy dissipation	R_{fp}	0.1 N/m^5
Recirculation pump energy dissipation	R_{rp}	0.1 N/m^5
Conductivity capacitor	C_k	565 m^5/N
Hydraulic resistance	$R_{forward}$	70 N/m^5
Capacitance of the tubular reservoir	C_{tr}	1.5 m^5/N
Hydraulic resistance	R_{return_l}	15 N/m^5
Hydraulic resistance	R_{return_s}	8 N/m^5
Hydraulic resistance	$R_{return_{AES}}$	5 N/m^5
Capacitance of the membrane module	C_{memb}	0.6 m^5/N
Brine capacitor	C_{brine}	8 m^5/N
Feed pump nominal pressure	p_{fp}	1 N/m^2
Recirculation pump nominal pressure	p_{rp}	160 N/m^2

$$e_{RO14} : q_{fp}(t) = \int \dot{q}_{fp}(t) \qquad e_{RO17} : p_{memb}(t) = \int \dot{p}_{memb}(t)$$

$$e_{RO15} : p_{tr}(t) = \int \dot{p}_{tr}(t) \qquad e_{RO18} : e_{Cbrine}(t) = \int \dot{e}_{Cbrine}(t)$$

$$e_{RO16} : q_{rp}(t) = \int \dot{q}_{rp}(t) \qquad e_{RO19} : e_{Ck}(t) = \int \dot{e}_{Ck}(t)$$

$$e_{RO20} : R_{memb}(t) = 0.202\left(4.137 \times 10^{11}\left(\frac{e_{Ck}(t) - 12000}{165} + 29\right)\right).$$

In the original design and experiments conducted with a prototype RO system, the time spent in each mode was fixed by the system operators. However, to demonstrate our mode detection algorithm, we assume the transition times are unknown, and apply our algorithm to detect when mode transitions occur.

In the experiments, for the case study, we assume 1% uncertainty in the system parameters and Gaussian measurement noise, with variance = 0.01 × mean value of the signal.

13.8.2 Mode Detection for the RO System

Employing Algorithm 13.2 we extract $MSD_{RO1} = (\mathcal{E}_{RO1}, \mathcal{T}_{RO1}, \mathcal{X}_{RO1}, \Sigma_{RO1}, \mathcal{F}_{RO1})$, where $\mathcal{E}_{RO1} = \{e_{RO2}, e_{RO5}, e_{RO6}, e_{RO7}, e_{RO8}, e_{RO9}, e_{RO10}, e_{RO11}, e_{RO15}, e_{RO18}, e_{RO19}\}$,

$\mathcal{T}_{RO1} = \{\}$,
$\mathcal{X}_{RO1} = \{p_{mem}, p_{tr}, \dot{p}_{tr}, e_{Cbrine}, \dot{e}_{Cbrine}, e_{Ck}, \dot{e}_{Ck}, q_{rp}, q_{fp}\}$,
$\Sigma_{RO1} = \{\sigma_1, \sigma_2\}$, and
$\mathcal{F}_{RO1} = \{\}$,
to detect each discrete variable in the RO system.

By substituting the known variables in the set of determined equations in MSD_{RO1}, and after some algebraic manipulations, we derive the following linear equation for solving σ_1 and σ_2:

$$\begin{aligned} a_1(t)\sigma_1(t) + b_1(t)\sigma_2(t) &= c_1(t) \\ a_2(t)\sigma_1(t) + b_2(t)\sigma_2(t) &= c_2(t), \end{aligned} \tag{13.13}$$

where $a_1(t) = \frac{y_2(t)-y_1(t)}{R_{return_l}}$,
$b_1(t) = \frac{y_2(t)-y_1(t)}{R_{return_s}}$,
$c_1(t) = C_{tr}\dot{y}_1(t) - y_3(t) + \frac{\dot{y}_5(t)C_k 1.667\times 10^{-8}}{6e_{Cbrine}(t)+0.1}$,
$a_2(t) = \frac{y_2(t)-y_1(t)}{R_{return_l}} - \frac{y_2(t)}{R_{return_{AES}}}$,
$b_2(t) = \frac{y_2(t)-y_1(t)}{R_{return_s}} - \frac{y_2(t)}{R_{return_{AES}}}$,
$c_2(t) = (1.667 \times 10^{-8} C_{brine}(t))$
$\dot{y}_4(t) - \frac{y_2(t)}{R_{return_{AES}}}$.
Therefore, σ_1 and σ_2 can be computed as

$$\begin{bmatrix} \sigma_1(t) \\ \sigma_2(t) \end{bmatrix} = \begin{bmatrix} a_1(t) & b_1(t) \\ a_2(t) & b_2(t) \end{bmatrix}^{-1} \begin{bmatrix} c_1(t) \\ c_2(t) \end{bmatrix}. \tag{13.14}$$

To show the performance of our proposed mode detection approach, we performed a simulation study, where the RO system is operated for 1000 s and switched modes every 33 s. The system started in mode 1, switched to mode 2 after 33 s, switched to mode 3 after 66 s, and switched back to mode 1 at $t = 100$ s, and switching repeated in this manner at 33 s intervals. Note that Algorithm 13.2 guarantees mode detectability independent of the system faults. However, to show a possible fault does not affect the mode detection performance, we considered an abrupt fault $f_f = 0.5$ occurring at $t = 510$ s in the experiment. Figure 13.6 shows that Eq. (13.14) perfectly estimated the discrete variables, and therefore, the operating mode of the system. To overcome the effects of measurement noise, we applied a simple low-pass filter to the measured signals.

13.8.3 Fault Detection and Isolation in the RO System

Figure 13.2 shows that the hybrid systems FDI approach includes five steps: (1) mode detection; (2) HMSO generation for the detected mode; (3) selecting a minimal set of HMSOs for fault detection; (4) generating residuals from the selected HMSOs;

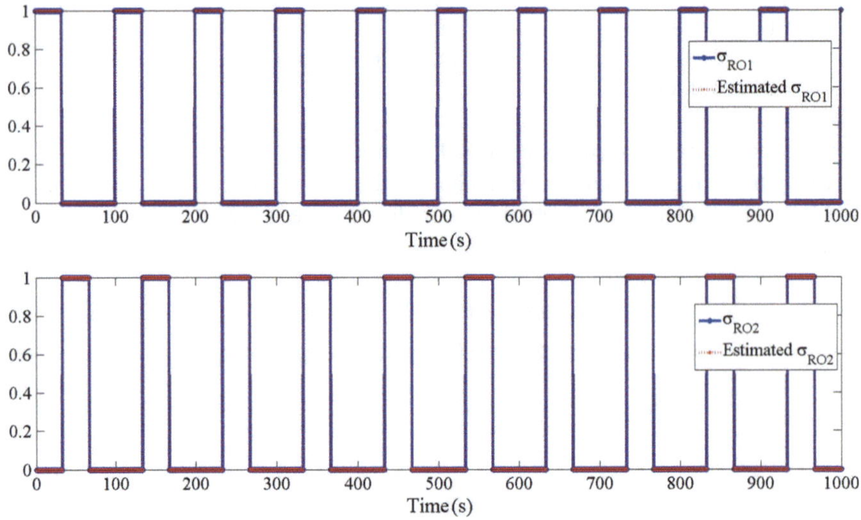

Fig. 13.6 Mode detection in the RO system in the presence of f_m

and (5) applying the generated residuals for FDI. We developed algorithms for mode detection, and selecting a minimal set of HMSOs (steps 1 and 3) in this paper. For HMSO generation and residual generation (steps 2 and 4), we used the fault diagnosis toolbox [18]. We discussed step 5 in Sect. 13.7.3.

In mode 1, this system has three possible faults $\mathcal{F} = \{f_f, f_r, f_m\}$. The fault diagnosis toolbox generates 98 HMSOs and Algorithm 13.3 selects 3 HMSOs, $HMSO_{RO1_1}$, $HMSO_{RO1_2}$, and $HMSO_{RO1_3}$, as a minimal set of HMSOs that can detect and isolate the faults. The fault diagnosis toolbox generates a residual from each HMSO.

In mode 2, the RO system has three possible faults: $\mathcal{F} = \{f_f, f_r, f_m\}$. The fault diagnosis toolbox generates 84 HMOSs and Algorithm 13.3 selects 3 HMSOs, $HMSO_{RO2_1}$, $HMSO_{RO2_2}$, and $HMSO_{RO2_3}$, as a minimal set of HMSOs that can detect and isolate the faults in the second mode. The fault diagnosis toolbox generates 3 residuals, one from each HMSO. In mode 3, the RO system has two possible faults, $\mathcal{F} = \{f_f, f_m\}$ (the recirculation pump is turned off). In this mode, the fault diagnosis toolbox generates 59 HMSOs and Algorithm 13.3 selects 2 HMSOs, $HMSO_{RO3_1}$, and $HMSO_{RO3_2}$, as a minimal set of HMSOs that can detect and isolate the faults. Table 13.5 represents the selected HMSOs and their associated faults in each mode. The fault diagnosis toolbox generates r_{RO3_1}, and r_{RO3_2} for the third mode. The set of residuals for each operating mode is available on GitHub.[2]

To detect and isolate the faults, we need to use a new set of residuals in each operating mode. Table 13.6 shows the selected hybrid residuals for FDI in each operating mode. For example, to detect f_f we have to use r_{RO1_1} when the system is

[2]https://rosystemresiduals.github.io/.

Table 13.5 Set of selected HMSOs for FDI in the RO system

HMSOs	Set of equations	Set of faults	Mode
$HMSO_{RO1_1}$	$e_{RO1}, e_{RO7}, e_{RO9}, e_{RO12}, e_{RO14}$	f_f	1
$HMSO_{RO1_2}$	$e_{RO3}, e_{RO6}, e_{RO8}, e_{RO10}, e_{RO11}, e_{RO13}, e_{RO16}, e_{RO19}$	f_r	1
$HMSO_{RO1_3}$	$e_{RO4}, e_{RO6}, e_{RO7}, e_{RO8}, e_{RO10}, e_{RO11}, e_{RO17}, e_{RO19}, e_{RO20}$	f_m	1
$HMSO_{RO2_1}$	$e_{RO1}, e_{RO7}, e_{RO9}, e_{RO12}, e_{RO14}$	f_f	2
$HMSO_{RO2_2}$	$e_{RO3}, e_{RO6}, e_{RO8}, e_{RO10}, e_{RO11}, e_{RO13}, e_{RO16}, e_{RO19}$	f_r	2
$HMSO_{RO2_3}$	$e_{RO4}, e_{RO6}, e_{RO7}, e_{RO8}, e_{RO10}, e_{RO11}, e_{RO17}, e_{RO19}, e_{RO20}$	f_m	2
$HMSO_{RO3_1}$	$e_{RO1}, e_{RO7}, e_{RO9}, e_{RO12}, e_{RO14}$	f_f	3
$HMSO_{RO3_2}$	$e_{RO3}, e_{RO4}, e_{RO7}, e_{RO8}, e_{RO11}, e_{RO17}, e_{RO20}$	f_m	3

Table 13.6 Selected residuals for fault detection and isolation

Selected residual	First mode	Second mode	Third mode
Detecting f_f	r_{RO1_1}	r_{RO2_1}	r_{RO3_1}
Detecting f_r	r_{RO1_2}	r_{RO2_2}	–
Detecting f_m	r_{RO1_3}	r_{RO2_3}	r_{RO3_2}
Isolating f_f from f_r	r_{RO1_1}	r_{RO2_1}	–
Isolating f_f from f_m	r_{RO1_1}	r_{RO2_1}	r_{RO3_1}
Isolating f_r from f_f	r_{RO1_2}	r_{RO2_2}	–
Isolating f_r from f_m	r_{RO1_2}	r_{RO2_2}	–
Isolating f_m from f_f	r_{RO1_3}	r_{RO2_3}	r_{RO3_2}
Isolating f_m from f_r	r_{RO1_3}	r_{RO2_3}	–

in the first operation mode, r_{RO2_1} when the system is in the second operation mode, and r_{RO3_1} when the system is in the third mode. Therefore, we define r_{f_f} to detect and isolate f_f.

$$r_{f_f}(t) = \begin{cases} r_{RO1_1}(t) & if \quad \sigma_1(t) = 1 \\ r_{RO2_1}(t) & if \quad \sigma_2(t) = 1 \\ r_{RO3_1}(t) & otherwise. \end{cases} \tag{13.15}$$

By running multiple simulations, we showed that the hybrid residuals derived from Table 13.6 were successful in detecting and isolating faults: f_f, f_r, and f_m in the different operation modes. However, we only present the scenario where the system executed the cycle shown in Fig. 13.6 and an abrupt efficiency decrease occurred in the feed pump $f_f = 0.5$ at $t = 510$ s.

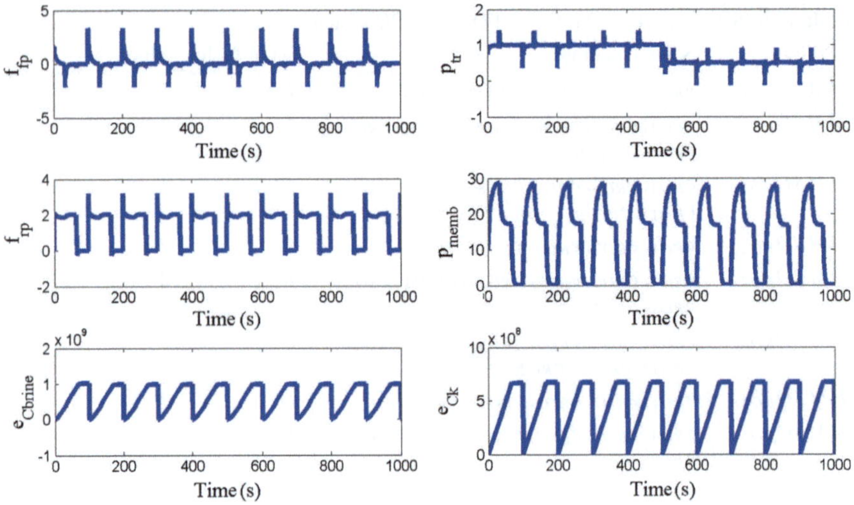

Fig. 13.7 Continuous state variables in the RO system

Fig. 13.8 f_f detection

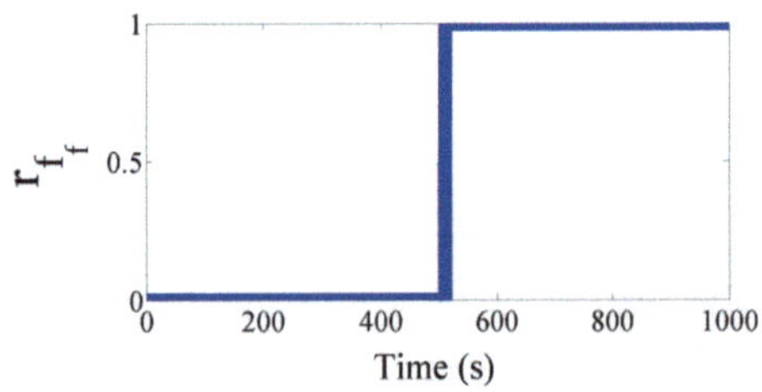

Figure 13.7 shows the state variables in this case study. Figure 13.8 shows that the hybrid residuals, r_{f_f} were successful in detecting and isolating f_f. For all of our detection and isolation experiments, we used the Z-test with 95% confidence level to evaluate residuals in each operating mode.

13.9 Conclusions

We have presented a new approach to the problem of fault detection and isolation in hybrid systems in this chapter that applies to complex hybrid systems that have a large number of operating modes. Our proposed approach consists of two algorithms: (1) mode detection and (2) fault detection and isolation in each operating mode. The contribution of this work is that we do not need to pre-compile the MSOs and residuals for every possible mode of the hybrid system, which can be computationally intractable. Therefore, the algorithm does not have to pre-enumerate all the possible modes, which is exponential in number of discrete variables in the model. Instead, our approach recomputes the HMSO's after a mode change, and updates the diagnoser model.

Unlike previous work [21], where we formulated HMSO selection as an exponential problem in the number of HMSOs in each mode, Algorithm 13.3 selects a minimal set of residuals with $O(l_{fm}^2 r_m)$ time complexity, where l_{fm} is the number of faults in mode m, and r_m is the number of HMSOs in that mode. The approach presented in this chapter adopts a greedy search, and, therefore, our algorithm does not guarantee the minimum number of HMSOs for each mode. However, in the running example and in the case study the algorithm selected the minimum number of HMSOs and in general, we expect the number of selected HMSOs to be close to the optimal solution.

It is expected that solving the mode detection and fault detection and isolation as two interrelated but separated problems requires more redundancies, and, therefore, more measurements when compared to the approaches that address both problems simultaneously, (e.g., [16, 34]). This is because we have to extract sets of just determined equations for mode identification that are independent of the system faults. However, recent developments in manufacturing inexpensive and robust sensors and efficient processors make our approach feasible for complex, modern systems.

In this work, we used mixed causality (integral and derivative) to solve for discrete variables for mode detection, and to generate residuals for fault detection and isolation. Using mixed causality improves mode detection and FDI performances. Limiting our approach to integral causality can reduce the number of solvable equations and this may make a set of discrete variables or a set of system faults undetectable [17]. However, in the presence of measurement noise, derivative computation can be error prone. In that case, restricting mode detection and residual generation to integral causality naturally increases the robustness of the approach.

To handle issues of robustness in a more general way, we have to consider the effect of model uncertainty and sensor noise in mode detection and FDI. In previous work [22, 23], we used sensitivity analysis [36] to define a detectability ratio measure and a isolability ratio measure to quantify the performance of a residual in fault detection and isolation. In future work, we will extend the previous work in robust residual selection to hybrid systems. Toward this end, we will develop a quantitative measure for robust mode detection and apply detectability ratio and isolability ratio measures for residual selection for robust FDI in each operating mode.

References

1. Ackerson, G., Fu, K.: On state estimation in switching environments. IEEE Trans. Autom. Control **15**(1), 10–17 (1970)
2. Babaali, M., Egerstedt, M.: Observability of switched linear systems. In: International Workshop on Hybrid Systems: Computation and Control, pp. 48–63. Springer (2004)
3. Bayoudh, M., Travé-Massuyès, L., Olive, X.: Coupling continuous and discrete event system techniques for hybrid system diagnosability analysis. In: 18th European Conference on Artificial Intelligence, pp. 219–223, Amsterdam, The Netherlands (2008)

4. Bayoudh, M., Travé-Massuyès, L., Olive, X.: On-line analytic redundancy relations instantiation guided by component discrete-dynamics for a class of non-linear hybrid systems. In: 48th IEEE Conference on Decision and Control, Held Jointly with the 28th Chinese Control Conference (2009)
5. Bemporad, A., Morari, M.: Control of systems integrating logic, dynamics, and constraints. Automatica **35**(3), 407–427 (1999)
6. Biswas, G., Khorasgani, H., Stanje, G., Dubey, A., Deb, S., Ghoshal, S.: An approach to mode and anomaly detection with spacecraft telemetry data. Int. J. Progn. Health Manag. (2016)
7. Biswas, G., Manders, E.J., Ramirez, J., Mahadevan, N., Abdelwahed, S.: Online model-based diagnosis to support autonomous operation of an advanced life support system. Habitat. Int. J. Hum. Support Res. **10**(1), 21–38 (2004)
8. Biswas, G., Simon, G., Mahadevan, N., Narasimhan, S., Ramirez, J., Karsai, G., et al.: A robust method for hybrid diagnosis of complex systems. In: Proceedings of the 5th Symposium on Fault Detection, Supervision and Safety for Technical Processes, pp. 1125–1131, Washington D.C., USA (2003)
9. Bregon, A., Alonso, C., Biswas, G., Pulido, B., Moya, N.: Hybrid systems fault diagnosis with possible conflicts. In: Proceedings of the 22nd International Workshop on Principles of Diagnosis, pp. 195–202, Murnau, Germany (2011)
10. Cocquempot, V., El Mezyani, T., Staroswiecki, M.: Fault detection and isolation for hybrid systems using structured parity residuals. In: 5th Asian Control conference, 2004, vol. 2, pp. 1204–1212. IEEE (2004)
11. Cocquempot, V., Staroswieck, M., Mezyani, T.E.: Switching time estimation and fault detection for hybrid system using structured parity residuals. In: IFAC Conference Safeprocess, pp. 2045–2055, Washington D.C., USA (2003)
12. Daigle, M.J., Roychoudhury, I., Biswas, G., Koutsoukos, X.D., Patterson-Hine, A., Poll, S.: A comprehensive diagnosis methodology for complex hybrid systems: a case study on spacecraft power distribution systems. IEEE Trans. Syst. Man Cybern. Part A Syst. Hum. **5**(40), 917–931 (2010)
13. Derler, P., Lee, E.A., Vincentelli, A.S.: Modeling cyber-physical systems. Proc. IEEE **100**(1), 13–28 (2012)
14. Domlan, E.A., Ragot, J., Maquin, D.: Active mode estimation for switching systems. In: American Control Conference, 2007, ACC'07, pp. 1143–1148. IEEE (2007)
15. Flaugergues, V., Cocquempot, V., Bayart, M., Pengov, M.: Structural analysis for FDI: a modified, invertibility-based canonical decomposition. In: Proceedings of the 20th International Workshop on Principles of Diagnosis, DX09, pp. 59–66, Stockholm, Sweden (2009)
16. de Freitas, N., Dearden, R., Hutter, F., Morales-Menendez, R., Mutch, J., Poole, D.: Diagnosis by a waiter and a mars explorer. Proc. IEEE **92**(3), 455–468 (2004). https://doi.org/10.1109/JPROC.2003.823157
17. Frisk, E., Bregon, A., Aslund, J., Krysander, M., Pulido, B., Biswas, G.: Diagnosability analysis considering causal interpretations for differential constraints. IEEE Trans. Syst. Man Cybern. Part A Syst. Hum. **42**(5), 1216–1229 (2012)
18. Frisk, E., Krysander, M., Jung, D.: A toolbox for analysis and design of model based diagnosis systems for large scale models. In: IFAC World Congress, Toulouse, France (2017)
19. Hofbaur, M.W., Williams, B.C.: Hybrid estimation of complex systems. IEEE Trans. Syst. Man Cybern. Part B (Cybern.) **34**(5), 2178–2191 (2004)
20. Hutter, F., Dearden, R.: Efficient on-line fault diagnosis for non-linear systems. In: Proceedings of the 7th International Symposium on Artificial Intelligence, Robotics and Automation in Space (2003)
21. Khorasgani, H., Biswas, G.: Structural fault detection and isolation in hybrid systems. IEEE Trans. Autom. Sci. Eng. **15**(4), 1585–1599 (2018). https://doi.org/10.1109/TASE.2017.2749447
22. Khorasgani, H., Jung, D.E., Biswas, G., Frisk, E., Krysander, M.: Off-line robust residual selection using sensitivity analysis. In: International Workshop on Principles of Diagnosis (DX-14), Graz, Austria (2014)

23. Khorasgani, H., Jung, D.E., Biswas, G., Frisk, E., Krysander, M.: Robust residual selection for fault detection. In: IEEE 53rd Annual Conference on Decision and Control (CDC), pp. 5764–5769. IEEE, Los Angeles, CA (2014)
24. Korte, B., Vygen, B.K.: Combinatorial Optimization. Springer, Heidelberg (2002)
25. Krysander, M., Aslund, J., Nyberg, M.: An efficient algorithm for finding minimal overconstrained subsystems for model-based diagnosis. IEEE Trans. Syst. Man Cybern. Part A Syst. Hum. **38**(1), 197–206 (2008)
26. Levy, R., Arogeti, S.A., Wang, D.: An integrated approach to mode tracking and diagnosis of hybrid systems. IEEE Trans. Ind. Electron. **61**(4), 2024–2040 (2014)
27. Li, X.R., Bar-Shalom, Y.: Multiple-model estimation with variable structure. IEEE Trans. Autom. Control **41**(4), 478–493 (1996)
28. Low, C.B.D.W., Arogeti, S., Zhang, J.B.: Causality assignment and model approximation for quantitative hybrid bond graph-based fault diagnosis. In: Proceedings of the 17th IFAC World Congress, vol. 41, pp. 10,522–10,527, Seoul, Korea (2008)
29. Low, C.B.D.W., Arogeti, S., Zhang, J.B.: Monitoring ability analysis and qualitative fault diagnosis using hybrid bond graph. In: Proceedings of the 17th IFAC World Congress, vol. 41, pp. 10,516–10,521, Seoul, Korea (2008)
30. Mazor, E., Averbuch, A., Bar-Shalom, Y., Dayan, J.: Interacting multiple model methods in target tracking: a survey. IEEE Trans. Aerosp. Electron. Syst. **34**(1), 103–123 (1998)
31. de Mortain, F., Subias, A., Travé-Massuyès, L., de Flaugergues, V.: Towards active diagnosis of hybrid systems leveraging multimodel identification and a markov decision process. IFAC-PapersOnLine **48**(21), 171–176 (2015)
32. Mosterman, P.J.: An overview of hybrid simulation phenomena and their support by simulation packages. In: International Workshop on Hybrid Systems: Computation and Control, pp. 165–177. Springer, Berg en Dal, The Netherlands (1999)
33. Mosterman, P.J., Biswas, G.: A theory of discontinuities in physical system models. J. Frankl. Inst. **335**(3), 401–439 (1998)
34. Narasimhan, S., Biswas, G.: Model-based diagnosis of hybrid systems. IEEE Trans. Syst. Man Cybern. Part A **27**(3), 348–361 (2007)
35. Narasimhan, S., Brownston, L.: HyDE-a general framework for stochastic and hybrid model based diagnosis. Proc. DX **7**, 162–169 (2007)
36. Petzold, L.R., Ascher, U.M.: Computer methods for ordinary differential equations and differential-algebraic equations. SIAM (1998)
37. Pulido, B., González, C.A.: Possible conflicts: a compilation technique for consistency-based diagnosis. IEEE Trans. Syst. Man Cybern. Part B (Cybern.) **34**(5), 2192–2206 (2004)
38. Sampath, M., Sengupta, R., Lafortune, S., Sinnamohideen, K., Teneketzis, D.C.: Failure diagnosis using discrete-event models. IEEE Trans. Control Syst. Technol. **4**(2), 105–124 (1996)
39. Torrisi, F.D., Bemporad, A.: HYSDEL-a tool for generating computational hybrid models for analysis and synthesis problems. IEEE Trans. Control Syst. Technol. **12**(2), 235–249 (2004)
40. Travé-Massuyes, L., Escobet, T., Olive, X.: Diagnosability analysis based on component-supported analytical redundancy relations. IEEE Trans. Syst. Man Cybern. Part A Syst. Hum. **36**(6), 1146–1160 (2006)

Chapter 14
Constraint-Driven Fault Diagnosis

Rafael M. Gasca, Ángel Jesús Varela-Vaca and Rafael Ceballos

14.1 Introduction

Constraint-Driven Fault Diagnosis (CDD) is based on the concept of *constraint suspension* [6], which was proposed as an approach to fault detection and diagnosis. In this chapter, its capabilities are demonstrated by describing how it might be applied to hardware systems. With this idea, a model-based fault diagnosis problem may be considered as a Constraint Satisfaction Problem (CSP) in order to detect any unexpected behavior and Constraint Satisfaction Optimization Problem (COP) in order to identify the reason for any unexpected behavior because the parsimony principle is taken into account.

In order to automate the CDD, in an efficient way, different techniques for solving these CSP/COPs could be considered. The first to be considered is related to the mathematical properties of the semiring CSP [26], where efficient diagnosis solutions are obtained decomposing the problem into trees applying dynamic programming. The second one is related to the solving of overconstrained CSPs and identification of conflicts [1, 20, 24]. The third approach [14, 15] uses a known symbolic technique called as Gröbner basis, that allows obtaining automatically, in certain cases, the Analytic Redundancy Restrictions which are used in Fault Detection and Identification (FDI) paradigm. The last efficient technique is related to specific clustering techniques that allow decomposing the original model-based diagnosis in several simpler problems and the computational complexity is reduced in a significant way.

R. M. Gasca · Á. J. Varela-Vaca · R. Ceballos (✉)
Department of Computer Languages and Systems,
Universidad de Sevilla, Seville, Spain
e-mail: ceball@us.es

R. M. Gasca
e-mail: gasca@us.es

Á. J. Varela-Vaca
e-mail: ajvarela@us.es

T. Escobet et al. (eds.), *Fault Diagnosis of Dynamic Systems*,
https://doi.org/10.1007/978-3-030-17728-7_14

In this chapter, the CDD is applied to typical hardware diagnostic problems such as multiplier–adder circuits and heat exchangers.

14.2 Constraint Programming in a Nutshell

Constraint Programming is based on the automatic resolution of CSPs, which are problems where an assignment of values to variables must be found in order to satisfy a finite number of constraints.

The model-based approach for automating the fault detection problems may be expressed as a Boolean Satisfiability Problem (SAT) or CSP instances. Therefore, the solving process consists of trying to run an SAT or CSP solver, respectively, and to determine whether they have any solution. In the case that the problems have no solution, the automation of the fault diagnosis problems consist of solving a MAX-SAT or MAX-CSP (COP) in order to implement constraint suspension.

A good guideline or rule to choose SAT or CSP representation can be related to the natural expression of the hardware models for using in model-based diagnosis approach. Another rule could be related to the computational efficiency, because either SAT solvers perform better than CSP solvers or where CSP solvers perform better than SAT solvers.

14.2.1 CSP and COP

On top of constraint technology lies the concept of the CSP [10]. In general, a CSP is composed of a set of variables, a domain for each variable, and a set of constraints. Each constraint is defined over some subset of the original set of variables and limits the combinations of values that the variables in this subset can take. The goal is to find one assignment to the variables such that the assignment satisfies all the constraints. In some kind of problems, the goal is to find all such assignments [23]. The constraints are then the relationships between the various choices for the variables.

Definition 14.1 A Constraint Satisfaction Problem (CSP) represents a reasoning framework consisting of variables, domains and constraints $\langle \mathcal{V}, \mathcal{D}, \mathcal{C} \rangle$. The set of variables $\mathcal{V} = \{v_1, v_2, \ldots, v_n\}$ could have the following domains $\mathcal{D} = \{d_1, d_2, \ldots, d_n\}$ and a constraint c_k $(v_{k_1}, \ldots, v_{k_n})$ is the set of assignment of the subset of variables in the corresponding domains that belongs to Cartesian Product $d_{k_1} \times \cdots \times d_{k_n}$. The solutions of a CSP is the set of assignment of the variables $\mathcal{V}$ in their corresponding domains that satisfy all constraints $\mathcal{C}$.

A typical example of CSP is the map-coloring problem, where given a geographical map split into regions, a color has to be assigned to all the regions dealing with adjacent regions must have different colors. An example map with eight regions is depicted in Fig. 14.1. The formalization of this CSP could be as follows:

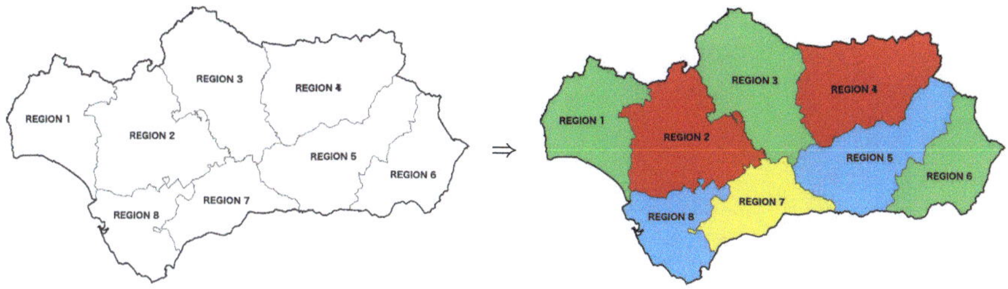

Fig. 14.1 Map-coloring problem

$$\mathcal{V} = \{R_1, R_2, R_3, R_4, R_5, R_6, R_7, R_8\}. \tag{14.1}$$

$$\mathcal{D} = \{red, green, blue, yellow\}. \tag{14.2}$$

$$\mathcal{C} = \{R_1 \neq R_2, R_1 \neq R_8, R_2 \neq R_3, R_2 \neq R_8, R_2 \neq R_7, R_3 \neq R_4, R_3 \neq R_5, R_7 \neq R_5, R_5 \neq R_4, R_5 \neq R_6\}. \tag{14.3}$$

The problem is composed of eight variables that represent each region, the domains are the colors, and the constraints establish the inequality conditions that the color of one region is different from the adjacent regions.

One application of the CSP might be the determination of the color assignments to regions, for instance, as given in Fig. 14.1. Another application might be given a solution to determine whether the solution is satisfiable or to determine the fault diagnosis pointing out the constraint and variables that make the CSP unsatisfied. This type of problem is modeled as MAX-CSP, thus a Constraint Optimization Problem (COP).

Definition 14.2 A Constraint Optimization Problem (COP) is a CSP in which the solutions optimize (minimize or maximize) an objective function, f over a subset of variables $\mathcal{V}$ of the CSP.

Following the map-coloring example, if a color assignment is given as observational model such as the right map as shown in Fig. 14.2. By applying the previous CSP, we can check the unsatisfiability since the regions R_1 and R_8 have the same color $green$. In this case, reified constraints can be applied by means of a COP (i.e., MAX-CSP) in order to automatically determine the faults. The representation of the problem might be as follows:

$$\mathcal{V} = \{R_1, R_2, R_3, R_4, R_5, R_6, R_7, R_8, RR_1, RR_2, RR_3, RR_4, RR_5, RR_6, RR_7, RR_8, RR_9, RR_{10}\}. \tag{14.4}$$

$$\mathcal{D}_{R_x} = \{red, green, blue, yellow\}. \tag{14.5}$$

$$\mathcal{D}_{RR_x} = \{0, 1\}. \tag{14.6}$$

$$\mathcal{C} = \{R_1 = green, R_2 = red, R_3 = green, R_4 = red, R_5 = blue,$$

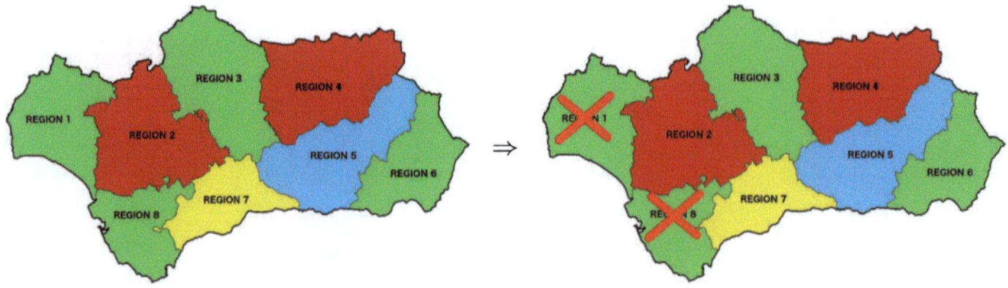

Fig. 14.2 Fault diagnosis of an assignment

$$R_6 = green, R_7 = yellow, R_8 = green. \quad (14.7)$$

$$RR_1 = 1, RR_2 = 1, RR_3 = 1, RR_4 = 1, RR_5 = 1, RR_6 = 1,$$
$$RR_7 = 1, RR_8 = 1, RR_9 = 1, RR_{10} = 1, \quad (14.8)$$

$$RR_1 == R_1 \neq R_2, RR_2 == R_1 \neq R_8, RR_3 == R_2 \neq R_3,$$
$$RR_4 == R_2 \neq R_8, RR_5 == R_2 \neq R_7, RR_6 == R_3 \neq R_4,$$
$$RR_7 == R_3 \neq R_5, RR_8 == R_7 \neq R_5, RR_9 == R_5 \neq R_4,$$
$$RR_{10} == R_5 \neq R_6\}. \quad (14.9)$$

$$\mathcal{F} = maximize(RR_1 + RR_2 + RR_3 + RR_4 + RR_5 + RR_6 +$$
$$RR_7 + RR_8 + RR_9 + RR_{10}). \quad (14.10)$$

The COP tries to maximize the sum of the reified constraints to be satisfied. The reified constraints are represented as "Boolean" variables RR_x with zero or one as domain. The RR_x variables have been forced to 1 value (cf. $RR_x = 1$). These values force the accomplishment of the constraints such as $RR_1 == R_1 \neq R_8$. However, this constraint cannot be satisfiable since the values of R_1 and R_8 are the same, *green*. This contradiction makes the COP unsatisfiable due to the assignment of *green* color for the adjacent regions *Region 1* and *Region 8* as shown in right map in Fig. 14.2.

The techniques used in solving the CSPs depend on the kind of constraints being considered. Constraints are often used on a finite domain, to the point that CSP is typically identified with problems based on constraints on a finite domain. Such problems are usually solved via techniques that combine propagation and search, in a particular way of backtracking or local search. Constraint propagation is a method used on such problems, but the majority of them are incomplete. In general, they may solve the problem or check if it is unsatisfiable, but not always. For this reason, these methods are also used in conjunction with a search algorithm to make possible the solving of these problems. Other considered kinds of constraints are on real or rational numbers, solving problems on these constraints is done via specific constraint propagation and search algorithms.

One of the main difficulties in CSP resolution is the appearance of local inconsistencies. Local inconsistencies are values of the variables that cannot take part in the solution because they do not satisfy any consistency property. Therefore if any consistency property is forced, we can remove all the values which are inconsistent with regard to the property. But it can be possible that some values which are consistent with regard to a property are inconsistent with regard to another property at the same time. Global consistency implies that all values which cannot take part in a solution can be removed. The constraints of a CSP generate local inconsistencies because they are combined. If the search algorithm does not store these inconsistencies, it will waste time and effort trying to carry out instantiations which have already been tested.

Different orders of consistency can be considered [22] according to the size of subnetworks taken into account: 1-consistency for networks involving one variable, 2-consistency for two variables, and in general k-consistency for networks involving k variables. An algorithm for computing k-consistency for discrete CSP [12] had been proposed whose runtime is exponential in k.

As it has been shown, since CSP can be used to detect constraint inconsistencies, it can be also be used within the model-based diagnosis paradigm. To diagnose a system within the CSP framework one must first represent the model of the system structure and the correct behavior. The structural elements of the system being diagnosed are modeled as CSP variables, and the behavioral relationships among the constituent components are expressed as constraints.

The diagnosis schema employs a CSP engine, whose outputs are the expected diagnoses based on the discrepancies between the predictions of the CSP representation and the observations from running the system. The diagnosed system is found faulty if some constraints cannot be satisfied, in which case the violated constraints are the precise reasons for the cause of the fault [11].

14.2.2 SAT and MAX-SAT

The Boolean Satisfiability Problem (SAT) is the problem of deciding whether there is a variable assignment that satisfies a given propositional expression. The boolean expressions with Boolean variables have no quantifiers. In 3-SAT, the maximum number of literals for the clause is 3 and in 2-SAT, the maximum number of literals for the clause is 2. It is solved efficiently by using path searches in graphs. For certain problems of fault detection, the SAT can be used and the MAX-SAT for fault diagnosis. An application of SAT and MAX-SAT is going to be illustrated in the Model-based Software Debugging chapter.

The solving process of these problems is carried out by means of SAT Solver or either translating MAX-SAT to MAX-CSP that is trivial. Translating CSP to SAT is not trivial, because it is necessary to convert multivalued variables into a set of boolean assignment variables and translate the constraints into clauses over assignment variables. SAT solvers may be complete such as solvers based on the Davis–Putnam–Logemann–Loveland (DPLL) algorithm, and incomplete SAT solvers based on local search and hybrid SAT solvers.

14.3 Automation of Constraint-Driven Fault Diagnosis

In order to efficiently automate the constraint-driven fault diagnosis, a set of concepts is necessary to consider for a system with *m* components (physical components):

Definition 14.3 *System Model (SM)* is a finite set of linear or nonlinear (polynomial) equality constraints ($\mathcal{P}$), which determine the hardware system behavior. This is done by means of the relations between the system non-observable variables ($\mathcal{V}_{nob}$) and the system observable variables ($\mathcal{V}_{ob}$), which are directly obtained from sensors that are supposed to work correctly for hardware systems. The representation of an SM is a tuple ($\mathcal{P}$, $\mathcal{V}_{ob}$, $\mathcal{V}_{nob}$).

An example used in the bibliography with respect to model-based diagnosis [5–7, 16, 22] is the polybox system previously introduced in Chap. 2 (cf. Logical Case Studies). This system is composed of three multipliers and two adders such as the shown in Fig. 14.3. We will have the following System Model (SM): $\langle \mathcal{P} : \{a * c = x, b * d = y, c * e = z, x + y = f, y + z = g\}, \mathcal{V}_{ob}\{a, b, c, d, e, f, g\}, \mathcal{V}_{nob} : \{x, y, z\}\rangle$.

Definition 14.4 *Context Set (CS)* is a subset of components of the system and its associated constraints. The possible context set will be 2^m.

Definition 14.5 *Context Network (CN)* is a lattice of subsets of the component of the system, ordered by "is subset of", according to the way proposed by ATMS [21]. The number of possible contexts for a system of *n* components is $2^m - 1$.

Definition 14.6 *Observational Model (OM)* is defined by a tuple of values for the observable variables. In the example model in Fig. 14.3, two observation models could be

$$OM_1 \equiv \{a = 3, b, = 2, c = 2, d = 3, e = 3, f = 10, g = 12\}. \quad (14.11)$$

$$OM_2 \equiv \{a = 3, b = 2, c = 2, d = 3, e = 3, f = 10, g = 10\}. \quad (14.12)$$

Definition 14.7 *Diagnosis Problem* is defined by a tuple composed of a System Model (SM) and an Observational Model. The solution to this problem is a set of possible failed components of the system.

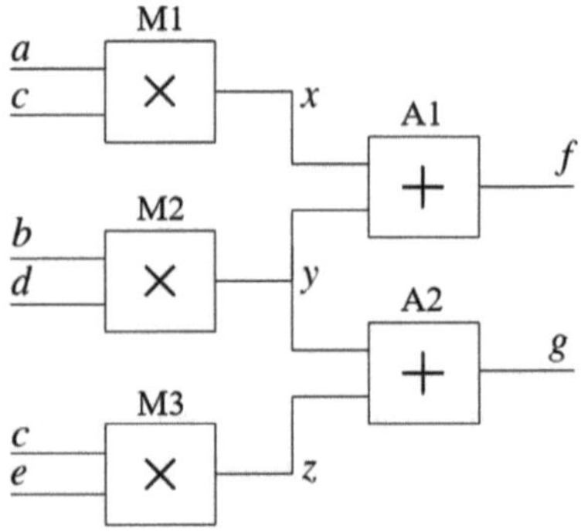

Fig. 14.3 Model of three multipliers and two adders

A set of computational techniques for the constraint-driven fault diagnosis are presented in the following subsections.

14.3.1 Constraint Suspension for Diagnostic Reasoning

In 1984, Davis [6] presents a new technique for diagnostic reasoning called constraint suspension. The reasoning from first principles is proposed using different languages for describing the structure and the component behavior. According to these languages, a constraint network is obtained and then the goal is to derive the components that could be responsible for a concrete fault, given the observations of the system. In the faulty case, the constraint network is inconsistent and the possible consistency is achieved by retracting some constraints (component behavior) of the network when we take into account the observational model. In order to improve the efficiency, this work proposed the concept of *discrepancy detection*.

14.3.2 Semiring CSP for Diagnostic Reasoning

Semiring CSP [2] is a framework for soft constraints. Soft constraints establish preference levels changing the definition of hard constraints. In these soft constraints, the assignments are associated with different interpretations such as weight, cost, preference, etc.

Model-based diagnosis may be solved as a COP over lattices. Sachenbacher and Williams [26] presents a framework as semiring CSPs. The mathematical properties of a semiring CSP allow to obtain efficient solution decomposing model-based problems into trees and later applying dynamic programming. The method allows to perform diagnosis over the general class of lattice preference structures and it is compared with the diagnosis algorithms SAB [9] and TREE* [27] for tree-structured problems that are special cases.

14.3.3 Overconstrained CSPs and Identification of Conflicts

Solving CSP process assigns values to variables satisfying all of the set of constraints, but it is impossible when a fault is present in the diagnosis problems. In the constraint programming area, these problems are called as overconstrained CSP and do not have any solution. We need to consider some ways of relaxing or weakening the specification of detection fault problem (in our case of the description of the hardware system model) until solutions can be found. Therefore, the systematic or local search does not find solutions for these overconstrained CSPs. Nevertheless, the number of satisfied constraints could be maximized for the detection fault problem. There are in the bibliography four ways of weakening an overconstrained CSP:

- Extending the domain of a variable,
- Extending the definition of a constraint,
- Eliminating any variable,
- Suspending any constraint.

The idea, related to fault diagnosis, is that given an infeasible set of constraints C, the goal is obtain explanations of infeasibility. For this reason, we must find out:

- All Minimal Unsatisfiable Subsets (MUSes) of the problem where all proper subsets satisfiable or,
- All Maximal Satisfiable Subsets (MSSes) of the problem.

The definition and strong relationship between MUSes and MSSes of a determined overconstrained problem has been analyzed in a previous work [25]. An adequate and specific solver could compute all Minimally Unsatisfiable Subsets (MUSes) of a given overconstrained CSP. For the following overconstrained CSP:

$$\mathcal{V} = \{a, b, c, d, e, f, g, x, y, z\}. \tag{14.13}$$

$$\mathcal{D} = \{2, 3, 3, 2, 2, 10, 12, [0, 100], [0, 100], [0, 100]\}. \tag{14.14}$$

$$\begin{aligned} \mathcal{C} = \{c_1 = a \times c = x, \quad & c_2 = b \times d = y, c_3 = c \times e = z, \\ & c_4 = x + y = f, c_5 = y + z = g\}. \end{aligned} \tag{14.15}$$

The set of constraints $\{c_1, c_2, c_4\}$ and $\{c_1, c_3, c_4, c_5\}$ are minimally unsatisfiable subsets for this problem. They will eventually be used for obtaining the solution of the fault diagnosis problem. A CSP is satisfiable iff it contains no MUSes.

Also, we could determine MSSes or CoMSSes, that is the complement of MSSes. Every CoMSS could be obtained as an irreducible hitting set of the collection of all MUSes. In the above example, CoMSSes are $\{c_1\}$, $\{c_4\}$, $\{c_2, c_3\}$, $\{c_2, c_5\}$. Therefore, hitting sets provide a transformation from MUSes to CoMSSes and they provide the possible solutions for the fault diagnosis problem. This transformation removes those constraints making the overconstrained CSP unsatisfiable.

Also, in the literature, Baker et al. [1] propose the Diagnosis of Over-determined CSPs (DOC), which identifies the set of least-important constraints that should be relaxed to the initial overconstrained CSP has a solution. If the solution is unacceptable for a user, the DOC selects next-best sets of least-important constraints until an acceptable solution has been generated. QUICKXPLAIN [20] improves this method by decomposing the complete explanation into subproblems of the same size.

For overconstrained CSP in specific domains, different works have been proposed in the bibliography. In overconstrained instances of the Disjunctive Temporal Problem (DTP), Liffiton et al. [24] describe the Musilitis algorithm for finding MUSes of overconstrained DTP. Also, for numeric overconstrained CSPs (NCSPs), the paper [13] proposes a set of methods for efficiently deriving all Numeric MUSes (NMUS) using the Neighborhood-based Structural Analysis on overconstrained NCSP. In such work, different bottom-up derivation strategies are described by taking into account the concept of the neighborhood for the different types of NCSPs. Depending on the structural aspects of these problems, the search methods are different.

14.3.4 Symbolic Techniques

In engineering problems, the models are almost always a set of linear and polynomial constraints. In order to automate and improve the fault diagnosis a new model can be derived, which is made up of a single constraint using a symbolic technique such as Gröbner basis. First, it eliminates the non-observable variables and obtains new constraints of the different set of components of the system. We build a context network with these polynomial constraints in the node and propose a methodology that is composed of an incremental algorithm that avoids the recalculation of previous constraints. A standard framework is used to obtain the minimal diagnosis. The obtained results are similar to those shown in the bibliography, but they are obtained by means of a more efficient and automatic way. This approach may be very useful for on-board diagnosis [14, 15].

Before presenting the methodology to carry out the fault diagnosis process for this model, we need to present the concept of Gröbner basis. A Gröbner basis is a set of multivariate polynomials that have desirable algorithmic properties. An algorithm permits to transform the set of polynomials into a Gröbner basis. It generalizes previous techniques such as Gaussian elimination for solving linear systems of equations, the Euclidean algorithm for computing the greatest common divisor of two univariate polynomials, and the Simplex Algorithm for linear programming. The introduction for Gröbner basis is presented in several works [3, 8, 19].

According to theoretical considerations, a set of multivariate polynomials $\mathcal{P} = 0$ does not have solutions if and only if 1 is in its Gröbner basis [3]. Therefore,

- If a polynomial model is overconstrained and have redundant constraints, the calculation of Gröbner basis eliminated them.
- If the polynomial model is overconstrained and inconsistent, one of the constraints obtained is $1 = 0$, that determine the inconsistency of the model.

Table 14.1 Minimal conflict contexts of the model in Fig. 14.3 without observational model

Minimal conflict contexts	Associated constraints	Truth value
M_1, M_2, A_1	$a * c + b * d - f = 0$	Not evaluated
M_2, M_3, A_2	$b * d + c * e - g = 0$	Not evaluated
M_1, M_3, A_1, A_2	$a * c - c * e - f + g = 0$	Not evaluated

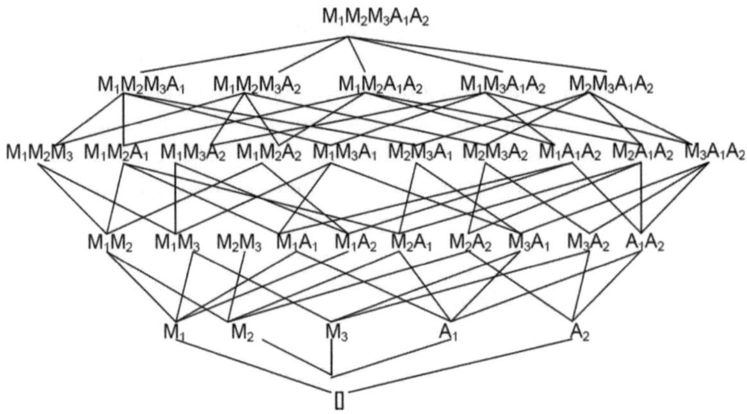

Fig. 14.4 Context network for the model of three multipliers and two adders

- If a set of polynomials is underconstrained, the Gröbner basis obtained is very useful for the possible resolution.

The parameters of the function *GröbnerBasis* are the set of polynomial constraints, set of observable variables, set of non-observable variables. These functions allow the calculation of the Gröbner basis. For example in Fig. 14.4, given the context for the components $\{M_1, M_2, M_3, A_1, A_2\}$, the function GröbnerBasis $(\{a * c - x, b * d - y, c * e - z, x + y - f, y + z - g\}, \{a, b, c, d, e, f, g\}, \{x, y, z\})$ obtains the following result:

$$\{b * d + c * e - g = 0, a * c - c * e - f + g = 0\}. \tag{14.16}$$

The application of the *GröbnerBasis* function to all contexts of the system represented in Fig. 14.3 permits to obtain the new context network represented in Table 14.1 and Fig. 14.5. In this figure, the contexts with only one component are not represented because all the contexts with two components of the Gröbner function are empty. Only the contexts with no-empty Gröbner functions are relevant and useful for fault diagnosis.

According to this new context network, the methodology for solving the fault diagnosis problem for the system given an observation model is the following:

- Searching the Minimal Conflict Contexts: An algorithm of Breadth-First Search (BFS) of the graph represented in Fig. 14.5 obtain for this system:
- Considering the observational model OM_1 the previous table is completed. We obtain the minimal conflict contexts (ARR in FDI) that are evaluated as false, resulting as shown in Table 14.2.
- Obtaining the minimal diagnoses. A standard algorithm to obtain the hitting set is used. For the previous observational models, the results are presented in Table 14.3.

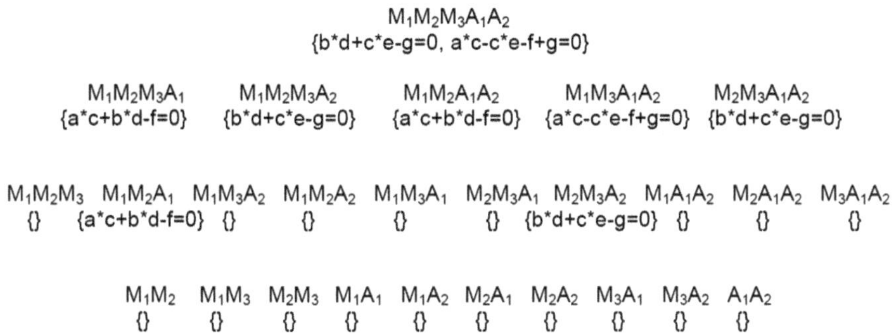

Fig. 14.5 Context network with Gröbner Basis for the model of three multipliers and two adders

Table 14.2 Minimal conflict contexts of the model in Fig. 14.3 with observational model OM_1

Minimal conflict contexts	Associated constraints	Truth value
M_1, M_2, A_1	$a*c+b*d-f=0$	False
M_2, M_3, A_2	$b*d+c*e-g=0$	True
M_1, M_3, A_1, A_2	$a*c-c*e-f+g=0$	False

Table 14.3 Minimal diagnoses of the model in Fig. 14.3

Observational model	Minimal diagnosis
OM_1	{A1}, {M1}, {A2, M2}, {M2, M3}
OM_2	{M2}, {A1, A2}, {A1, M3}, {A2, M1},{M1, M3}

As an exercise, the reader can try to build the Minimal Conflict Context for OM_2 similarly to the Table 14.2. Finally, compare the results with the diagnosis in Table 14.3.

14.3.5 Improvements for Determining the Minimal Conflicts Contexts

For determining the Minimal Conflicts Contexts in an efficient way, in this subsection two improvements are proposed [4]: a structural pretreatment in order to reduce drastically the number of nodes of context network, and a reduction of the number of contexts where Gröbner Bases algorithm is applied. These two improvements can be done in an offline process.

14.3.5.1 Clusters of Components

The first step is the partition of the system into independent subsystems, in such a way that the Minimal Conflicts Contexts of the system can be obtained as the union of all Minimal Conflicts Contexts of all subsystems. These subsystems are smaller than the whole system, and therefore the computational complexity for detecting conflicts is reduced.

Definition 14.8 *Cluster of components.* A set of components $\mathcal{C}$ is a cluster of component only if:

1. For all non-observable inputs and outputs of each component of $\mathcal{C}$, these inputs and outputs are always linked only with components of $\mathcal{C}$.
2. It does not exist another set $\mathcal{C}' \subseteq \mathcal{C}$ which satisfies the first condition.

The first part the definition guarantees that we are able to detect a conflict in a cluster of components without information about other clusters. The second part of the definition guarantees that the size of clusters will be as small as possible. For each cluster, an independent context network is obtained. The set of conflicts for each cluster can be obtained by independent processes (also in a parallel way), and the union of the conflicts of all the clusters will be the complete set of conflicts of the system.

Example There is just one cluster in the polybox example, and the context network contains $2^5 - 1 = 31$ nodes. But, for example, if the visibility of variable y is changed from non-observable to observable, there will be three clusters: $\{M_1, A_1\}$, $\{M_2\}$ and $\{M_3, A_2\}$, and there will be three context networks with a total of $(2^2 - 1) + (2^1 - 1) + (2^2 - 1) = 7$ nodes.

14.3.5.2 Relevant Contexts

In order to obtain an equivalent context network but without non-observable variables, Gröbner Basis algorithm is applied. The goal is reducing the number of nodes (of the context network) to be processed, and therefore reducing the computational complexity. The idea is to select which contexts are important for detecting conflicts. These contexts are named relevant contexts.

Definition 14.9 *Relevant contexts.* RC is a relevant context if applying Gröbner Basis algorithm (GB), one or more obtained constraints cannot be obtained by applying the same algorithm to other sub-contexts $RC' \subseteq RC$, and $\forall\, RC' \subseteq RC$: $GB(RC') \subseteq GB(RC)$.

In order to know if a context is relevant or irrelevant, the first step is removing for each context all the constraints which include only observable variables. In this case, it has no sense to apply Gröbner Basis algorithm. The next step is removing the constraints that contain at least one non-observable variable, which appears only

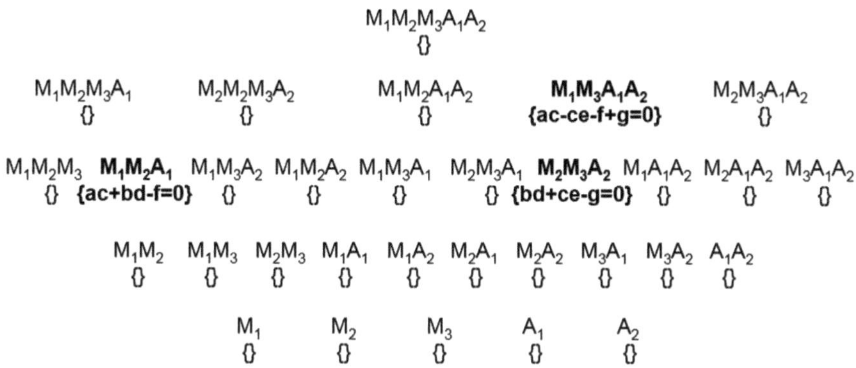

Fig. 14.6 Context network of the polybox example. Relevant contexts are shown in bold

one time in the set of constraints associated with each context because Gröbner Basis algorithm is not able to remove these variables. Finally, the context is relevant, if after the first and second step, there is at least one constraint for each component of the context.

Gröbner Basis algorithm is not applied to irrelevant contexts, because the set of obtained constraints will be empty or included in the set of obtained constraints of relevant contexts. Irrelevant contexts are not necessary for obtaining the complete set of conflicts of a system.

Example In Fig. 14.6, the context network for the polybox example is shown. Gröbner Basis algorithm is applied only to the relevant contexts: $M_2M_3A_1A_2$, $M_1M_2A_2$, and $M_2M_3A_2$.

14.3.5.3 Determination of Minimal Conflict Contexts

A constraint-driven algorithm is proposed in order to obtain the Minimal Conflict Contexts (MCCs), based on the following definition:

Definition 14.10 *Context Analytical Redundancy Constraint (CARC).* A constraint derived from SM where only observable variables are related. The set of CARCs is the union of all the constraints obtained by applying Gröbner Basis algorithm for each relevant context.

A relevant context RC is an MCC if it has not an empty set of constraints and it satisfies one of the following predicates:

- All its sub-contexts are not MCCs.
- At least one CARC of the context *RC* is not included in any of its sub-contexts which are MCCs.

In order to determine which relevant contexts are MCCs, the graph of relevant contexts is traversed from the leaf nodes to the root, in such a way that only upper relevant contexts that satisfy the definition of MCC will be taken into account.

Example In the polybox example, all relevant contexts are MCCs. There are three CARCs: $ac - ce - f + g = 0$ (context $M_2M_3A_1A_2$), $ac + bd - f = 0$ (context $M_1M_2A_2$), $bd + ce - g = 0$ (context $M_2M_3A_2$).

The set of minimal diagnoses is obtained by applying a standard hitting set algorithm to the MCCs, which are evaluated to false when an observational model is applied. The evaluation of the MCCs and the hitting set algorithm are done in an online process (when an observational model is known), but all previous steps (as the process for obtaining the MCCs) can be done in an offline process and just one time for all possible observational models.

14.4 Application to a Case of Study

In Fig. 14.7, a system of heat exchangers is shown. This system [18] consists of six heat exchangers, three flows f_i come in at different temperatures t_i. There are three different subsystems, each one formed by two exchangers: E1, E2, E3, E4, and E5, E6. Each of the six exchangers and each of the eight nodes of the system are considered as components in order to be able to verify their correct behavior. In [17], the complete SM of the example is shown.

The system of heat exchangers has five clusters: $\{N_{11}\}$, $\{N_{13}\}$, $\{N_{12}, N_{21}, N_{22}, E_1, E_2\}$, $\{N_{14}, N_{23}, N_{24}, E_5, E_6\}$, and $\{E_3, E_4\}$. For example, $\{E_3, E_4\}$ is a cluster because all the connections (inputs and outputs) to other components outside the cluster are observable. There are some non-observable connections inside the cluster, but do not connect to components outside the cluster. For example, the outputs f_{32} and t_{32} of the component E_3 are not observables, but they are inputs for the component E_4.

For each cluster, an independent context network is obtained. This pretreatment reduces the number of nodes of the context network, for example from $2^{14} - 1$ to $(2^1 - 1) + (2^1 - 1) + (2^5 - 1) + (2^5 - 1) + (2^2 - 1) = 67$.

Gröbner Basis algorithm is not applied for irrelevant contexts. For example, context N_{12} is irrelevant because all variables of their constraints are observable. The context $N_{12}N_{21}E_1E_2$ is irrelevant because there exists another context, $N_{12}E_1E_2$, without component N_{21}, that generates the same set of constraints. Another example is E_1E_2, it is an irrelevant context because applying Gröbner Basis is not possible to obtain any constraint.

In Fig. 14.8, the context network for the components cluster $\{N_{12}, N_{21}, N_{22}, E_1, E_2\}$ is shown. Gröbner Basis algorithm is applied only to the relevant contexts: $N_{12}E_1E_2$, $N_{21}N_{22}E_1E_2$, and $N_{12}N_{21}N_{22}E_1E_2$. In Table 14.4, the differences between applying Gröbner Basis algorithm to all contexts (as in [14, 15]) or only to the relevant contexts for each cluster are shown.

In Fig. 14.9, all CARCs grouped by clusters are shown. In order to determine which relevant context is MCC, we have to traverse the graph of the context network

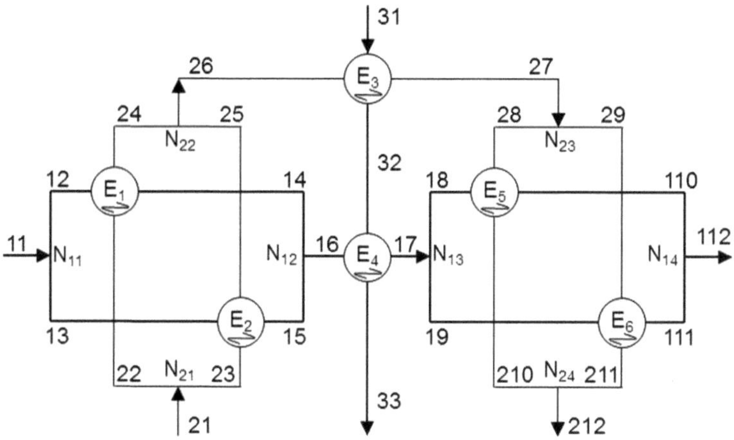

The normal behaviour of the system can be described by means of these constraints:

$\sum_i f_i = 0$: mass balance at each node

$\sum_i f_i \cdot t_i = 0$: thermal balance at each node

$\sum_{in} f_i \cdot t_i - \sum_{out} f_j \cdot t_j = 0$: enthalpy balance for each heat exchanger

C.	Constraints
N_{11}	$\mathbf{f_{11}}\text{-}\mathbf{f_{12}}\text{-}\mathbf{f_{13}}$
	$\mathbf{f_{11}\cdot t_{11}}\text{-}\mathbf{f_{12}\cdot t_{12}}\text{-}\mathbf{f_{13}\cdot t_{13}}$
N_{12}	$f_{14}+f_{15}\text{-}\mathbf{f_{16}}$
	$f_{14}\cdot t_{14}+f_{15}\cdot t_{15}\text{-}\mathbf{f_{16}\cdot t_{16}}$
N_{13}	$\mathbf{f_{17}}\text{-}\mathbf{f_{18}}\text{-}\mathbf{f_{19}}$
	$\mathbf{f_{17}\cdot t_{17}}\text{-}\mathbf{f_{18}\cdot t_{18}}\text{-}\mathbf{f_{19}\cdot t_{19}}$
N_{14}	$f_{110}+f_{111}\text{-}\mathbf{f_{112}}$
	$f_{110}\cdot t_{110}+f_{111}\cdot t_{111}\text{-}\mathbf{f_{112}\cdot t_{112}}$
N_{21}	$\mathbf{f_{21}}\text{-}f_{22}\text{-}f_{23}$
	$\mathbf{f_{21}\cdot t_{21}}\text{-}f_{22}\cdot t_{22}\text{-}f_{23}\cdot t_{23}$

C.	Constraints
N_{22}	$f_{24}+f_{25}\text{-}\mathbf{f_{26}}$
	$f_{24}\cdot t_{24}+f_{25}\cdot t_{25}\text{-}\mathbf{f_{26}\cdot t_{26}}$
N_{23}	$\mathbf{f_{27}}\text{-}f_{28}\text{-}f_{29}$
	$\mathbf{f_{27}\cdot t_{27}}\text{-}f_{28}\cdot t_{28}\text{-}f_{29}\cdot t_{29}$
N_{24}	$f_{210}+f_{211}\text{-}\mathbf{f_{212}}$
	$f_{210}\cdot t_{210}+f_{211}\cdot t_{211}\text{-}\mathbf{f_{212}\cdot t_{212}}$
E_1	$\mathbf{f_{12}}\text{-}f_{14}$
	$f_{22}\text{-}f_{24}$
	$\mathbf{f_{12}\cdot t_{12}}\text{-}f_{14}\cdot t_{14}+f_{22}\cdot t_{22}\text{-}f_{24}\cdot t_{24}$

C.	Constraints
E_2	$\mathbf{f_{13}}\text{-}f_{15}$
	$f_{23}\text{-}f_{25}$
	$\mathbf{f_{13}\cdot t_{13}}\text{-}f_{15}\cdot t_{15}+f_{23}\cdot t_{23}\text{-}f_{25}\cdot t_{25}$
E_3	$\mathbf{f_{26}}\text{-}\mathbf{f_{27}}$
	$\mathbf{f_{31}}\text{-}f_{32}$
	$\mathbf{f_{26}\cdot t_{26}}\text{-}\mathbf{f_{27}\cdot t_{27}}+\mathbf{f_{31}\cdot t_{31}}\text{-}f_{32}\cdot t_{32}$
E_4	$\mathbf{f_{16}}\text{-}\mathbf{f_{17}}$
	$f_{32}\text{-}\mathbf{f_{33}}$
	$\mathbf{f_{16}\cdot t_{16}}\text{-}\mathbf{f_{17}\cdot t_{17}}+f_{32}\cdot t_{32}\text{-}\mathbf{f_{33}\cdot t_{33}}$

C.	Constraints
E_5	$\mathbf{f_{18}}\text{-}f_{110}$
	$f_{28}\text{-}f_{210}$
	$\mathbf{f_{18}\cdot t_{18}}\text{-}f_{110}\cdot t_{110}+f_{28}\cdot t_{28}\ \text{-}f_{210}\cdot t_{210}$

C.	Constraints
E_6	$\mathbf{f_{19}}\text{-}f_{111}$
	$f_{29}\text{-}f_{211}$
	$\mathbf{f_{19}\cdot t_{19}}\text{-}f_{111}\cdot t_{111}+f_{29}\cdot t_{29}\ \text{-}f_{211}\cdot t_{211}$

$$\mathcal{V}_{observable}=\{f_{11},f_{12},f_{13},f_{16},f_{17},f_{18},f_{19},f_{112},f_{23},f_{21},f_{26},f_{27},f_{212},f_{31},f_{33},t_{11},t_{12},t_{13},t_{16},t_{17},t_{18},t_{19},t_{112},t_{21},t_{26},t_{27},t_{212},t_{31},t_{33}\}$$

$$\mathcal{V}_{nonObservable}=\{f_{14},f_{15},f_{21},f_{22},f_{22},f_{23},f_{24},f_{25},f_{28},f_{29},f_{210},f_{211},f_{32},t_{14},t_{15},t_{110},t_{111},t_{22},t_{23},t_{24},t_{25},t_{28},t_{29},t_{210},t_{211},t_{32}\}$$

Fig. 14.7 System of heat exchangers, components, equations, and variables

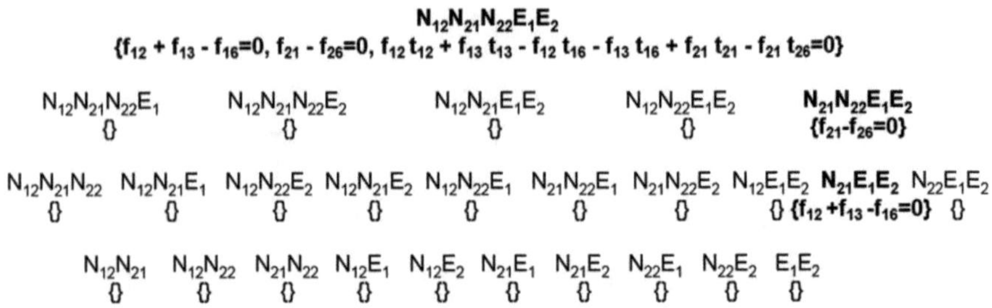

Fig. 14.8 Context network of the cluster $\{N_{12}, N_{21}, N_{22}, E_1, E_2\}$. Relevant contexts are shown in bold

Table 14.4 Improvement by obtaining clusters of components and relevant contexts

	No reduction	Using clusters	Using clusters and Relevant C
Number of nodes of the context net	$2^{14} - 1$	67	67
Calls to Gröbner Basis	$2^{14} - 1$	67	7
Number of obtained constraints	64	14	14

Index	Cluster	Constraints
1	1	$f_{11} - f_{12} - f_{13} = 0$
2	1	$f_{11} t_{11} - f_{12} t_{12} - f_{13} t_{13} = 0$
3	2	$f_{17} - f_{18} - f_{19} = 0$
4	2	$f_{17} t_{17} - f_{18} t_{18} - f_{19} t_{19} = 0$
5	3	$f_{12} + f_{13} - f_{16} = 0$
6	3	$f_{21} - f_{26} = 0$
7	3	$f_{12} t_{12} + f_{13} t_{13} - f_{12} t_{16} - f_{13} t_{16} + f_{21} t_{21} - f_{21} t_{26} = 0$
8	4	$f_{18} + f_{19} - f_{112} = 0$
9	4	$f_{27} - f_{212} = 0$
10	4	$f_{18} t_{18} + f_{19} t_{19} - f_{18} t_{112} - f_{19} t_{112} + f_{27} t_{27} - f_{27} t_{212} = 0$
11	5	$f_{26} - f_{27} = 0$
12	5	$f_{16} - f_{17} = 0$
13	5	$f_{31} - f_{33} = 0$
14	5	$f_{16} t_{16} - f_{17} t_{17} + f_{26} t_{26} - f_{27} t_{27} + f_{31} t_{31} - f_{31} t_{33} = 0$

Fig. 14.9 Context analytical redundancy constraints

from the leaf nodes to the root, in such a way that, if any of the upper contexts do not validate the definition of MCC, it cannot be considered as an MCC. In the heat exchangers example, all the relevant contexts are MCC. In Fig. 14.10, all MCCs of the system are shown.

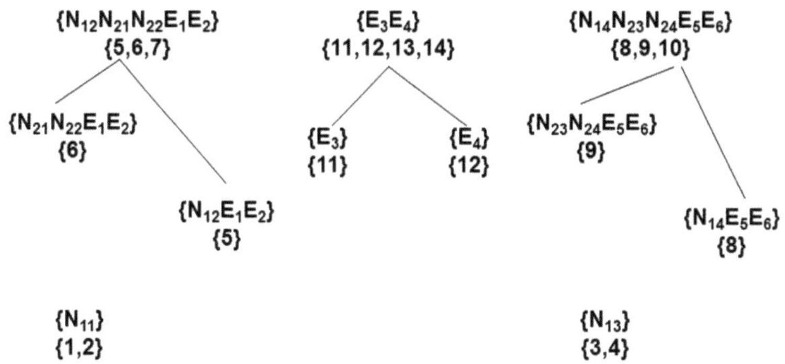

Fig. 14.10 Minimal conflicts contexts network of the system

14.5 Conclusions

The paradigm of Constraint-driven Fault Diagnosis is an adequate alternative to the automation of the fault diagnosis problems for hardware systems. These problems are represented as CSPs/COPs or SAT/MaxSAT in order to detect and identify the possible cause of the faults. The solving process of CSPs/COPs is based on applying consistency and search techniques. The high complexity of the exhaustive search algorithm for certain CSPs/COPs requires the application of decomposition or symbolic techniques in order to reduce this complexity.

Acknowledgements This work has been partially funded by the Ministry of Science and Technology of Spain (TIN2015-63502-C3-2-R) and the European Regional Development Fund (ERDF/FEDER).

References

1. Bakker, R.R., Dikker, F., Tempelman, F., Wogmim, P.M.: Diagnosing and solving over-determined constraint satisfaction problems. In: Proceedings of the 13th International Joint Conference on Artificial Intelligence, IJCAI'93, vol. 1, pp. 276–281. Morgan Kaufmann Publishers Inc., San Francisco, CA, USA (1993)
2. Bistarelli, S., Montanari, U., Rossi, F.: Semiring-based constraint satisfaction and optimization. J. ACM **44**(2), 201–236 (1997)
3. Bose, N.K.: Gröbner bases: an algorithmic method in polynomial ideal theory. In: Multidimensional Systems Theory and Applications, pp. 89–127. Springer Netherlands, Dordrecht (1995)
4. Ceballos, R., Gómez-López, M.T., Gasca, R.M., Del Valle, C.: Determination of possible minimal conflict sets using components clusters and grobner bases. In: Proceedings of the 15th International Workshop on Principles of Diagnosis, DX04, Carcassonne, France, pp. 21–26 (2004)
5. Cordier, M., Dague, P., Dumas, M., Lévy, F., Montmain, J., Staroswiecki, M., Travé-Massuyès, L.: A comparative analysis of AI and control theory approaches to model-based diagnosis. In:

ECAI 2000, Proceedings of the 14th European Conference on Artificial Intelligence, Berlin, Germany, pp. 136–140, 20–25 Aug 2000
6. Davis, R.: Diagnostic reasoning based on structure and behavior. Artif. Intell. **24**(1), 347–410 (1984)
7. Dechter, R., Pearl, J.: The anatomy of easy problems: a constraint-satisfaction formulation. In: Proceedings of the 9th International Joint Conference on Artificial Intelligence, Los Angeles, CA, USA, pp. 1066–1072, Aug 1985
8. Donald, B.R., Kapur, D., Mundy, J.L.: Symbolic and numerical computation for artificial intelligence (1992)
9. El Fattah, Y., Dechter, R.: Diagnosing tree-decomposable circuits. In: Proceedings of the 14th International Joint Conference on Artificial Intelligence, IJCAI'95, pp. 1742–1749. Morgan Kaufmann Publishers Inc., San Francisco, CA, USA (1995)
10. Freuder, E., Mackworth, A.: Constraint-Based Reasoning. A Bradford Book. MIT Press (1994)
11. Freuder, E., Wallace, R.: Partial constraint satisfaction. Artif. Intell. **58**, 21–70 (1992)
12. Freuder, E.C.: Synthesizing constraint expressions. Commun. ACM **21**(11), 958–966 (1978)
13. Gasca, R.M., Del Valle, C., Gómez-López, M.T., Ceballos, R.: NMUS: structural analysis for improving the derivation of all MUSes in overconstrained numeric CSPs. In: Borrajo, D., Castillo, L., Corchado, J.M. (eds.) Current Topics in Artificial Intelligence, pp. 160–169. Springer, Berlin, Heidelberg (2007)
14. Gasca, R.M., Ortega, J., Toro, M.: Diagnosis dirigida por restricciones simbólicas para modelos polinómicos. In: Diagnosis, Razonamiento Cualitativo y Sistemas Socieconómicos, pp. 71–78. Carlos Alonso y J. A. Ortega (2001)
15. Gasca, R.M., Ortega, J.A., Toro, M.: Diagnosis basada en modelos polinomicos usando técnicas simbólicas. Inteligencia Artificial, Revista Iberoamericana de Inteligencia Artificial **5**(14), 68–77 (2001)
16. Genesereth, M.R.: The use of design descriptions in automated diagnosis. Artif. Intell. **24**(1–3), 411–436 (1984)
17. Gómez-López, M.T., Ceballos, R., Gasca, R.M., Del Valle, C.: Applying constraint databases in the determination of potential minimal conflicts to polynomial model-based diagnosis. In: Constraint Databases, Proceedings of the 1st International Symposium on Applications of Constraint Databases, CDB'04, Paris, pp. 75–89, 12–13 June 2004
18. Guernez, C.: Fault detection and isolation on non linear polynomial systems. In: 15th IMACS World Congress on Scientific, Computation, Modelling and Applied Mathematics (1997)
19. Hoffmann, C.M.: Geometric and Solid Modeling: An Introduction. Morgan Kaufmann Publishers Inc., San Francisco, CA, USA (1989)
20. Junker, U.: QUICKXPLAIN: preferred explanations and relaxations for over-constrained problems. In: Proceedings of the 19th National Conference on Artifical Intelligence, AAAI'04, pp. 167–172. AAAI Press (2004)
21. de Kleer, J.: An assumption-based TMS. Artif. Intell. **28**(2), 127–162 (1986)
22. de Kleer, J., Mackworth, A.K., Reiter, R.: Characterizing diagnoses and systems. Artif. Intell. **56**(23), 197–222 (1992)
23. Kumar, V.: Algorithms for constraint satisfaction problems: a survey. AI Mag. **13**(1), 32–44 (1992)
24. Liffiton, M.H., Moffit, M.D., Pollack, M.E., Sakallah, K.A.: Identifying conflicts in overconstrained temporal problems. In: Proceedings of the Nineteenth International Joint Conference on Artificial Intelligence, IJCAI'05, pp. 199–204 (2005)
25. Liffiton, M.H., Sakallah, K.A.: On finding all minimally unsatisfiable subformulas. In: Proceedings of the 8th International Conference on Theory and Applications of Satisfiability Testing, SAT 2005. Lecture Notes in Computer Science. Springer (2005)
26. Sachenbacher, M., Williams, B.: Diagnosis as semiring-based constraint optimization. In: Proceedings of the 16th European Conference on Artificial Intelligence, ECAI'04, pp. 873–877. IOS Press, Amsterdam (2004)
27. Stumptner, M., Wotawa, F.: Diagnosing tree-structured systems. Artif. Intell. **127**(1), 1–29 (2001)

Chapter 15
Model-Based Software Debugging

Rafael Ceballos, Rui Abreu, Ángel Jesús Varela-Vaca and Rafael M. Gasca

15.1 Introduction

The complexity and size of software systems have rapidly increased in recent years, with software engineers facing ever-growing challenges in building and maintaining such systems. In particular, testing and debugging, that is, finding, isolating, and eliminating defects in software systems still constitute a major challenge in practice [47].

Debugging is an iterative process, where hypothesis generation, hypothesis selection, and hypothesis confirmation form the central tasks [7]. However, selection and exploration of good hypotheses remain difficult, and programmers often rely on intuition and spend considerable effort on pursuing seemingly promising hypotheses that ultimately do not lead to the true fault [37]. Similar to physicians, who often apply multiple tests to arrive at a diagnosis, software engineers have multiple debugging tools to analyze a program failure at their hands. In the software domain, debugging tools can leverage information obtained from (i) the execution of the program and from (ii) formal analysis of the behavior of a program or its model. While a rich set of tools has been developed to ease the burden of gathering information about a program and its execution(s), complementary approaches to use and analyze this

R. Ceballos (✉) · Á. J. Varela-Vaca · R. M. Gasca
University of Seville, Seville, Spain
e-mail: ceball@us.es

Á. J. Varela-Vaca
e-mail: ajvarela@us.es

R. M. Gasca
e-mail: gasca@us.es

R. Abreu
IST, University of Lisbon and INESC-ID, Lisbon, Portugal
e-mail: rui@computer.org

T. Escobet et al. (eds.), *Fault Diagnosis of Dynamic Systems*,
https://doi.org/10.1007/978-3-030-17728-7_15

information must be leveraged to target a broader spectrum of faults and compensate the limitations of individual techniques.

Model-Based Software Debugging (MBSD) is a technique that leverages concepts from program slicing. In MBSD, a diagnosis is obtained by the logical inference from the static model of the system, combined with a set of run-time observations. There are several types of models, derived from the source code and test cases that are then used by the technique to reason about observed failures. Despite the accuracy of this technique, in most cases, the computational effort required to create a model of a large program forbids the use of model-based approaches in real-life applications [29, 40, 42–44, 52]. Regardless of the complexity of the approach, there have been developments into making the real-life applications more tractable to model-based debugging:

- One option is the combination of MBSD with a more lightweight and accurate technique that uses coverage to reason about observed failures [40] Yet another field of application is end-user programming, such as spreadsheets, where programs tend to be smaller [31].
- Another option is based on the combination of different paradigms, such as Design by Contract. This methodology is named Software Diagnosis Based on Constraints Models [20, 54]. The goal is to isolate and identify faults in assertions (Design by Contract) and/or in sentences (source code). Design by Contract was proposed in [41], in order to improve the Object-Oriented software quality.

The remainder of this section elaborates on techniques to isolate software faults, in particular, model-based approaches.

15.2 Background and Problem Statement

Improving software quality has been a long-standing issue in academia, leading to considerable advances in automated fault detection, localization, and correction. While fault detection strives to expose defects in a given program, fault localization aims to identify particular program fragments that may be responsible for a failure, and fault correction aims to repair the program.

Effective (automated) debugging assistance can be provided by supporting the developer in one or more of the aforementioned tasks. Recent advances in program analysis and verification techniques have led to mature mathematical frameworks and tools focused on specific purposes, but their individual utility for general debugging remains limited to specific programs, execution environments, and problem contexts. As a result, the overall debugging process remains complex and much time is devoted to analyzing (possibly irrelevant) results gathered from different tools.

For example, techniques that are solely based on a program, such as slicing [49] or invariant learning [28], are often ineffective in locating faults that stem from functionality that has been “forgotten” in the implementation or that is not covered by tests. Conversely, techniques based on abstract specifications (if available), such as abstract state machines [56], can locate regions in a program where specification and program show different behaviors but may suffer from difficulties obtaining sufficiently detailed, correct specifications that would allow to confine the cause to a specific small region. Only by a combination of complementary techniques can such complex faults be located effectively.

However, many formal methods suffer from spurious explanations that are caused by approximations and abstractions introduced by the formalism [13], while others may return results that are too large to be useful [46]. Experience with automated debugging and checking tools shows that the remaining results are often dismissed if the fault has not been located after examining the first few candidate explanations [36]. Therefore, discriminating between true explanations and those that are caused by approximations in the analysis is essential. Statistics-based techniques are among the more popular automated fault localization techniques. By correlating information about which program fragments have been exercised in a set of multiple program execution traces (also called program spectra) with information about successful and failing executions, Spectrum-based Fault Localization (SFL) and other statistics-based approaches yield a list of suspect program fragments sorted by their likelihood to be at fault.

Since this technique is efficient in practice, this and other dynamic analysis techniques are attractive for large modern software systems [52]. Overall, statistical techniques are lightweight, but are rather dependent on the availability of a suitable test suite.

Machine learning techniques also feature prominently among automated program analysis tools. In this context, learning is applied to infer from execution traces models that describe the program's intended behavior. For example, Daikon [26] is an invariant detection tool built with the intention of supporting program evolution by helping programmers to better understand the code. It analyzes the values of variables encountered in executions of the program and retains only those Boolean expression over program variables that are satisfied in all executions. For example, if in all observed executions, the value of variable x was less than a variable y at some point in the execution, the invariant $x < y$ will be reported. The same approach has since been applied to debugging, where violations of inferred invariants are used to detect errors [28]. However, existing work does little to help determine the origin of the fault once a failure has been detected, and the learning algorithms may produce results that are too numerous or too specific to be of value to the programmer.

Better results than those obtained from methods based on dynamic analysis alone can often be achieved if a model of the correct program behavior is available [1]. Model-Based Software Debugging (MBSD) techniques have been advocated as powerful debugging aid that isolate faults in complex programs [40]. By comparing the execution of a program to what is anticipated by its programmer, model-based reasoning techniques separate those parts of a program that may contain a fault from those that cannot fully explain the observed symptoms. Compared to spectrum-based localization, model-based analysis yields better accuracy due to precise reasoning about the possible effects of each program fragment but suffers from poor scalability.

Better and precise results can be achieved if the model-based methodology for diagnosing bugs takes into account Design by Contract (DbC) methodology because the correct behavior is available for automatic reasoning [21]. DbC is the model of the correct behavior that the source code must satisfy. In [50], two measures in order to validate the benefits of using DbC are proposed: robustness and diagnosability. The robustness is the degree to which the software is able to recover from internal faults that would otherwise have provoked a failure. DbC enables the development of more reliable and robust software applications and the control of abnormal situations. Diagnosability expresses the effort required in the localization of a fault as well as the precision allowed by a test strategy on a given system. The results show that the robustness improves rapidly with only a few contracts, and for improving diagnosability the quantity of contracts is less important than their quality. In [14], it is shown that contracts are useful for fault isolation if they are defined during analysis. By using contracts, the fault isolation and diagnosability are significantly improved in object-oriented code (it implies a wider distribution of functions).

In this chapter, we use the terminology introduced by Avizienis et al. [8]:

- A *failure* is an event that occurs when delivered service deviates from correct service.
- An *error* is a system state that may cause a failure.
- A *fault* (defect/bug) is the cause of an error in the system.

In the context of this chapter, faults are bugs in the code of a software program.

Definition 15.1 A software program Π is formed by a sequence M of one or more *components* (e.g., statements). Components can be of several levels of granularities, such as classes, methods, and statements.

Failures and errors are symptoms caused by faults in the program. Fault localization aims at isolating the root cause of observed symptoms. The fault localization techniques considered in this chapter consider the existence of a test suite revealing the software faults.

Definition 15.2 A test case t is a (i, o) tuple, where i is a collection of input settings or variables for determining whether a software system works as expected or not, and o is the expected output. If $\Pi(i) = o$ the test case passes, otherwise fails.

Definition 15.3 A test suite $T = \{t_1, \ldots, t_N\}$ is a collection of test cases that are intended to test whether the program follows the specified set of requirements. The cardinality of T is the number of test cases in the set $|T| = N$.

15.3 Software Diagnosis Based on Constraints

A methodology based on constraints for diagnosing software is proposed in [21]. The main idea is the transformation of contracts and source code into a model based on constraints for locating faults or defects in source code and assertions. These faults are wrongly designed assertion or statements, for example, variations in Boolean conditions or in assignment statements. Other types of errors, as syntax errors, memory access violations, or infinite loops, could be considered in future work.

15.3.1 Constraints-Based Model

A diagnosis is a hypothesis about what are the changes to do in a program in order to obtain a correct behavior. A component has an abnormal behavior [35] if the outputs are different from the expected results. For example, a multiplier component is abnormal if the output of the multiplier is different to the multiplication of its inputs. For each component c, a Boolean variable $AB(c)$ stores if c is abnormal or not, in order to know whether a component is abnormal and can be a part of the minimal diagnosis. The goal of this methodology is detecting semantic defects in source code or assertions, and these defects are modeled as components with abnormal behavior.

In a software program, blocks of source code are linked in order to obtain the specified behavior. Each statement of a source code can be considered as a component, with inputs and outputs (results). An executed program is a set of linked blocks of code. The order of these blocks can be represented as a *Control Flow Graph* (CFG). The CFG is a directed graph that represents a set of sequential blocks and decision statements. A *Path* is a possible sequence of statements of the CFG. In order to detect and isolate defects in the design of programs, the CFG and program contracts are transformed into a model based on constraints. Testing techniques select the observational models which are most significant for detecting failures in programs. A test case is designed for executing a particular program path and determining whether a program works as expected or not.

When a program has executed the order of the assertions and statements is very important. It is necessary to maintain this order when the source code and contracts are transformed into constraints. For this reason, the statements of the CFG are translated into a Static Single Assignment (SSA) form. This translation maintains the execution sequence when the program is transformed into constraints. In SSA form, only one assignment is allowed for each variable in the whole program. For example, the statements *b = x*q;...b = b+3;... {Post:b =...}* is transformed to *b1 = x1*q1;...b2 = b1 + 3;...{Post:b2 =...}*.

The System Model (SM) is a finite set of constraints which determine the software behavior. It will be obtained by transforming the set of statements and assertions of a program to constraints (in SSA form). A test case is an Observational Model that can be applied to an SM.

The subset $D \subseteq SM$ is a diagnosis if $SM' \cup TC$ is satisfied, where $SM' = SM - D$. The minimal diagnoses imply to modify the smallest number of program statements or assertions. A diagnosis is a set of components with abnormal behavior and the goal is to minimize this set. In order to maximize the number of components with normal behavior, the idea is solving a MAX-CSP (CSP and MAX-CSP had been introduced in Chap. 14).

```
/**
 *  @inv getBalance()>= 0
 *  @inv getInterest >= 0
 */
public interface Account {

    /**
     *  @pre income > 0
     *  @post getBalance() >= 0
     */
    public void deposit (double income);

    /**
     *  @pre withdrawal > 0
     *  @post getBalance() ==
     *      getBalance()@pre - withdrawal
     */
    public void withdraw (double withdrawal);

    public double getBalance ();

}
```

```
public class AccountImp implements Account {

    private double interest;

    private double balance;

    public AccountImp()  {
       this.balance = 0;
    }

    public void deposit (double income) {
       this.balance = this.balance + income;
    }

    public void withdraw (double withdrawal) {
       this.balance = this.balance - withdrawal;
    }

    public double getBalance() {
       return this.balance;
    }
}
```

Fig. 15.1 Contract and source code for the bank account class example

The first goal is detecting the inconsistencies between test cases and contracts, and then between test cases, contracts and source code. These inconsistencies are detected, for example, if the SM does not satisfy a Test Case. If there are inconsistencies, the second goal is isolating the inconsistencies between test cases and contracts, and then between test cases, contracts, and source code. These are explained in more detail in the sections below.

15.3.2 Diagnosing DbC Defects

In Fig. 15.1, the class *AccountImp* is shown. It is an example that simulates a bank account. There are methods for depositing and withdrawing money. Assertions are checked in two different ways: first without test cases, and then with test cases.

- Without test cases. Two kinds of checks are proposed:
  - Checking if all the invariants of a class can be satisfied together.
  - Checking if the precondition and postcondition of a method are feasible with the invariants of its class.
- With test cases. The idea is applying test cases to the sequence {invariants + precondition + postcondition + invariants} for each method.

Example 1 In Fig. 15.2, the checking of method *withdraw* is shown. The initial balance must be 0 units and, when a nonnegative amount is withdrawn, the balance must preserve the value 0. The balance must be equal or greater than zero when the method finishes because of the invariant, but the postcondition implies that *balance* = *balance@pre* – *withdrawal*, that is, $0 - withdrawal > 0$, and this is impossible if the *withdrawal* is positive. There is a problem with the precondition since it is not

DbC:	*CSP*:	*Max-CSP*:
Inv.: balance >= 0 Pre.: withdrawal > 0 Post.: balance = balance@pre - withdrawal Inv.: balance >= 0	C_1: balance@pre >= 0 C_2: withdrawal > 0 C_3: balance = balance@pre - withdrawal C_4: balance >= 0 Var = {balance@pre, withdrawal,balance} Dom = {0,[0, 100],0}	C_1: AB(Inv) ∨ (balance@pre >= 0) C_2: AB(Pre) ∨ (withdrawal > 0) C_3: AB(Post) ∨ (balance = balance@pre - withdrawal) C_4: AB(Inv) ∨ (balance >= 0) Var = {balance@pre,withdrawal, balance,AB(Inv),AB(Pre),AB(Post)} Dom = {0,[0, 100],0,[*false*,*true*], [*false*,*true*],...,[*false*,*true*]}
Test case: Inputs={balance@pre = 0, withdrawal > 0} Outputs= {balance = 0} Source code: Method Withdraw		

Fig. 15.2 CSP for detecting and isolating defects by using a test case for method *Withdraw*

strong enough to stop the program execution when the withdrawal is not equal or greater than the balance of the account.

15.3.3 Diagnosing Source Code Defects

After checking DbC, then the source code of the program is checked by using test cases. A System Model is obtained by transforming DbC assertions (preconditions, postconditions, invariants) and statements. DbC assertions are directly transformed into constraints (into SSA form). The source code outputs must satisfy these constraints because they correspond to the correct behavior.

In order to transform the statements to constraints (System Model), the source code is divided into basic blocks: sequential blocks (declarations, assignments, and method calls), conditional blocks, and loop blocks. For example, an assignment *Ident* = *Exp*; is transformed into an equality constraint in a CSP, and for the MAX-CSP the abnormal behavior is added: AB(S_{Asig}) ∨ *Ident* = *Exp*. After execution of S_{Asig}, the equality between the assigned variable and assigned expression must be satisfied, or in another case, this statement (assignment) has an abnormal behavior (the statement contains a defect).

In Fig. 15.3, the polybox example is transformed into five statements (source code). Polybox example had been introduced in Chap. 2. The program cannot reach the correct output because the second statement is an adder (bug) instead of a multiplier. By transforming the source code, a CSP and Max-CSP are obtained. For checking if there are failures, a test case is applied to the CSP. The test case does not satisfy the CSP. There is almost one semantic defect that generates an unexpected output. In order to identify this defect, the same test case is applied to the Max-CSP. By solving this Max-CSP, the obtained minimal diagnosis is $\{S_2\}$. The diagnosis is minimal because the goal of the MAX-CSP is to find an assignment of the AB variables that enable the maximum number of components with normal behavior.

Source Code:	*CSP*:	*Max-CSP*:
S_1: int x = a * c	C_1: x = a × c	C_1: $AB(S_1) \vee (x = a \times c)$
S_2: int y = b + d	C_2: y = b + d	C_2: $AB(S_2) \vee (y = b + d)$
S_3: int z = c * e	C_3: z = c × e	C_3: $AB(S_3) \vee (z = c \times e)$
S_4: int f = x + y	C_4: f = x + y	C_4: $AB(S_4) \vee (f = x + y)$
S_5: int g = y + z	C_5: g = y + z	C_5: $AB(S_5) \vee (g = y + z)$
	Var = {a,b,c,d,e,f,g,x,y,z}	Var = {a,b,c,d,e,f,g,x,y,z,$AB(S_1)$,$AB(S_2)$,..., $AB(S_5)$}
	Dom = {2,3,3,2,2,12,12, [0,100],[0,100],[0,100]}	Dom = {2,3,3,2,2,12,12,[0,100],[0,100],[0,100], [*false*,*true*],[*false*,*true*],...,[*false*,*true*]}
Test case: Inputs={a = 3, b = 2, c = 2, d = 3, e= 3} Outputs= {f = 12, g = 12} Source code: S1 .. S5		

Fig. 15.3 CSP for detecting and isolating defects by using a test case for the Toy Program.

Example 1

```
{Pre: x>0 ∧ y>0}
 public int dif(int x,int y){
    int max,min,s,z;
(S1)   if (x>=y){
(S2)      min=y;
(S3)      max=x;
       }else{
(S4)      min=x;
(S5)      max=y;}
     {Assert: max>=min }
(S6)   z=max-min;
(S7)   s=0;
(S8)   while (z>0){
(S9)     s=s+z;
(S10)    z=z-1;}
(S11)  return s;
{Post: s=∑_{a=1}^{|x-y|} a}
```

Example 2

```
{Pre: i>=0 ∧ i<=n ∧ p>0}
 public int rec(int n,
       int i,int p){
       int s;
(S1)     if (i==n)
(S2)       s=1;
         else{
(S3)       p=2*p;
(S4)       s=this.rec(n,i+1,p);
(S5)       s=s+p;}
(S6)     return s; }
{Post:s=1+p∑_{y=1}^{n-i} 2^y}
```

Fig. 15.4 Source code examples that include conditional statements, loops, and method calls

In Fig. 15.4 two more examples are shown. For conditional statements, as in example 1, the predicate *P* applied to each statement store if each statement is part of the *path* or not. When we have a statement S_x, if $P(S_x)$ is true then S_x belongs to the *path* of the executed program; otherwise, it does not belong to the path. For a conditional statement [22], generated constraints include the transformation of the condition and the block of statements included in each case. Only one path is possible, and this path will depend on the condition. The transformation to constraints of the conditional statement allows to detect and isolate defects in conditions and to set the correct path for obtaining the correct behavior. A loop statement can be transformed into a sequence of conditional statements. The number of iterations depends on the test case. For each iteration, the equivalent nested conditional statement is transformed into constraints. Maintaining the order is important for obtaining the same result as the loop statement. The invariant of the loop is also transformed into constraints of the model.

For method calls and return statements, as in example 2 (Fig. 15.4), is possible to substitute the method call by the precondition and postcondition of the called method. The constraints obtained from the postcondition give us information about the correct behavior. If there is no contract, another option is to substitute the method call by the statements of the method [23].

15.4 Spectrum-Based Reasoning for Software Debugging

This section introduces a lightweight, reasoning technique that reasons over abstractions of program traces, called program spectra, to produce a diagnostic report for observed failures in program executions. The technique is known as Spectrum-based Fault Localization (SFL, for short), and is among the best fault localization techniques [42, 52].

15.4.1 Program Spectra

A program spectrum, introduced by Reps et al. in 1997 to address the Year 2000 problem[1] [32], is a characterization of the execution of a program execution on a set of inputs. These set of inputs can be, for instance, test cases in a test suite. Note that in the following execution, transaction, and test case are used interchangeably.

Program spectra are information collected at run-time, hence it provides a view on the dynamic behavior of a program. A program spectrum is represented as a vector of M counters or flags, where M is the number of *software components*. Various different program spectra exist [30]; e.g., path-hit spectra, data-dependence-hit spectra, and block-hit spectra are among the most common ones. The spectra commonly used in spectrum-based fault localization is component-hit spectra, a type of spectra that merely indicates whether a component was involved in program execution.

Simlar to code coverage tools [55], the source code needs to be instrumented to collect which components were covered in each execution. In addition to the program spectra, information whether that particular spectra corresponds to a failing or a passing execution is also collected in a so-called error vector. In the following, we will refer to the collected information as the (A, e) tuple, where

- $a_{mn} = 1$ if component $1 \leq m \leq M$ was involved in transaction $1 \leq n \leq N$, and 0 otherwise;
- $e_n = 1$ if transaction $1 \leq n \leq N$ failed, and 0 if passed.

[1]The Year 2000 problem is also known as Y2K problem, Y2K bug, or simply Y2K.

	obs			
	c_1	c_2	c_3	e
t_1	1	1	0	1
t_2	0	1	1	1
t_3	1	0	0	1
t_4	1	0	1	0

Fig. 15.5 Hit-spectra matrix

15.4.2 Modus Operandi of Fault Localization

Next, we will illustrate how spectrum-based reasoning to fault localization works using a running example. Consider the hit-spectra matrix in Fig. 15.5 (containing a set of component[2] observations *obs* and transaction outcomes *e*), with 4 transactions and 3 components.

Spectrum-based reasoning [4, 6, 25] consists on the following:

1. Generate sets of components (candidates) that would explain the observed erroneous behavior.
2. Rank the candidates according to their probability.

Candidate Generation

A diagnostic candidate d is a set of components that are said to be valid if at least one component in d is exercised in all failed transactions. That is,

$$\forall_{n \in 0..N} : e_n = 1 \Rightarrow \exists_{m \in 0..M} : a_{mn} = 1 \wedge d_m \in d$$

We are only interested in minimal candidates,[3] as they can subsume others of higher cardinality. There may be several minimal candidates d_k for a particular spectrum, which constitutes a collection of minimal candidates D.

In our example, the collection of minimal diagnostic candidates that can explain the erroneous behavior are

- $d_1 = \{c_1, c_2\}$
- $d_2 = \{c_1, c_3\}$

Note that the problem of computing the set of minimal candidates is equivalent to computing minimal hitting sets [18].

Candidate Ranking

For each candidate d, the posterior probability is calculated using the naïve Bayes rule[4]

[2]As said in the previous section, by component we mean the unit by which we gather coverage. Basically, components are the columns in the hit-spectra matrix and can represent, e.g., every statement in the source code.

[3]A candidate d is said to be minimal if no valid candidate d' is contained in d.

[4]Probabilities are calculated assuming conditional independence throughout the process.

$$\Pr(d \mid (A, e)) = \Pr(d) \cdot \prod_{n}^{N} \frac{\Pr((A_i, e_i) \mid d)}{\Pr(A_n)}, \quad (15.1)$$

where $\Pr(obs_i)$ is a normalizing term that is identical for all candidates; hence, this term is not considered for ranking purposes.

In order to define $\Pr(d)$, let p_j denote the prior probability that a component c_j is at fault.[5] The prior probability for a candidate d is given by

$$\Pr(d) = \prod_{n \in d} p_n \cdot \prod_{n \notin d} (1 - p_n). \quad (15.2)$$

$\Pr(d)$ estimates the probability that a candidate, without further evidence, is responsible for erroneous behavior—that is, the prior probability of being faulty. It is also used to make to make larger candidates (in terms of cardinality) less probable. In order to bias the prior probability taking observations into account, $\Pr(A_i, e_i \mid d)$ is used. Let g_j (referred to as component goodness) denote the probability that a component c_j performs nominally

$$\Pr((A_i, e_i) \mid d) = \begin{cases} \prod_{j \in (d \cap A_i)} g_j & \text{if } e_n = 0 \\ 1 - \prod_{j \in (d \cap A_i)} g_j & \text{otherwise} \end{cases} \quad (15.3)$$

In cases where values for g_n are not available (which is the case for software components), they can be estimated by maximizing $\Pr((A, e) \mid d)$ (Maximum Likelihood Estimation (MLE) for the naïve Bayes classifier) under parameters $\{g_n \mid n \in d\}$. This MLE-based approach is the BARINEL approach and will be detailed in the next section [6]).

Considering our example, the probabilities for both candidates are

$$\Pr(d_1 \mid (A, e)) = \overbrace{\left(\frac{1}{1000} \cdot \frac{1}{1000} \cdot \left(1 - \frac{1}{1000}\right)\right)}^{\Pr(d)} \times \overbrace{\underbrace{(1 - g_1 \cdot g_2)}_{t_1} \times \underbrace{(1 - g_2)}_{t_2} \times \underbrace{(1 - g_1)}_{t_3} \times \underbrace{g_1}_{t_4}}^{\Pr((A,e)|d)} \quad (15.4)$$

$$\Pr(d_2 \mid A, e) = \overbrace{\left(\frac{1}{1000} \cdot \frac{1}{1000} \cdot \left(1 - \frac{1}{1000}\right)\right)}^{\Pr(d)} \times \overbrace{\underbrace{(1 - g_1)}_{t_1} \times \underbrace{(1 - g_3)}_{t_2} \times \underbrace{(1 - g_1)}_{t_3} \times \underbrace{g_1 \cdot g_3}_{t_4}}^{\Pr((A,e)|d)} \quad (15.5)$$

By performing a MLE for both functions it follows that $\Pr(d_1 \mid (A, e))$ is maximized for $g_1 = 0.47$ and $g_2 = 0.19$, and $\Pr(d_2 \mid A, e)$ is maximized for $g_1 = 0.41$ and $g_3 = 0.50$. Applying the goodness values to both expressions, it follows that

[5] In the context of development-time fault localization, we often approximate p_j as $1/1000$, i.e., 1 fault for each 1000 lines of code.

$\Pr(d_1 \mid (A, e)) = 1.9 \times 10^{-9}$ and $\Pr(d_2 \mid A, e) = 4.0 \times 10^{-10}$, entailing the ranking (d_1, d_2).

15.4.3 *The* BARINEL *Approach to Compute Goodness*

In the previous section, we have described the *modus operandi* of spectrum-based reasoning to fault localization. A key issue in the outlined approach is to compute the component goodness g_j of each component, as these values influence the posterior probabilities of the diagnostic candidates ($\Pr(d_k)$). The approach was first introduced in [6], as is named BARINEL.

15.4.3.1 Component Goodness Estimation

As mentioned before, the key idea underlying the BARINEL approach is that it computes the g_j for each candidate's faulty components that *maximizes the probability* $\Pr((A, e)|d_k)$ *of the observations* (A, e) *occurring, conditioned on candidate* d_k, yielding statistically perfect estimators for g_j. Following equation (15.3), BARINEL bias the prior probability taking observations into account conditioned on a particular diagnostic candidate d_k using $\Pr(A_i, e_i \mid d_k)$:

$$\Pr((A_i, e_i) \mid d_k) = \begin{cases} \prod\limits_{n \in d_k \wedge a_{mn}=1} g_n & \text{if } e_n = 0 \\ 1 - \prod\limits_{n \in d_k \wedge a_{mn}=1} g_n & \text{if } e_n = 1 \end{cases}$$

Hence, the component goodnesses (g_n) are computed by maximizing $Pr((A_i, e_i)|d_k)$ according to

$$\arg\max_{g_j | j \in d_k} \Pr(e|d_k).$$

Note that this approach implies *optimum* g_n values may differ per diagnostic candidate for the exact same set of components. For instance, suppose a system with $M = 4$ components and the following double- and triple-fault candidates. Let the following be the g_n that optimally explain the observed failures and passes for the double and triple fault $\{0.15, 0.8, 0.4, 1\}$ and $\{0.7, 1, 1, 0.3\}$, respectively. Note that g_n differ for the same component n (e.g., 0.15 vs. 0.7).

The BARINEL approach generalizes over both persistent and intermittent faults. Each component $n \in d_k$ is associated with a component goodness $g_n \in [0, 1]$, which represents a generalization over the classical "normal/abnormal" entries:

- **0** – persistently failing to
- **1** – *healthy*, i.e., faulty but not yielding observed failures.

15.4.3.2 Algorithm

The approach, named BARINEL, is described in detail in Algorithm 15.1 and has three main phases [6]. BARINEL, taking as input (A, e), starts to generate a set of diagnostic candidates $D = \{d_1, \ldots d_k, \ldots, d_{|D|}\}$ using a low-cost, heuristic, and optimized MHS algorithm called STACCATO. The STACCATO algorithm is guided to return a set of limited diagnostic candidates that captures all significant probability mass[6] [2, 19].

Algorithm 15.1: Diagnostic Algorithm: BARINEL (© 2009 IEEE. Reprinted, with permission, from [4]).

Inputs: Activity matrix A, error vector e,
Output: Diagnostic Report D

1: $\gamma \leftarrow \epsilon$
2: $D \leftarrow$ STACCATO((A, e)) {Compute MHS}
3: **for** all $d_k \in D$ **do**
4:   expr $\leftarrow$ GENERATEPR($(A, e), d_k$)
5:   $i \leftarrow 0$
6:   $\Pr[d_k]^i \leftarrow 0$
7:   **repeat**
8:     $i \leftarrow i + 1$
9:     **for** all $j \in d_k$ **do**
10:       $g_j \leftarrow g_j + \gamma \cdot \nabla \mathrm{expr}(g_j)$
11:     $\Pr[d_k]^i \leftarrow$ EVALUATE(expr, $\forall_{j \in d_k} g_j$)
12:   **until** $|\Pr[d_k]^{i-1} - \Pr[d_k]^i| \leq \xi$
13: **return** SORT(D, Pr)

In the second phase, $\Pr(d_k|(A, e))$ is computed for each $d_k \in D$ (lines 3 to 14). The function GENERATEPR derives the symbolic formula for $\Pr((A, e)|d_k)$. To illustrate this function, suppose the following observations:

c_1	c_2	e	$\Pr(e_i\|\{1, 2\})$
1	0	1	$1 - g_1$
1	1	1	$1 - g_1 \cdot g_2$
0	1	0	g_2
1	0	0	g_1

According to the BARINEL reasoning, the probability of obtaining (A, e) given $d_k = \{1, 2\}$ equals

$$\Pr((A, e)|d_k) = g_1 \cdot g_2 \cdot (1 - g_1) \cdot (1 - g_1 \cdot g_2)$$

[6]For an efficient implementation of STACCATO, refer to https://github.com/npcardoso/MHS2 or http://mhs2.algorun.org/.

Next, the component goodness values are calculated such that they maximize $\Pr((A, e)|d_k)$. This is calculated by applying a gradient ascent procedure [9] (lines 9–11).

Finally, the diagnoses are ranked according to $\Pr(d_k|(A, e))$, which is computed by EVALUATE according to the posterior Bayes update (line 12):

$$\Pr(d_k|(A, e)) = \frac{\Pr((A, e)|d_k)}{\Pr(A)} \cdot \Pr(d_k)$$

where $\Pr(d_k)$ is the prior probability that d_k is the true fault explanation, $\Pr(A)$ is a normalization factor, and $\Pr((A, e)|d_k)$ is the probability that (A, e) is observed assuming d_k correct.

BARINEL's MLE for single fault diagnostic candidates has been proven to be the intuitive way to estimate the true intermittency parameter (i.e., component goodness values) [6]. Consider the following (A, e) as well as the probability of that occurring (Pr), being g_1 the true intermittency parameter:

c_1	e	$\Pr(e_i\|d_k)$
1	0	g_1
1	0	g_1
1	1	$1 - g_1$

MLE estimates g_1 to be $\frac{2}{3}$. Showing that g_1 maximizes the probability of this particular (A, e) to occur, proves that this is a perfect estimate. As $\Pr(e|\{1\})$ is given by $\Pr(e|\{1\}) = g_1^2 \cdot (1 - g_1)$, the value of g_1 that maximizes $\Pr(e|\{1\})$ is indeed $\frac{2}{3}$. Furthermore, it has also been shown that these estimators yield optimal diagnostic reports, if considering single faults only [6, 42].

There are success stories of applying BARINEL in practice [27, 57] and other areas of research (e.g., [15, 16, 43–45]). This is the case because the time/space complexity of our approach is low—hence, being amenable to large software systems. The complexity is essentially the same as other lightweight approaches modulo a constant factor on the account of the gradient ascent procedure, which exhibits rapid convergence [6]. The approach is available within the GZoltar toolset at http://www.gzoltar.com [17].

15.4.4 Results

To assess the performance improvement of our approach, we generate synthetic observations based on sample (A, e) generated for various values of N, M, and number of injected faults C (cardinality). Essentially, the simulator[7] samples component activity from a Bernoulli distribution with parameter r, i.e., the probability a compo-

[7] Simulator is available at https://github.com/SERG-Delft/sfl-simulator.

nent is involved in a row equals r. For the C faulty components $c_j \in C$ we also set g_j. Thus, the probability of a component being involved *and* generating a failure equals $r \cdot (1 - g)$. A row i in A generates an error ($e_i = 1$) if at least 1 of the C components generates a failure (noisy-or model). Measurements for a specific (N, M, C, r, g) scenario are averaged over $1,000$ sample matrices, yielding a coefficient of variance of approximately 0.02.

We compare the accuracy of our Bayesian framework with the classical framework [35] in terms of a diagnostic performance metric C_d, that denotes the cost of diagnosis (that is, the percentage of statements that a developer needs to inspect before finding finding the actual components at fault) [48]. Given a diagnosis $D = \{d_1, \ldots, d_k, \ldots, d_K\}$, the computation of C_d proceeds as follows: (i) the diagnostic ranking is mapped into a component ranking according to $\Pr(j) = \sum_{k=1}^{K} \Pr(d_k) \cdot h_k[j] / \sum_{k=1}^{K} \Pr(d_k)$, (ii) the ranking is traversed; inspected healthy components contribute to C_d.

For instance, consider a 4-component program with a unique diagnosis $d_1 = \{c_1, c_2, c_4\}$ with an associated $g = \{0.70, 0.20, 1, 0.15\}$, and c_1, c_2 faulty. The first component to be verified/ replaced is the non-faulty c_4, as its goodness is the lowest. Consequently, C_d is increased with $\frac{1}{4}$ to reflect that it was inspected in vain.

Our experiments for the different (N, M, C, r, g) scenarios, lead us to conclude the following:

- For a sufficient large number of executions (N), BARINEL produces optimal diagnosis, being able to correctly pinpoint the true faulty components. One of the reasons for this observation is that the chance that non-faulty components are still within the MHS is low. Furthermore, for single faults ($C = 1$), BARINEL yields optimal diagnosis.
- For small g_j (that is, faulty components are very likely to yield observed failures), C_d converges more quickly than for large g_j as executions involving faulty components are much more likely to fail. For large g_j, BARINEL requires many more observations (larger N) to rank the true faulty components higher.
- More observations (N) are needed to pinpoint true faulty components as the number of faults C increase. The reason is because failure behavior can be caused by much more components, reducing the correlation between failure and a particular component involvement.
- BARINEL is superior to other related approaches for $C \geq 2$. In particular, the other approaches steadily deteriorate for increasing C.

We refrain from detailing the results obtained with BARINEL in many different contexts, instead we have decided to outline the main findings. A detailed description of results—including simulator and real software systems results—can be found in related research papers [3, 4, 6].

15.5 Software Configuration Errors Diagnosis

Software Product Line (SPL) [10, 11, 34] is a new paradigm in the Software Engineering field, which provides the basis for the development of products. This paradigm is based on the identification of a set of core features and their relations in the development of products. SPL methods consist of the process of analyzing related products in order to identify their common and variable features. The main method for the domain analysis of an SPL is based on feature models and Feature-Oriented Domain Analysis [34] represents one of the most used techniques for the domain analysis of feature models.

Feature models (hereinafter FM) describe a model that define features and their relations. FMs enable to reason about certain properties, for instance, the potential number of valid products (set of valid configurations as all valid combination of a selection of features); and whether a particular configuration (selection of features) constitutes a valid product. There are several types of models to design FMs [10]. Although the notation proposed by Czarnecki [24] is the most used in the literature, an example of this notation is shown in the in Fig. 15.6. This notation enables four type of relations between a parent and its child features:

- **Mandatory** relation indicates that a child feature is required as shown in Fig. 15.6 where *computer* feature requires the mandatory sub-feature of *video*, *computer* $\leftrightarrow$ *video*.
- **Optional** relation indicates that a child feature is optional as shown in Fig. 15.6 where *computer* feature implies an optional sub-feature of *audio*, *computer* $\rightarrow$ *audio*.
- **Alternative** relation indicates that one of the sub-features must be selected. In general, $a_1, a_2, \ldots, a_n$ alternative sub-features of b, $a_1 \wedge a_2 \wedge \cdots \wedge a_n \leftrightarrow b_{\bigvee_{i<j}(a_i \vee \cdots \vee a_j)}$. In Fig. 15.6 where the *video* feature implies the selection of one *VGA* or *HDMI* feature, *video* $\leftrightarrow$ *VGA* $\vee$ *HDMI*.

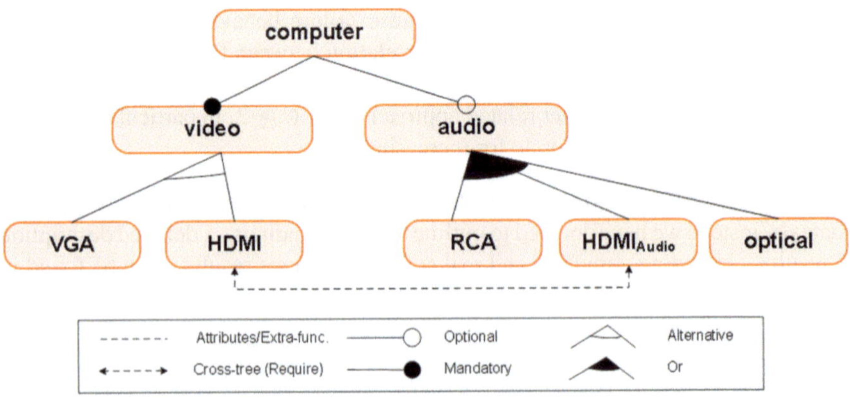

Fig. 15.6 Feature model example

- **Or** relation indicate that at least one of the sub-features must be selected. In general, $a_1, a_2, \ldots, a_n$ sub-features of b, $a_1 \vee a_2 \wedge \cdots \wedge a_n \leftrightarrow b$. In Fig. 15.6, $audio \leftrightarrow (RCA \vee optical \vee HDMI_{Audio}) \wedge \neg(RCA \wedge optical \wedge HDMI_{Audio}))$.

In addition, other relations called cross-tree constraints are allowed. These cross-tree constraints enable to represent constraints that relate features without a direct parent–child relation. The most common are the inclusion and exclusion relations:

- Feature A *requires* feature B, for instance, in Fig. 15.6, $HDMI \leftrightarrow HDMI_{Audio}$.
- Feature A *excludes* feature B, for instance, ¬(A ∧ B)).

This graphical notation lacks mechanisms to define certain relations and information of the features. There are some extensions such as proposed by [11, 24] that enable the specification of attributes and extra-functionalities for features. These extensions enable characteristics of features that can be measured to be provided and to include the facility to express relations between these characteristics (extra-functionalities).

In order to determine a software configuration, fault detection or diagnosis feature models can be transformed into formal models. The formal models can be also used to extract information related to the product. This information can vary from: number of configurations (number of all valid configurations), filters (selection of specific characteristics for the features), all valid products (all valid products with certain features), validation (check if a selection of characteristics represent a valid configuration), optimum products (determine the best products according to a criterion), variability (inspect the relation between a set of potential products and certain products), commodity (the relation between certain products and the total of products).

SAT and constraint programming can be used as a formal model to perform feature model analysis since this approach handles integer domains that are used for attributes and optimization functions. Feature models might be transformed into Constraint Satisfaction Problem (CSP), SAT problem, Constraint Optimization Problem (COP), MAX-SAT problem [12, 51] depending on the reasoning question to achieve. Figure 15.7 depicts a transformation example of a feature model to an SAT and MAX-SAT problem.

As aforementioned, other information can be obtained such as the determination whether a configuration is valid or invalid. For a better understanding of these two example configurations related to Fig. 15.6 are given in next as follows:

$$C1 : \{computer, video, HDMI, audio, HDMI_{Audio}\} \tag{15.6}$$

$$C2 : \{computer, video, HDMI, audio, RCA\} \tag{15.7}$$

The first example shows a valid configuration in so far as it has {*computer, video, HDMI,*} $HDMI_{Audio}$ features and the values for C_S, and C_k are valid values. In this case, the configuration is valid since *computer* and *video* are mandatory and the alternative *HDMI* has been selected. Regarding *audio*, $HDMI_{Audio}$ is chosen as requirement of *HDMI* video feature, thus the configuration is valid. Nevertheless, the second

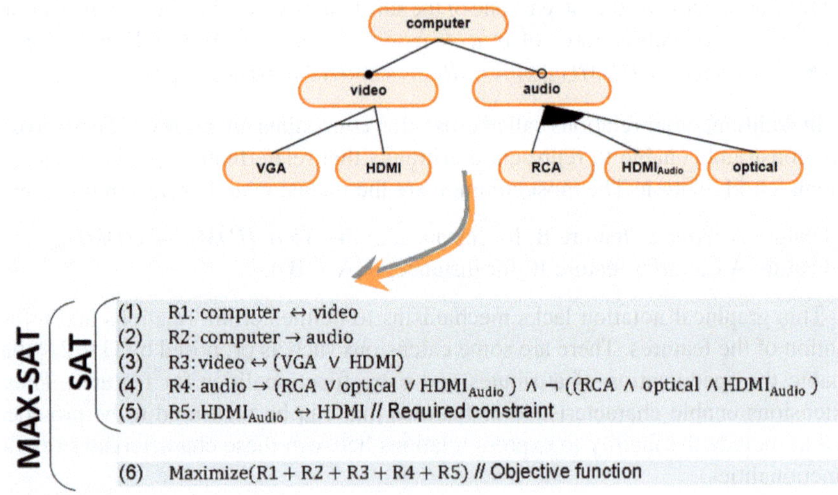

Fig. 15.7 Transformation of a feature model to a SAT

configuration is invalid since the *RCA* feature is in the configuration, however, *HDMI* feature requires $HDMI_{Audio}$ feature to be chosen as a required constraint. This invalid configuration might be diagnosed in order to determine why and what are responsible for this invalid configuration or to determine a faulty feature model composition.

From the point of view of model-based diagnosis, it can be applied in two different ways: (1) diagnose (i.e., detection of faulty software configurations) the feature model in order to identify malformed structures; thus no legal configurations can be computed due to overconstrained structured or bad constructions; (2) given a configuration, diagnose why it is illegal with regard to the feature model. If we assume that feature model is correct and represent a product line, the main problem of diagnosis is in the second case.

Formally, in this context, the System Model *(SM)* is defined as a feature model FM_i and a specific configuration C as an Observational Model *(OM)* which is unsatisfiable. These two components define the System Configuration Error (*SCE*) problem to be diagnosed. The diagnosis strives toward the identification of inconsistencies in the configuration, and specifically the set of features and attributes that produce the inconsistency. This identification might be solved by means of fault diagnosis theory. Following the theory of consistency-based diagnosis proposed in [35] and given in previous chapters, the problem has been formalized as the fault diagnosis of a *SCE* problem in the following definitions.

// Configuration:

(1) computer, video, HDMI, audio, RCA = true

// Reified constraints:

(2) $R1 = computer \leftrightarrow video$
(3) $R2 = computer \rightarrow audio$
(4) $R3 = video \leftrightarrow (VGA \vee HDMI)$
(5) $R4 = audio \rightarrow (RCA \vee optical \vee HDMI_{Audio}) \wedge \neg(RCA \wedge optical \wedge HDMI_{Audio})$
(6) $R5 = HDMI_{Audio} \leftrightarrow HDMI$

// Objective function:

(7) $Maximize(R1 + R2 + R3 + R4 + R5)$

Fig. 15.8 MAX-SAT to diagnose the a problem

Definition 15.4 (*Fault Diagnosis of an SCE*). A fault diagnosis of an *SCE* is a set of features and attributes $\Delta \subseteq C$ such that $\Delta = \{\Delta_{F_j} \cup \Delta_{F_l} \cup \cdots \cup \Delta_{F_m}\}$.

$$FM_i \cup (C - \Delta) \vdash \top \tag{15.8}$$

At this point, the system has been diagnosed by identifying the Δ as responsible for the inconsistency in the SCE.

In order to do the diagnosis, constraint suspension can be applied. In this case, the problem is translated into a MAX-SAT problem in which *SM* where all the constraints from the original problem are reified, thus is translated to a Boolean constraint; the variables from the configuration, *C*, are established as a part of the model, and the objective function is established to the satisfaction of the maximum number of the reified constraints. An example of MAX-SAT is shown in Fig. 15.8, it is important to highlight that features that are unestablished in the model are assumed *false* by default.

The MAX-SAT can be solved and it retrieves the information about fault diagnosis as the minimal number of constraints that cannot be satisfied. The solution of this programme is the fault diagnosis, and therefore Δ as defined previously. In the particulars of the example, the fault diagnosis is $\Delta = R5$ which means that $R5$ cannot be satisfied. The explanation of the fault diagnosis is also obtained for the solver in so far as *HDMI* feature is *true* but $HDMI_{Audio}$ is *false* since it is not within the configuration but *RCA*.

Although a very simple problem has been used as a proof of concept, the real utility of fault detection and diagnosis in software product configuration is demonstrated when the configurations of the system is too complex and these reasoning techniques help to easily and automatically determine the errors made in the configurations.

15.6 Conclusions

In this chapter, we have presented a model-based diagnosis approach to software debugging. We have introduced three different approaches. A model-based diagnosis approach which uses dynamic information, namely abstraction of program traces, to generate a (dynamic, sub-) model of the program under analysis. The model, along with the set of traces for pass/fail executions is used to reason about the observed failures. In contrast to most approaches to software fault diagnosis, which present diagnosis candidates as single explanations [5, 33, 38], our approach also contains multiple fault explanations in the diagnostic ranking (typical of model-based approaches [39, 53]).

A constraint model-based diagnosis approach is also proposed. The source code, test cases, and contracts (assertions, precondition, and postcondition) are taken into account in order to obtain a more precise diagnosis. The approach diagnoses defects in contracts and source code. And finally, an approach for diagnosing a feature model in order to identify malformed structures is proposed. Given a configuration, this methodology is able to diagnose why it is illegal with regard to the feature model.

Acknowledgements This work has been partially funded by the Ministry of Science and Technology of Spain (TIN2015-63502-C3-2-R) and the European Regional Development Fund (ERDF/FEDER). This material is based upon work supported by the ERDF's COMPETE 2020 Programme under project No. POCI-01-0145-FEDER-006961 and FCT under project No. UID/EEA/50014/2013.

References

1. Abreu, R., Mayer, W., Stumptner, M., van Gemund, A.J.: Refining spectrum-based fault localization rankings. In: Proceedings of the 2009 ACM symposium on Applied Computing, pp. 409–414. ACM (2009)
2. Abreu, R., Van Gemund, A.J.: A low-cost approximate minimal hitting set algorithm and its application to model-based diagnosis. SARA **9**, 2–9 (2009)
3. Abreu, R., Zoeteweij, P., Van Gemund, A.J.: An observation-based model for fault localization. In: Proceedings of the 2008 International Workshop on Dynamic Analysis: Held in Conjunction with the ACM SIGSOFT International Symposium on Software Testing and Analysis (ISSTA 2008), pp. 64–70. ACM (2008)
4. Abreu, R., Zoeteweij, P., Van Gemund, A.J.: Spectrum-based multiple fault localization. In: Proceedings of the 2009 IEEE/ACM International Conference on Automated Software Engineering, pp. 88–99. IEEE Computer Society (2009)
5. Abreu, R., Zoeteweij, P., van Gemund, A.J.C.: On the accuracy of spectrum-based fault localization. In: TAICPART-MUTATION '07: Proceedings of the Testing: Academic and Industrial Conference Practice and Research Techniques - MUTATION, pp. 89–98. IEEE Computer Society, Washington, DC, USA (2007)
6. Abreu, R., Zoeteweij, P., Van Gemund, A.J.C.: A new bayesian approach to multiple intermittent fault diagnosis. In: Proceedings of the 21st International Joint Conference on Artificial Intelligence, IJCAI'09, pp. 653–658 (2009)
7. Araki, K., Furukawa, Z., Cheng, J.: A general framework for debugging. IEEE Softw. **8**(3), 14–20 (1991)

8. Avižienis, A., Laprie, J.C., Randell, B., Landwehr, C.E.: Basic concepts and taxonomy of dependable and secure computing. IEEE Trans. Dependable Secur. Comput. **1**(1), 11–33 (2004)
9. Avriel, M.: Nonlinear Programming: Analysis and Methods. Courier Corporation, North Chelmsford (2003)
10. Batory, D.: Feature models, grammars, and propositional formulas. In: Proceedings of the 9th international conference on Software Product Lines, SPLC'05, pp. 7–20. Springer, Berlin (2005)
11. Benavides, D., Segura, S., Ruiz-Cortés, A.: Automated analysis of feature models 20 years later: a literature review. Inf. Syst. **35**(6), 615–636 (2010)
12. Benavides, D., Trinidad, P., Ruiz-Cortés, A.: Automated reasoning on feature models. In: LNCS, Advanced Information Systems Engineering: 17th International Conference, CAISE 2005, p. 2005. Springer (2005)
13. Brat, G., Drusinsky, D., Giannakopoulou, D., Goldberg, A., Havelund, K., Lowry, M., Pasareanu, C., Venet, A., Visser, W., Washington, R.: Experimental evaluation of verification and validation tools on martian rover software. Form. Methods Syst. Des. **25**(2–3), 167–198 (2004)
14. Briand, L.C., Labiche, Y., Sun, H.: Investigating the use of analysis contracts to support fault isolation in object oriented code. In: Proceedings of the International Symposium on Software Testing and Analysis, ISSTA 2002, Roma, Italy, 22–24 July 2002, pp. 70–80 (2002)
15. Campos, J., Abreu, R., Fraser, G., d'Amorim, M.: Entropy-based test generation for improved fault localization. In: Proceedings of the 28th IEEE/ACM International Conference on Automated Software Engineering, pp. 257–267. IEEE Press (2013)
16. Campos, J., Arcuri, A., Fraser, G., Abreu, R.: Continuous test generation: enhancing continuous integration with automated test generation. In: Proceedings of the 29th ACM/IEEE international conference on Automated Software Engineering, pp. 55–66. ACM (2014)
17. Campos, J., Riboira, A., Perez, A., Abreu, R.: Gzoltar: an eclipse plug-in for testing and debugging. In: Proceedings of the 27th IEEE/ACM International Conference on Automated Software Engineering, pp. 378–381. ACM (2012)
18. Cardoso, N., Abreu, R.: A distributed approach to diagnosis candidate generation. In: Portuguese Conference on Artificial Intelligence, pp. 175–186. Springer (2013)
19. Cardoso, N., Abreu, R.: An efficient distributed algorithm for computing minimal hitting sets. In: Proceedings of the 25th International Workshop on Principles of Diagnosis, DX, vol. 14, p. 23 (2014)
20. Ceballos, R., Gasca, R.M., Borrego, D.: Constraint satisfaction techniques for diagnosing errors in design by contract software. ACM SIGSOFT Softw. Eng. Notes **31**(2) (2006)
21. Ceballos, R., Gasca, R.M., Valle, C.D., Borrego, D.: Diagnosing errors in dbc programs using constraint programming. In: Current Topics in Artificial Intelligence, 11th Conference of the Spanish Association for Artificial Intelligence, CAEPIA 2005, Santiago de Compostela, Spain, Revised Selected Papers, pp. 200–210 (2005)
22. Ceballos, R., Gasca, R.M., Valle, C.D., Rosa, F.D.L.: A constraint programming approach for software diagnosis. In: Proceedings of the Fifth International Workshop on Automated Debugging, AADEBUG 2003, pp. 187–197 (2003)
23. Ceballos, R., Gasca, R.M., Valle, C.D., Toro, M.: Max-csp approach for software diagnosis. In: Advances in Artificial Intelligence - IBERAMIA 2002, 8th Ibero-American Conference on AI, pp. 172–181 (2002)
24. Czarnecki, K., Helsen, S., Eisenecker, U.: Formalizing cardinality-based feature models and their specialization. In: Software Process: Improvement and Practice, p. 2005 (2005)
25. De Kleer, J.: Diagnosing multiple persistent and intermittent faults. In: Proceedings of the 21st International Joint Conference on Artificial Intelligence, IJCAI'09, pp. 733–738 (2009)
26. Ernst, M.D., Perkins, J.H., Guo, P.J., McCamant, S., Pacheco, C., Tschantz, M.S., Xiao, C.: The daikon system for dynamic detection of likely invariants. Sci. Comput. Program. **69**(1–3), 35–45 (2007)
27. Gouveia, C., Campos, J., Abreu, R.: Using HTML5 visualizations in software fault localization. In: 2013 First IEEE Working Conference on Software Visualization (VISSOFT), pp. 1–10. IEEE (2013)

28. Hangal, S., Lam, M.S.: Tracking down software bugs using automatic anomaly detection. In: Proceedings of the 24th International Conference on Software Engineering, ICSE 2002, pp. 291–301. IEEE (2002)
29. Hao, D., Zhang, L., Zhang, L., Sun, J., Mei, H.: Vida: Visual interactive debugging. In: IEEE 31st International Conference on Software Engineering, ICSE 2009, pp. 583–586. IEEE (2009)
30. Harrold, M.J., Rothermel, G., Sayre, K., Wu, R., Yi, L.: An empirical investigation of the relationship between fault-revealing test behavior and differences in program spectra. STVR J. Softw. Test., Verif., Reliab. **3**, 171–194 (2000)
31. Hofer, B., Riboira, A., Wotawa, F., Abreu, R., Getzner, E.: On the empirical evaluation of fault localization techniques for spreadsheets. In: ICSE '13: Proceedings of the 2013 International Conference on Software Engineering, pp. 68–82. Springer (2013)
32. Jones, C.: The Year 2000 Software Problem: Quantifying the Costs and Assessing the Consequences. ACM Press/Addison-Wesley Publishing Co, Boston (1997)
33. Jones, J.A., Harrold, M.J., Stasko, J.: Visualization of test information to assist fault localization. In: ICSE '02: Proceedings of the 24th International Conference on Software Engineering, pp. 467–477. IEEE (2002)
34. Kang, K.: Feature-oriented domain analysis (FODA): feasibility study. Technical Report CMU/SEI-90-TR-21 - ESD-90-TR-222, Carnegie Mellon University, Software Engineering Institute (1990)
35. de Kleer, J., Mackworth, A.K., Reiter, R.: Characterizing diagnoses and systems. Artif. Intell. **56**(2–3), 197–222 (1992)
36. Kremenek, T., Ashcraft, K., Yang, J., Engler, D.: Correlation exploitation in error ranking. In: ACM SIGSOFT Software Engineering Notes, vol. 29, pp. 83–93. ACM (2004)
37. Lieberman, H.: The debugging scandal and what to do about it. Commun. ACM **40**(4), 26–30 (1997)
38. Liu, C., Yan, X., Fei, L., Han, J., Midkiff, S.P.: Sober: Statistical model-based bug localization. In: Proceeding of the ESEC/FSE-13. ACM, Lisbon, Portugal (2005)
39. Mayer, W., Stumptner, M.: Models and tradeoffs in model-based debugging. In: 18th International Workshop on Principles of Diagnosis. Nashville, TN, USA (2007)
40. Mayer, W., Stumptner, M.: Evaluating models for model-based debugging. In: Proceedings of the 2008 23rd IEEE/ACM International Conference on Automated Software Engineering, pp. 128–137. IEEE Computer Society (2008)
41. Meyer, B.: Applying design by contract. IEEE Comput. **25**(10), 40–51 (1992)
42. Pearson, S., Campos, J., Just, R., Fraser, G., Abreu, R., Ernst, M.D., Pang, D., Keller, B.: Evaluating and improving fault localization. In: ICSE '17: Proceedings of the 39th International Conference on Software Engineering, pp. 609–620. IEEE Press (2017)
43. Perez, A., Abreu, R., D'Amorim, M.: Prevalence of single-fault fixes and its impact on fault localization. In: 2017 IEEE International Conference on Software Testing, Verification and Validation (ICST), pp. 12–22. IEEE (2017)
44. Perez, A., Abreu, R., van Deursen, A.: A test-suite diagnosability metric for spectrum-based fault localization approaches. In: Proceedings of the 39th International Conference on Software Engineering, pp. 654–664. IEEE Press (2017)
45. Perez, A., Abreu, R., HASLab, I.T.: Leveraging qualitative reasoning to improve SFL. In: IJCAI, pp. 1935–1941 (2018)
46. Renieres, M., Reiss, S.P.: Fault localization with nearest neighbor queries. In: Proceedings of the 18th IEEE International Conference on Automated Software Engineering, 2003, pp. 30–39. IEEE (2003)
47. Shepard, T., Lamb, M., Kelly, D.: More testing should be taught. Commun. ACM **44**(6), 103–108 (2001)
48. Steimann, F., Frenkel, M., Abreu, R.: Threats to the validity and value of empirical assessments of the accuracy of coverage-based fault locators. In: Proceedings of the 2013 International Symposium on Software Testing and Analysis, pp. 314–324. ACM (2013)
49. Tip, F.: A Survey of Program Slicing Techniques. Centrum voor Wiskunde en Informatica, Amsterdam (1994)

50. Traon, Y.L., Ouabdesselam, F., Robach, C., Baudry, B.: From diagnosis to diagnosability: axiomatization, measurement and application. J. Syst. Softw. **65**(1), 31–50 (2003)
51. Varela-Vaca, Á.J., Gasca, R.M.: Towards the automatic and optimal selection of risk treatments for business processes using a constraint programming approach. Inf. Softw. Technol. **55**(11), 1948–1973 (2013)
52. Wong, W.E., Gao, R., Li, Y., Abreu, R., Wotawa, F.: A survey on software fault localization. IEEE Trans. Softw. Eng. **42**(8), 707–740 (2016)
53. Wotawa, F., Stumptner, M., Mayer, W.: Model-based debugging or how to diagnose programs automatically. In: Hendtlass, T., Ali, M. (eds.) Developments in Applied Artificial Intelligence, pp. 746–757. Springer, Berlin (2002)
54. Wotawa, F., Weber, J., Nica, M., Ceballos, R.: On the complexity of program debugging using constraints for modeling the program's syntax and semantics. In: Current Topics in Artificial Intelligence, 13th Conference of the Spanish Association for Artificial Intelligence, CAEPIA. Selected Papers, pp. 22–31 (2009)
55. Yang, Q., Li, J.J., Weiss, D.: A survey of coverage based testing tools. In: Proceedings of the 2006 International Workshop on Automation of software test, AST '06, pp. 99–103. ACM, New York, NY, USA (2006)
56. Yilmaz, C., Williams, C.: An automated model-based debugging approach. In: ASE '07: Proceedings of the Twenty-Second IEEE/ACM International Conference on Automated Software Engineering, pp. 174–183. ACM (2007)
57. Zoeteweij, P., Abreu, R., Golsteijn, R., Van Gemund, A.J.: Diagnosis of embedded software using program spectra. In: 14th Annual IEEE International Conference and Workshops on the Engineering of Computer-Based Systems (ECBS'07), pp. 213–220. IEEE (2007)

Chapter 16
Diagnosing Business Processes

Diana Borrego and María Teresa Gómez-López

16.1 Introduction

Most of the organizations related to industrial manufacturing are process-oriented. In order to facilitate their daily processes, enterprises use software, automating the execution order of their more frequent activities. Business Process Management Systems (BPMSs) have helped Industry to orchestrate their tasks and recover every used information, combining knowledge from information technology area and management science to improve operational business processes [2]. Business Process Models (BPMs) are formed of tasks combined by means of a workflow that lets the achievement of a particular business goal [21]. The BPMs can be enriched with rules to describe the values of the data that flows. For example, in the manufacturing process of an aerospace company, the process for aircraft assembly is able to be developed using a BPMS, because it is a frequent process where faults can turn up.

BPMs can be described using different perspectives, i.e., control flow, data flow, business rules, resources, and time. BPMs can or cannot work as expected, being necessary to diagnose them. The challenge is that business processes are designed to coordinate different elements related to an organization, then several of them are possible sources of a fault. For example, diagnosis can be related to the execution order of the activities, the input data introduced, the person who executes an activity, the behavior of an activity, etc.

Classical model-based diagnosis cannot be directly applied to BPMs, since they have special characteristics that are analyzed in this chapter. Aspects, such as the great quantity of data used, how the information is shared between different instances of a

D. Borrego (✉) · M. T. Gómez-López
Languages and Computer Systems Department, Universidad de Sevilla, Sevilla, Spain
e-mail: dianabn@us.es

M. T. Gómez-López
e-mail: maytegomez@us.es

T. Escobet et al. (eds.), *Fault Diagnosis of Dynamic Systems*,
https://doi.org/10.1007/978-3-030-17728-7_16

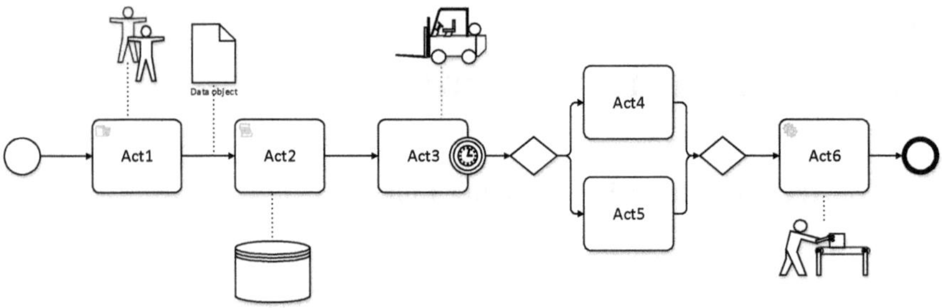

Fig. 16.1 Example of business process

process, or the heterogeneity of the used information or the shared resources, make necessary to tackle the diagnosis from others points of view.

An example of a general business process is depicted in Fig. 16.1, where a set of activities are combined to develop an objective in a company. The business process can include: who is executing an activity, what data objects are flowing in the instances, which resource is used in each activity, or the database whose data are read and written during the executions.

According to the business process life cycle (i.e., Design, Analysis, Configuration, Enactment and Evaluation [21]), Design, Analysis, and Evaluation are the most typical sources of faults, thereby where the diagnosis tends to be applied. These phases can be supported by a Process-Aware Information System (PAIS), that is, a software system that helps to manage and execute operational processes. PAISs include the possibility to model various perspectives, such as people, control flow, data layer or resources, and it is involved in every phase of the business processes.

The chapter is organized as follows: Sect. 16.2 introduces definitions about business processes and the main characteristics of BPMs to be diagnosed in their different phases. Section 16.3 explains the specific diagnosis techniques that can be used to tackle the challenge found in each phase of the business process life cycle. Finally, Sect. 14.5 concludes the chapter.

16.2 Problem Statement

Business Process Management (BPM) offers a set of mechanisms that facilitate the modeling of processes by nontechnical experts with a high level of automation [9]. Business Process Management Systems (BPMSs) are a set of software tools and methodologies focused on the management and recovery of business processes execution. The obtained information from the instantiations of the processes can be used to detect and diagnose incorrect behaviour or to improve the process executions. The use of this widespread recovery information enables enterprises to detect in a more efficient and effective way a malfunction in a system. Since BPMs combine various systems related and choreographed, a more complete and complex diagnosis can be faced. BPM fits well with the necessities of factories [15] since every day they perform a set of well-defined processes to produce their products, where heterogeneous data are recovered.

16.2.1 Business Process Description

Business processes help organizations to describe and monitor their daily activities to be improved. The most relevant definitions are shown below:

Definition 16.1 (*Business Process*) A set of activities that are performed in coordination in an organization and technical environment. These activities jointly achieve a business goal [21].

The quality of a business process is able to be defined as both the capacity to produce quality outcomes or to produce the outcome in a quality way. Quality is usually measured by aspects, such as accuracy or security [14]. Verification and diagnosis of business processes help to ascertain the quality of the process, especially analyzing suitability and accuracy aspects in relation to the expected behaviour:

- **Accuracy**. The capability of a business process to provide the expected results with a certain degree of precision. The verification and diagnosis of business processes ensure obtaining the results in accordance with the expected behaviour, ensuring the attainment of the expected results.
- **Suitability**. The capacity to provide the appropriate solution to get the specified business goals.

Definition 16.2 (*Business Process Model (BPM)*) It is formed by a set of activities, a set of control flow gateways (AND, OR, XOR) that describes the relationship between the activities, and a set of Data Objects that flows in the business process [21]. BPMs can be activity oriented or artifact oriented.

Definition 16.3 (*Business Process Management*) It can be considered as an extension of classical Workflow Management (WFM) that includes concepts, methods, and techniques to support the design, administration, conformance, enactment, and analysis of business processes [21].

Definition 16.4 (*Business Process Management System*) It is a generic software system that is driven by explicit process representations to coordinate the enactment of business processes [21].

Definition 16.5 (*Business Data Constraint (BDC)*) They are a subset of business compliance rules that represent the compliance relation between the values of data during a business process instance. It is a numerical constraint that involves business process variables that can be related to the data flow or the stored information. BDCs can be associated as invariant to the business process or as pre- or post-conditions [13].

16.2.2 Business Process Phases

The whole business process life cycle includes various phases: In the **Design phase**, business processes are modeled, where typically livelock and deadlock analyses are performed [10] to avoid the incorrect synchronization of the project choreography. Also, data read and written must be analyzed, in order to avoid missing, redundant, and conflict data [19]. In the **Analysis phase**, validation and verification by simulating are performed. In order to cover the possible instances of a BPM, both control flow and data flow perspectives must be combined. A fully automated diagnosis of the correctness of BPMs implies to include the semantics of activities by means of business rules. What are the most relevant data to be monitored in a business process is not always a trivial problem. In the **Configuration phase**, the most appropriate points of monitoring are analyzed to obtain the level of diagnosability needed in each case. A correct BPM is not able to work correctly using incorrect input data. In the **Enactment and Evaluation phase**, processes are monitored and analyzed to evaluate and diagnose both BPMs (workflow and rules) [17] and the used data for each instance [13].

16.2.3 Model-Based Diagnosis and Business Processes

In order to diagnose if a system is working as expected, it is necessary to compare the designed model with the observations (Observational Model).

16.2.3.1 Model

The model in business processes tends to be formed of two parts: Activity-oriented business model and Business Data constraints.

Activity-oriented business model that tends to be described by using the standard BPMN [16] that permits the inclusion in the model of several artifacts to describe activities workflow, used data objects, choreography between processes, etc. The activity-oriented workflow describes the possible order of execution of the activities.

To include in the model how the variables can evolve during the execution of business process instances, business data constraints (Definition 16.5) describe the relation between the data values during the instantiation of a process. BDCs are presented as constraints with numerical variables (integer, float, and natural) by following the next grammar [13], where it is important to highlight that the variables participating in the constraints can come from the database or from the data flow:

```
BusinessDataConst := Atomic_Const BOOL_OP BusinessDataConst
 | Atomic_Const | 'NOT' Constraint
BOOL_OP := 'AND' | 'OR' | '→'
Atomic_Const := function PREDICATE function
```

```
function := Variable FUNCTION_SYMBOL function
  | Variable | Constant
Variable := Table'.'Attribute | Attribute | DataFlowVariable |
Constant
PREDICATE := '=' | '<' | '≤' | '>' | '≥'
FUNCTION_SYMBOL := '+' | '−' | '*' | '/'
```

BDCs are able to describe the pre- and post-conditions of the variables before and after an activity is executed. For example, {A_OutputSignal ≥ B_InputSignal + C_InputSignal AND A_OutputSignal ≤ C_InputSignal}.

16.2.3.2 Observational Model

In a Process-Aware Information System (PAIS), the information is typically stored in a relational database, and therefore the tuples of the tables compose the Observational Model (OM). Since the variables involved in a BDC have different origins, it is possible that attributes from various tables are related in a single BDC. The location of the data in various tables is due to the necessity to follow the *Normal Forms* defined in relational database theory. The normalization rules are designed to prevent update anomalies and data inconsistencies. Since data is stored in various tables, to ascertain the full tuple of values for an OM, a *denormalization process* needs to be carried out [13]. To understand how the OM can be extracted from a relational model, the example of Fig. 16.2 is used. It includes two entities (A and B) with a relation by means of a foreign key that relates each tuple of entity A with a tuple of entity B. If a join is executed between both entities, relation *Entity A&B* is obtained. This new relation represents three different OMs, where finally a diagnosis process can be executed. The different aspect, in this case, is that each OM is not independent of the others. For example, attributes *d* in first and second tuples come from a single value in the original entities. It means that if *d1* is an explanation of malfunction for a diagnosis, it has to cause a malfunction for the second tuple of the OM too. It is not always related to the value of the variable itself since, for example, the attribute *b* in the second and third tuples has the same value *b2*, but they represent two different variables that can take the same value. In the last case, the explanation of a malfunction in the second OM does not imply the same explanation for the last tuple.

16.2.4 *Special Characteristics*

Business processes have special characteristics that provoke the necessity of adapting classical diagnosis techniques. Both the great amount of data generated and the multisource heterogeneous information cause the need for adaptation of the classi-

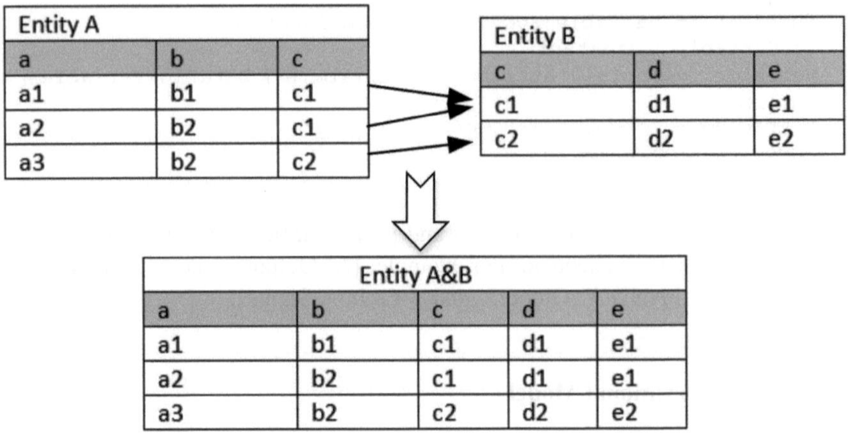

Fig. 16.2 Example of creation of OM

cal diagnosis for business processes. Some of the most relevant characteristics are explained in the following:

- **Shared data among different process executions**. Business processes tend to be modeled when a set of activities in a defined order are frequently repeated in a company. Each process execution is called instance. These instances use data that flow in the process or is stored in an external database. Derived from this external storage, the same data can be used in different instances, as explained in Sect. 16.2.3.2. It gives rise to when a value of a data is determined as incorrect in a diagnosis evaluation, it is necessary to take into account that the diagnosis explanation must be conformance with every instance where the data is involved, and if the data is incorrect or correct in an instance, it must be incorrect or correct for every instances where this data participates.
- **Different sources of fault**. Business processes are able to coordinate every participant in an organization into the same model. This has the advantage of integrating several possibilities in a diagnosis study, but also makes more complex the detection of the possible responsible for a malfunction. Several causes can be the origin: the involved persons, data, activities, resources, etc. This heterogeneity of possible faults makes more complex the isolation of the responsible for a malfunction.
- **Rules and structures combined**. Models in business processes are formed by the activities described in the *Business Process Model* of Definition 16.2, and the Business Data Constraints of Definition 16.5. This causes that, when a diagnosis is done, the whole process has not been analyzed, since it depends on the point of analysis in the business process instance.
- **Choreography between several processes**. Processes usually work in a coordinated way, although it brings about more complex models that can imply more faults. It is derived from several persons working and interacting in different parts of the models, increasing the possibilities of introducing malfunctions.

- **High changeability of business rules**. Business rules are frequently changed, since rules can change over time. This is because of compliance rules let business experts describe how the process works, not being necessary to modify the workflow.

16.3 Proposed Approach

The mentioned characteristics affect the diagnosis in different ways according to the involved phase. In following subsections, each phase is analyzed and how the diagnosis can be tackled derived from the special characteristics is explained.

16.3.1 Design

Only a reduced percentage of software projects succeed. The primary reason for these failures is poor conceptual modeling. To detect and avoid this, the aim of verification analysis methods is to check the correctness of BPMs, concerning the detection of syntax errors or violations in control flow and data flow perspectives.

16.3.1.1 Workflow

Most modeling and verification approaches in the literature only consider the control flow perspective of BPMs. These contributions typically deal with errors in the modeling of the flow, detecting deadlocks, livelocks (infinite loops), lack of synchronization, and dangling reference, most of them checking an important correctness criterion known as the soundness property, which guarantees proper termination of the business processes.

The most popular and effective notation for modeling business process at the conceptual level is the Business Process Modeling Notation (BPMN). It is a notation used to characterize the identification and the specification of the business processes. However, the general modeling includes the arbitrariness and lacks the strictness. For example, the congestion of business flows. If the congestion is not properly controlled, the software may not meet the requirements.

As explained in van der Aalst in [1], process chains are considered to be sound if they fulfill a series of conditions: (i) for every process A which is reachable from a start process, there exists a sequence leading from process A to the terminate process; (ii) the terminate process is the process that has no subsequent process; and (iii) for each process A, there is at least one occurrence sequence from the start process to the terminate process that involves process A.

In detail, the congestion can lead to the structural problem such as livelock. A livelock problem is an infinite execution of a process. In this problem, some of the

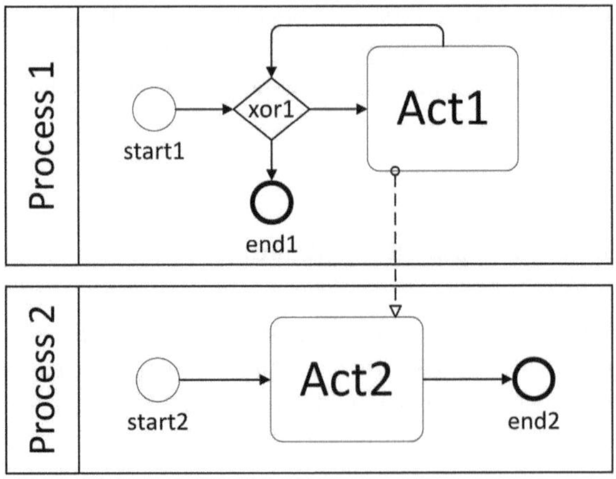

Fig. 16.3 Livelock and deadlock example

processes may run successfully but some of the processes trap in an endless loop of execution.

Likewise, deadlock takes place when a business process requires the execution of previous processes to be completed, but at least one of them is never executed. Such errors inhibit the BPM to work correctly.

As an example, Fig. 16.3 shows a simple BP model, where two processes collaborate by means of exchanging a message (dashed arrow). Depending on the evaluation of the exclusive gateway in Process 1, different traces can be obtained, which is possible to get stuck in a livelock or a deadlock. Such possible traces are

$T1 : (start1, start2) \rightarrow (xor1, Act2) \rightarrow (Act1, end2) \rightarrow (xor1) \rightarrow (end1) \Rightarrow$
$CORRECT\ TRACE$

$T2 : (start1, start2) \rightarrow (xor1, Act2) \rightarrow (end1, Act2) \rightarrow (Act2) \Rightarrow DEADLOCK$

$T3 : (start1, start2) \rightarrow (xor1, Act2) \rightarrow (Act1, end2) \rightarrow (xor1) \rightarrow (Act1) \rightarrow (xor1)$
$\Rightarrow LIVELOCK$

16.3.1.2 Data Flow

On the other hand, besides the structural verification regarding conflicts in the control flow perspective, when the data flow perspective is not correctly designed within the BPM, it can cause errors or conflicts, referred to as data anomalies in [18], classified into three types in [19] as given below:

1. Missing data. It takes place when a data item is accessed before it is initialized, for four different causes: (i) a data is never assigned an initial value; (ii) an activity which uses a data item is executed before the activity which initializes

it; (iii) the initialization and access are performed in parallel branches, thereby the data item may not have been initialized when it is read; and (iv) the data item is used but not initialized under certain workflow routing conditions.
2. Redundant data. A data item that does not contribute to the production of the final output data of the process is produced, causing inefficiency. The main cause is that a data item is produced by an activity, but never used by any activity executed after it in the process instance.
3. Conflicting data. There exist different versions of the same data item. It is impossible to decide which version of the data item should be considered.

In addition, it is necessary to consider verification of data flow errors, specifying variables read and written per activity, and even detailing the effect of each activity by means of pre- and post-conditions.

16.3.2 Analysis

The verification of BPMs allows to determine the correctness of business processes, because a business process cannot be correct, if: (i) one or more of its activities can never get executed for any possible value of the data input or (ii) there is not any possible process instance for a certain value of the input data. In these cases, despite the BPM can be correct regarding the control flow perspective, it is not correct from the perspective of the data flow.

In order to diagnose faults in data semantics, an effective means to express data constraints is to annotate activities in a BPM with pre- and post-conditions that specify the effect on the data state for each activity. Annotating process' activities with pre- and post-conditions facilitates compliance checking to ensure that business processes are properly designed. It is an effective means to capture such dependencies between control flow and data flow. Moreover, pre- and post-conditions can capture requirements on the data flow that any execution of the business process has to comply with.

This subsection presents an approach for diagnosing the correctness of the relationship between values in semantic BPMs, containing activities whose effects are formally specified using pre- and post-conditions. An activity can start if the execution of the BPM has reached the activity and its precondition is satisfied. Upon completion, the activity delivers data that satisfies its post-condition. An execution of the business process can reach an activity whose pre-condition is not satisfied. In that case, the execution gets stuck in the activity and fails.

To detect these errors, we distinguish between two different notions of correctness to diagnose such data flow errors [5]:

- May-correctness. A BPM is may-correct if every activity can be reached and executed at least once, so there is an execution in which the activity is done. The fulfillment of this correctness ensures that there is at least one valuation of the data inputs which makes an activity executable.

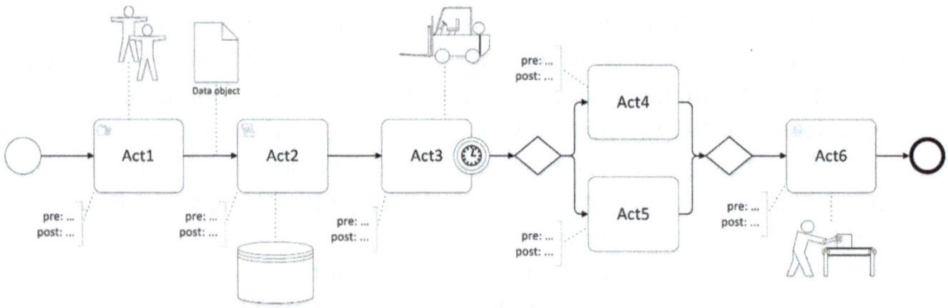

Fig. 16.4 Business process with annotated activities

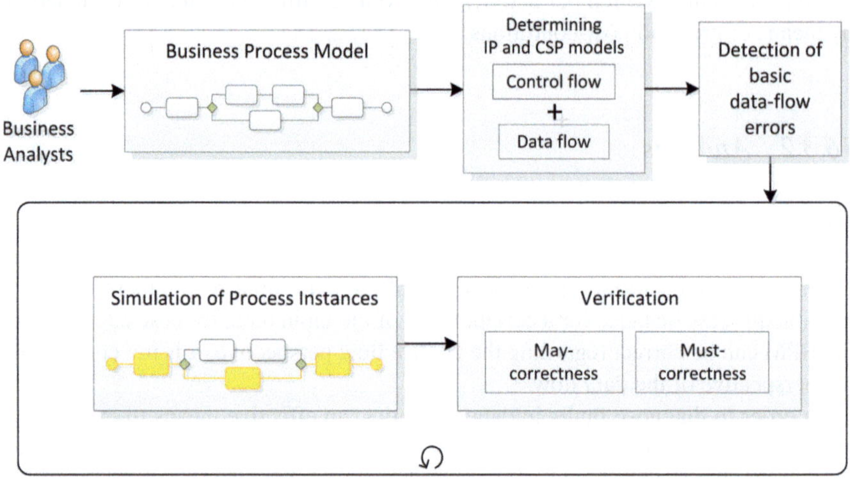

Fig. 16.5 Diagram of the verification process

- Must-correctness. A BPM is must-correct if every possible execution that reaches an activity satisfies the pre-condition of the activity. The fulfillment of this correctness ensures that there is no valuation of the data inputs which makes an activity non-executable.

To illustrate the idea, Fig. 16.4 shows a business process whose activities are annotated with pre- and post-conditions. Those conditions should be satisfied in accordance with the valuations of the data flowing and changing during the execution. As a concrete example, if activities *Act1* and *Act2* in Fig. 16.4 assign the valuation $\{a = 30, b = 100, c = 30, d = 200\}$, and the pre-condition of *Act3* is $pre(Act3) = 3 * a + b + c < d$, it would not be possible to execute *Act3* since $3 * 30 + 100 + 30 \not< 200$, so that the process would get stuck at *Act3*.

Diagram in Fig. 16.5 illustrates the diagnosis process, which is performed in the analysis phase, at design time, using artificial intelligence techniques to compute the execution instances allowed by a BPM. In detail:

1. The diagnosis process starts from BPMs with correct control flow definitions.
2. For diagnosis, the BPM is translated into two models: (i) an Integer Programming model (IP model), to determine the different instances of execution of the business process, and (ii) the pre- and post-conditions of the activities are modeled as constraints in a Constraint Satisfaction Problem (CSP), following a Backus–Naur Form (BNF) grammar in order to avoid any ambiguity.
3. Next, basic data flow errors are determined. These data flow errors at the basic level can cause unexpected process interruptions and should, therefore, be avoided. In detail:

   - Conflicts among postconditions: It is possible to find different postconditions that seem to be conflicting regarding the value of some variable. However, all of them could be satisfied since the value of a variable can be modified several times during the execution of a process, therefore making possible that those different valuations satisfy each postcondition as each activity is executed. In order to resolve conflicts among post-conditions, it becomes necessary to convert the variables and constraints of the CSP model into Static Single Assignment (SSA) form [3, 8]. As a result, in a CSP in SSA form, each variable is assigned a value by only one activity.
   - Missing data: When a variable is read by an activity, so referenced in its pre-condition, but not written in any preceding activity. To identify this kind of errors, the CSP is enriched with constraints that check the presence of at least a direct path from an activity writing a data and any activity reading it, so that the writing precedes the reading.
   - Conflicting data: When the same variable is written in two parallel activities. In that case, one activity overwrites the value of the variable written earlier by the other activity. The identification of this kind of errors implies the use of an algorithm to iterate over all activities writing a data, testing whether there exists any trace triggering any pair of them.

   Such data flow errors are at a more basic level than violations of must and may correctness. Therefore, these data flow errors need to be detected and resolved before may- and must-correctness can be diagnosed.
4. Finally, from the final obtained CSP, it is possible to simulate executions to check if the pre-condition of the final activity is satisfied in a scenario where all pre- and post-conditions of previous activities are satisfied, for both types of correctness:

   - May-correctness: There is at least one valuation of the data inputs which makes an activity executable.
   - Must-correctness: There is no valuation of the data inputs which makes an activity non-executable.

16.3.3 Configuration

Since there is no single entity which provides an overall view of the complete data flow model of a business process, it is desirable to improve the monitoring of business processes to be fully aware of possible deviations from expected behaviour.

Since the monitoring of a business process, and later diagnosis, is based on observations providing information about the behavior of the process, for the later diagnosis process to be successful, diagnosability analysis becomes a configuration time requirement. The diagnosability level of the business process depends upon the observations. If the diagnosis is based on few observations, or if observations are not allocated at the most convenient places, it is very difficult to distinguish which parts of the business process are failing. Both the number of observations and the location where they are performed enable the cause of the error to be precisely located. In general, diagnosis systems not only incorporate identification and isolation of faults, but also the monitoring.

The allocation of the test points is, therefore, a crucial task, since if the test points are not correctly allocated, then the observational model may be rendered useless for the isolation of faults, and hence the diagnosis process would fail to determine the activities which are not behaving as they were modeled.

Then, why not allocate test points at every possible location? Weighting the pros and cons, we would get a complete monitoring, but on the other hand, it would cause problems regarding confidentiality, privacy, and security, and it would also highly increase the necessary communications and cost.

So, test points should be located depending on certain requirements: (i) cost limitations, (ii) optimization of the diagnosis time, or (iii) expected diagnosability level.

As an example, for the business process in Fig. 16.6, the allocation of a test point as depicted gives rise to two clusters of activities (shadowed and non-shadowed activities, respectively).

Diagram in Fig. 16.7 shows the steps of the test points allocation process, which is performed in the configuration phase, using artificial intelligence techniques to compute the allocation depending on the chosen objective. In detail:

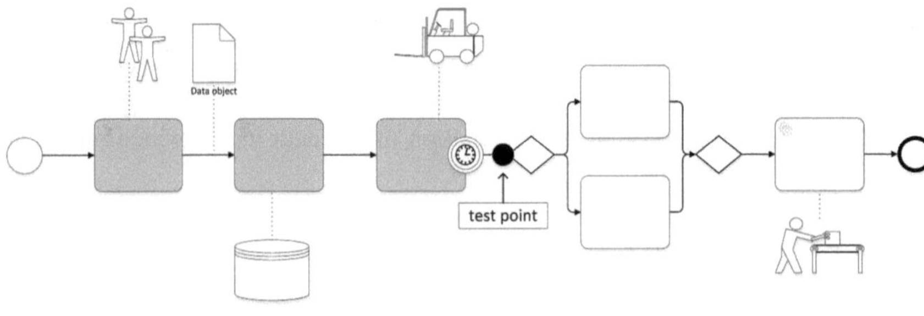

Fig. 16.6 Business process with test point

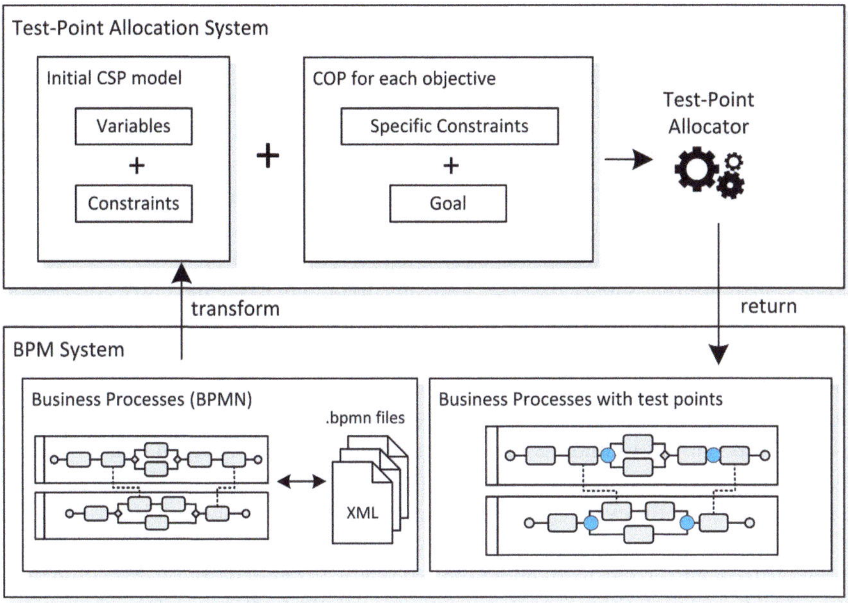

Fig. 16.7 Diagram of the process of allocation of test points

1. The process of allocation starts from BPMs with correct control flow and data flow definitions.
2. Then, the business process is modeled as an initial CSP, with no test points, so that all the activities belong to the same cluster.
3. Next, and according to the objective to achieve (cost, execution time, or diagnosability level) the initial CSP is enriched with specific constraints and goals for each objective, so that it becomes a COP. In detail, for each objective:

   - Cost limitations: It implies to limit the number of test points to allocate, with the goal of maximizing the number of clusters.
   - Minimization of the diagnosis process execution time: It begins with a fixed number of balanced clusters, with the goal of minimizing the number of test points to allocate.
   - Diagnosability level: From a maximum number of activities per cluster, the goal is to minimize the number of test points to allocate.

4. The allocation of test points is performed (test-point allocator) through solving the corresponding COP, getting different clusters of activities.

16.3.4 Enactment and Evaluation

A process correctly modeled and deployed can work incorrectly, since the use of incorrect data can cause improper working. Business Data Constraints can be used to validate data correctness during instantiation according to the Business Rules. Incorrect data can be included by users in the system or generated by activities during the process execution. The quantity of data can be huge and stored in external repositories, such as on a relational database. Therefore, the data involved in a business process instance that satisfy the BDCs are able to belong to the data flow (depending on the instance) and can be stored in a database (persistence layer).

Model-based diagnosis needs to compare the expected model and the observed model. Expected model is represented by a workflow that combines the activities and the Business Data Constraints. But the observational model must be extracted from external repositories to find the misalignment between the observed trances and the model [12]. Business processes usually are supported by relational databases that contain the traces for every instance. During a process execution, it is possible that a correctly designed model works incorrectly because of data. For this reason, it is necessary to analyze how to diagnose the process at different points at execution time. Figure 16.8 depicts the elements that are combined: BPMs, Business Data Constraints associated with activities or to the whole process, and external databases. BPM and Business Data Constraints conform the model of the system. The involved data stored in the database and flowing in the data flow are extracted to create the observational model as explained in [13].

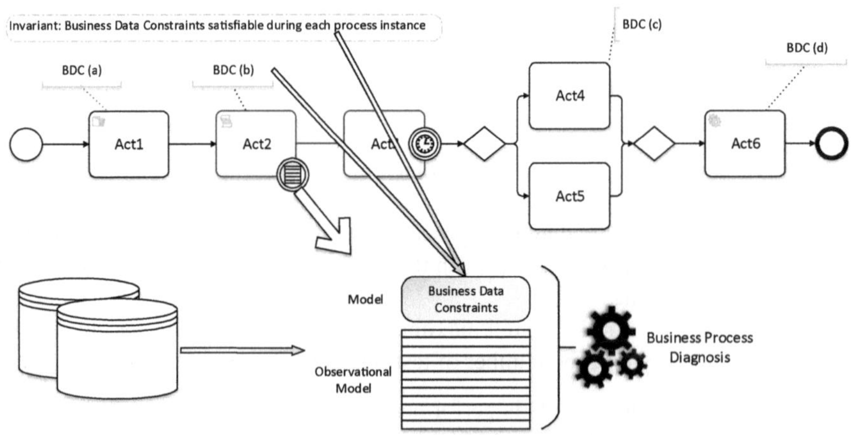

Fig. 16.8 Model and observational model of business process diagnosis at runtime

16.3.4.1 Constraint Programming to Find and Isolate Incorrect Values of Variables

As explained in the previous chapter, a Constraint Satisfaction Problem (CSP) consists of the assignment of a set of values to a set of variables which are involved in a set of constraints. A solution of a CSP is a total assignment satisfying all constraint. If no solution is found for a CSP, it means that the constraints are not satisfied between them, and the responsible component for this situation must be found. This is similar to the process of diagnosing in a system. For example, to find out the incorrect variable in a minimal diagnosis, we must eliminate some constraints, letting the CSP be satisfiable. To know which are the variables, finding an assignment which satisfies as many constraints as possible is a problem that can be solved using Max-CSP. The Maximal Constraint Satisfaction Problem (Max-CSP) consists of finding a total assignment which satisfies the maximum number of constraints. Max-CSP is an NP-hard problem and generally is more difficult to solve than the CSP problem [20]. Max-CSPs have already been used in model-based diagnosis [7] as a type of Constraint Optimization Problem (COP), as a CSP with an optimization function (maximization of minimization). Other different algorithms have also been proposed in order to improve the algorithmic determination of all minimal unsatisfiable subsets using notions of independence of constraints and incremental constraint solvers [4] and structural analysis [11].

Most times the problem is not related to incorrect restrictions nor a malfunction of the activities, very often the problem is due to incorrect input data. If we want to model a business process as a set of values for a set of business rules, where the error is the valuation of a variable for a tuple in a database, the constraints that can be relaxed to obtain a correct behavior will be the values of the variables. It means, to find the minimum number of variables that whether they are not instantiated, all the constraints are satisfiable.

Not all the variables can be incorrect for a direct human task, since some of them only depend on the values of input variables. For example, for the constraint $c = a + b$, a, and b are introduced by an user and c is derived from the values of a and b. It implies that if a or b are incorrect, then c will be incorrect too. For this situation, two types of variables are defined:

Definition 16.6 (*Input Variable*) Variable whose value is introduced by the user in an activity of a business process.

Definition 16.7 (*Derived Variable*) Variable whose value depends on the value of other variables (input or derived variables). It means that its value is not introduced as an input variable.

Once the input and derived variables are determined, depending on the business process, the goal of a diagnosis process is to detect the possible minimal incorrect variables set.

Definition 16.8 (*Possible Minimal Incorrect Variables Set*) Set of input variables ($\mathcal{S}$) that if they are not instantiated, the set of business rules are satisfiable. This set is minimal iff there is not a subset $\mathcal{S}\sqcap\lfloor\mathcal{S} \subset \mathcal{S}$ where $\mathcal{S}\sqcap\lfloor\mathcal{S}$ is a possible minimal incorrect variables set.

The Max-CSP shown below is created to obtain this maximum set of variables in (1) with the value associated by the user. In order to obtain this minimal combination, reified constraints in (8) are used. A reified constraint relies on a variable that denotes its truth value. It is, therefore, necessary to add new variables to the CSP (in (2)), whose domain is reduced to values 0 (false value) and 1 (true value). The Max-CSP is

$$type\ InputVar_1, \ldots, InputVar_n \tag{16.1}$$

$$type\ InputVarBool_1, \ldots, InputVarBool_n \tag{16.2}$$

$$type\ DerivedVar_1, \ldots, DerivedVar_m \tag{16.3}$$

$$type\ varMaximize \tag{16.4}$$

$$InputVar_1 = [min\_domain_1, max\_domain_1] \tag{16.5}$$

$$\ldots$$

$$InputVar_n = [min\_domain_n, max\_domain_n]$$

$$DerivedVar_1 = [min\_domain\_derived_1, max\_domain\_derived_1] \tag{16.6}$$

$$\ldots$$

$$DerivedVar_m = [min\_domain\_derived_m, max\_domain\_derived_m]$$

$$Business\_Rule_1 \tag{16.7}$$

$$\ldots$$

$$Business\_Rule_k$$

$$Constraint_1 = (InputVarBool_1 = [InputVar_1\_Instance]) \tag{16.8}$$

$$\ldots$$

$$Constraint_n = (InputVarBool_n = [InputVar_n\_Instance])$$

$$varMaximize = Constraint_1 + \ldots + Constraint_n \tag{16.9}$$

$$maximize(varMaximize) \tag{16.10}$$

16.3.4.2 Hybrid Faults in Business Process Diagnosis

One of the challenges described in Sect. 16.2 is the continuous evolution of the rules in a business process. This is an advantage of flexibility in the system, but it is also a potential source of faults, because business constraints can be introduced incorrectly or being inconsistent according to the legacy data. Thereby, a great quantity of data and rules can be involved in a malfunction, then several possible diagnoses can be found. In order to reduce these possibilities, it is necessary to include weights associated with the various possibilities [6]. It is even more complex because of the high quantity of information and rules evolving in the processes at runtime. Derived from this complexity, a hybrid diagnosis needs to be performed regarding the likelihood of faults in data versus business rules. In order to achieve the correct diagnosis, it is fundamental to attain the best assumption concerning the degree of likelihood. For example, if the possible minimal explanation of a fault is that one data value is incorrect or one BDC is incorrect, which is the more likely minimal diagnosis? We could consider that data are more frequently introduced, and by more nonexpert user, then probably data is the incorrect element. But in the case that the possible explanation are four variables or one BDC? It could depend on the number of introduced input data. If 2,000 input data were introduced, 4 incorrect faults are a very low percentage, but if only 5 data have been introduced, it is more likely that the BDC is the incorrect element rather than 4 out of 5 values.

When both data and BDCs can be incorrect, the Max-CSP explained in the previous subsection must be adapted, since both variables and BDCs must be associated with a reified constraint that can be true or false. It is also necessary to include different likelihoods for each BDC, since it will depend on the number of variables associated to each one of them. For this case, a Min-CSP instead of a Max-CSP is needed.

In order to declare the **Variables of the problem**, a new variable is added to the Min-CSP for each variable obtained in BDC^i_j as explained in Sect. 16.3.4.1 following the syntax: *type* var^1_k, ..., var^m_k.

Furthermore, in order to provide the Min-CSP with the ability to distinguish between different sources of faults (i.e., data and/or BDCs), also reified constraints are used. There are different reified constraints, some associated to each BDC (BDC_i) and some others to each BDC_i for each tuple j (BDC^j_i). The Boolean variable $rBDC_i$ will be true to denote the correctness of BDC_i for any tuple of values and $rBDC^j_i$ for each different tuple. The Min-CSP has the following form:

//Boolean variables associated to reified constraint to ascertain the satisfiability of each BDC:

$$Bool\ rVar^1_k, \ldots, rVar^m_k \tag{16.11}$$

$$Bool\ rBDC_i, rBDC^1_i, \ldots, rBDC^n_i \tag{16.12}$$

//Reified constraints to represent the correctness of the variables assignment:

$$rVar^j_k = \neg(var^j_k = value^j_k) \tag{16.13}$$

//Variable to represent the reified BDCs for each tuple:

$$rBDC_i^1 = \neg(BusinessRule_i instantiatedbytuple1) \tag{16.14}$$

$$\dots$$

$$rBDC_i^n = \neg(BusinessRule_i instantiatedbytuplen) \tag{16.15}$$

The reified variables are assigned to negated constraints since the objective function is to minimize the number of elements with incorrect behavior (noncompliant). To know when a fault in a BDC is less likely than in a data value, it is necessary to include in the Min-CSP the following BDCs:

$$rBDC_i = rBDC_i^1 + \dots + rBDC_i^n = \sum_j^n rBDC_i^j \geq minLik_i \tag{16.16}$$

These constraints incorporate the likelihood concept into the CSP, by using the parameter minLik_i.

Definition 16.9 ($minLik_i$) This is a parameter that represents the minimum number of faults (noncompliant instances of a BDC_i) that is determined as the threshold to represent that there is a malfunction in a BDC_i [6].

This value depends on the number of data related to each BDC. For example, if there are 5 tuples where BDC_i is involved, $minLik_i$ can take a value between 1 and 5. If at least the $minLik_i$ threshold number of instances is not satisfiable, then BDC_i is considered as a part of the minimal diagnosis. How the values of each $minLik_i$ is determined is detailed in [6].

The **Objective function** to minimize is defined as follows:

$$minimize(rVar_k^1 + \dots + rVar_k^m + \dots + rBDC_1 * minLik_1 + \dots + rBDC_q * minLik_q) \tag{16.17}$$

Each rBDC_i has a weighting that multiplies it that is proportional to each parameter $minLik_i$. This objective implies finding the minimal hybrid diagnosis.

Finally, it is important to include in the Min-CSP the following constraint, one for each BDC_i:

$$(rBDC_i^1 + \dots + rBDC_i^n = 0) \vee (rBDC_i^1 + \dots + rBDC_i^n \geq minLik_i) \tag{16.18}$$

Previous constraint contemplates two scenarios: (i) $\sum_j^n \text{rBDC}_i^j$ is equal to 0 being BDC_i correct for every tuple of data or (ii) $\sum_j^n \text{rBDC}_i^j$ is equal to or greater than minLik_i, representing that the BDC_i has a defect that is observed for the inconsistency of some (depending on the minLik) tuples. The inclusion of $\geq$ avoids intermediate values between 0 and minLik_i. This constraint strengths that a BDC_i is incorrect iff another explanation will imply that more than MinLik input data were incorrect.

16.4 Conclusions

Business Process Models facilitate the description, implementation, and deployment of the activities of an organization. As other systems, both design and execution can be incorrect, being necessary the use of diagnosis techniques. Since industrial business processes have special characteristics, classic model-based diagnosis must be adapted to be applicable in this context. In this chapter, how the special characteristics can affect the most relevant business process phases have been presented, at Design, Analysis, Configuration, and Enactment and Evaluation phases. Regarding the Design and Analysis phases, both workflow and data management must be included in the business process diagnosis. In relation to the configuration phase, it is relevant to decide where a business process can be monitored to obtain the wished level of diagnosability. During enactment and evaluation phase, the model and data must be combined to detect incorrect behaviors at instantiation time.

Acknowledgements This work has been partially funded by the Ministry of Science and Technology of Spain (TIN2015-63502-C3-2-R) and the European Regional Development Fund (ERDF /FEDER). We would like to thank SACU of the University of Seville for the valuable information that has contributed toward the development of the ideas in this chapter.

References

1. van der Aalst, W.: Formalization and verification of event-driven process chains. Inf. Softw. Technol. **41**(10), 639–650 (1999). https://doi.org/10.1016/S0950-5849(99)00016-6
2. van der Aalst, W.M.P.: Process-aware information systems: design, enactment, and analysis. Wiley Encyclopedia of Computer Science and Engineering. Wiley, New Jersey (2008)
3. Alpern, B., Wegman, M.N., Zadeck, F.K.: Detecting equality of variables in programs. In: Proceedings of the 15th ACM SIGPLAN-SIGACT Symposium on Principles of Programming Languages, pp. 1–11. San Diego, California (1988)
4. de la Banda, M.G., Stuckey, P.J., Wazny, J.: Finding all minimal unsatisfiable subsets. In: PPDP '03: Proceedings of the 5th ACM SIGPLAN International Conference on Principles and Practice of Declaritive Programming, pp. 32–43. ACM Press (2003)
5. Borrego, D., Eshuis, R., Gómez-López, M.T.: Diagnosing correctness of semantic workflow models. Data Knowl. Eng. **87**, 167–184 (2013). https://doi.org/10.1016/j.datak.2013.04.008
6. Ceballos, R., Borrego, D., Gómez-López, M.T., Gasca, R.M.: Hybrid diagnosis applied to multiple instances in business processes. In: Enterprise, Business-Process and Information Systems Modeling - 17th International Conference, BPMDS 2016, 21st International Conference, EMMSAD 2016, Held at CAiSE 2016, Ljubljana, Slovenia, 13–14 June 2016, Proceedings, pp. 212–227 (2016)
7. Ceballos, R., Gasca, R.M., Valle, C.D., Toro, M.: Max-CSP approach for software diagnosis. In: IBERAMIA, pp. 172–181 (2002)
8. Cytron, R., Ferrante, J., Rosen, B.K., Wegman, M.N., Zadeck, F.K.: Efficiently computing static single assignment form and the control dependence graph. ACM Trans. Program. Lang. Syst. **13**, 451–490 (1991)
9. Dumas, M., Rosa, M.L., Mendling, J., Reijers, H.A.: Fundamentals of Business Process Management. Springer, Berlin (2013). https://doi.org/10.1007/978-3-642-33143-5

10. Falcioni, D., Polini, A., Polzonetti, A., Re, B.: Livelock and deadlock detection for pa inter-organizational business processes. In: Kő, A., Leitner, C., Leitold, H., Prosser, A. (eds.) Advancing Democracy, Government and Governance (2012)
11. Gasca, R.M., Valle, C.D., Gómez-López, M.T., Ceballos, R.: Nmus: Structural analysis for improving the derivation of all muses in overconstrained numeric csps. In: CAEPIA, pp. 160–169 (2007)
12. Gómez-López, M.T., Borrego, D., Carmona, J., Gasca, R.M.: Computing alignments with constraint programming: The acyclic case. In: Proceedings of the International Workshop on Algorithms & Theories for the Analysis of Event Data 2016 Satellite event of the conferences: 37th International Conference on Application and Theory of Petri Nets and Concurrency Petri Nets 2016 and 16th International Conference on Application of Concurrency to System Design ACSD 2016, Torun, Poland, 20–21 June 2016, pp. 96–110 (2016). http://ceur-ws.org/Vol-1592/paper07.pdf
13. Gómez-López, M.T., Gasca, R.M., Pérez-Álvarez, J.M.: Compliance validation and diagnosis of business data constraints in business processes at runtime. Inf. Syst. **48**, 26–43 (2015). https://doi.org/10.1016/j.is.2014.07.007
14. Heravizadeh, M., Mendling, J., Rosemann, M.: Dimensions of business processes quality (QoBP). In: Business Process Management Workshops, BPM 2008 International Workshops, Milano, Italy, 1–4 September 2008, pp. 80–91 (2008). https://doi.org/10.1007/978-3-642-00328-8_8
15. Kalpic, B., Bernus, P.: Business process modelling in industry - the powerful tool in enterprise management. Comput. Ind. **47**(3), 299–318 (2002). https://doi.org/10.1016/S0166-3615(01)00151-8
16. OMG: Object Management Group, Business Process Model and Notation (BPMN) Version 2.0. OMG Standard (2011)
17. Rozinat, A., van der Aalst, W.M.P.: Conformance checking of processes based on monitoring real behavior. Inf. Syst. **33**(1), 64–95 (2008). https://doi.org/10.1016/j.is.2007.07.001
18. Sun, S.X., Zhao, J.L., Sheng, O.R.L.: Data flow modeling and verification in business process management. In: AMCIS Association for Information Systems (2004)
19. Sun, S.X., Zhao, J.L., Nunamaker, J.F., Sheng, O.R.L.: Formulating the data-flow perspective for business process management. Inf. Syst. Res. **17**(4), 374–391 (2006)
20. Wallace, R.J.: Directed arc consistency preprocessing. In: Constraint Processing. Selected Papers, pp. 121–137. Springer, London, UK (1995)
21. Weske, M.: Business Process Management: Concepts, Languages. Architectures. Springer-Verlag New York Inc., Secaucus, NJ, USA (2007)

Chapter 17
Fundamentals of Prognostics

Anibal Bregon and Matthew J. Daigle

17.1 Introduction

Systems Health Management (SHM) is an engineering discipline aiming to improve the design and operation of systems in the presence of degradations and/or faults. Prognostics has arisen as a promising technology that has quickly become essential to manage the safety and cost of complex systems [25, 32, 38]. Prognostics and Health Management (PHM) solutions [13] have been successfully applied within the engineering systems community to prolong the life of assets, optimize maintenance activities, and guarantee a safe operation. The success stories of PHM for engineering systems are numerous. Ranking among the most important examples, we can find aerospace systems [3, 15], electrical vehicles [23], batteries [27, 40], and industrial systems [17, 18], to mention a few.

In a general way, prognostics can be defined as "the problem of predicting the future state of a system". In PHM, the idea of prognostics is typically linked to the prediction of a future state that is related to a previous fault situation. The classical definition of prognostics given by the PHM Society[1] is as follows:

Definition 17.1 (*Prognostics*) Estimation of the remaining useful life of a component or subsystem.

In this definition, a core concept within the prognostics literature appears, the *Remaining Useful Life* (RUL). The RUL accounts for the time from the current instant to the time of failure (also known as *End of Life*, EOL). However, in general,

A. Bregon (✉)
Departamento de Informática, Universidad de Valladolid, Segovia, Spain
e-mail: anibal@infor.uva.es

M. J. Daigle
NIO USA, Inc., 3200 N 1st St, San Jose, CA, USA
e-mail: matthew.daigle@nio.io

[1] http://www.phmsociety.org/.

T. Escobet et al. (eds.), *Fault Diagnosis of Dynamic Systems*,
https://doi.org/10.1007/978-3-030-17728-7_17

we do not want to predict when a critical failure occurs, but rather that point at which a repair or replacement is necessary, which should be before any damage occurs, that is, irreparable. Ideally, there is a predefined *failure threshold*, at which point components or subsystems can be repaired or replaced right before the critical failure occurs. Hence, we can distinguish between the time from the current instant to the time of critical failure, known as *Estimated Time to Failure* (ETTF), and the time to the failure threshold, that will be the RUL. The recent revision of the International Organization for Standardization (ISO) 13381-1:2015 [1] includes the previous idea, defining prognostics, ETTF, and RUL as follows:

Definition 17.2 (*Prognostics—ISO 13381-1:2015*) Analysis of the symptoms of faults to predict future condition and residual life within design parameters.

Definition 17.3 (*Estimated Time to Failure*) Estimation of the period from the current point in time to the point in time where the monitored.

Definition 17.4 (*Remaining Useful Life*) Remaining time before system health falls below a defined failure threshold.

However, as we will see later in this chapter, the prediction of the future system state and RUL does not necessarily need to be linked with the existence of a fault situation. Consider the well-known case study of battery prognostics in systems such as electric cars or electric aircraft. For these systems, we are interested in knowing with precision the remaining charge in the batteries so that we do not run out of power in the middle of the highway or during a mission. In this case, there is no fault occurring in the system, just the nominal discharge process of the batteries. Hence, a prognostics solution should be general enough to provide estimations of future events, independently whether they are triggered by a previous faulty situation or not. Also, these definitions oversimplify the prognostics problem, and the reality of complex systems is much more harsh, since it has a strong dependency on future operating conditions (input commands, environment, and unforeseen loads) and different sources of uncertainty. In Sect. 17.2, we present an extended definition of prognostics including all these ideas that will be the basis for the model-based prognostics formulation that we present in Sect. 17.3. Then, Sect. 17.4 shows how model-based diagnosis applies to the prognostics of a centrifugal pump. The chapter finishes in Sect. 17.5 with a few conclusions and a brief description of challenges and open problems in prognostics.

17.1.1 Goals of Prognostics

As we have mentioned, prognostics plays a key role within the SHM technology stack. Among the goals of prognostics, we can emphasize the following:

- Increase safety and reliability: Used together with diagnostic techniques (like most of the solutions shown in previous chapters), prognostics enables making decisions either to fix the diagnosed fault before it becomes critical, or to accurately decide if the system can keep working within safe conditions.

- Decrease collateral damage: By avoiding cascading effects that harm healthy components or subsystems.
- Reduce operating costs: The costs of repairs and maintenance can be reduced by eliminating unnecessary scheduled maintenance (e.g., the cost of changing system components can be highly decreased if those components are changed based on a varying future health prediction that highly depends on the current health, usage, and working environment, rather than having periodic inspections at fixed time intervals).
- Improve availability: Similarly to the previous point, a more realistic maintenance scheduling will help to avoid unnecessary downtime and improves utilization of assets.
- Improve planning: A mission can be optimally planned or replanned online in faulty situations for a better use of the available resources.

The next section will present the fundamental ideas underlying the prognostics process and prognostics methods.

17.2 Prognostics Fundamentals

In this section, we present the basic idea of the prognostics process and then discuss the importance of uncertainty within this process. We describe the different sources of uncertainty that can arise and how they can be included within the prognostics process. We then present the different prognostics methods, namely, data-driven, physics-based and hybrid prognostics approaches, and give details on each one of them, discussing their advantages and disadvantages.

17.2.1 Understanding the Prognostics Process

Figure 17.1 shows the basic idea underlying the prognostics process. In the figure, the time t_P represents the current time (also called prediction time) where the prognostics algorithms are run. To that end, the current health state of the system at t_P is estimated, and with that, the prediction algorithm predicts the evolution of the damage in the component until the failure threshold. The time instant where the damage evolution crosses the failure threshold is the time of EOL, t_{EOL}, and the difference between t_{EOL} and t_P is the Remaining Useful Life (RUL). As we have already mentioned, the failure threshold is somehow a conservative approach, such that it is far enough from the no return point of the critical failure level. The exact value of the failure threshold varies depending on the particular system and the required safety or cost constraints. Some prognostics solutions consider here the so-called *hazard zone* where a probability distribution determines the percentage of danger around the failure zone. Higher probabilities will mean the failure threshold is less conservative and lower probabilities that the failure threshold is set in a more conservative position.

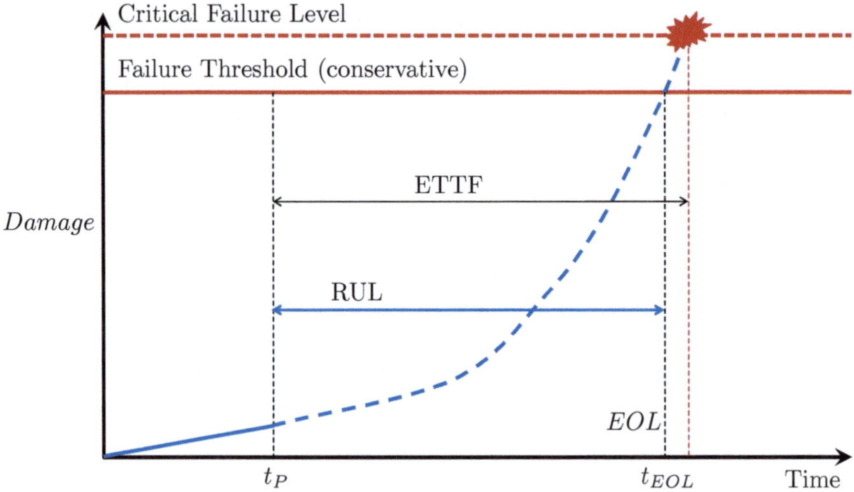

Fig. 17.1 Prognostics process

In an idealized world, accurate and precise estimation following this process would be possible. However, real life is much more complicated, and this theoretical solution faces many problems related to uncertainty when dealing with engineering systems (e.g., we need some idea of future loads or unforeseen environmental events). Figure 17.2 shows two different situations caused by a misrepresentation of the uncertainty that results in unreliable prognostics results. First, in green, the system can overestimate the true time of EOL, thus determining an RUL much shorter than the real one, which might affect negatively in the cost, for example, by having unnecessary maintenance operations or by shortening the mission time, among other consequences. And second, in orange, the system can underestimate the true time of EOL, thus having catastrophic safety consequences. The next section discusses the different sources of uncertainty in a prognostics system and how they can affect the prognostics process.

17.2.2 Uncertainty

In engineering systems, we hardly have access to the ground truth, and instead we have a set of noisy measurements, that we use as input to a model, which also has a high level of uncertainty. The determination of an accurate and appropriate failure threshold is also difficult and can compromise the quality of the RUL estimation. Finally, the lack of knowledge about the future loads and environmental conditions increases the uncertainty even more. Uncertainties in prognostics arise from a diverse set of sources. In general, we can consider uncertainties in the measurements, model, and system inputs [29–31, 36].

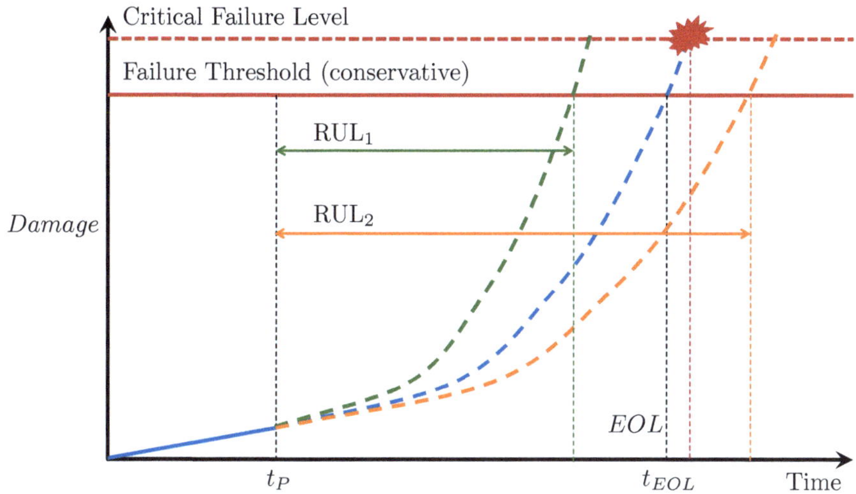

Fig. 17.2 Prognostics process with different RUL computations

- **Measurements**: Sensor measurements exhibit different issues that are usually sources of uncertainty. First, even the best and most expensive sensors provide noisy measurements, and this noise can have a big influence in all the prognostics process, since they are used for many tasks, going from the estimation of the current health of the system to the parameter tuning for the fault degradation models. Second, sensor coverage, especially in aerospace systems, is usually limited and does not cover all the components in the system. And third, preprocessing processes carried out on the sensors measurements usually imply some kind of loss of information that can also affect the prognostics results.
- **Model**: Several problems arise related to the modeling of the systems. First, when doing prognostics, we need a model of the system and some kind of fault propagation model. Creating such a model is not an easy task, due to the unknown dynamics of evolution of the faults and, consequently, the existing fault propagation models, if any, have a relatively low fidelity. Second, the existence of unmodeled phenomena also generates uncertainties in the prognostics process since all the behavior of the system, both in nominal and fault situations, cannot be modeled. And third, since models have to be tuned using a (usually) small amount of real noisy data, numerical errors will arise when computing the predictions with such models.
- **Present state estimate**: Current state estimation of the system at the prediction time includes additional uncertainty to the prognostics process.

Additional to the aforementioned sources of uncertainty, one of the most challenging problems for the prognostics calculations is related with the operating environment on the system. In particular, the anticipation of future loads and future operating conditions are two of the big problems that any prognostics solution has to face and is usually the most significant source of uncertainty [30, 31]. In reality, it is practically impossible to accurately anticipate for future loads, especially for large periods of time, so any prognostics system has to account for this kind of uncertainty and carry out periodical repetitions on the prognostics calculations to account for the new loads in the system since the previous prediction point.

Based on these ideas of uncertainty, the next section revisits the idea of prognostics and proposes a new more general and complete definition.

17.2.3 Revisiting Prognostics

To sum up, we can see prognostics as a prediction process, and for that we need the following:

- To know the current health state of the system (in reality, we can say that we need some estimation of the current system state, since, as we have mentioned before, we want to perform prognostics for general situations where the so-called *failure threshold* is related to some future event not necessarily linked to a fault).
- To anticipate some idea of the future load and future operating conditions of the system.
- To include some kind of model to account for remaining sources of uncertainty.

Taking this into account, we redefine the concept of prognostics as follows:

Definition 17.5 (*Prognostics*) Prognostics estimates the current system state and, conditional on future load and uncertainty, predicts at what time the component will reach a determinate state.

Where this state can be defined, for example, as a failure state or as a state where the system no longer operates within its given specifications. Figure 17.3 illustrates the idea of the prognostics process when carried out following this revisited idea. Instead of failure threshold, we now consider a threshold (T_e) that is computed as a functions of the system state. Instead of having an EOL point, we now consider an event (e). Finally, the prediction problem is carried out by estimating the current state, which is now represented as a probability distribution (to account for the initial state uncertainty), and propagated, along with the distribution of future loads, noise, etc..., until the threshold is reached. Next section will provide a detailed view on the mathematical formulation of the problem of prognostics.

17.2.4 Prognostics Methods

In engineering systems, prognostics has been approached by a large number of techniques. In general, these techniques can be classified into data-driven, physics-based, and hybrid approaches [25, 38]. The *data-driven* solutions use large amounts of measured data to learn some kind of fault degradation model. The physics-based solutions

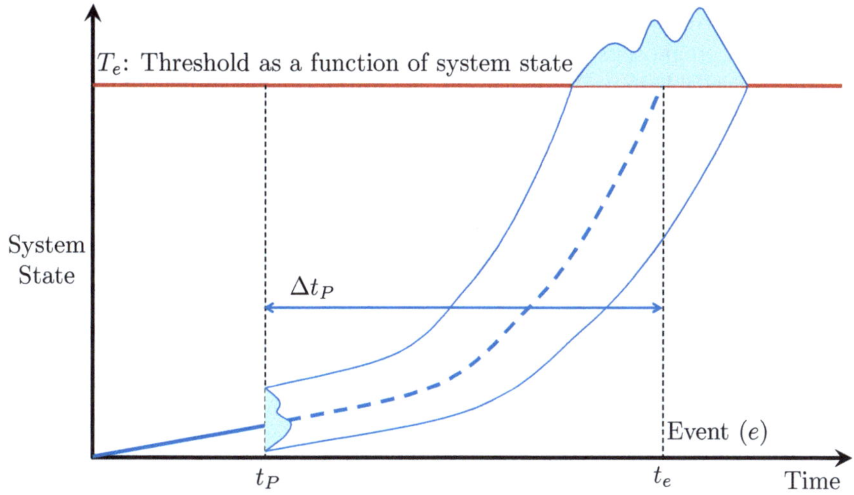

Fig. 17.3 Prognostics process revisited

use mainly physical models of the system behavior and fault degradation. Although all approaches use models of some kind, these sort of solutions using models derived from first principles are typically referred to as *model-based* solutions. Finally, the *hybrid* solutions combine the data-based and the model-based solutions in a number of ways to improve the prognostics performance.

Next, we present a brief description of each one of the prognostics methods and discuss their advantages and disadvantages. For a more detailed description and references on the large number of prognostics solutions, we refer the reader to some of the main works on prognostics methods in the literature, such as [13, 20, 25, 34, 38].

17.2.4.1 Data-Driven Prognostics

Data-driven prognostics solutions make use of previously observed data to carry out the prognostics predictions. It starts off with raw data that is processed to extract features that indicate the behavior of the system and its future evolution. In the case of prognostics problems where the prediction state is related to a faulty state, the extracted features will give some indication on the fault growth. Predictive models are then trained with this data to provide estimations of the RUL. Many different approaches have been used in the literature for data-driven prognostics, going from regression tools to probabilistic solutions (such as hidden Markov models or dynamic Bayesian networks), to artificial-intelligence-related techniques (such as neural networks or support vector machines). An extensive survey on statistical techniques for prognostics is given in [34] and for artificial intelligence techniques in [33].

The data-driven solutions to prognostics provide a fast and relatively simple way to implement prognostics solutions (there exists a very large corpus of well-proven techniques available to use). Also, using data-driven techniques help to gain understanding of the physical behaviors when large amounts of data are available. On the other hand, data-driven approaches also face many flaws, for example, that the physical cause–effect relations are not considered when building the prognostics models or, more important, that large amounts of data are usually required for the solutions to provide satisfactory prognostics results.

17.2.4.2 Model-Based Prognostics

Model-based prognostics approaches use physics-based models to describe the operation of a system or component and to capture the knowledge about how it fails. The model needs to be able to explain the fault, but also model how the fault will evolve along time. This is particularly difficult because fault effects and its evolution on the systems are not always known and usually difficult to model mathematically. Moreover, cause–effect phenomenon is not always directly observable. For example, structural cracks are observable faults, but internal resistance changes in batteries caused by aging are not directly observable. Although model-based solutions to prognostics have not been as popular as the data-driven ones, there exist multiple works that have successfully used model-based techniques to solve a variety of prognostics problems [10, 21, 28]. Among the most popular solutions, we can distinguish those using Particle filters [2, 11, 24] or any kind of Kalman filter [4, 19].

The model-based solutions to prognostics provide results that are intuitive based on modeled cause–effect relationships. Also, reusability is favored with model-based techniques since, once the model is created, only parameter calibration will be needed to adapt it to different scenarios. However, creating such models is also its main flaw. First, because it is not trivial to create accurate models of the systems degradation, and even if it is possible, the computational burden of these models is sometimes quite high, what makes their use impractical for real-time prognostics. And second, even if high fidelity models are available, a relatively small amount of real data (or expert knowledge) is necessary for parameter tuning.

17.2.4.3 Hybrid Solutions

The hybrid prognostics methods combine data-driven and model-based solutions in different intelligent manners such that the performance of the prognostics process is improved. There are many proposals for hybrid solutions. Among the most popular solutions, one can be to use the model to make initial predictions and then refine such predictions using data. Another solution uses data to learn models to estimate the health state of the systems (since typically there is more experimental data on diagnosis than prognostics) and then make the predictions using a model-based technique. Practitioners have also combined techniques in a way that the knowledge of one technique helps to efficiently develop the other one. For example, use knowledge about the physical behavior to focus the data-driven learning techniques, or use the understanding from the data to develop more accurate models. Liao and Kottig [20] provide a review of hybrid prognostics for RUL prediction of engineered systems.

Developing hybrid solutions has a great advantage for prognostics because, by working in a complementary way, we will not require excessively large volumes of data or high fidelity models, both difficult to get for prognostics. However, this could also be a double-edged sword since we will always need some kind of physical knowledge of the system and some kind of experimental data.

17.3 Model-Based Prognostics

In this section, we present a mathematical formulation of the problem of prognostics, and algorithms for solving the prognostics under uncertainty within the model-based prognostics paradigm.

17.3.1 Problem Formulation

As defined in the previous section, at its core, prognostics is a problem of predicting future states of the system. Mathematically, we are interested in how the *state* of the system, $\mathbf{x}$, will evolve as a function of time. In particular, we are interested in determining whether a particular subset of the state space will be reached, corresponding to the occurrence of some event of interest, e.

Let $\mathcal{X} = \Re^{n_x}$ represent the state space of the system (where n_x is the dimension of the state vector), and let $\mathcal{X}_e \subset \mathcal{X}$ represent the set of states of interest. In failure prognostics, e denotes failure, $\mathcal{X}_e$ represents the set of failure states, and $\mathcal{X} \backslash \mathcal{X}_e$ represents the set of nominal or nonfailure states.

The separation between $\mathcal{X}_e$ and $\mathcal{X} \backslash \mathcal{X}_e$ may be defined by a Boolean function that classifies states into either category. In a one-dimensional state space, this usually takes the form of a threshold, and hence this function has historically been called the *threshold function*.

For a state $\mathbf{x} \in \mathcal{X}$, the threshold function, T_e, is formulated as

$$o_e(k) = T_e(k, \mathbf{x}(k), \mathbf{u}(k)), \tag{17.1}$$

where $k \in \mathbb{N}$ is the discrete time variable, $\mathbf{u}(k) \in \Re^{n_u}$ is the input vector, and $o_e(k) \in \mathbb{B}$ is a Boolean variable indicating if $x(k) \in \mathcal{X}_e$. Typically, o_e is defined only as a function of the state, but in general, it may also be time dependent and input dependent.

Given this basic framework, there are several kinds of questions that can be answered, which are given as follows:

1. Will e occur at some future time instant?
2. At what time will e occur?
3. How long until e will occur?

Since prognostics deals with forecasting into the future, the answers to these questions will be uncertain and represented as probability distributions.

In failure-based prognostics, the first question relates to computing the probability of failure, the second relates to computing the EOL, and the third the RUL.

The inputs to the prediction problem are the following [29]:

1. a time of prediction, k_o;
2. a time horizon of prediction, $[k_o, k_h]$;
3. the initial state probability distribution, $p(\mathbf{x}_o(k_o))$;
4. the future input trajectory distribution, $p(\mathbf{U}_{k_o,k_h})$, where $\mathbf{U}_{k_o,k_h} = [\mathbf{u}(k_o), \mathbf{u}(k_o+1), \ldots, \mathbf{u}(k_h)]$ and $\mathbf{u}(k)$ are the system inputs; and
5. the future process noise trajectory distribution, $p(\mathbf{V}_{k_o,k_h})$, where $\mathbf{V}_{k_o,k_h} = [\mathbf{v}(k_o), \mathbf{v}(k_o+1), \ldots, \mathbf{v}(k_h)]$ and $\mathbf{v}(k)$ are the system process noise.

These inputs may change for every new time of prediction.

Note that the following notation is adopted. For a vector $\mathbf{a}$, a trajectory of that vector over a time interval $[k_o, k_h]$ is denoted by the matrix $\mathbf{A}_{k_o,k_h}$, and $\mathbf{A}_{k_o,k_h}(k) = \mathbf{a}(k)$, for $k \in [k_o, k_h]$.

The core problem of prognostics is to predict the future states of the system within a time interval:

Problem 17.1 Given a time interval $[k_o, k_h]$, an initial state $p(\mathbf{x}(k_o))$, process noise $p(\mathbf{V}_{k_o,k_h})$, and future inputs $p(\mathbf{U}_{k_o,k_h})$, determine $p(\mathbf{X}_{k_o,k_h})$.

A more specific problem is to compute which of these states belong to $\mathcal{X}_e$ for a given event e, and determine the associated probability of reaching such a state at a particular time:

Problem 17.2 Given a event e, a time interval $[k_o, k_h]$, an initial state $p(\mathbf{x}(k_o))$, process noise $p(\mathbf{V}_{k_o,k_h})$, and future inputs $p(\mathbf{U}_{k_o,k_h})$, determine the probability that some state $\mathbf{x} \in \mathcal{X}_e$ will be reached at a given time $k \in [k_o, k_h]$.

For example, if e represents failure, then we would be computing the probability of failure occurring at a particular time.

A related problem is to compute the probability of reaching a state with a given label at *any* time within a given time interval:

Problem 17.3 Given a event e, a time interval $[k_o, k_h]$, an initial state $p(\mathbf{x}(k_o))$, process noise $p(\mathbf{V}_{k_o,k_h})$, and future inputs $p(\mathbf{U}_{k_o,k_h})$, determine the probability of reaching some state $\mathbf{x} \in \mathcal{X}_e$ for any time $k \in [k_o, k_h]$.

Algorithm 17.1: `Simulate.`

Inputs: $k_o, k_h, \mathbf{x}(k_o), \mathbf{U}_{k_o,k_h}, \mathbf{V}_{k_o,k_h}$
$\mathbf{X}_{k_o,k_h}(k_o) \leftarrow \mathbf{x}(k_o)$
for $k = k_o$ **to** $k_h - 1$ **do**
  $\mathbf{X}_{k_o,k_h}(k+1) \leftarrow \mathbf{f}(\mathbf{X}_{k_o,k_h}(k), \mathbf{U}_{k_o,k_h}(k), \mathbf{V}_{k_o,k_h}(k))$
return $\mathbf{X}_{k_o,k_h}$

For example, if e represents failure and $k_h - k_o$ is 1 hour, then we would be computing the probability that the system will fail within the next hour.

The final problem is to compute *when* a state with a given label will be reached. For a event e, we define k_e as follows:

$$k_e(k) = \min\{k' : k' \geq k \text{ and } \mathbf{x}(k) \in \mathcal{X}_e\}, \tag{17.2}$$

i.e., k_e is the earliest time point at which the system state is assigned the event e. This is a generalization of EOL; if e represents failure, then k_e is EOL and $k_e - k_o$ is RUL. The problem of computing k_e is then defined as follows:

Problem 17.4 Given a event e, a time interval $[k_o, k_h]$, an initial state $p(\mathbf{x}(k_o))$, process noise $p(\mathbf{V}_{k_o,k_h})$, and future inputs $p(\mathbf{U}_{k_o,k_h})$, determine $p(k_e)$.

17.3.2 Model-Based Prognostics

In a model-based paradigm, we go about answering these questions by computing the evolution of some initial state $\mathbf{x}$ through a model. The state evolves according to a state equation as given below:

$$\mathbf{x}(k+1) = \mathbf{f}(\mathbf{x}(k), \mathbf{u}(k), \mathbf{v}(k)), \tag{17.3}$$

where $\mathbf{v}(k) \in \Re^{n_v}$ is the process noise vector and $\mathbf{f}$ is the state update function.

With this, we can then define the basic prediction algorithm, shown as Algorithm 17.1. This algorithm computes the future states simply by simulating the state using the state equation. Here, we assume that the trajectory of future inputs, $\mathbf{U}$, and the trajectory of process noise, $\mathbf{V}$, are known.

Similarly, Algorithm 17.2 predicts the state ahead and determines when the threshold evaluates to true. If so, the first time point at which this occurs is returned. If the threshold never evaluates to true within the finite prediction horizon, then ∞ is returned.

Algorithm 17.2: `ComputeK`.

Inputs: $k_o, k_h, \mathbf{x}(k_o), \mathbf{U}_{k_o,k_h}, \mathbf{V}_{k_o,k_h}$
$\mathbf{X}_{k_o,k_h}(k_o) \leftarrow \mathbf{x}(k_o)$
for $k = k_o$ **to** $k_h - 1$ **do**
  if $T_e(\mathbf{X}_{k_o,k_h}(k), \mathbf{U}_{k_o,k_h}(k))$ **then**
    return k
  $\mathbf{X}_{k_o,k_h}(k+1) \leftarrow \mathbf{f}(\mathbf{X}_{k_o,k_h}(k), \mathbf{U}_{k_o,k_h}(k), \mathbf{V}_{k_o,k_h}(k))$
if $T_e(\mathbf{X}_{k_o,k_h}(k), \mathbf{U}_{k_o,k_h}(k))$ **then**
  return k
else
  return ∞

Algorithm 17.3: `StatePrediction`.

Inputs: $k_o, k_h, p(\mathbf{x}(k_o)), p(\mathbf{U}_{k_o,k_h}), p(\mathbf{V}_{k_o,k_h})$
$\{(\mathbf{x}(k_o)^i, \mathbf{U}^i_{k_o,k_h}, \mathbf{V}^i_{k_o,k_h})\}_{i=1}^N \leftarrow$ `GenerateSamples` $(N, p(\mathbf{x}(k_o)), p(\mathbf{U}_{k_o,k_h}), p(\mathbf{V}_{k_o,k_h}))$
for $i = 1$ **to** N **do**
  $\mathbf{X}^i_{k_o,k_h} \leftarrow$ `Simulate`$(k_o, k_h, \mathbf{x}(k_o)^i, \mathbf{U}^i_{k_o,k_h}, \mathbf{V}^i_{k_o,k_h})$
return $\{\mathbf{X}^i_{k_o,k_h}\}_{i=1}^N$

17.3.3 Uncertainty

Of course, assuming that future inputs and process noise are known is not a practical assumption. Since these are trajectories of future values, they are unknown. At best, we know only the distributions of future values.

Now, we assume that we know the distributions of future trajectories of the inputs and process noise, along with the distribution of the initial state at the time of prediction.

Algorithm 17.3 computes the probability distribution of future state trajectories, represented as discrete samples, using Algorithm 17.1. Using `GenerateSamples`, realizations for the different inputs to Algorithm 17.1 are generated, and Algorithm 17.1 is called for each. Here, `GenerateSamples` may refer to any appropriate sampling algorithm, e.g., Monte Carlo sampling, Latin Hypercube sampling, etc.; we omit the details here. The probability distribution $p(\mathbf{X}_{k_o,k_h})$ is defined by the set of samples of state trajectories that are produced. Here, we denote this set using $\{\mathbf{X}^i_{k_o,k_h}\}_{i=1}^N$, where the i superscript denotes sample i and N is the total number of samples. The value of N depends on the sampling algorithm, and must be sufficiently large so that the statistics of the result captured the accuracy demanded by the application, subject to computational constraints.

Similarly, we can write an algorithm that computes the probability distribution of the time at which the threshold evaluates to true. This algorithm is shown in Algorithm 17.4

Algorithm 17.4: `KPrediction`.

1: **Inputs:** $k_o, k_h, p(\mathbf{x}(k_o)), p(\mathbf{U}_{k_o,k_h}), p(\mathbf{V}_{k_o,k_h})$
2: $\{(\mathbf{x}(k_o)^i, \mathbf{U}^i_{k_o,k_h}, \mathbf{V}^i_{k_o,k_h})\}_{i=1}^N \leftarrow$ `GenerateSamples` $(N, p(\mathbf{x}(k_o)), p(\mathbf{U}_{k_o,k_h}), p(\mathbf{V}_{k_o,k_h}))$
3: **for** $i = 1$ **to** N **do**
4: $\quad k^i \leftarrow$ `ComputeK`$(k_o, k_h, \mathbf{x}(k_o)^i, \mathbf{U}^i_{k_o,k_h}, \mathbf{V}^i_{k_o,k_h})$
5: **return** $\{k^i\}_{i=1}^N$

17.3.4 Modeling Uncertainty

This leaves another problem unresolved, which is, how do we come up with the needed probability distributions in order to make a prediction?

For determining the initial state distribution, there are several methods available. If measurements related to the system state are known, then these can be potentially inverted to determine the state. In other circumstances, methods such as Kalman filters can be used to estimate the state only. In this setting, such an algorithm gives an estimate for the system state at any time point, and this probability distribution becomes the starting point for any prediction.

For the process noise trajectory, typically some assumption about the process noise (e.g., an assumed state covariance matrix) is made. It is practical to assume that the distribution of the noise at time k is independent of the distribution at $k-1$. As such, we can generate trajectories of process noise by simply sampling independently at all time points in the prediction horizon.

The future input trajectories can be the most difficult to represent, depending on the environment in which the system is run. In many cases, these trajectories can be indirectly represented through a set of *surrogate* variables. That is, the trajectories are parameterized by a set of variables, and we sample these variables to sample the trajectories. For example, the future input may be linear, and we represent this by sampling the slope of the line. It then falls to estimating the distribution of the slope.

17.3.5 Distributed Prognostics

As shown in the above algorithms, prognostics can be computationally very expensive. There have been two fundamental approaches to addressing this challenge. One approach is to parallelize the computations, for example, using a GPU [12]. Consider the algorithms above. Several *independent* samples are simulated ahead to determine the future system states. Since these samples are independent, they can be computed in parallel.

A second approach is to decompose the prognostics problem itself into independent subproblems, as in [8]. Such an approach is useful when the system under prognostics is very large, and the event of interest is defined through several independent conditions, such that the event is said to occur when any of the conditions are met. In this case, we can define one prognostics problem for each condition, solve them independently, then merge the results to determine the overall event distribution. Such an approach can be achieved through formal approaches such as structural model decomposition, as in [8].

17.4 Case Study

As an example for prognostics modeling, let us consider a centrifugal pump, originally presented in [11]. In order to apply model-based prognostics, we must develop a model of the system that includes identifying the state vector **x**, the input vector **u**, the output vector **y**, the state equation **f**, the output equation **h**, and the threshold function T_e. This model is open source and available as part of the Prognostics Model Library [6].

In the following, we present the model in continuous time. The model may then be converted to a discrete time representation for implementation within prognostics algorithms.

17.4.1 Nominal Model

Centrifugal pumps deliver fluids in a system. A schematic is shown in Fig. 17.4 and its parameters are given in Table 17.1. An electric motor rotates the impeller, causing a pressure difference that pulls fluid in through the inlet, and across the outlet. Bearings, lubricated by oil, minimize friction along the shaft. The pump state includes $\omega(t)$, the rotational velocity of the pump; $Q(t)$, the discharge flow; $T_t(t)$, the thrust bearing temperature; $T_r(t)$ the radial bearing temperature; and $T_o(t)$, the oil temperature.

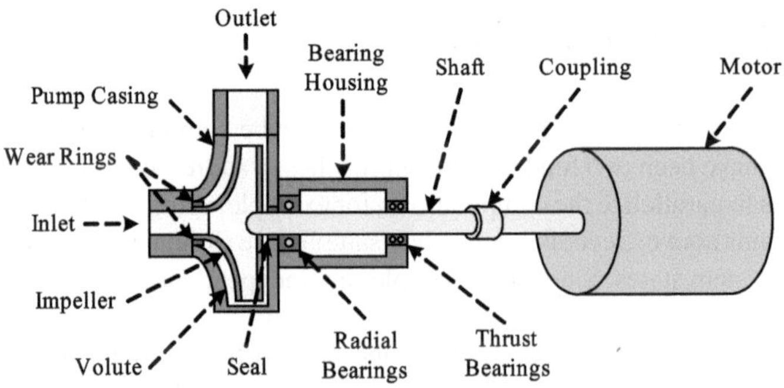

Fig. 17.4 Centrifugal pump (© 2014 IEEE. Reprinted, with permission, from [8])

Table 17.1 Nominal Pump Parameters (© 2014 IEEE. Reprinted, with permission, from [8])

Parameter	Value
$\omega(0)$	376 rad/s
J	50 kg m^2
r	8.0×10^{-3} N m s
n	3 phases
p	1 pole pair
R_1	$3.6 \times 10^{-1}\,\Omega$
R_2	$7.6 \times 10^{-2}\,\Omega$
$L_1 + L_2$	6.3×10^{-4} H
$Q(0)$	0 m^3/s
a_0	1.5×10^{-3} kg m^2
a_1	5.8 kg/m
a_2	9.2×10^3 kg/m^4
$b_0(0)$	12.7 kg/m
b_1	1.8×10^4 kg/m^4
b_2	0 kg/m^7
c	8.2×10^{-5} m$^{7/2}$/kg$^{1/2}$
c_l	1.0×10^{-10} m$^{7/2}$/kg$^{1/2}$
J_Q	5.0 s^{-1}
$T_o(0)$	290 K
J_o	8.0×10^3 K/J/s
$H_{o,1}$	1.0 W/K
$H_{o,2}$	3.0 W/K
$H_{o,3}$	1.5 W/K
$T_r(0)$	290 K
J_r	2.4 K/J/s
$r_r(0)$	1.8×10^{-6} N m s
$H_{r,1}$	1.8×10^{-3} W/K
$H_{r,2}$	2.0×10^{-2} W/K
$T_t(0)$	290 K
J_t	7.3 K/J/s
$r_t(0)$	1.4×10^{-6} N m s
$H_{t,1}$	3.4×10^{-3} W/K
$H_{t,2}$	2.6×10^{-2} W/K

The rotational velocity of the pump is described using a torque balance,

$$\dot{\omega}(t) = \frac{1}{J}\left(\tau_e(t) - r\omega(t) - \tau_L(t)\right), \tag{17.4}$$

where J is the lumped motor/pump inertia, τ_e is the electromagnetic torque provided by the motor, r is the lumped friction parameter, and τ_L is the load torque. We assume the pump is driven by an induction motor, in which torque is produced only when there is a slip, s, between the synchronous speed of the supply voltage, ω_s and the mechanical rotation, ω:

$$s(t) = \frac{\omega_s(t) - \omega(t)}{\omega_s(t)}. \tag{17.5}$$

The expression for the torque τ_e is derived from an equivalent circuit representation for a three-phase induction motor [22]:

$$\tau_e(t) = \frac{npR_2}{s(t)\omega_s(t)} \frac{V(t)^2}{(R_1 + R_2/s(t))^2 + (\omega_s(t)L_1 + \omega_s L_2)^2}, \tag{17.6}$$

where R_1 is the stator resistance, L_1 is the stator inductance, R_2 is the rotor resistance, L_2 is the rotor inductance, n is the number of phases (typically 3), p is the number of magnetic pole pairs, and V is the applied root mean square (rms) motor voltage. The dependence of torque on slip creates a feedback loop that causes the rotor to follow the rotation of the magnetic field. Rotor speed is controlled by changing ω_s, e.g., through the use of a variable-frequency drive, which will also change V to manage power usage.

Load torque τ_L is a polynomial function of the pump flow rate and the impeller rotational velocity [16, 39], which is given below:

$$\tau_L(t) = a_0\omega(t)^2 + a_1\omega(t)Q(t) - a_2Q(t)^2, \tag{17.7}$$

where a_0, a_1, and a_2 are coefficients derived from the pump geometry [16].

The rotation of the impeller creates a pressure difference from the inlet to the outlet of the pump, which drives the pump flow, Q. The pump pressure is computed as

$$p_p(t) = b_0\omega(t)^2 + b_1\omega(t)Q(t) - b_2Q(t)^2, \tag{17.8}$$

where b_0, b_1, and b_2 are coefficients derived from the pump geometry. The parameter b_0 is proportional to impeller area A [5]. Flow through the impeller, Q_i, is computed using the pressure differences as follows:

$$Q_i(t) = c\sqrt{|p_s(t) + p_p(t) - p_d(t)|}sign(p_s(t) + p_p(t) - p_d(t)), \tag{17.9}$$

where c is a flow coefficient, p_s is the suction pressure, and p_d is the discharge pressure.

To account for fluid inertia, the discharge flow is described by

$$\dot{Q}(t) = \frac{1}{J_Q}(Q_o(t) - Q_i(t)), \tag{17.10}$$

where J_Q is the flow inertia.

Pump temperatures are often monitored as indicators of pump condition. The oil heats up due to the radial and thrust bearings and cools to the environment:

$$\dot{T}_o(t) = \frac{1}{J_o}(H_{o,1}(T_t(t) - T_o(t)) + H_{o,2}(T_r(t) - T_o(t)) - H_{o,3}(T_o(t) - T_a(t))), \tag{17.11}$$

where J_o is the thermal inertia of the oil and the $H_{o,i}$ terms are heat transfer coefficients. The thrust bearings heat up due to the friction between the pump shaft and the bearings, and cool to the oil and the environment:

$$\dot{T}_t(t) = \frac{1}{J_t}(r_t\omega(t)^2 - H_{t,1}(T_t(t) - T_o(t)) - H_{t,2}(T_t(t) - T_a(t))), \tag{17.12}$$

where J_t is the thermal inertia of the thrust bearings, r_t is the friction coefficient for the thrust bearings, and the $H_{t,i}$ terms are heat transfer coefficients. The radial bearings behave similarly:

$$\dot{T}_r(t) = \frac{1}{J_r}(r_r(t)\omega(t)^2 - H_{r,1}(T_r(t) - T_o(t)) - H_{r,2}(T_r(t) - T_a(t))), \tag{17.13}$$

where J_r is the thermal inertia of the radial bearings, r_r is the friction coefficient for the radial bearings, and the $H_{r,i}$ terms are heat transfer coefficients.

The overall input vector $\mathbf{u}$ is given by

$$\mathbf{u}(t) = \left[p_s(t)\ p_d(t)\ T_a(t)\ V(t)\ \omega_s(t)\right]^T. \tag{17.14}$$

The available pump sensors form the measurement vector $\mathbf{y}$ given by

$$\mathbf{y}(t) = \left[\omega(t)\ Q(t)\ T_o(t)\ T_t(t)\ T_r(t)\right]^T. \tag{17.15}$$

Figure 17.5 shows nominal pump operation, with typical parameter values (see [11]. The input voltage and line frequency are varied to control the pump speed. The pump rotation creates fluid flow and heats up the bearings.

17.4.2 Damage Modeling

The life of the pump is driven by its efficiency and its temperatures. There are upper temperature limits for each of the three temperatures (leading to irreversible damage), and a lower limit on pump efficiency, η which is defined as $\eta(t) = \frac{V(t)I(t)}{(p_d(t)-p_s(t))Q(t)}$ for nominal inputs ($I(t)$ is rms motor current). The threshold function evaluates to true when any of these limits are violated.

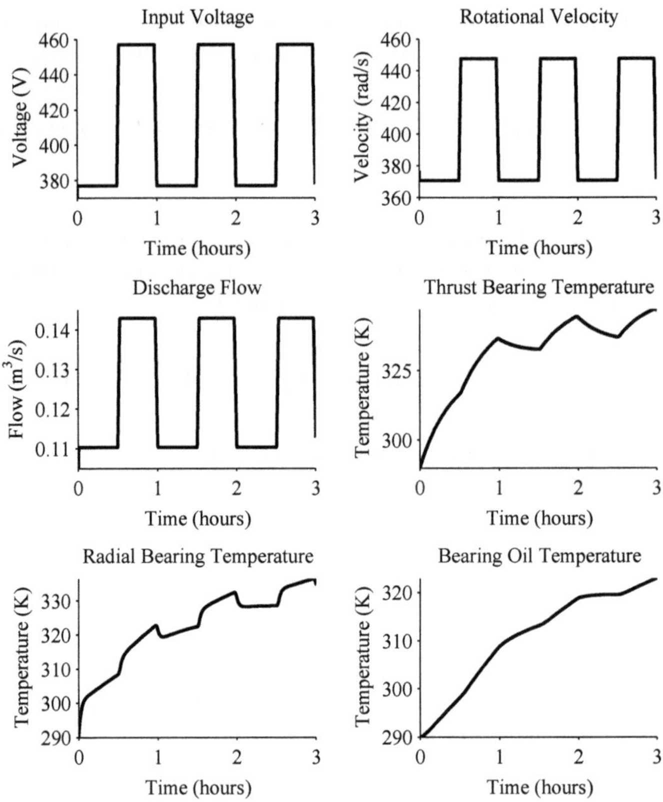

Fig. 17.5 Nominal pump operation (© 2014 IEEE. Reprinted, with permission, from [8])

The most significant damage mechanism for pumps is impeller wear, which is represented as a decrease in the effective impeller area A [5, 37]. Since the impeller area is proportional to b_0, a decrease in impeller area causes a decrease in the pump pressure, and, hence, the pump efficiency. We use the erosive wear equation [14] to describe how the impeller area changes over time. The erosive wear rate is proportional to fluid velocity times friction force. Fluid velocity is proportional to volumetric flow rate and friction force is proportional to fluid velocity. We lump the proportionality constants into the wear coefficient w_A to obtain [11] the following:

$$\dot{A}(t) = -w_A Q_i(t)^2. \tag{17.16}$$

Because A is proportional to b_0, then $\dot{b}_0(t) = k\dot{A}(t) = -kw_A Q_i(t)^2$, so we estimate b_0 and $w_{b_0} = kw_{A(t)}$.

Another significant damage mechanism for pumps is bearing wear, which is captured as an increase in the friction coefficient. Sliding and rolling friction generate wear of material which increases the coefficient of friction [9, 11, 14]:

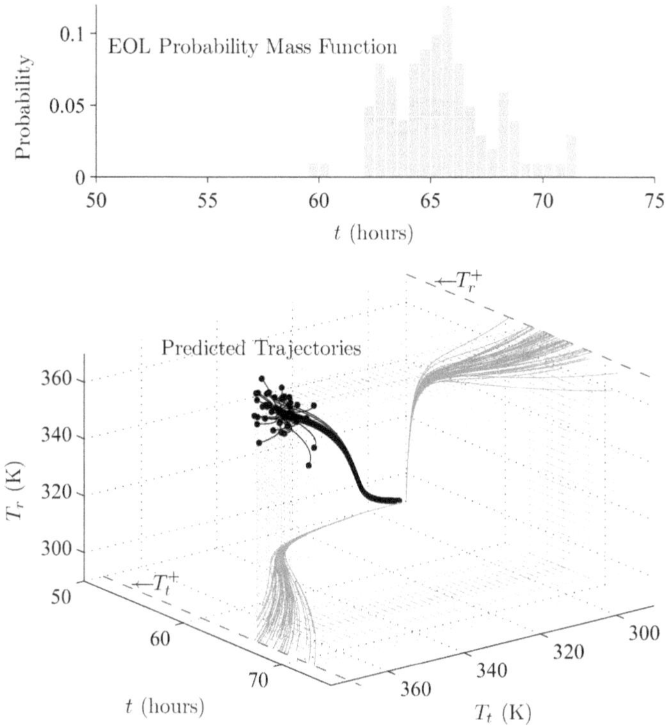

Fig. 17.6 Simultaneous prediction of thrust bearing and radial bearing wear in the pump. The damage trajectories are coming out of the page, increasing in T_t, increasing in T_r, and increasing in t

$$\dot{r}_t(t) = w_t r_t(t) \omega(t)^2, \tag{17.17}$$

$$\dot{r}_r(t) = w_r r_r(t) \omega(t)^2, \tag{17.18}$$

where w_t and w_r are the wear coefficients. The slip compensation provided by the electromagnetic torque generation masks small changes in friction, so it is only with very large increase that a change in ω will be observed. Changes in friction manifest more strongly in the bearing temperatures, eventually driving them to the temperature limits.

So, the full state vector becomes

$$\mathbf{x}(t) = \left[\omega(t)\ Q(t)\ T_o(t)\ T_t(t)\ T_r(t)\ b_0(t)\ r_t(t)\ r_r(t)\right]^T. \tag{17.19}$$

That is, these additional states are *damage* variables that evolve in time by *damage progression equations* and must be estimated online.

The *wear parameters* are also unknown, but assumed to be constant in time, i.e.,

$$\boldsymbol{\theta}(t) = \left[w_{b_0}\ w_t\ w_r\right]^T. \tag{17.20}$$

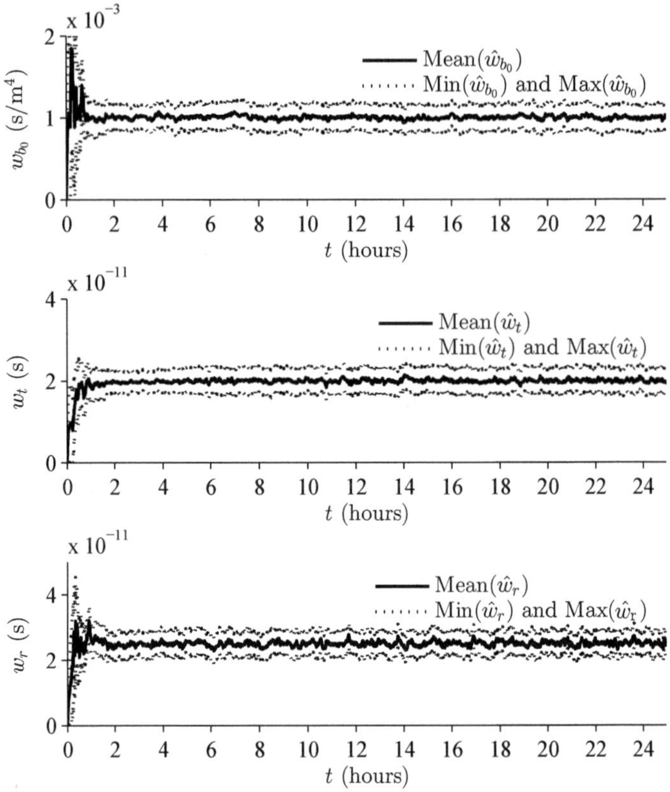

Fig. 17.7 Estimation results for $w_{b_0} = 1 \times 10^{-3}$, $w_t = 2 \times 10^{-11}$, and $w_r = 2.5 \times 10^{-11}$ (©2014 IEEE. Reprinted, with permission, from [8])

17.4.3 Uncertainty Modeling

As described in Sect. 17.3.3, in order to perform prognostics, we must define the uncertainty in the state at the time of prediction, the future inputs, and the process noise. For the states, a state observer is used to jointly estimate states and parameters online. At a given prediction time, the initial state is obtained from the filter.

For the process noise, time-invariant variances are assumed for each state. These same variances are used when sampling new states using the state equation.

For the future inputs, we consider a use case where the pump undergoes a fixed loading schedule, and the ultimate goal is to predict how many loading cycles (each lasting about an hour) the pump can be used before maintenance is needed. Since the loading schedule is fairly fixed, some small variances are added to the duration of the loading segment and the magnitude of the pump speed within the loading segment within a given loading schedule. To sample a loading schedule, we need to sample only the durations and pump speeds for each segment within a schedule.

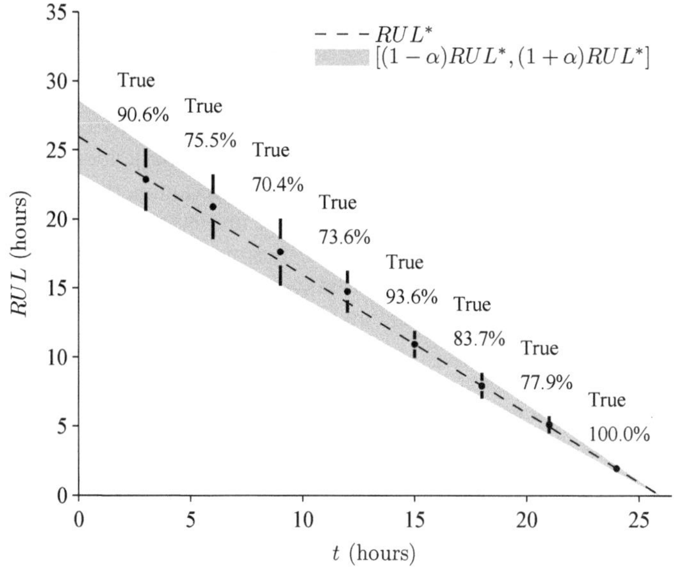

Fig. 17.8 Prognostics results for $w_{b_0} = 1 \times 10^{-3}$, $w_t = 2 \times 10^{-11}$, and $w_r = 2.5 \times 10^{-11}$ with $\alpha = 0.1$ and $\beta = 0.5$ (© 2014 IEEE. Reprinted, with permission, from [8])

In the case of the pump, the future inputs can have a significant impact on the predictions because they effect the damage progression rates. For example, an increased rotation speed will cause bearing friction to increase at a faster rate, and will cause an increased pump flow, which, in turn, will cause impeller wear to increase at a faster rate.

17.4.4 Results

Figure 17.6 shows results from the prediction of thrust bearing and radial bearing wear for $N = 100$ (not all trajectories are shown in the lower plot). There is initially a small uncertainty in the initial values, so the distribution of samples is very tight. The distribution of the wear parameters, w_t and w_r, however, is relatively large. Thus, the individual trajectories spread significantly as EOL is approached.

Estimation and prediction results are shown in Figs. 17.7 and 17.8, respectively (from [8]). In the estimation results, we can see the estimation of the wear parameters as a function of time. Here, an unscented Kalman filter was used to jointly estimate the system state and the wear parameters. Once the estimates converge, the health state is well known and the predictions become accurate. Prognostics results are shown in the form of an α-λ plot, which computes the relative accuracy of the predictions as EOL is approached.

17.5 Conclusions, Challenges and Open Problems

In this chapter, we have presented the fundamental concepts of prognostics. At the beginning of the chapter, we provide the reader with a general view and main definitions and concepts of prognostics. We then present the basic ideas underlying the prognostics process, discuss on how to deal with uncertainty, the main problem of prognostics, and then give a brief overview of the prognostics methods. Finally, we go into the details of the model-based prognostics method and present a framework to carry out prognostics using physics-based system models. The section ends up presenting a centrifugal pump as the case study and shows how to perform model-based prognostics for such system.

Although the development of accurate and efficient prognostics methodologies and its application to a variety of engineering systems have seen a huge increase in the latest years, there is still a long way to go. Many challenges arise when developing prognostics solutions to a new domain [35]. First, having some sort of experimental data for new domains is usually difficult, but this is especially relevant for prognostics, since, on one hand, prognostics techniques have a strong dependency of measurement data, and, on the other hand, systems are not usually run until the failure point, and hence it is difficult to have large run-to-failure datasets. But the alternative of using model-based techniques is also nontrivial since developing degradation models of systems is not an easy task. Dealing with uncertainty and carrying out a cost–benefit analysis are also quite challenging problems for prognostics. Dealing with uncertainty is always ranked top of the main challenges when developing prognostics solutions. The different possible sources of uncertainty and the difficulty to deal with them (especially the future inputs uncertainty) have a big influence on the accuracy of the prognostics results. One previous task that is not always carried out is the cost–benefit analysis of implementing prognostics solutions. As we have seen in this chapter, the benefits of integrating prognostics in engineering systems are many, but also the development, installation, and implementation costs.

Among the many open problems in prognostics, we can underline the uncertainty management, validation and verification [26], development of hybrid prognostics solutions [20], and system-level prognostics [7]. The latest is especially relevant since most prognostics solutions only deal with a component or small subsystems and do not provide prognostics decisions at a system level.

Acknowledgements The authors would like to thank Dr. Indranil Roychoudhury for his contributions on several stages of this work.

References

1. ISO 13381-1:2015 - Condition Monitoring and Diagnostics of Machines - Prognostics - Part 1: General Guidelines. Standard, International Organization for Standardization, Geneva, CH (2015)

2. An, D., Choi, J.H., Kim, N.H.: Prognostics 101: a tutorial for particle filter-based prognostics algorithm using matlab. Reliab. Eng. Syst. Saf. **115**, 161–169 (2013). https://doi.org/10.1016/j.ress.2013.02.019
3. Balaban, E., Narasimhan, S., Daigle, M., Roychoudhury, I., Sweet, A., Bond, C., Gorospe, G.: Development of a mobile robot test platform and methods for validation of prognostics-enabled decision making algorithms. Int. J. Progn. Health Manag. **4**(1) (2013)
4. Baraldi, P., Mangili, F., Zio, E.: A kalman filter-based ensemble approach with application to turbine creep prognostics. IEEE Trans. Reliab. **61**(4), 966–977 (2012)
5. Biswas, G., Mahadevan, S.: A hierarchical model-based approach to systems health management. In: Proceedings of the 2007 IEEE Aerospace Conference (2007)
6. Daigle, M.: Prognostics Model Library (2016). https://github.com/nasa/PrognosticsModelLibrary
7. Daigle, M., Bregon, A., Roychoudhury, I.: A distributed approach to system-level prognostics. In: Annual Conference of the Prognostics and Health Management Society 2012, pp. 71–82, Minneapolis, MN (2012)
8. Daigle, M., Bregon, A., Roychoudhury, I.: Distributed prognostics based on structural model decomposition. IEEE Trans. Reliab. **63**(2), 495–510 (2014)
9. Daigle, M., Goebel, K.: Model-based prognostics under limited sensing. In: 2010 IEEE Aerospace Conference, Big Sky, MT, USA (2010)
10. Daigle, M., Saha, B., Goebel, K.: A comparison of filter-based approaches for model-based prognostics. In: Proceedings of the 2012 IEEE Aerospace Conference, pp. 1–10, Big Sky, MT, USA (2012)
11. Daigle, M.J., Goebel, K.: Model-based prognostics with concurrent damage progression processes. IEEE Trans. Syst. Man Cybern. Syst. **43**(3), 535–546 (2013). https://doi.org/10.1109/TSMCA.2012.2207109
12. Gorospe, G., Daigle, M., Sankararaman, S., Kulkarni, C., Ng, E.: GPU accelerated prognostics. In: Annual Conference of the Prognostics and Health Management Society 2016, St. Petersburg, FL, USA (2017)
13. Gouriveau, R., Medjaher, K., Zerhouni, N.: From Prognostics and Health Systems Management to Predictive Maintenance 1: Monitoring and Prognostics. Wiley-ISTE, New York (2016)
14. Hutchings, I.M.: Tribology: Friction and Wear of Engineering Materials. CRC Press, Boca Raton (1992)
15. Johnson, S.B., Gormley, T., Kessler, S., Mott, C., Patterson-Hine, A., Reichard, K., Scandura, P.: System Health Management with Aerospace Applications. Wiley-Blackwell, Oxford (2011)
16. Kallesøe, C.: Fault detection and isolation in centrifugal pumps. Ph.D. Thesis, Aalborg University (2005)
17. Kapur, K.C., Pecht, M.: Reliability Engineering. Wiley-Blackwell, Oxford (2014)
18. Kwon, D., Hodkiewicz, M., Fan, J., Shibutani, T., Pecht, M.: Iot-based prognostics and systems health management for industrial applications. IEEE Access **4**, 3659–3670 (2016). https://doi.org/10.1109/ACCESS.2016.2587754
19. Lall, P., Lowe, R., Goebel, K.: Extended kalman filter models and resistance spectroscopy for prognostication and health monitoring of lead-free electronics under vibration. IEEE Trans. Reliab. **61**(4), 858–871 (2012)
20. Liao, L., Kottig, F.: Review of hybrid prognostics approaches for remaining useful life prediction of engineered systems, and an application to battery life prediction. IEEE Trans. Reliab. **63**(1), 191–207 (2014). https://doi.org/10.1109/TR.2014.2299152
21. Luo, J., Pattipati, K.R., Qiao, L., Chigusa, S.: Model-based prognostic techniques applied to a suspension system. IEEE Trans. Syst. Man Cybern. Part A Syst. Hum. **38**(5), 1156–1168 (2008)
22. Lyshevski, S.E.: Electromechanical Systems, Electric Machines, and Applied Mechatronics. CRC Press, Boca Raton (1999)
23. Mi, C., Masrur, M.A., Gao, D.W.: Diagnostics, Prognostics, Reliability, EMC, and Other Topics Related to HEVs, Chap. 8, pp. 177–195. Wiley-Blackwell, Oxford (2011)

24. Orchard, M., Vachtsevanos, G.: A particle filtering approach for on-line fault diagnosis and failure prognosis. Trans. Inst. Meas. Control **31**(3–4), 221–246 (2009)
25. Roemer, M.J., Byington, C.S., Kacprzynski, G.J., Vachtsevanos, G., Goebel, K.: Prognostics. In: System Health Management with Aerospace Applications, pp. 281–295. Wiley-Blackwell, Oxford (2011)
26. Roemer, M.J., Dzakowic, J., Orsagh, R.F., Byington, C.S., Vachtsevanos, G.: Validation and verification of prognostic and health management technologies. In: 2005 IEEE Aerospace Conference, pp. 3941–3947 (2005). https://doi.org/10.1109/AERO.2005.1559699
27. Saha, B.: A Bayesian Approach to Battery Prognostics and Health Management, Chap. 7, pp. 151–174. Wiley-Blackwell, Oxford (2016)
28. Saha, B., Goebel, K.: Modeling Li-ion battery capacity depletion in a particle filtering framework. In: Proceedings of the Annual Conference of the Prognostics and Health Management Society 2009, San Diego, CA (2009)
29. Sankararaman, S., Daigle, M., Goebel, K.: Uncertainty quantification in remaining useful life prediction using first-order reliability methods. IEEE Trans. Reliab. **63**(2), 603–619 (2014)
30. Sankararaman, S., Goebel, K.: Remaining Useful Life Estimation in Prognosis: An Uncertainty Propagation Problem, p. 4901. American Institute of Aeronautics and Astronautics, Reston (2013). https://doi.org/10.2514/6.2013-4901
31. Sankararaman, S., Goebel, K.: Uncertainty in prognostics and systems health management. Int. J. Progn. Health Manag. **6**(10) (2015)
32. Saxena, A., Celaya, J., Saha, B., Saha, S., Goebel, K.: Metrics for Offline Evaluation of Prognostic Performance. Int. J. Progn. Health Manag. **1**(1) (2010)
33. Schwabacher, M., Goebel, K.: A survey of artificial intelligence for prognostics. In: AAAI Fall Symposium - Technical Report, pp. 107–114, Arlington, Virginia (2007)
34. Si, X.S., Wang, W., Hu, C.H., Zhou, D.H.: Remaining useful life estimation - a review on the statistical data driven approaches. Eur. J. Oper. Res. **213**(1), 1–14 (2011). https://doi.org/10.1016/j.ejor.2010.11.018
35. Sun, B., Zeng, S., Kang, R., Pecht, M.G.: Benefits and challenges of system prognostics. IEEE Trans. Reliab. **61**(2), 323–335 (2012). https://doi.org/10.1109/TR.2012.2194173
36. Tang, L., Kacprzynski, G.J., Goebel, K., Vachtsevanos, G.: Methodologies for uncertainty management in prognostics. In: 2009 IEEE Aerospace Conference, pp. 1–12 (2009). https://doi.org/10.1109/AERO.2009.4839668
37. Tu, F., Ghoshal, S., Luo, J., Biswas, G., Mahadevan, S., Jaw, L., Navarra, K.: PHM integration with maintenance and inventory management systems. In: Proceedings of the 2007 IEEE Aerospace Conference, pp. 1–12. IEEE (2007)
38. Vachtsevanos, G., Lewis, F.L., Roemer, M., Hess, A., Wu, B.: Intelligent Fault Diagnosis and Prognosis for Engineering Systems. Wiley, New York (2011)
39. Wolfram, A., Fussel, D., Brune, T., Isermann, R.: Component-based multi-model approach for fault detection and diagnosis of a centrifugal pump. In: Proceedings of the 2001 American Control Conference, vol. 6, pp. 4443–4448 (2001)
40. Xi, Z., Jing, R., Lee, C., Hayrapetyan, M.: Recent Research on Battery Diagnostics, Prognostics, and Uncertainty Management, Chap. 8, pp. 175–216. Wiley-Blackwell, Oxford (2016)

Chapter 18
Electronics Prognostics

Chetan S. Kulkarni and Jose Celaya

Nomenclature

ϵ_R	Relative dielectric constant
ϵ_O	Permittivity of free space
t_o	Oxide thickness
V_e	Dispersion volume at time t
V_O	Initial electrolyte volume
j_{eo}	Evaporation rate (h^{-1} cm^{-2})
t	Time in minutes
ρ_E	Electrolyte resistivity
P_E	Correlation factor related to electrolyte spacer porosity and average liquid pathway
r_c	Radius of capacitor capsule
h_c	Height of capacitor capsule
V_{paper}	Volume of paper
A_s	Effective oxide surface area (L × H)
w_e	Volume of ethyl glycol molecule
V_c	Total capacitor capsule volume
d_A	Thickness of anode strip
d_C	Thickness of cathode strip
d_s	Thickness of spacer strip
A_s	Effective oxide surface area.
M_1	Electrical lumped parameter model
M_2	Simplified lumped parameter model

C. S. Kulkarni (✉)
SGT, Inc., NASA Ames Research Center, Moffett Field, Mountain View, CA 94035, USA
e-mail: chetan.s.kulkarni@nasa.gov

J. Celaya
Schlumberger, Software Technology Innovation Center, Menlo Park, CA 94025, USA
e-mail: jcelaya@slb.com

T. Escobet et al. (eds.), *Fault Diagnosis of Dynamic Systems*,
https://doi.org/10.1007/978-3-030-17728-7_18

M_3	Basic lumped parameter model
D_1	Capacitance degradation model
D_2	ESR degradation model
D_3	Volume degradation model
C_I	Ideal capacitance value for an ideal capacitor
C_R	Real capacitor value for a nonideal capacitor model
R_E	Equivalent series resistance of the capacitor
$C_l(k)$	Capacitance percentage loss at time t_k
T_i	Validation test on capacitor i
M_i	Nominal model for a component or system
D_i	Degradation model for a capacitor
R_L	Load resistance on electrical overstress system
V_L	Load voltage on electrical overstress system
V_o	Electrical overstress voltage in aging system
Z_I	Ideal capacitor impedance
Z	Capacitor impedance for nonideal capacitor model $\mathcal{M}_1$

18.1 Introduction

This book chapter proposes the use of a model-based prognostics approach for electronics components. Components such as electrolytic capacitors, MOSFETs, and IGBTs are critical components in electronics systems in safety critical domains such as aeronautics, medical, etc. These devices are known to have lower reliability than other electronic components that are used in power supplies of avionics equipment and electrical drivers of electromechanical actuators of control surfaces. The field of prognostics for these is concerned with the prediction of Remaining Useful Life (RUL) of the components and systems. This notion of condition-based health assessment leverages the knowledge of the device physics to model the degradation process, which is then used to estimate remaining useful life as a function of current state of health and future operational and environmental conditions.

The prognostics methodology is based on a Bayesian tracking framework and dynamic degradation models developed from electrical overstress accelerated aging tests. A validation methodology is presented to assess the efficacy of the method using available run-to-failure data. The novelty of the approach consists in its ability to periodically estimate remaining useful life. This estimation process is condition-based in the sense that periodic measurements of the component under consideration are used in the estimation process. The contribution of this work is in the area of dynamic degradation model development and implementation of a model-based prognostics methodology for electronic components.

Low reliability and their criticality in avionics systems make switching components and electrolytic capacitors important candidates for a health management solution. In addition to this, degradation at component level could lead to cascading faults at subsystem and system level. In order to mitigate the effects of capacitor failures in critical to safety systems, we introduce here, a condition-based prognostics methodology. This methodology provides with the ability to identify degradation effects and to estimate the remaining life of the components periodically. This method will further allow for prognostics-based decision-making for optimal maintenance scheduling of the system or for implementation of mitigation strategies in case of contingencies during operation.

18.1.1 Motivation

The development in the field of prognostics methodologies has become essential as more electrical and electronics systems are being used to replace traditional systems in several safety critical applications. The development of prognostics methods for electronics presents several challenges due to a wide variation in components being used in a system, a continuous development in electronics technologies, and a general lack of understanding of how electronics systems and devices fail. Traditional reliability techniques in electronics tend to focus on understanding the time to failure for a batch of components of the same type by running multiple experiments and making probabilistic estimates from the accumulated data. There has been a push to understand, in more depth, how faults progress as a function of loading and environmental conditions. Furthermore, just until recently, it was assumed that electronic components and systems do not indicate any precursors to failure.

Avionics systems perform critical functions in aircraft greatly escalating the ramification of an in-flight malfunction [5, 19]. These systems combine physical processes, computational hardware and software, and present unique challenges for fault detection and isolation. A systematic analysis of these conditions is very important for analysis of aircraft safety and also to avoid catastrophic failures during flight.

Power supplies and specifically DC–DC converter are key systems of modern avionics systems. Any degradation or faults in these converter units propagate to the Global Positioning System (GPS) and navigation subsystems as cascading effects thus affecting the overall operations of the aircraft. Capacitors and MOSFETs (metal oxide field effect transistor) are the key components present in DC–DC converters, which degrade and fail consistently [17]. Some of the more widespread fault effects, such as a ripple voltage surge at the output can cause glitches in the GPS position and velocity output, and this in turn, if not rectified can propagate and distort the navigation resultant solution.

18.1.2 Research Challenges

Physics-based degradation modeling evaluate fundamental processes, such as mechanical, chemical, electrical, and thermal, that govern the behavior of a component and the effect of operational and nonoperational stressors on component performance. The methodology derives a cumulative damage function due to various failure mechanism states of the component, the relevant physical processes that govern component behavior, and its operating conditions. A failure mode is defined by the causes for the failure [27]. It can also be defined as the way in which a component, subsystem, or system fails to meet or deliver the intended functionality and behaviors. A single component may have multiple failure modes, each of which can be derived based on an underlying physics-based degradation phenomenon associated with the component [36]. A physics-based model is laborious to derive, but high fidelity empirical methods have been developed using selective accelerated life testing to observe the failure mode effects in components [27].

Capturing the fundamental physics-based degradation phenomenon of each component or system is key to generate accurate prediction models and estimate remaining useful life of a system or component. There are several important steps and challenges to overcome before these approaches can be applied to real-world applications, behavior analysis, performance-driven predictions, and decision-making. The most important issues currently being addressed in the Prognostics and Health Management community include system parameter monitoring, failure precursors selection, sensors selection and placement to ensure tracking of relevant system behavior and operating conditions, in-situ data acquisition, feature extraction for refining the degradation models, and behavior prediction from the collected data [3, 10]. It is a challenging task to define and develop procedures for using the developed models to predict the overall system performance along with cascading failures. Specifically implementing PHM methodologies for electronic systems include methodologies for tracking degradation in components and observe its behavior on connected systems and subsystems.

The outline of the rest of the chapter presented is as follows: Section 18.2 discusses physics-based modeling approach and its implementation to prognostics framework. Section 18.3 introduces electrolytic capacitors, their basic structure, degradation mechanisms, and derived first principles models. Section 18.4 presents the accelerated aging experimental setup along with specific tests developed. Section 18.5 discusses implementation and the resulting first-principles models for electrolytic capacitor degradation. Section 18.6 presents results and validation tests. Section 18.7 ends the chapter with a discussion, conclusions, and directions for future work.

18.2 Physics-Based Modeling for Prognostics

The section discusses approaches for physics-based degradation modeling in electronics and electrical systems. The section further briefly discusses a generalized prognostics methodology and implementation framework and ends with the prognostics approach implemented for electronics components.

18.2.1 Physics-Based Degradation Modeling

The Physics-Based Degradation Modeling (PBDM) approach implements a first principles framework for prognostics, as opposed to traditional approaches that employ data-driven empirical models. The derived models are dynamic which evolve over time as the device/system degrades due to operation. The derived modeling techniques are effective for estimating the impact on lifetime due to certain failure mechanisms [14]. Failure horizon estimating, i.e., remaining useful life and the evolution of degradation through this time interval depending upon the current health state and operating condition of the component/system.

In safety critical systems, determining fault initiation and propagation is key to accurate and effective fault prognosis. The approach uses first-principles physics-based methods to temporal models of fault propagation and failure. Recent materials research study has focused on microstructural characterization of materials and modeling frameworks that effectively and reliably describes anomaly initiation and propagation as a function of time [14]. These models, which address the fundamental material properties, generally are not suitable for real-time treatment due to computational complexity. In such cases, more abstract models, also called lumped models, are usually derived from the microstructural characterizations and can be used to estimate degradation and system performance [9].

Typically, a device/subsystem does not degrade rapidly under normal operating conditions, where operations are restricted to manufacturer-specified threshold limits. In such cases, tests may have to be conducted over longer periods of time (e.g., months and even years) to study and detect varying failure modes. Accelerated Stress Tests (AST) have been recognized to be a valuable in gathering data on estimates of the reliability and quality of electronics components [14]. These are carried out under enhanced adverse environmental conditions to accentuate the damage accumulation rate due to physical or environmental phenomena [35]. In the Physics-Based Degradation Modeling (PBDM) approach, it is very important to identify, understand, and model the underlying failure mechanisms to derive general models of the aging phenomena which are not adequately highlighted in traditional techniques.

The following activities are implemented for PHM methodology:

- using system structure and physical laws to derive a first-principles operational model of the device/system;
- identifying failure modes based on stressors, which then helps to establish the underlying root cause and derive temporal models of component degradation [37];
- deriving applicable degradation/failure mechanism models;
- updating and enhancing the physics-based models using experimental data and stochastic modeling methods [14];
- using the derived physics-based models and prognostic algorithms for predicting the RUL of the system.

18.2.2 Prognostics Methodology

Model-based prognostics methodology specifically for electrolytic capacitors is discussed in this section. The methodology relies on accelerated aging experiments to observe and identify degradation behavior and further derive time-dependent degradation models. The proposed methodology is presented in Fig. 18.1. The methodology is based on time-dependent empirical degradation models derived from accelerated aging test data.

Accelerated Aging: In this test commercial, off the shelf capacitors are electrically overstressed, in order to observe and record the degradation process along with identifying performance conditions close to failure criteria in a short time frame. Electrochemical-Impedance Spectroscopy (EIS) method is used periodically during the accelerated aging test to characterize the frequency response of the capacitor's impedance. Several measurements are available through the aging time, including measurements at pristine condition before the aging starts and measurements after failure condition.

System Identification: A lumped parameter model (M_1) of the nonideal capacitor impedance is assumed. This impedance model includes a capacitance element and an Equivalent Series Resistance (ESR) parasitic element. The EIS measurements along with the impedance model structure are used in a systems identification setting to estimate the model parameters available throughout the aging test. This results in time-dependent capacitance and ESR measurements trajectories reflecting capacitor degradation.

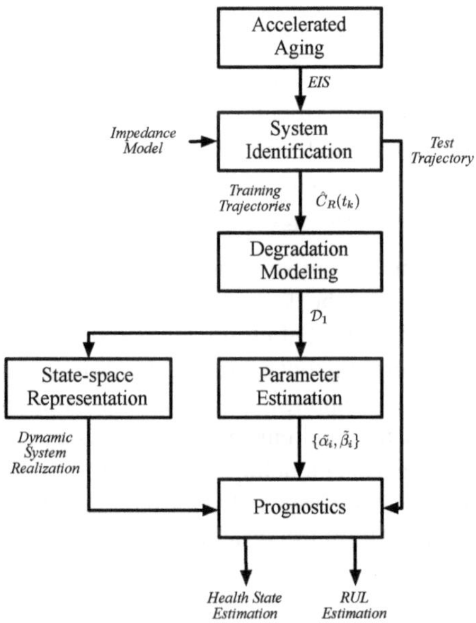

Fig. 18.1 Methodology for capacitor prognostics

Degradation Modeling: An empirical degradation model based on the observed degradation process during the accelerated life test is being developed. The objective of the model is to generate a parametrized model of time-dependent capacitance degradation as generated by the system identification step.

Parameter Estimation: The parameters of the degradation model are estimated using nonlinear least squares regression. The parameters are assumed to be static during the prognostics estimation.

Prognostics: A Bayesian framework is employed to estimate (track) the state of health of the capacitor based on measurement updates of key capacitor parameters. A Kalman filter algorithm framework is used to track the state of health and the degradation model is used to make predictions of remaining useful life once no further measurements are available.

18.2.3 Model-Based Prognostics Framework

A generalized model-based prognostics algorithm based on Kalman filter and a physics-inspired empirical degradation model is presented.

The prognostic methodology implemented in this work is divided into two parts: (1) first part consists of a tracking framework where the developed algorithm tracks the health state based on inputs and present operating conditions, (2) in the second part, the algorithm predicts RUL of the capacitor. The prediction is based on measurement data from the accelerated aging experiments. The complete methodology is depicted in Fig. 18.2 and consists of the following:

1. State tracking (Kalman filter): The capacitance loss C_l is defined as the state variable to be estimated and the degradation model is expressed as a discrete time dynamic model in order to estimate capacitance loss as new measurements become available. Direct measurements of the capacitance are assumed for the filter.
2. Health state forecasting: It is necessary to forecast the state variable once there are no more measurements available at time or RUL prediction t_p. This is done by evaluating the degradation model through time using the state estimate at time t_p as initial value.
3. Remaining life computation: RUL is computed as the time between time of prediction t_p and the time when then forecasted state crosses the failure threshold value.

This process is repeated at varying values of t_p through the life of the component under consideration.

18.2.3.1 Kalman Filter for State Estimation

A state-space dynamic model is being implemented for filtering. The state variable x_k at time t_k is defined as the percentage loss in capacitance denoted as $C_l(k)$. Since

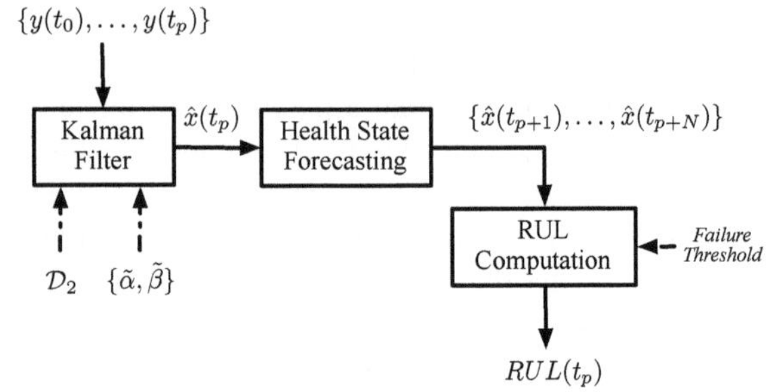

Fig. 18.2 Model-based prognostics methodology

system measurements are presented as percentage loss in capacitance, output equation is given by $y_k = hx_k$, where the value of h is equal to one. The following system structure is implemented for filtering and prediction steps using the Kalman filter:

$$\begin{aligned} x_k &= A_k x_{k-1} + B_k u + v, \\ y_k &= h x_k + w, \end{aligned} \tag{18.1}$$

where,

$$\begin{aligned} A_k &= (1 + \Delta_k), \\ B_k &= -\alpha\beta\Delta_k, \\ h &= 1, \\ u &= 1. \end{aligned} \tag{18.2}$$

There is inconsistent delta in time increment measurements Δ_k, since measurements were taken at nonuniform sampling rate. Thus, some of the parameters of the model in Eq. (18.1) will change through time. Further, v and w are normal random variables with zero mean and Q and R variance, respectively. A detailed description of the algorithm can be found in [33], a description of how the algorithm is used for forecasting can be found in [11] and an example of its usage for prognostics can be found in [32].

18.2.3.2 Future State Forecasting

Implementing Kalman filter in a RUL forecasting algorithm framework requires evolution of the state without updating the error covariance matrix and the posterior of the state vector. The last update is done at the time of the last measurement t_l. The n-step forecasting equation for the Kalman filter is presented in (18.3) the following equation:

$$\hat{x}_{l+n} = A^n x_l + \sum_{i=0}^{n-1} A^i B. \tag{18.3}$$

A constant Δ_t is used in the forecasting mode, and hence subscripts from parameters A and B are omitted.

Table 18.1 Parameter estimation results

Validation test	Test capacitor	Training capacitor	α (95% CI)	β (95% CI)	σ_v^2
T_2	#2	#1, #3–#6	**0.0162** (0.0160, 0.0164)	**−0.8398** (−1.1373, −0.5423)	1.8778
T_3	#3	#1, #2, #4–#6	**0.0162** (0.0160, 0.0164)	**−0.8287** (−1.1211, −0.5363)	1.9654
T_4	#4	#1–#3, #5, #6	**0.0161** (0.0159, 0.0162)	**−0.8217** (−1.1125, −0.5308)	1.8860
T_5	#5	#1–#4, #6	**0.0162** (0.0161, 0.0164)	**−0.7847** (−1.1134, −0.4560)	2.1041
T_6	#6	#1–#5	**0.0169** (0.0167, 0.0170)	**−1.0049** (−1.2646, −0.7453)	2.9812

18.2.3.3 Noise Models

The model noise variance Q is estimated from model regression residuals for each test case and presented in Table 18.1. This variance is being used for model noise in the Kalman filter implementation. Measurement noise variance R required in the filter implementation and computed from direct measurements of the capacitance with the EIS equipment, and the observed variance is 4.99×10^{-7}.

18.2.4 Research Approach

Prognostics deals with determining the health state of components, and projecting this state into the future to make End Of Life (EOL) and RUL estimations. These approaches are crucial for improving system safety, reliability, and availability. The key task employed in model-based approaches is to derive first-principles physics models that capture knowledge about the system, components, and their underlying degradation mechanism [12, 31]. Faults and degradations appear as parameter value changes in the model which provides the mechanism for tracking system behavior under degraded conditions [7, 24].

The implementation of prognostics modeling process is illustrated in Fig. 18.3. The work is focused on developing degradation models of components, where fault parameters evolve in time and are governed by their operating conditions. Experimental studies on the device are combined with electrical and mechanical configuration information to device physics-based model of behavior phenomena as can be seen in Fig. 18.3. Identifying the failure precursors and developing accurate models taking into consideration the basic understanding of the first principles of operation is an important challenge that is addressed in this research work. Early detection and analysis leads to more accurate estimation of parameter changes, and hence a bet-

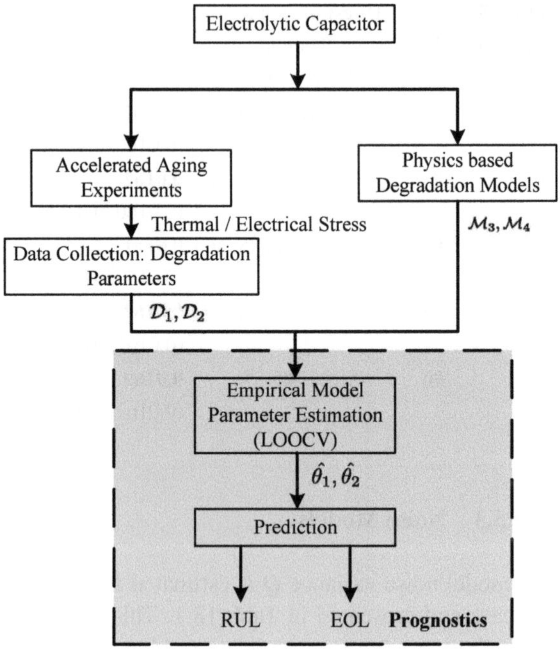

Fig. 18.3 Research methodology

ter prediction and end-of-life estimates of the component. The derived state-space models are then implemented in a Bayesian framework for prognostics.

18.3 Degradation in Electrolytic Capacitors

Capacitor health is heavily affected by their operational conditions, such as voltage, current, frequency, and ambient temperatures among others. Degradation of these devices observed in DC–DC power converters and signal filters leads to increase resistance and decrease in capacitance parameters respectively. Continued capacitor degradation leads to larger noise in converter output due to the inefficient filtering, and hence the average voltage drops below specified threshold value. This further affects the health and performance of downstream components and systems and in some cases may damage them.

18.3.1 Physical Interpretation of the Degradation Process

There are a number of factors that cause electrolytic capacitors to degrade and eventually fail. Continuous gradual degradation in operational functionality over a period of time results in the failure of the component. A sudden complete loss of function is termed a *catastrophic* failure, i.e., a short or open circuit in the capacitor. Gradual degradation results in a increase in the equivalent series resistance and/or decrease in capacitance over time.

Degradation can be primarily attributed to electrolyte evaporation, leakage current, or increase in internal pressure due to gas released by chemical reactions [15, 20, 26]. An ideal capacitor would offer no resistance to the flow of current at its leads. However, the electrolyte that fills the space between the plates and the electrodes develops certain equivalent internal series resistance. Figure 18.4 shows the structure of an electrolytic capacitor in detail. Heath dissipation due to ESR in the capacitor leads to increase in internal temperature and thus causing electrolyte evaporation.

ESR and capacitance represent health state of any capacitor device and hence are the two key failure precursors. The values were calculated by characterizing the capacitors at regular time intervals. The two parameters degrade due to different failure mechanisms which are observed through the experimental measurements. Literature studies [22] on capacitor degradation discuss direct relationship between electrolyte decrease with increase in ESR and decrease in capacitance.

The values are estimated using a system identification method and deriving a basic lumped parameter model consistent with the capacitance and the ESR as shown in Fig. 18.5. Parameter estimation is done using capacitor impedance frequency response measurement data. It is quite a complex and difficult process to estimate ESR and capacitance parameters values through in-situ voltage and current measurements through the accelerated aging test.

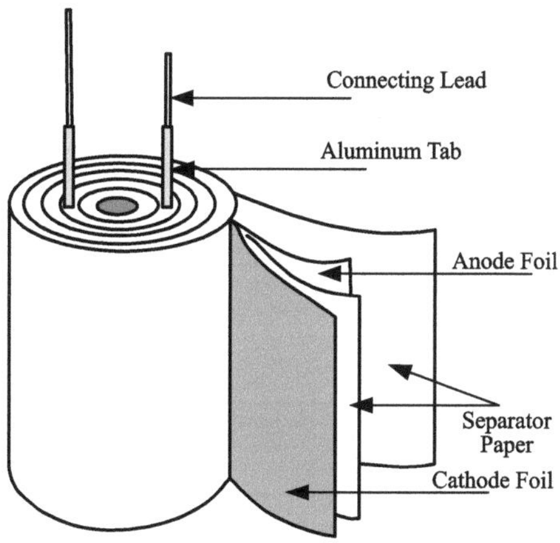

Fig. 18.4 Electrolytic capacitor structure

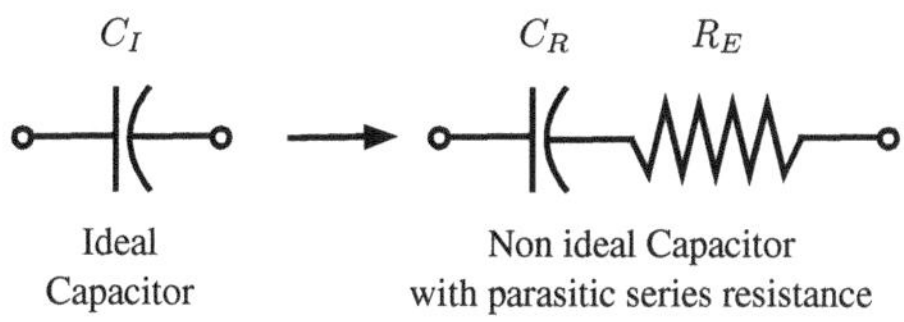

Fig. 18.5 Lumped parameter model (M_1) for a real capacitor

18.3.2 Degradation Mechanisms

Failures typically can be divided as (1) catastrophic failures, where there is complete loss of functionality primarily, i.e., short and open circuits and (2) degradation failures, where a gradual deterioration of parameters is observed due to accumulated internal damage. The fish-bone diagram in Fig. 18.6 summarizes the comprehensive set of different failure modes in electrolytic capacitors [1, 2, 13, 16, 29, 38] which is being presented. The diagram identifies relationship between root causes and failure modes observed in capacitors. The approach presented tracks behavior of the device untill complete failure is observed in the unit. The rest of the section describes the degradation mechanisms and their impact on the electrical parameters of the capacitors.

Literature studies [6, 25] determine deterioration in electrolyte quality as the primary causes for degradation as decrease in electrolyte volume due to evaporation and weakening of the oxide layer due to increased core temperature.

18.3.3 Degradation Models

As discussed in Fig. 18.6, the fish-bone diagram is an exhaustive list of key underlying phenomena that cause capacitor degradation. The derived approximate internal lumped parameter electrical circuit model of the capacitor determines that capacitance *C* and resistance *ESR* are the primary determinants of capacitor health. We implement a physics modeling approach that links degradation of *C* and *ESR* param-

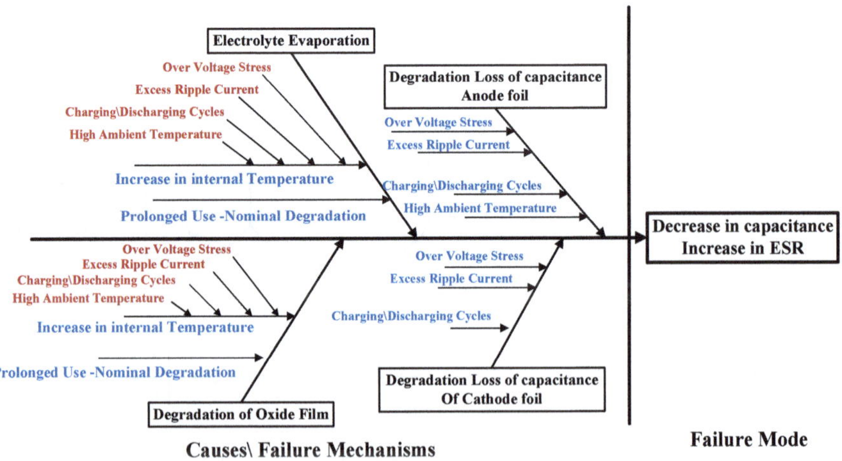

Fig. 18.6 Fish-bone diagram of failure mechanisms in aluminum electrolytic capacitor

eters of the capacitor to physical phenomena associated with electrolyte evaporation and oxide layer deterioration.

18.3.3.1 Capacitance Model

The input impedance of a capacitor network is defined in terms of combined lumped series and parallel impedance of the derived circuit model [34]. Capacitance represents the ability of a unit to store electrical charge and is calculated from first principles understanding of the configuration and the geometry of the conductors and dielectric properties of the insulator. For a foil-type electrolytic capacitor with a cylindrical geometry, the total lumped capacitance [34] is given by

$$C = \frac{2\epsilon_R \epsilon_0 A_s}{d_C}. \tag{18.4}$$

18.3.3.2 ESR Model

An ideal capacitor offers no internal resistance to the current flow. Using first principles, internal resistance R_E is computed as a function of the effective oxide surface area, A_s [34], as given below:

$$R_E = \frac{1}{2}\left(\frac{\rho_E d_C P_E}{A_s}\right). \tag{18.5}$$

Since $R_E \approx ESR$, increase in R_E due to evaporation of electrolyte leads to a decrease in A_s and subsequent increase in ESR as given below:

$$ESR = \frac{1}{2}\left(\frac{\rho_E d_C P_E}{A_s}\right). \tag{18.6}$$

18.3.3.3 Electrolyte Volume

Decrease in C due to electrolyte evaporation [4, 18, 23, 28, 30] is given by

$$V_e(t) = V_{e0} - (A_s j_{eo} w_e) \times t. \tag{18.7}$$

Then using (18.7), we get

$$A_s = \frac{V_{e0} - V_e(t)}{j_{eo}\, w_e\, t}. \tag{18.8}$$

18.3.4 Time-Dependent Degradation Models

Time-dependent degradation models for C and ESR are derived using physics-based model of electrolyte evaporation. Referring (18.4) and (18.8), *capacitance degradation model* can be presented as follows:

$$D_1 : C(t) = \left(\frac{2\epsilon_R\epsilon_0}{d_C}\right)\left(\frac{V_{e0} - V_e(t)}{j_{eo}\, t\, w_e}\right), \tag{18.9}$$

where V_{e0} [18, 23] and rest of parameters are defined specifically for a capacitor type, i.e., structure, manufacturing specifications, etc.

Similarly, change in ESR is represented by increase in electrolyte resistance (R_E), and is computed from (18.5) and (18.7), which is presented as follows:

$$D_2 : ESR(t) = \frac{1}{2}\left(\rho_E\, d_C\, P_E\right)\left(\frac{j_{eo}\, t\, w_e}{V_{e0} - V_e(t)}\right). \tag{18.10}$$

In model, D_2, rate of evaporation j_{eo}, correlation factor, P_E related to electrolyte spacer porosity, and average electrolyte pathway are temperature dependent. Electrolyte evaporation due to high temperature increases correlation factor P_E as the average pathway of the liquid decreases. Electrolyte evaporation under thermal stress storage conditions is attributed to increased temperature of the capacitor capsule. A thermal model [19, 23] explains the phenomenon of heat travel from the surface of the body to the core of the capacitor body.

18.4 Accelerated Aging Experiments

Accelerated life test experiments are often carried out in prognostics research to observe and assess effects of long-term degradation processes in a short time frame. In addition, the tests allow identification and study of different failure mechanisms and their effect on observable signals and parameters. The section presents discussion on accelerated aging methodology and an analysis on degradation patterns induced due to the aging process. A brief description of the aging setup in the first half of the section is followed by analysis of the observed degradation mechanisms. The precursors to failure are identified along with the physical processes that contribute to the degradation in the capacitor component.

18.4.1 Experimental Setup

The objective of the electrical overstress experiment is to observe and study effects of high voltage on capacitor degradation where units were subjected to high voltage stress using developed hardware system. Square waveforms are simulated using a

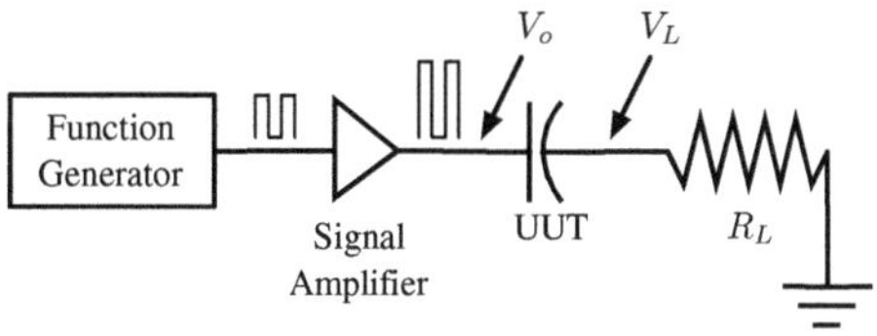

Fig. 18.7 Block diagram of the experimental setup

function generator as input to the developed hardware charge and discharge of the capacitors units in continuous cycles.

The capacitors charge and discharge simultaneously in the early stages of the experiment indicating a consistent time constant, and as time progresses, the cycles go out of sync as the units degrade. Even though all the units under test are subjected to similar operating conditions, change in respective ESR and capacitance values is different. Charging and discharging of each capacitor under test is monitored along with the input and output voltages of the capacitor. Figure 18.7 shows the block diagram for the developed hardware setup. Additional details on the accelerated aging system are presented in [21].

A set of six capacitors is considered for the EOS experimental setup. Electrolytic capacitors of 2200 μF capacitance, with a maximum rated voltage of 10 V, current rating of 1 A, and maximum operating temperature of 105 °C are used in this study. These were the recommended capacitors by the manufacturer for DC–DC converters. The electrolytic capacitors under test were characterized before the start of the experiment at room temperature.

Measurements on each unit were recorded every 8–10 h of the total 180 plus hours of experiment time at 25 °C to capture the change in C and ESR due to degradation. At each measurement step, the voltage source was shut down, capacitors were discharged completely, and the characterization procedure was carried out. This was done for all the six capacitors under test. Details regarding the aging experiment results and analysis of the measured data refer to [8, 21].

18.4.2 System Identification for Real Capacitor Model

C and ESR values are estimated with an assumption of a series combination of resistance and capacitance, from impedance frequency response measurements using an SP-150 Biologic SAS Electro-Impedance Spectroscopy (EIS) instrument.

An ideal capacitor has complex impedance represented as $Z_I = 1/sC_I$, where C_I is the ideal capacitance value. The complex impedance of derived model M_1 is given by

$$Z = R_E + \frac{1}{sC_R}, \tag{18.11}$$

where R_E is the equivalent series resistance and C_R is the real capacitance.

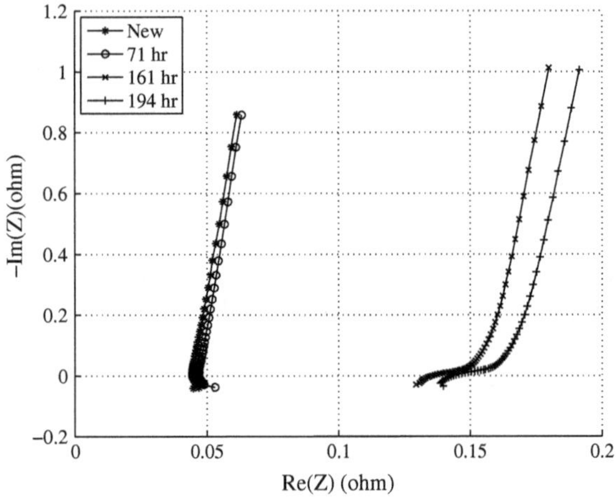

Fig. 18.8 Electro-impedance measurements at different aging times

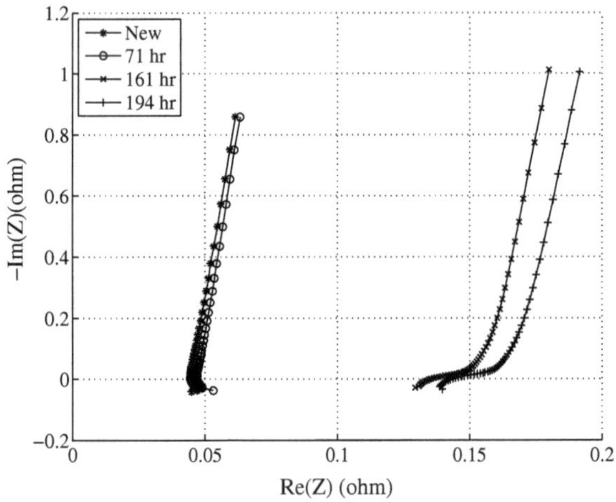

Fig. 18.9 Degradation of capacitor performance, percentage ESR increase as a function of aging time

EIS measurements characterize the electrical performance of the capacitor, Fig. 18.8 shows the Nyquist plots of the impedance measurements for capacitor #1 at pristine condition and after accelerated aging at intervals of 71, 161, and 194 h. The degradation can be observed as the Nyquist plot shifts to the right as a function of aging time due to change in R_E. These measurements are further used to estimate parameters of impedance model M_1 using (18.11). The parameter estimation performed using the EIS instrument software (EC lab) calculates the optimal values for $C_R{}^*$ and $R_E{}^*$.

The parameter estimation iteration cycle is performed every time an EIS measurement is taken resulting in updated values for C_R and R_E at different points of time. The average pristine condition ESR value is measured to be 0.056 mΩ and average capacitance of 2123 μF, respectively, for the set of capacitors under test.

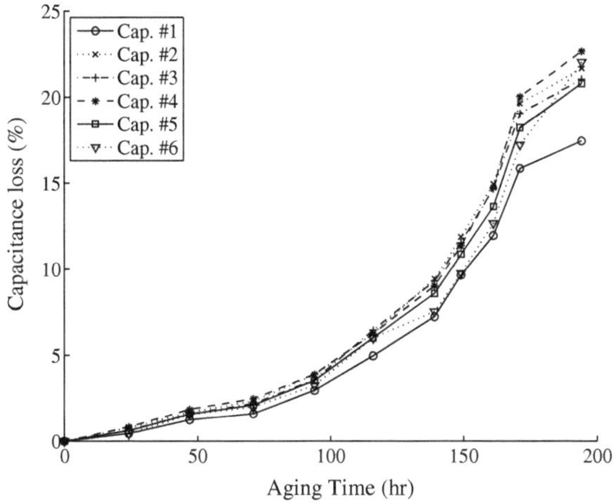

Fig. 18.10 Degradation of capacitor performance, percentage capacitance loss as a function of aging time

Figure 18.9 shows percentage increase in the ESR for respective capacitors under test, which is calculated from the impedance measurement step. Similarly, Fig. 18.10 shows the percentage decrease in capacitance as the respective capacitor degrades over the EOS stress cycle. As per MIL standards [26], a capacitor is considered unhealthy if under electrical operation its ESR increases by 280–300% of its initial value and/or the capacitance decreases by 20% below its initial condition value. The plots in Fig. 18.9 show increase in average ESR by 54–55% and decrease in capacitance by more than 20% (the threshold mark for a healthy capacitor) over the same operational time. (see Fig. 18.10). Since the capacitance loss crossed the threshold, it was selected as a precursor of failure in model development of component degradation.

18.5 Degradation Modeling for Prognostics

Degradation model is key element of the model-based prognostics framework implementation. A model is constructed based on the empirical evidence of the degradation process along with time evolution measured through measured data, particularly, capacitance loss as described in Fig. 18.10.

18.5.1 Nominal Model

A nonideal capacitor model M_1 is used to represent capacitors, a low-pass filter implementation is demonstrated as an example in Fig. 18.11. In the circuit, input voltage V_i is voltage to be filtered and V_o is the filtered output voltage across the

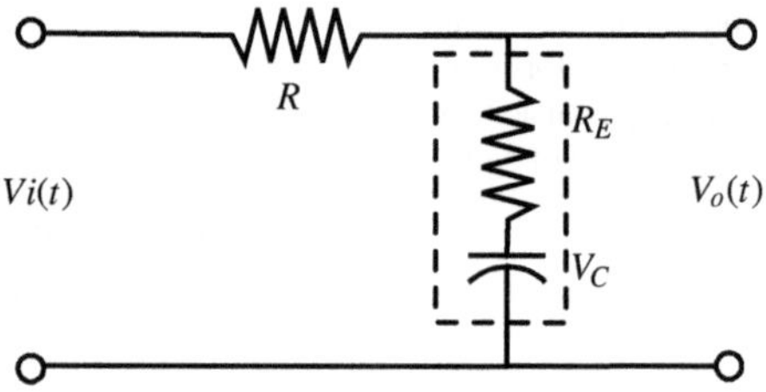

Fig. 18.11 Low-pass filter model (M_2)

capacitor. An assumption is made, where $v(t) = V_o(t)$ and $u(t) = V_i(t)$ in the low-pass system circuit, for M_2 is shown in Fig. 18.11. A state-space realization (M_2) of the dynamic system is given by

$$\begin{aligned} \dot{z}(t) &= \frac{-1}{C_R(R+R_E)} z + \frac{1}{C_R(R+R_E)} u(t), \\ v(t) &= \left[1 - \frac{R_E}{R+R_E}\right] z + \frac{R_E}{R+R_E} u(t), \end{aligned} \tag{18.12}$$

where $z(t) = V_C(t)$ is the state variable representing the capacitor voltage C_R, R_E, and R are system parameters. Furthermore, C_R and R_E are parameters that will change through time as the capacitor degrades.

Model M_1 represents nominal dynamics of a low-pass filter with a non-ideal capacitor. Since the degradation model parameters are reflected in M_1, it cannot be used directly in the algorithm. To develop a complete degradation model incorporating time evolution of R_E or C_R are needed in order to enhance M_1 for model-based prognostics.

18.5.2 Empirical Model

Capacitance percentage loss is used as failure precursor to develop the degradation model. This model correlates aging time to capacitance percentage loss. C_l is considered to be percentage loss of capacitance due to degradation as shown in Fig. 18.10. Degradation model D of the capacitance parameter in the real capacitor model M_1 is given by

$$D_1 : C_l(t) = e^{\alpha t} + \beta, \tag{18.13}$$

where α and β are degradation model parameters that will be estimated from the experimental data of accelerated aging experiments.

A leave-one-out cross-validation technique is to test the prognostics algorithm, which results in six leave-one-out test cases for validation of the results. A non-linear least squares regression algorithm is used to estimate the model parameters. Parameter estimation results are presented in Table 18.1, along with estimate and 95% confidence interval for α and β.

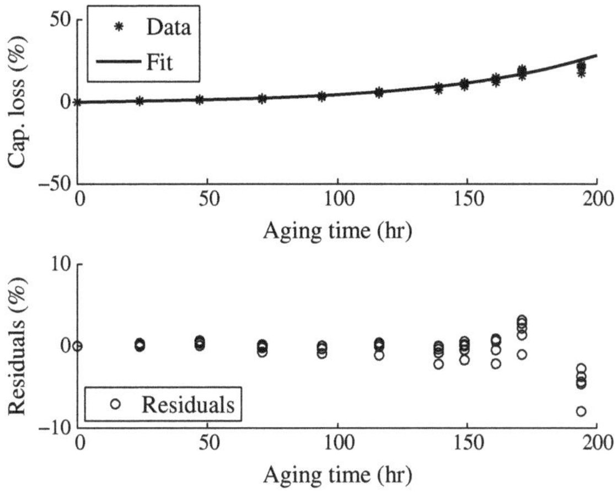

Fig. 18.12 Estimation results for the empirical degradation model

Figure 18.12 shows the estimation results for test case T_6. The experimental data are presented together with results from the exponential fit function. It can be observed from the residuals that the estimation error increases with time. This is to be expected since the last data point measured for all capacitor units falls slightly off the concave exponential model.

The developed degradation model with static parameters is used in the Bayesian tracking framework. This enables the framework to overcome limitation of the degradation model to represent operational behavior close to the failure threshold given the tracking framework ability to compensate the estimation when measurements are unavailable.

18.5.3 State-Space Realization for Tracking

A Bayesian tracking framework using the Kalman filter technique is implemented incorporating the developed degradation model. The methodology requires a state-space dynamic model to associate degradation level at time t_k to degradation level at time t_{k-1}. The following steps are followed to derive a state-space model for D_1, (1) the model nonlinear exponential behavior is represented as a first-order differential equation, represented by time evolution of $C_l(t)$, (2) obtain a discrete time state-space model D_2.

From Eq. (18.13), we have that $C_l(t) = e^{\alpha t} + \beta$, taking the first derivative with respect to time and substituting $e^{\alpha t} = C_l(t) - \beta$ from Eq. (18.13), we have

$$\dot{C}_l = \frac{dC_l(t)}{dt} = \alpha C_l(t) - \alpha\beta. \tag{18.14}$$

The state-space model taking the finite difference approximation for $\dot{C}_l$ using discrete time index k and time interval Δk, we have

$$\frac{C_l(k) - C_l(k - \Delta k)}{\Delta k} = \alpha C_l(k - \Delta k) - \alpha\beta \text{ and}$$

$$C_l(k) = (1 + \alpha\Delta_k)C_l(k - \Delta t) - \alpha\beta\Delta_k.$$

A Bayesian tracking framework implemented on the derived model estimates the value of the capacitance loss through time at available measurements steps.

18.6 Prediction of Remaining Useful Life Results

RUL prediction results for $T6$ are presented in Fig. 18.13. It is observed that the residual error increased with aging time, which is expected based on the results observed in model estimation process.

Figure 18.14 presents RUL prediction results for *ESR* and C measurements at time $t_p = 161$ (h). The failure threshold is considered at 20% decrease in C. EOL is defined as the time at which the capacity percentage loss forecast trajectory crosses the EOL threshold and RUL is EOL minus 161 h.

Figure 18.15 presents estimated capacitance loss and EOL prediction at different points during the aging time. Predictions are made at each measurement step. It is observed that the predictions become better as they are made closer to the EOL. This is due to the additional information available to update the estimates as it nears EOL.

An α-λ performance metric for implemented prognostics framework is presented in Fig. 18.16 for validation test T_6. The central solid line represents ground truth and

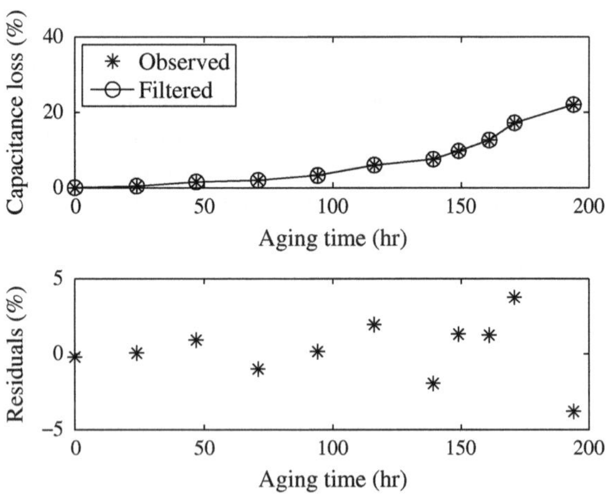

Fig. 18.13 Results for the Kalman filter implementation for $T6$

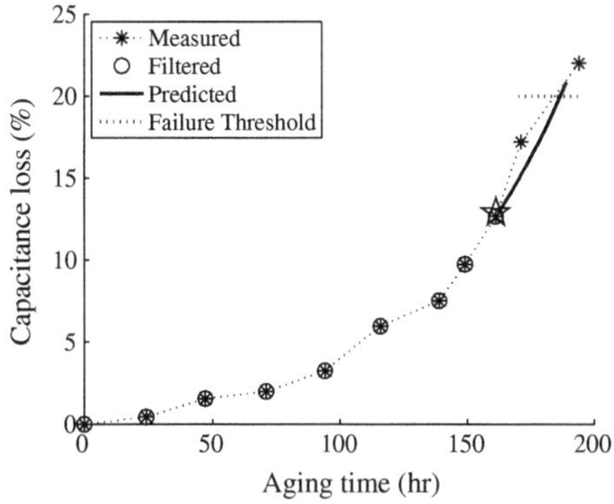

Fig. 18.14 RUL prediction at $t_p = 149$ (h)

shaded region corresponds to a 30% ($\alpha = 0.3$) error bound in the RUL prediction. The metric specifies prediction to be within the defined error bound halfway between first prediction and EOL ($\lambda = 0.5$). The metric allows the user to visualize change in RUL prediction performance changes as additional data is available closer to EOL.

18.6.1 Validation Tests

Table 18.2 summarizes results for RUL predictions respective measurement time steps. The last column indicates the RUL prediction error. The magnitude of the error decreases as the prediction time gets closer to EOL. The decrease is not monotonic which is to be expected when using a tracking framework to estimate health state because the last point of estimation is used to start the forecasting process.

Table 18.3 shows Relative Accuracy (RA) metric performance in Eq. (18.15). The metrics allows for an assessment of the percentage accuracy relative to the ground-truth value, e.g., a 100 represents perfect accuracy. The RA is presented for all the test cases for different prediction times. The last column in Table 18.3 represents median RA for tests cases at particular prediction time step. The RA values decrease considerably for $t_p = 171$ which is consistent with earlier observations, indicating the algorithm with a fixed-parameter model is not able to capture exponential behavior around time step 171 h. This limitation in the model can be improved by enhancing the degradation model or implementing an online estimation of degradation model parameters using Kalman or Particle filters.

$$RA = 100\left(1 - \frac{RUL^{*} - RUL^{'}}{RUL^{*}}\right) \tag{18.15}$$

Fig. 18.15 T_6: Health state estimation and forecasting of capacitance loss (%) at variable time steps t_p; $t_p = [0, 24, 47, 71, 94, 116, 139, 149, 161, 171]$

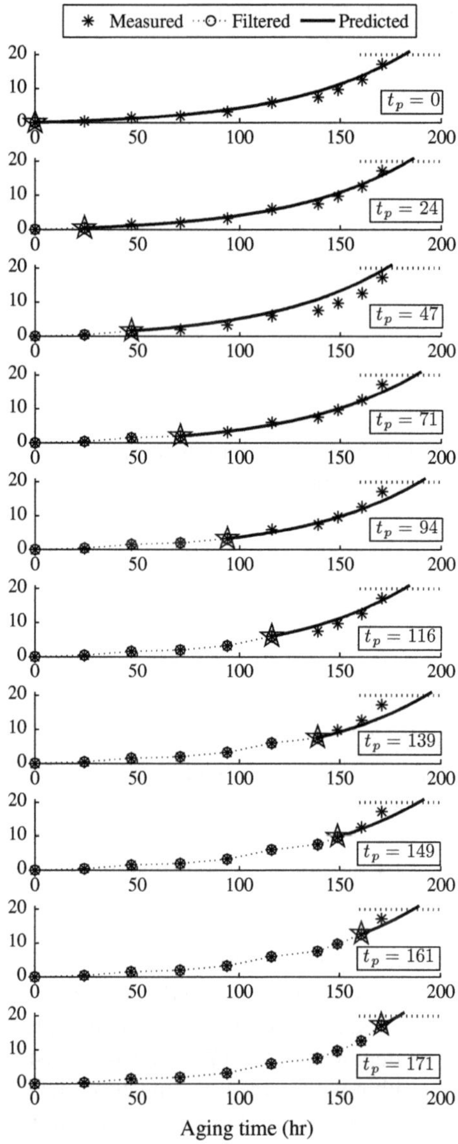

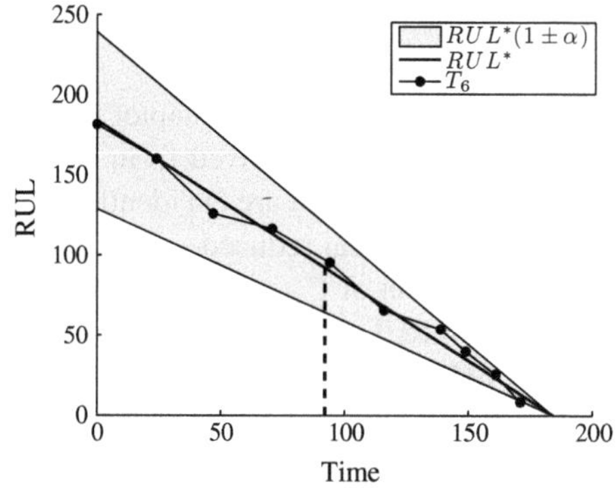

Fig. 18.16 Results based on α-λ performance metric

Table 18.2 Summary of RUL forecasting results

t_p	RUL^*	RUL'_{T2}	RUL'_{T3}	RUL'_{T4}	RUL'_{T5}	RUL'_{T6}
24	151.04	158.84	164.88	158.76	167.76	159.89
47	128.04	131.32	134.08	128.35	135.32	125.91
71	104.04	117.01	119.88	115.37	122.63	116.41
94	81.04	92.69	96.64	93.09	97.6	95.42
116	59.04	67.28	65.39	67.77	69.5	65.71
139	36.04	44.01	44.72	46.88	49.4	53.75
149	26.04	30.67	32.41	33.55	35.92	39.95
161	14.04	17.23	18.28	18.2	22.64	25.6
171	4.04	1.07	2.89	N/A	5.52	8.45

Table 18.3 Validation based on relative accuracy metric

t_p	RA_{T2}	RA_{T3}	RA_{T4}	RA_{T5}	RA_{T6}	$\widetilde{RA}$
24	94.8	95.5	91.9	96.9	99.7	**95.5**
47	97.4	99.3	96.4	96.7	91.7	**96.7**
71	87.5	91.9	84.5	94.1	97.1	**91.9**
94	85.6	90	78.9	94.8	94.2	**90**
116	86	99.1	76.5	98	96.2	**96.2**
139	77.8	95.8	53.1	96.7	81.1	**81.1**
149	82.1	98.4	46.9	94.8	86.6	**86.6**
161	77.2	87.3	16.6	87.5	89.8	**87.3**
171	26.6	26.4	N/A	34.8	63.7	**30.7**

18.7 Conclusion

The work discussed in this chapter presents an RUL prediction framework based on degradation model derived from accelerated aging measurement data. The key contributions presented are (1) identification of the lumped parameter model for capacitor as a practical reduced-order model for prognostics algorithm development; (2) identification of *C* and *ESR* model parameters as failure precursors; (3) implementation of RUL prediction algorithm framework based on the Kalman filters; and (4) the prediction of RUL for capacitors as new measurements time step. The performance of the developed model is found to be satisfactory under the current study, based on the RUL prediction performance metric results.

Implementation of the developed framework extends the technology readiness level of prognostics implementation to components used in critical systems. The results presented are hinged on accelerated life test measurement data as well as the accelerated life timescale. Next steps are planned to focus on development of functional mapping to construe accelerated timescale factor to real-time operational conditions timescale, where the degradation process dynamics are slower.

References

1. Albella, J.M., Gomez-Aleixandre, C., Martinez-Duart, J.M.: Dielectric characteristics of miniature aluminium electrolytic capacitors under stressed voltage conditions. J. Appl. Electrochem. **14**(1), 9–14 (1984)
2. Alwitt, R.S., Parler, S.G.: Heat generation during pulse operation of prototype aluminum electrolytic capacitors. J. Appl. Electrochem. **25**(6), 533–542 (1995)
3. Balaban, E., Saxena, A., Narasimhan, S., Roychoudhury, I., Goebel, K., Koopmans, M.: Airborne electro-mechanical actuator test stand for development of prognostic health management systems. In: Proceedings of the Annual Conference of the Prognostics and Health Management Society, Portland, OR (2010)
4. Bengt, A.: Electrolytic Capacitors Theory and Applications. RIFA Electrolytic Capacitors (1995)
5. Bhatti, U., Ochieng, W.: Failure modes and models for integrated GPS/INS systems. J. Navig. **60**(2), 327–348 (2007)
6. Brettle, J., Jackson, N.F.: Failure mechanisms in solid electrolytic capacitors. Electrocompon. Sci. Technol. **3**, 233–246 (1977)
7. Celaya, J., Kulkarni, C., Biswas, G., Goebel, K.: A Model-Based Prognostics Methodology for Electrolytic Capacitors Based on Electrical Overstress Accelerated Aging (2011)
8. Celaya, J., Kulkarni, C., Biswas, G., Goebel, K.: Towards Prognostics of Electrolytic Capacitors. In: AIAA 2011 Infotech@Aerospace Conference. AIAA, St. Louis, MO (2011)
9. Celaya, J., Kulkarni, C., Biswas, G., Goebel, K.: Towards a model-based prognostics methodology for electrolytic capacitors: A case study based on electrical overstress accelerated aging. Int. J. Progn. Health Manag. **3**(004) (2012)
10. Celaya, J., Saxena, A., Wysocki, P., Saha, S., Goebel, K.: Towards prognostics of power mosfets: accelerated aging and precursors of failure. In: Proceedings of the Annual Conference of the Prognostics and Health Management Society, Portland, OR (2010)
11. Chatfield, C.: The Analysis of Time Series: An Introduction, 6th edn. Chapman and Hall/CRC, Boca Raton (2003)

12. Daigle, M., Goebel, K.: A model-based prognostics approach applied to pneumatic valves. Int. J. Progn. Health Manag. Soc. **2**(8), 1–16 (2011)
13. Gomez-Aleixandre, C., Albella, J.M., Martinez-Duart, J.M.: Gas evolution in aluminum electrolytic capacitors. J. Electrochem. Soc. **131**(3), 612–614 (1984)
14. Hansen, P., McCluskey, P.: Failure models in power device interconnects. In: European Conference on Power Electronics and Applications, pp. 1–9, Aalborg, Denmark (2007)
15. IEC: 60384-4-1 fixed capacitors for use in electronic equipment. Technical Report (2007-03)
16. Ikonopisov, S.: Theory of electrical breakdown during formation of barrier anodic films. Electrochim. Acta **22**(10), 1077–1082 (1977)
17. Kulkarni, C., Biswas, G., Bharadwaj, R., Kim, K.: Effects of Degradation in DC-DC Converters on Avionics Systems: A Model Based Approach, pp. 8–13 (2010)
18. Kulkarni, C., Biswas, G., Celaya, J., Goebel, K.: Prognostic Techniques for Capacitor Degradation and Health Monitoring. In: The Maintenance and Reliability Conference (MARCON), Knoxville, TN (2011)
19. Kulkarni, C., Biswas, G., Koutsoukos, X.: A prognosis case study for electrolytic capacitor degradation in DC-DC converters. In: Proceedings of the Annual Conference of the Prognostics and Health Management Society, San Diego, CA (2009)
20. Kulkarni, C., Biswas, G., Koutsoukos, X., Goebel, K., Celaya, J.: Experimental studies of ageing in electrolytic capacitors. In: Proceedings of the Annual Conference of the Prognostics and Health Management Society, Portland, OR (2010)
21. Kulkarni, C., Biswas, G., Koutsoukos, X., Goebel, K., Celaya, J.: Integrated diagnostic/prognostic experimental setup for capacitor degradation and health monitoring. In: IEEE AUTOTESTCON, Orlando, FL (2010)
22. Kulkarni, C., Biswas, G., Koutsoukos, X., Goebel, K., Celaya, J.: Physics of failure models for capacitor degradation in dc-dc converters. In: The Maintenance and Reliability Conference (MARCON), Knoxville, TN (2010)
23. Kulkarni, C., Celaya, J., Biswas, G., Goebel, K.: Prognostic modeling and experimental techniques for electrolytic capacitor health monitoring. In: The 8th International Workshop on Structural Health Monitoring (IWSHM). Stanford University, Stanford, CA (2011)
24. Kulkarni, C., Celaya, J., Biswas, G., Goebel, K.: Physics based degradation models for capacitor prognostics under thermal overstress conditions. Int. J. Progn. Health Manag. **4**(005) (2013)
25. Luo, J., Namburu, M., Pattipati, K., Qiao, L., Kawamoto, M., Chigusa, S.: Model-based prognostic techniques [maintenance applications]. In: Proceedings AUTOTESTCON 2003. IEEE Systems Readiness Technology Conference, pp. 330–340, Anaheim, CA (2003). https://doi.org/10.1109/AUTEST.2003.1243596
26. MIL-C-62F: General specification for capacitors. Fixed Electrolytic (2008)
27. Nasser, L., Curtin, M.: Electronics reliability prognosis through material modeling and simulation. In: Proceedings of IEEE Aerospace Conference, Big Sky, MT, USA (2006)
28. Rodner, S.C., Wedin, P., Bergstram, L.: Effect of electrolyte and evaporation rate on the structural features of dried silica monolayer films. Langmuir **18**(24), 9327–9333 (2002). https://doi.org/10.1021/la020259q
29. Roederstein, V.: Aluminum capacitors - general information. Document - **25001** (2007)
30. Rusdi, M., Moroi, Y., Nakahara, H., Shibata, O.: Evaporation from water ethylene glycol liquid mixture. Langmuir Am. Chem. Soc. **21**(16), 7308–7310 (2005)
31. Saha, B., Goebel, K.: Model adaptation for prognostics in a particle filtering framework. Int. J. Progn. Health Manag. Soc. **2**(6), 1–12 (2011)
32. Saha, B., Goebel, K., Christophersen, J.: Comparison of prognostic algorithms for estimating remaining useful life of batteries. Trans. Inst. Meas. Control **31**(3–4), 293–308 (2009)
33. Stengel, R.F.: Optimal Control and Estimation. Dover Books on Advanced Mathematics (1994)
34. Tasca, D., O'Donnell, H., Stokes, S.: Pulse Power Response and Damage Characterization of Capacitors. In: Proceedings of the EOS/ESD Symposium, pp. 174–191. ESD Assn, Las Vegas, Rome, NY (1981)
35. Upadhyayula, K., Dasgupta, A.: Guidelines for Physics-of-Failure Based Accelerated Stress Testing. In: Annual Reliability and Maintainability Symposium, pp. 345–357, Anaheim, CA (1998)

36. Vichare, N., Pecht, M.: Prognostics and health management of electronics. IEEE Trans. Compon. Packag. Technol. **29**(1), 222–229 (2006)
37. Watson, G.F.: Mil reliabiliy: a new approach. IEEE Spectr. **29**(8), 46–49 (1992)
38. Wit, H.D., Crevecoeur, C.: The dielectric breakdown of anodic aluminum oxide. Phys. Lett. A **50**(5), 365–366 (1974)

Index

T. Escobet et al. (eds.), *Fault Diagnosis of Dynamic Systems*,
https://doi.org/10.1007/978-3-030-17728-7

MIX
Papier aus verantwortungsvollen Quellen
Paper from responsible sources
FSC® C105338

If you have any concerns about our products,
you can contact us on
ProductSafety@springernature.com

In case Publisher is established outside the EU,
the EU authorized representative is:
Springer Nature Customer Service Center GmbH
Europaplatz 3, 69115 Heidelberg, Germany

Printed by Libri Plureos GmbH
in Hamburg, Germany